Leaf Shapes and Forms
Palmate
Linear
Cordate
Oblong
Obovate
Doubly Pinnate
Ovate
Pinnate
Reniform
Oval
D1429166

Dendrobium nobile
Eucalyptus leucoxylon
Abutilon megapotamicum
Lilium regale
Petrea kohautiana
Rosa 'First Love'
Clerodendrum thomsoniae
Papaver orientale
'Mrs Perry'
Passiflora caerulea

Frances Perry

M.B.E., V.M.H., F.L.S.

Flowers of the World

Illustrated by

Leslie Greenwood

F.L.S., F.R.S.A.

Foreword by

The Lord Aberconway

V.M.H., President of The Royal Horticultural Society

Consultant Taxonomist

C. D. Brickell

B.Sc., Director, Royal Horticultural Society Garden, Wisley, England

American Consultant

George Kalmbacher

Taxonomist & Curator of the Herbarium, Brooklyn Botanic Garden, U.S.A.

Hamlyn

London . New York . Sydney . Toronto

Acknowledgements

Published by
The Hamlyn Publishing Group Limited
London · New York · Sydney · Toronto
Hamlyn House, Feltham, Middlesex, England
Distributed in the USA
by Crown Publishers, Inc.

ISBN 0 600 01634 x

Photoset by BAS Printers Ltd., Wallop, Hants

Printed in Italy by Officine Grafiche A. Mondadori, Verona

Designed by
David Warner

Dr J. S. Beard, Sydney Botanic Garden (previously Perth), Australia.
P. Black, Black & Flory Ltd., Langley, Buckinghamshire.
Blackmore & Langdon Ltd., Bath Somerset.
A. Blombery, Sydney, Australia.
F. W. Buglass.

Cambridge University Botanic Garden (R. W. Younger, Superintendent).
Thomas Carlile (Loddon Nurseries) Ltd. Twyford, Berkshire.
L. Cheang, Director, Botanic Garden, Penang.

Mrs. E. du Plessis, Kirstenbosch Botanic Gardens, Cape Town, S Africa.

E. T. Ekanayaka, Peradeniya Botanic Gardens, Ceylon.
Elm Garden Nurseries, Claygate, Surrey.
H. J. Everitt.
J. R. B. Evison, N.D.H., F.L.S., V.M.H., The Director of Parks, County Borough of Brighton.

Miss V. Finnis.

Miss M. Giasi, Librarian, Brooklyn Botanic Gardens, New York, USA.
H. Gilpin, Christchurch Botanic Garden, New Zealand.

Mrs N. Hart, Wendover, Buckinghamshire.
G. Hermon Slade, Sydney, Australia.
Miss B. Havergall, M.B.E., V.M.H., N.D.H., Waterperry Horticultural School, Oxford.

Kings Park and Botanic Garden, Perth, Western Australia.
L. Kuzmin, Deputy Director, Botanic Garden, Moscow, USSR.

The late Max Lewis, Singapore.

W. G. Mackenzie, F. L. S., V.M.H., Chelsea Physic Garden, London.
L. Maurice Mason, V.M.H.
Mrs. V. Mellor, Summertown, Oxford.

Norwood Hall Institue of Horticulture, London Borough of Ealing (J. Warren, N.D.H., Principal).

Oxford Botanic Garden (J. K. Burras, Superintendent).

Perry's Hardy Plant Farm, Enfield, Middlesex.

Regents Park, London (Mr. Stephenson, Superintendent).
Thomas Rochford & Sons Ltd., Hoddesdon, Hertfordshire.
The Royal Botanic Garden, Edinburgh (Dr H. R. Fletcher, D.SC., PH.D., F.R.S.E., V.M.H., Regius Keeper).
The Royal Botanic Garden, Kew (Sir George Taylor, D.SC., V.M.H., Director and his staff), including the Library and Herbarium at Kew.
The Royal Botanic Gardens, Sydney, New South Wales, Australia (K. Mair, Director and Chief Botanist).
The Royal Botanic Garden, Wakehurst, Sussex.

The Royal Horticultural Society's Garden, Wisley, Ripley, Surrey (F. S. Knight, F.L.S., V.M.H., and his staff).
Professor B. Rycroft, Kirstenbosch Botanic Gardens, Cape Town, S Africa.

Miss E. Scholtz, Brooklyn Botanic Garden, New York, USA.
Dr W. T. Stearn, V.M.H., Natural History Museum, South Kensington, London.
Suttons, Reading, Berkshire.

N. G. Treseder, N.D.H., Treseders' Nurseries (Truro) Ltd., Cornwall.
Reginald Try, Windsor, Berkshire.

Yong Fan Chin, Curator, Singapore Botanic Garden.

Brassolaeliocattleya
June Moore

Table of Contents

Strelitzia reginae

Foreword

The Lord Aberconway

Of gardening books there seems no end. Their purchase by ardent garden lovers is a continual assurance of the vitality and popularity of a wonderful pastime.

Of the hundreds of books on gardening that appear in a year, some are beautiful, some are learned; some are popular but ephemeral, some have a perennial quality.

One type of book has a universal and lasting appeal: a book which records flower portraits faithfully reproduced in their natural colours, accompanied by a text which tells enough, but not too much, of their story.

That is what this book does.

The illustrations have rarely been surpassed for accuracy and colour, even in old works of the past. Mr Leslie Greenwood, the artist, exhibits frequently at R.H.S. shows, where thousands have seen and admired his work.

The text is by Frances Perry, and no-one could have done it better. It not only describes the flowers illustrated and others, but fills in for each the background of history and legend, and tells of the uses of the plants.

To give a double assurance that the nomenclature is accurate and modern, Mr Christopher Brickell, the Director of Wisley, has acted as taxonomist.

This is less a gardening book than a lovely work about flowers. It will give enjoyment throughout the world for, as the Chinese proverb says, 'Habits and customs differ, but all people have the love of flowers in common'. To all, therefore, I commend this book.

Aberconway

Preface

Frances Perry and Leslie Greenwood

On joyful occasions flowers seem to have a close affinity with happiness. This is perhaps one of the reasons we use them so lavishly at gala events like carnivals, triumphal entries or weddings. But they also give pleasure in small quantities—witness the hospital bunch, the mothering day posy and the congratulatory bouquet.

We also look to flowers to aid us in expressing grief or in the alleviation of distress. Albeit silently they can speak without tongues, their fragility and sweetness bringing comfort to millions. This practice is as old as history and as widespread as the globe.

In a different context millions of people grow flowers for the pure joy of witnessing the annual renewal of plant life—a miracle which never palls—and brighten their homes and gardens with blossoms from all corners of the globe.

Others take holidays in the flower strewn lands of Greece, California, SW Australia, or tramp the Alps or Himalayas—simply in order to find plants in their native habitat and discover hundreds they never knew existed.

Some can do none of these things. Their life is more circumscribed and to these we offer flower paintings—as true to life as the artist can make them.

Plants display tremendous diversity. Each continent produces its distinct varieties; the elements affect their nature and the physical properties of the soil. There are land plants and water plants, desert dwellers and others which can withstand salt spray, wind or snow. Some live out their lives clinging precariously to a cliff face or perched on another plant; a few rob their host in order to survive or form with them a symbiotic partnership. Certain species have carnivorous tastes and set traps for insects to satisfy these needs, and there are plants sensitive to the touch or with huge or very small organs or powers of mimicry, particularly in orchids.

The green world of Nature is full of miracles and with so many plants from which to choose we have only been able to skim the surface. The limits on space and the availability of suitable material to paint have been major criteria, although there have also been subjects we liked personally and just *had* to include. With but few exceptions the illustrations were drawn from life by the artist, and the author in the course of a long horticultural career and travels through the five continents, met up with or has grown most of the plants described.

Inevitably a work of this nature demands assistance and goodwill from many people and we have been very fortunate in this respect. In particular we feel a deep indebtedness to Christopher Brickell, who has steered us through sundry taxonomic difficulties, and gone to untold trouble in this respect. Amid a welter of 'authorities' we chose J. C. Willis as our main guide, as explained by Mr Brickell in his introduction. We are also grateful to Peter Hunt at Kew Gardens for checking over the chapter on Orchids, and to sundry people in various parts of the world for showing or telling us about their native flora or providing specimens for illustration purposes. There are so many of these that their names have been listed on page 4.

Again, both artist and author wish to express their appreciation for much help received from the publishers and in particular Mrs Janet Liebster, the Art Director Mr Roger Denning, Mr Bill Brott and Mr Norman Humphreys. These have steered, pruned and pushed in the kindest possible way and altogether given us the kind of help 'makers of books' pray for. We thank them greatly.

Finally since, as Ruskin said, 'flowers seem intended for the solace of ordinary humanity' we hope this book of flowers will convey to others some of the pleasure we ourselves have experienced during its compilation.

Introduction

C. D. Brickell

Without plant life man could not exist. Many may think this an over-simplified statement, and look on plants as part of their surroundings, to be used as food and medicine; to be admired if they are beautiful; or simply as extraneous matter to be bulldozed out of the way to make room for the next block of houses or a chemical plant. The simple fact is that life as we know it depends completely on series of interacting food-chains which have at their source some plant.

The diversity of plants is astonishing. They vary from microscopic, one-celled algae like *Pleurococcus*, to the gigantic Redwoods, three hundred feet or more high. In between there is a bewildering assortment of seaweeds, fungi, lichens, mosses, ferns, Orchids, insect eaters, and a mass of other flowering plants. Close to 250,000 different species of flowering plants are now known—and man continually tries to improve on nature so that hybridists have produced many thousands more selected for their economic or ornamental uses in forestry, agriculture, horticulture and medicine.

The development of primitive agriculture allowed man to settle in groups with a division of labour where some were able to grow plants for food whilst others worked to produce clothing, furniture and the many other adjuncts to living in permanent or semi-permanent communities. In different areas of the world the agriculture, and through this the evolving civilization, was based on limited groups of species found wild in the area concerned. The Russian botanist Vavilov defined eight major centres of development, with various subdivisions, and in these areas large numbers of local races of crop plants arose due to deliberate or natural selection, mutation and other factors. The limited range of species within each region resulted in a high degree of development of the crops concerned and consequently food production methods became more efficient. Once methods of cultivation had developed sufficiently to produce an excess of food in the community, the inhabitants could indulge a liking for flowers and herbs and grow them on a smaller scale for pleasure or for their uses in medicine and cooking. Gradually this developed by many diverse ways into the art of gardening as we know it today.

Naturally our knowledge of the early methods of gardening is scanty and based mainly on legend. Religious writings in particular give us an idea of the plants grown for food and for ornament. It is to Theophrastus, however, well named the Father of Botany that we owe so much of our knowledge of the plants grown and the methods used of growing them in the Mediterranean region four centuries before Christ. His *Enquiry into Plants* is a remarkable account of the plants known to the Greek world and it is evident that he himself was a practical gardener. In it he tells us numerous anecdotes of the uses to which various plants could be put, and relates many legends which had inevitably grown up around them.

The Romans advanced considerably the cultivation of plants for decorative purposes but with the passing of the Roman Empire ornamental gardening in Europe remained almost at a standstill. Vegetables and fruit, together with many plants grown for herbal and medicinal purposes, were maintained particularly in monastery gardens. No doubt many of the herbs were grown for the pleasant scents they imparted in the somewhat insanitary living conditions of the times. For food, too, they would probably have been in considerable demand adding flavour or spiciness to otherwise dull concoctions.

The Herbals from that of Dioscorides, a Greek of the first century A.D., to the *Herball* of John Gerard in 1597 (an adaptation of the writings of Robert Dodoens) provide the main sources of our knowledge of the plants grown in Europe in the sixteen hundred years after Christ's birth and it is very evident that the healing properties, reputed or actual, were of more concern than the ornamental value. Almost all the plants grown would have been native to Europe or the Mediterranean countries as the rich floras of the remainder of the world could not be drawn upon. Conrad Gesner in *Horti Germaniae* (1561) listed over a thousand plants from gardens in Germany, many evidently selected forms of wild European species, some with doubled flowers, others varying in flower colour from the normal populations.

From Gerard's time onwards our information on plants and gardening increased considerably. The *Paradisus* of John Parkinson, the Tradescants and Philip Miller with his invaluable *Gardeners Dictionary* (1731) set the pace for the botanical works of Linnaeus, whose system of classification forms the basis on which the structure of plant naming is built.

On the other side of the world in China and Japan the art of horticulture was highly developed by the ancient civilizations of those countries. The cultivation of Chrysanthemums, Camellias and Paeonies, and the artificial dwarfing of trees, today known as bonsai were well advanced. But it was not until Kaempfer, Thunberg (a pupil of Linnaeus) and Von Siebold sent back information on Japanese plants and the Frenchman Pierre d'Incarville and later Reeves and Fortune opened the floodgates of China's floral wealth, that knowledge and both wild and cultivated plants from those areas reached Europe in any quantity. Since the initial explorations many botanists and plant collectors have sent back an incredible diversity of plants for our gardens, particularly from China with which country the names of such eminent collectors, as Henry, Forrest, Wilson, Farrer, Kingdon Ward, Farqes, Delavay Soulié, and Père David will always be associated.

It is impossible here even to touch on the development of gardening following the enormous influx of plants brought about by increased trading and travel from the 16th century onwards. And so to the present day bypassing such milestones as the 'Tulipomania' which swept Holland in the 1630s, the founding of the (now Royal) Horticultural Society in 1804, and the era of William Robinson and Gertrude Jekyll, arch exponents of natural gardening in the late nineteenth and early twentieth centuries.

In a book such as this, covering a wide range of tropical, subtropical and temperate plants, it is necessary to limit very considerably the number of species mentioned. Frances Perry has included the major families of plants and from those families selected individual genera, species and cultivars for their spectacular flowers, for their general interest or for the frequency with which they are met. Of necessity, this must be a personal choice as it would be quite impracticable to include in this book more than a small fraction of the enormous number of plants grown throughout the world.

Botanical plant names to the gardener are wearisome and often annoying. It frequently seems that the Latin names of the plants one knows best have been, or are just about to be changed by some interfering botanist. The horticulturist who all his gardening life has known, for example, the Shrimp Plant, *Beloperone guttata*, is first told it should be referred to as *Drejerella guttata*, and before this name has become currently used, even by botanic gardens, it is replaced by *Justicia brandegeana*. There are very good reasons for alterations of this kind. Botany is not an exact science. Our overall knowledge of plants, even in areas reasonably well botanized, is relatively small and much detailed research is still required; the relationships of species and often genera are constantly argued about by botanists. The Shrimp Plant was placed in the genus *Beloperone* by the botanist Brandegee. Recent research indicated that its morphological relationships were much closer to those of other species in the genus *Drejerella*; and yet further research, based on pollen characters, showed that the Shrimp Plant and several related species were best placed in the genus *Justicia*. This is an example of altering the classification of a plant in the light of additional knowledge of its relationships.

The opposite course is sometimes taken. Groups maintained as separate entities by some botanists may be merged by others who, on close scrutiny, often of more adequate material, consider the differences originally used to separate the groups to be insufficient. Thus the genus *Negundo* was at one time considered distinct from *Acer* but is nowadays almost always merged with it. Similarly, *Clematis* is now held to include *Atragene*, in which genus such a well-known garden plant as *Clematis macropetala* would be placed if it were maintained as distinct.

Some changes are due to the need to correct misapplication of names or misidentifications in the past. And here gardeners are often as responsible as botanists for the confusion! A South African plant misidentified in our gardens for many years was *Osteospermum* (*Dimorphotheca*) *jucundum*, which is still to be seen under the names *Osteospermum* or *Dimorphotheca barberae*, a distinct but superficially very similar species.

Much more frustrating are nomenclatural changes due to strict applications of the principle of priority. *Viburnum fragrans* is now considered an invalid name for the superb winter-flowering shrub brought back to cultivation in Europe by Purdom from western China in 1911 and by Farrer in 1915. Unfortunately, the name *V. fragrans*, which was given to this species by the botanist Alexander von Bunge in 1833, is antedated by Loiseleur's use in 1824 of *V. fragrans* for a different plant, and under the rules of botanical nomenclature the earlier published name has priority. Bunge's use of the name must be rejected and in 1966 Dr W. T. Stearn reluctantly published the name *V. farreri* for this species. The curious rigidity of the rules is, in some respects, hard for gardeners (and many botanists) to understand. Under the Botanical Code conservation of a fairly large number of generic names is already accepted, and after reasoned argument others are added to the list following each Botanical Congress. It would not appear unreasonable for a similar procedure to be adopted for specific epithets, again after arguments had been put forward for consideration by a specialist committee and accepted or rejected as befitted the case. Few people will see reason in altering the botanical name of the Tiger Lily, universally known as *Lilium tigrinum*, to *L. lancifolium*, but under the rules as at present formulated this regrettably should be done.

Generally, of course, the effect of applying rules accepted internationally is extremely beneficial in stabilizing plant names, always bearing in mind that natural entities do not necessarily lend themselves to mathematical treatments!

The placing of plants into classified groups is a matter of much serious debate amongst botanists struggling towards the perfect system. Opinions differ considerably as to the classification of different groups—some separating, others merging related plants. In order to avoid needless controversy Willis' *Dictionary of the Flowering Plants and Ferns*, revised by H. K. Airy Shaw, 1966, was chosen as a standard. The names of the plant families and genera for this book, with one or two minor exceptions, accord with Willis, and as far as possible the nomenclature used follows the recommendations of the International Code of Botanical Nomenclature (1966) and the International Code of Nomenclature for Cultivated Plants (1969). This, in some cases, has meant using less commonly seen but more correct spellings of plant names. As an example, the Botanical Code recommends that specific epithets taken from the name of a man or woman which ends in *-er* should be formed by adding the letter *-i* (thus Fraser to *fraseri*), and not *-ii* which is more usual for names ending in consonants. Similarly, when such epithets are in adjectival form we have *Tulipa fosterana* not *fosteriana*, although the latter spelling is more common.

Where possible also, the most recent floras and monographs of genera have been consulted. Slight discrepancies occur in some instances between the names used for the same plant in floras which overlap partially. *Flora Europaea* (1964) retains *Anemone* and *Pulsatilla* as distinct genera. The Flora of Turkey (1965), on the other hand, merges the two under *Anemone*. In such cases the more normal horticultural usage has been preferred. Botanical taxonomic work is continually in progress and even as this book goes to press some of the names used will have been superseded.

The purpose of *Flowers of the World* is to outline the vast wealth of plant species throughout the world. Climatic conditions determine to a very great extent the type of plants we may grow in our gardens. Some plants are very tolerant and will grow in areas with very different climates. Others require more specific conditions to thrive. It is essential for some to have a period of cold or a rest period; others need definite periods at a specific temperature to form their flower buds; yet others will not tolerate any hint of frost; elsewhere water is a limiting factor; and so the fascinating, interwoven kaleidoscope goes on. There is a constant challenge to gardeners to try to grow plants in their gardens or greenhouses from other areas of the world. Now, as in the past, enthusiasts will go to immense lengths to provide just the right conditions in which they hope a particular species will reach perfection. We grow plants for many different reasons. For their beauty or their quaintness; for their botanical or scientific interest; for their value in landscaping; for human vanity in succeeding where others fail; for their herbal, medicinal or economic properties; or merely because we just like them to be part of our surroundings.

But the underlying theme is still there. Plants are inextricably bound up with man's survival and the development and maintenance of cultivated plants is a matter which should concern all of us.

The pressure of increasing population throughout the world is ever present. Using land for building according to short-sighted economic theories and laying waste vast areas for other purposes of the human race is liable, in the not too distant future, to end in man's starvation. We are all too ready to rely on 'someone else' to provide the food we need, forgetting that the two factors of increasing populations in almost every country of the world and decreasing land on which food can be grown will eventually meet. Our indifference to the extinction of a plant species or a cultivar of no apparent economic value is a contributory factor. Even the most unprepossessing of plants may be of use to man. It may contain chemicals which will contribute to controlling a human disease as is the case with *Catharanthus* (*Vinca*) *roseus*, the pretty Madagascar periwinkle, grown as an ornamental, but now of use in cancer research. Or it may be an obscure wheat cultivar, which, although poor-yielding, contains in its genes the potential to resist disease. If it is lost, this potential source of resistance is lost too.

And so unwittingly, often unknowingly, gardeners help maintain the balance between the human race and the plant life of the world. Any book which serves to popularize the cultivation of plants is thus contributing to this end. None of us knows the full extent to which plants can be used in man's service. The old lady with her own special brand of double Primrose or the Chinese peasant who has raised his particular strain of rice may not appear to contribute much to civilization by having grown them and maintained them—but they might just be preserving a potential cure for epilepsy or the means of combating a devastating plant disease—and we shall never know if they are ignored.

Glossary

A ACHENE a small one-seeded dry indehiscent fruit
ADPRESSED closely pressed together but not united
ADVENTITIOUS (of roots, buds, etc.) growing in an unusual place
AERIAL (of roots, except those direct from leaf cuttings) adventitious roots rising above ground and forming climbing, supporting or absorptive organs
ALBUMEN nutritive material stored in the seed
ALPINE 1. Growing above the tree-line on high mountains. 2. Suitable for growing in a rock garden
ALTERNATE (of leaves) placed singly at different heights on the stem or axis; one at a node
ANNUAL completing a life-cycle in only one year
ANTHER the part of the stamen containing the pollen
ARIL an extra covering or appendage on some seeds
AUTOTROPHIC able to build up food from simple inorganic compounds
AWN a bristle-like part or appendage
AXIL the upper angle between the leaf and the stem on which it is carried
AXIS the main stem or root or a branch in relation to its own branches or appendages

B BARB a fine hooked bristle
BERRY a fleshy fruit containing several seeds with no stony layer around them
BIENNIAL completing a life cycle within two years
BIGENERIC HYBRID a hybrid plant resulting from a cross involving parents from two different genera
BILOBED (of leaves) having two lobes
BIPARTITE divided into two parts
BIPINNATE (of leaves) leaflets on either side of a stalk (like a feather), each leaflet being sub-divided again in a similar way
BISEXUAL bearing both functional male and female organs in the same flower
BITERNATE (of leaves) three-parted with each division again three-parted
BLADE the expanded part of a leaf or petal
BRACT a modified leaf, sometimes brightly coloured, usually just below a flower or flower cluster
BRACTEOLE a secondary bract, often on a secondary axis such as the flower stalk
BUD a condensed, undeveloped shoot or flower
BUD SPORT usually a flower bud showing deviation from the normal type
BUDDING propagation by inserting a growth bud in a slit in the bark of another plant of the same or a different species (the stock) so that the tissues unite
BULB a modified stem surrounded by swollen fleshy scale leaves or leaf bases containing stored food
BULBIL a small bulb in the axil of a leaf or replacing a flower bud
BULBOUS 1. bulb-shaped. 2. perennating by means of a bulb, corm, tuber or rhizome
BUTTRESS ROOT a branch root arising on the stem above the soil and acting as additional support

C CALCIFUGE plants disliking lime in the soil
CALYX the outer whorl of the perianth composed of individual sepals
CAPITATE having a globular flower (or fruit) head
CAPSULE a dry dehiscent fruit containing many seeds and formed from two or more united carpels
CARPEL one of the units of a compound pistil; a simple pistil has one carpel consisting of a stigma, style and ovary
CELL a biological unit containing the nucleus; surrounded in plants by a cellulose wall
CHROMOSOME thread-like bodies found in the nucleus of a cell and containing the units of inheritance (genes)
CINCINNUS a cymose inflorescence with short alternating lateral branches
CIRCUMPOLAR having a natural distribution that includes both North America and Eurasia
CLAW the narrowed base of some petals or sepals
CLEISTOGAMY the production of self-pollinated flowers that do not open
CLEISTOGENE a plant which bears cleistogamous flowers
CLONE a group of plants propagated vegetatively from a single individual
COMPOUND (of leaves) having two or more leaflets
CONE a dense inflorescence or fruit body of more or less conical shape
CONSPECIFIC included in the same species
CORDATE (of leaves) heart-shaped (usually of the leaf-base)
CORM an enlarged fleshy usually underground stem-base often covered with thin fibrous scales
COROLLA the inner usually coloured members of the perianth, the petals
CORONA a crown-like structure on the inside of the corolla or perianth; or as an outgrowth of the stamen-tube
CORYMB a racemose flower head in which the lower stalks are elongated so that the top is more or less flat
CRENATE having scalloped edges
CROSS FERTILIZATION the fertilization of the ovules in the ovary with pollen from another flower or plant
CROSS POLLINATION the transference of pollen from the anthers of one flower to the stigma of a different flower
CROWN the point at which the root and stem of a plant merge; or the part of the stem at ground level
CULTIVAR an assemblage of clearly distinguished cultivated plants which when reproduced sexually or asexually retains its distinguishing characters; the horticultural equivalent of the botanical variety
CUTICLE the thin waxy external covering of plants forming a continuous skin over the epidermal cells
CYME an inflorescence where the main axis bears a single, terminal flower and further growth is by lateral branching
CYMOSE bearing cymes in which the central flower opens first, the remaining flowers opening in succession outwards to the periphery

D DECIDUOUS (of leaves) falling at the end of one season of growth
DECUMBENT (of stems) lying on the ground but with the tips upright
DEHISCENCE the method of opening of a seed-containing body or anther at maturity
DENTATE having sharp, coarsely-toothed indentations
DICOTYLEDON a class of seed-bearing plants typified by the two cotyledons (seed-leaves) of the seed embryo
DIOECIOUS having individual male and individual female flowers on different plants
DIPLOID having two sets of chromosomes per cell
DIVISION propagation of plants by dividing them into smaller pieces
DORSAL pertaining to the surface farthest from the axis, often the lower surface of a leaf
DRUPE a fleshy indehiscent fruit with the single seed enclosed in a stony/woody covering (endocarp)

E ENTIRE having a continuous margin without indentations
EPIDERMIS the outer single layer of cells covering vascular plants
EPIGYNOUS (of flowers) having the sepals, petals and stamens arising on or above the ovary
EPIPHYTE a plant that lives on the surface of another plant or other support, is not a parasite and is not usually attached directly to the soil
EXOTIC a plant that has been introduced from another region

F F_1 HYBRID first generation plants of selected parents; a group of plants maintained by crossing the same two parents each season
FAMILY (of plants) a group of related genera—used in classification
FARINA powder, meal (as on some *Primula* leaves)
FARINACEOUS containing starch; applied also to plants coated with farina
FASCIATION the fusing together of organs; much flattened, abnormal growths result
FASTIGIATE having erect branches close together in a shoot system, often forming a column
FEEDING (of plants) the application of fertilizer, manure or compost
FEMALE FLOWER a flower lacking fertile male organs but with fertile female organs
FENESTRA a small opening, resembling a window
FERTILE capable of producing seeds that are able to germinate
FERTILIZATION the union of the male and female gametes from the pollen grain and the ovule, respectively
FIBROUS containing many fibres, the strengthening tissues of plants
FISSURE a deep groove dividing a structure into lobes
FLORET small individual flowers that make up the flower head of a composite flower or other dense inflorescence
FOLLICLE a dry, dehiscent capsular fruit that has an opening on one side only, the product of a simple pistil
FORCE to induce growth and/or flowering artificially before the normal season
FRUIT the ripened pistil following fertilization; the seed-bearing organ

G GAMETES the sexual reproductive cells, contained in ovules and pollen grains
GENUS a group of closely related species, used in classification
GERMINATION the first stage in the development of plants from seed; sprouting; putting forth shoots
GLAND one or more cells that manufacture a secretion
GLAUCOUS grey-green; having a 'bloom' or whitish covering which may rub off
GLOBOSE almost spherical
GRAFTING the transference and attachment of one part of a plant (the scion) to another plant (the stock) so that the tissues unite; used for propagation
GYNOSTEMIUM a structure consisting of the united male and female organs of a flower (as in Orchids)

H HABIT the way in which a particular plant grows
HALF-HARDY (of perennials) plants requiring protection in winter, by lifting and storing in a frost-free place (of annuals) sown under glass and protected until frost danger has passed before planting
HALF-RIPE CUTTING a cutting taken from shoots or stems that have completed their growth but are not yet woody
HALF-RIPE SHOOT shoots which have completed their growth but are not yet woody
HAPLOID an organism or gamete having one set of chromosomes per cell, that is half the normal number
HARDY plants that will survive moderate winter frost
HEEL CUTTING a cutting of soft wood with a base of hard wood formed the previous year
HERB a seed plant with a soft non-woody stem; non-botanical name for culinary and medicinal plants
HERBACEOUS pertaining to herbs; (of borders) containing perennial plants which die down in autumn to basal crown buds from which fresh growth occurs following a dormant period
HERMAPHRODITE with both functional male and female organs in the same flower; bisexual
HETEROMORPHIC with more than one kind of form
HUMUS organic soil material consisting of decomposed plant and animal matter
HYBRID a plant resulting from a cross between parents genetically unalike; more commonly, the offspring from crossing two different species or their variants
HYGROSCOPIC capable of absorbing moisture

I IMBRICATE overlapping like tiles in regular order (of scales, bracts, petals)
INDEHISCENCE not opening, as of a seed pod or anther
INFERIOR (of ovaries) one that is seemingly below the calyx and fused with the receptacle
INFLORESCENCE the arrangement of flowers on the axis; a flower head
INTERNODE the part of the stem (or axis) between two nodes or joints
INVOLUCRE one or more whorls of bracts or small leaves forming a collar at the base of a flower or flower head
IRREGULAR (of flowers) asymmetrical; said of any whorls of flower parts (e.g. petals) in which the members are not all alike

K KEEL a boat-shaped petal formed from two front (anterior), often united, petals in legumes; ridge on a blade of grass

L LABELLUM lip; particularly the enlarged, lower (posterior) inner perianth segment of an orchid
LANCEOLATE (of leaves) long and thin, widening above the base and tapering at the top; lance-shaped
LATERAL positioned at the side, e.g. side shoots as opposed to the terminal shoot
LAYER to peg a shoot to the ground so that it will form adventitious roots and may be detached after a time as a separate plant
LIANA a woody, climbing plant
LIGULATE strap-shaped
LIME 1. a general term used for horticultural materials with a high calcium content; chalk or limestone; strictly referred to quicklime (calcium oxide). 2. a type of citrus fruit
LINEAR (of leaves) long and narrow, the sides parallel or almost so
LOAM a general term applied to soils of varying characteristics; particularly to those containing organic matter and of great fertility; and to those with medium content of sand and clay particles
LOBE part of a plant organ divided from the rest of that organ by a (usually) deep indentation
LOCULE a compartment in an organ such as an ovary

M MALE FLOWER a flower lacking fertile female organs but with fertile male organs
MARMORATE having an appearance like marble
MID-RIB the main rib or vein of a leaf, a continuation of the petiole
MONOCARPIC dying after bearing fruit once only
MONOCOTYLEDON a class of seed-bearing plants typified by the one cotyledon (seed leaf) of the seed embryo
MONOECIOUS having both individual male and individual female flowers on the same plant
MONOTYPIC GENUS a genus containing only one species
MORAINE a mass of crushed rocks at the base of a glacier; a rock-garden term for a specially contructed bed for alpine plants requiring well-drained conditions
MULCH a loose covering of straw, leaves, manure, etc., laid down to lessen moisture loss from soil around the roots of plants
MUTATION a spontaneously arising genetic change in a plant
MYCORRHIZA the association of a fungus with the roots of a higher plant

N NAKED 1. (of seeds) without a covering formed from the ovary 2. (of flowers) without a perianth
NECTARY a gland that secretes nectar, usually situated at the base of the petals or perianth
NODE the joint on the stem from which the leaves arise
NUT a dry, indehiscent, one-seeded fruit with hard woody walls
NUTLET the stone formed inside a drupe; one-seeded divisions of the carpel as in Labiatae and Boraginaceae

O OFFSET a small bulb produced by mature ones; small rooted pieces of herbaceous plants; a short runner ending in a plantlet
OPPOSITE (of leaves) two at a node on opposing sides of the axis (stem)
ORBICULAR round and flat; disc-shaped
OVARY the part of the female organ that encloses the ovules and develops into the fruit
OVATE (of leaves) egg-shaped and attached to the stem at the broader end
OVULE the structure containing the female gamete which develops into the seed after fertilization

P PAIRED (of leaves) two directly opposite each other on the stem
PALMATE (of leaves) divided or lobed like a hand with outstretched fingers
PANICLE a loosely, diversely branching racemose flower cluster
PARASITE a plant which obtain its food wholly or partially from another living plant known as the host
PARTHENOGENESIS the development of the seed embryo from an unfertilized ovule (female gamete)
PEDICEL a stalk of a single flower (or fruit) in a group
PEDUNCLE a stalk of a flower cluster or of a solitary flower
PELTATE (of leaves) attached to the stalk at a point away from the margin; shield-shaped
PERENNIAL having a life cycle that lasts more than two years
PERIANTH the outer whorls of a flower considered together; a collective term for the calyx and corolla
PERFECT (of flowers) containing functional pistils and stamens in the same flower
PERPETUAL FLOWERER a plant with a long flowering season
PETAL a perianth segment from the inner whorl, one of the members of the corolla, usually brightly coloured
PETALOID petal-like
PETIOLE the stalk supporting a leaf
PHYLLOCLADE a more or less flattened branch or stem functioning as a leaf
PINNATE (of leaves) leaflets on either side of a stalk (like a feather)
PINNATIFID (of leaves) with lobes, cut in a pinnate (rather than palmate) manner, up to halfway to the mid-rib
PISTIL the female organ of a flower consisting of one or more carpels
PITCHER a leaf or part of a leaf modified to catch insects
PITH the central soft, spongy part of the stem
PLANTLET a small plant at the end of a runner or on a leaf margin
PLICATE (of leaves) folded like a fan
POD a dehiscent, dry fruit
POLLEN the powder (spores or grains) borne by the anther, each grain containing the male gamete(s)
POLLINATION the transfer of pollen from anther to stigma (used incorrectly for fertilization)
POLYPLOID an organism having more than two sets of chromosomes per cell
PROBOSCIS the trunk-like structure on the head of some insects
PROPAGATE (of plants) to increase in number sexually (by seed) or asexually (by vegetative means—cuttings, layers, grafts, etc.)
PSEUDO-BULB the thickened bulb-like stem of some Orchids

R RACE often applied to a local population of a species, differing in minor points from other populations of the same species
RACEME a single, often elongated inflorescence with flowers usually borne on short pedicels along a common axis (stem) of indeterminate growth
RADIAL (of leaves or flowers) growing from a central point
RADICAL a leaf or stem arising directly from the root
RECEPTACLE the modified, often expanded, end part of the stem, on which some or all the flower parts are borne
RECURVED curved backwards
REFLEXED sharply curved or turned backwards
REGULAR (of flowers) radially symmetrical
RENIFORM kidney-shaped
RESIN the secretion products of certain plants; hard or gum-like in character
REVERT to show the features of an ancestor, used of a hybrid or highly selected plant
RHIZOME a frequently but not invariably horizontal stem often thickened which is at least partly underground and bears both roots and shoots
RHIZOMATOUS pertaining to or like a rhizome
ROOT the part of a plant that anchors it to the soil and (in most cases) absorbs moisture and nutrients
ROOTSTOCK the basal, persistent part of the stem or an erect herbaceous perennial, from which new roots and aerial shoots arise in the next growing season; a plant used as a stock in budding or grafting
ROSETTE an arrangement of leaves or other organs radiating from a central axis or crown
RUNNER a slender prostrate or creeping stem which roots at the nodes
RUNNING HABIT having prostrate or creeping stems that readily take root

S SAP the aqueous solution of mineral salts, sugars and other organic substances occurring in plant cells
SAPROPHYTE a plant which lives on dead or decaying organic matter
SCALE LEAVES small often appressed leaves protecting a bud or other organism
SCAPE a leafless flower stem that arises directly from the ground
SCREE a steep slope covered with loose stones; used in gardens for a stony or gritty well-drained area for growing alpine plants
SEED the ripened ovule, containing an embryo ready for germination under suitable conditions.
SEED-LEAF the first leaf produced by a germinating seed
SELF-FERTILE successful fertilization of the ovule(s) of a flower by pollen from the same flower or (more loosely) same plant
SELF-POLLINATION the transference of pollen from the anther of a flower to the stigma of the same flower or (more loosely) same plant
SELF-STERILE incapable of producing viable seeds when the ovary has been fertilized by pollen from the same flower or (more loosely) same plant
SEMI-DOUBLE a flower with more than the usual number of petals for the species but with not all the stamens and pistils petaloid
SEPAL a leaf-like member of the calyx
SEPTUM a partition dividing a cavity
SESSILE (of flowers or leaves) arising directly from the stem without a stalk
SHOOT the stem of a vascular plant derived from the plumule in the seed embryo
SHRUB a much-branched perennial woody plant usually without a central trunk
SILICULA a broad, often flat, dry, dehiscent fruit (a short pod) divided in two by a false septum
SIMPLE (of leaves) having an undivided blade, not split into leaflets
SINGLE (of flowers) having the normal number of petals for the species
SLIP a shoot removed from a plant to be used as a cutting
SPADIX an elongated raceme of sessile flowers surrounded by a spathe, the stem of the flower spike usually long and fleshy
SPATHE a bract or leaf-like structure protecting a spadix, sometimes coloured and flower-like
SPECIES the basic unit of classification; small natural populations of similar individuals separated from other populations by definable characters
SPECIES GROUP a group of species which may have many features in common
SPHAGNUM a bog moss which rots to form valuable type of peat
SPIKE an inflorescence with sessile flowers on an elongated, unbranched axis; commonly for any narrow elongated inflorescence
SPIKELET a secondary spike with only a few flowers
SPORT a mutation in the main vegetative body of a plant
SPUR a tube-like appendage on a flower
SPUR-BACK to cut out unwanted clusters of buds or growth
STAMEN the male pollen-bearing organ of a flower consisting of an anther borne on a filament
STAMINODE a sterile stamen, sometimes leaf-like or petaloid
STANDARD a tree or shrub with a tall erect stem and the lowest branches arising at a point several feet above the ground
STIGMA the part of the pistil that receives the pollen
STIPULE a leaf-like structure growing at the base of a leaf petiole
STOCK a plant that receives a graft
STOLON a creeping stem at or below ground level able to form a new plant at the tip; less commonly, a shoot that bends to the ground and roots at the tip

STOMA (pl. stomata) a minute opening or pore found mainly on the underside of the leaf

STRAIN a group of similar plants that has been raised and is maintained by careful selection

STRIKE to form roots

STYLE the part of the pistil between the ovary and the stigma

SUB-SHRUB a perennial in which the lower part is woody and the upper part is soft and herbaceous

SUBSPECIES a distinct sub-division of a species

SUB-TROPICAL nearly tropical; requiring a warm greenhouse or suitable for planting outside in frost-free areas

SUCKER a shoot arising from a subterranean stem or root

SUPERIOR (of ovaries) borne above the point of attachment for the perianth of stamens or surrounded by the receptacle but not fused to it

SYMBIOSIS the living together of two organisms of different species to their mutual benefit

T TAP-ROOT an elongated primary root bearing smaller secondary roots

TENDRIL a thin, specialized thread-like leaf or stem used by climbing plants to attach themselves to supports

TEPALS units of the perianth not clearly differentiated into calyx or corolla

TERNATE arranged in threes

TOP to remove the growing point

TOP DRESSING a layer of manure or other beneficial material around the base of a plant

TRANSPIRATION the loss of water usually as water vapour from the leaves

TRICARPELLATE having three carpels

TRILOBATE having three lobes

TRIPARTITE having three parts

TRIQUETROUS (of stems) having three angles

TRUSS a compact, moderate-sized cluster of flowers; a general horticultural term frequently used to describe ill-defined inflorescences

TUBE the fused part of the calyx or corolla

TUBER a short, congested, fleshy, often underground, stem acting usually as a food reserve

TYPE (HOLOTYPE) the specimen used by the author in describing a new plant, and upon which the name is based; horticulturally used for the original plant of a species known in cultivation as distinct from other forms which may be bred or introduced later

U UMBEL a racemose often flat-topped inflorescence with flower stalks arising from a common point like the spokes of an umbrella

UMBELLATE arranged in umbels

UNDER-SHRUB a low shrub

UNISEXUAL (of flowers) of one sex; male only or female only

V VARIEGATED having organs that contain one or more other colours in addition to the primary colour

VARIETY botanically a group of individuals distinguishable within a species but insufficiently so to be considered a distinct species; horticulturally, any variation of botanical or horticultural origin within a species; applied also to selected forms and hybrids

VEGETATIVE REPRODUCTION asexual; using vegetative parts from an individual to increase the stock of that individual (cuttings, grafting, budding, but *not* by seed)

VEIN a vascular strand in a leaf

VERTICILLATE arranged in whorls

VIVIPAROUS producing daughter plants or bulbils which develop while still attached to the parent plant

W WHORL a circle of three or more leaves or parts of flowers arising from one point

X XEROPHYTE a plant adapted to growth in very dry areas

Z ZYGOMORPHIC (of flowers) divisible into equal halves in one plane only (see also irregular)

Bibliography

Book of Spices, The by Frederick Rosengarten. Livingston Publishing Co., Macrea Smith Co., Philadelphia, 1969.

Carnations and all Dianthus by Montague Allwood. Allwood Bros., 1926.

Collins Guide to Bulbs by Patrick Synge. Collins, London, 1971.

Coming of the Flowers, The by A. W. Anderson. Williams & Norgate Ltd., 1950.

Companion Plants by Helen Philbrick and Richard Gregg. Stuart & Watkins, London, 1966.

Cultivated Aroids, The by M. B. Birdsey Gillich Press, Berkeley, USA, 1951.

Dictionary of Economic Plants by J. C. Th. Uphof. Weinheim (Bergstrasse), Hafner Publishing Co., New York, 1959.

Dictionary of the Flowering Plants and Ferns, A by J. C. Willis. H. K. Airy Shaw, Cambridge, 1966.

Early Uses of Californian Plants by Edward J. Balls. University of California Press, Berkeley, USA, 1965.

Encyclopaedia of Annual and Biennial Garden Plants, An by C. O. Booth. Faber & Faber Ltd., London.

English Rock Garden, The by Reginald Farrer. T. C. and E. C. Jack Ltd., London, 1918.

Ericas in Southern Africa by H. A. Baker and E. G. H. Oliver. Purnell & Sons (Pty) Ltd., Cape Town, S Africa, 1967.

Exotica 3 by Alfred Byrd Graf. Roehrs Co., Rutherford, USA, 1963.

Field Guide to Alpine Plants of New Zealand by J. T. Salmon. A. H. and A. W. Reed, Wellington, New Zealand, 1968.

Flora Brasilica by F. C. Hoehne. Romiti & Lanzard, 1940.

Flowering Plants of the Anglo-Egyptian Sudan, The by F. W. Andrews. 3 vols. Sudan Government Publication, 1952.

Flowering Trees and Shrubs in India by D. V. Cowen. Thacker & Co. Ltd., Bombay, India, 1969.

Flowering Trees of the Caribbean by Barnard and Harriett Pertchik. Rinehart & Co. Inc., New York, 1951.

Flowering Trees of the World by Edwin Menninger. Hearthside Press Inc., New York, 1962.

Flowers and Flower Lore by Hilderic Friend. W. Swann Sonnenschein & Co., London, 1884.

Flowers and their Histories by Alice M. Coats. Adam & Charles Black Ltd., London, 1956.

Flowers of the Brazilian Forests by Margaret Mee. Tryon Gallery, London, 1969.

Folklore and Odysseys of Food and Medicinal Plants by Ernst and Johanna Lehner. Tudor Publishing Co., New York, 1962.

Folklore and Symbolism of Flowers, Plants and Trees by E. and J. Lehner. Tudor Publishing Co., New York, 1960.

Gardening in the Lowlands of Malaya by R. E. Holttum. Straits Times Press, Singapore, 1953.

Garden Shrubs and their Histories by Alice Coats. Vista Books, London, 1963.

Gerard's Herball by Marcus Woodward from edition of Th. Johnson, 1636. Spring Books ed., London, 1964.

Guide to Native Australian Plants, A by A. M. Blombery. Angus & Robertson, Ltd., Sydney, Australia, 1967.

Hydrangeas, The by Michael Haworth-Booth from *Effective Flowering Shrubs*. Collins, London, 1951.

Illustrated Guide to Tropical Plants by E. J. H. Corner and K. Watanabe. Hirokawa Publishing Co., Tokyo, Japan, 1969.

Irises by Gwendolyn Anley. Collingridge, London, 1946.

Joy of the Ground by Marion Cran. Herbert Jenkins Ltd., London, 1929.

Ladies Flower Garden (Bulbs, Greenhouse Plants, Annuals, Ornamental Perennials), The by Mrs Loudon, 1840.

Living Plant, The by Knight and Step. Hutchinson & Co. Ltd., London, 1905.

Mountains of the Moon by Patrick Synge. Lindsay Drummond, 1937.

My Garden in Spring by E. A. Bowles. T. C. and E. C. Jack Ltd., London, 1914.

My Rock Garden by Reginald Farrer. Edward Arnold Ltd., 1907.

On the Eaves of the World by Reginald Farrer. Edward Arnold Ltd., 1917.

Orchids by Peter Black. Max Parrish & Co. Ltd., London, 1966.

Pioneers in Gardening by Miles Hadfield. Garden Book Club Ed., Foyles, London.

Plant Hunters in the Andes by T. Harper Goodspeed. Neville Spearman Ltd., London, 1961.

Plant Lore, Legends and Lyrics by Richard Folkard. London, 1884.

Plants for Man by Robert W. Schery. Allen & Unwin, London, 1954.

Plants of New Zealand by R. M. Laing and E. W. Blackwell. Whitcombe & Tombs Ltd., Christchurch, New Zealand, 1940.

Plants of the World by H. C. de Wit. Thames & Hudson Ltd., London, 1963.

Rochford Book of House Plants, The by T. Rochford and R. Gorer. Faber and Faber Ltd., London, 1961.

Rose in Britain, The by N. P. Harvey. Souvenir Press Ltd., London, 1951.

R. H. S. Dictionary of Gardening. 4 vols and supplement. Oxford University Press, Oxford.

Stalking the Wild Asparagus by Euell Gibbons. David McKay Co. Inc., New York, 1962.

Standard Cyclopedia of Horticulture, The by L. H. Bailey. Macmillan & Co. Ltd., London, 1937.

Trees and Shrubs Hardy in the British Isles by W. J. Bean. John Murray Ltd., London, 1950, new edition 1970.

Trees and Shrubs of New Zealand by A. L. Poole and N. M. Adams. Gov. Printers, Wellington, New Zealand, 1964.

Tropical Plants and their Cultivation by L. Bruggeman. Thames & Hudson Ltd., London, 1957.

Wild Flowers of the Cape Peninsula by Mary Maytham Kidd. Oxford University Press, Oxford, 1950.

Wild Flowers of the Transvaal by C. Letty. Howard Timmins, Cape Town, 1955.

Wild Flowers of the United States by H. W. Rickett. McGraw-Hill Book Co. Inc., New York, 1966 onwards.

World of Orchids, The by G. C. K. Dunsterville. W. H. Allen & Co., London, 1964.

Acanthaceae

250 genera and 2500 species

This is a large family of dicotyledonous plants, mostly of tropical distribution but also represented in the Mediterranean area, the United States and Australia.

The habit is varied for, while there are few trees, shrubs, climbers, herbaceous and marsh plants are well represented. Very many grow in damp places in tropical forests. The leaves are usually entire with thin blades and arranged in opposite pairs.

The flowers are frequently condensed in cymes (both terminal and in the leaf axils); but racemose inflorescences and also solitary flowers can occur. The bracts and bracteoles are often coloured, the latter more or less enclosing the flowers which are usually showy, often irregularly shaped and four-to five-petalled, in some cases one or more petals being elongated into protruding tongues. The calyx has four to five clefts and there are four or two (rarely five) stamens. The fruit is a capsule; in some genera (like *Ruellia*) the ripe seeds are scattered by means of a special ejection mechanism.

Acanthus is the hardiest genus, the majority of species coming from the Mediterranean areas of Greece and Turkey. Frost kills the foliage but the roots survive very low temperatures (or they can be protected by a mulch of leaves or litter) and send up new growths each spring. These are very permanent plants with striking architectural lines, especially in the foliage, which makes them highly suitable for key positions in such places as the corner of a border, against steps or in front of a large tree.

The foliage of *Acanthus spinosus*—or more possibly from the shape, *A. mollis*—is reputed to have inspired the traditional design on Corinthian columns in architecture. According to one legend, Callimachos, a Greek architect, was visiting the tomb of a young girl who had died on the eve of her wedding. There, standing on an *Acanthus* plant and left by a previous visitor, was a basket covered with a tile. Callimachos noticed that the leaves had been forced back by the tile into a decorative shape and adapted the motif to fit the pillars of a temple he was building at Corinth.

Both the Greeks and Romans made use of *A. mollis* (in the form of garlands) as an adornment for their buildings, their furniture and even their clothing. Virgil describes an *Acanthus* pattern design embroidered on the robe of Helen of Troy and in some legends it is one (of many plants) credited with furnishing the material for Christ's Crown of Thorns.

The leaves are deeply cut, sometimes spiny and of a rich deep green; the flowers are purple, green and white, borne on stout stems with a spiny bract beneath each blossom. *A. spinosus* var. *spinosissimus* has deeply divided sharply spined leaves. Some *Acanthus*—or Bear's Breeches as they are commonly known—have been used in cough medicine.

Acanthus as an architectural ornament

Favourite pot plants of temperate countries are species of *Jacobinia, Aphelandra, Crossandra* and *Ruellia*. Aphelandras in particular have received considerable attention in recent years from nurserymen as they make good house plants in centrally heated rooms. *Aphelandra squarrosa* 'Louisae', the Zebra Plant from South America, is especially fine with terminal four-sided spikes of deep yellow bracts edged with red, from which the yellow flowers emerge. Growing about 45 cms ($1\frac{1}{2}$ ft) high it has smooth stems furnished with rather drooping, dark green leaves which have their veins picked out in creamy-yellow. A cultivar called 'Brockfield' is an improvement as the leaves are held more upright.

Fittonias are ornamental foliage plants for damp shady borders in subtropical gardens, or for pot work in summer-shaded greenhouses with a moist atmosphere. The small rounded leaves are beautifully patterned with white or red veins, and although blooms appear these are secondary to the foliage.

Fittonia verschaffeltii has dark green leaves netted with carmine veins; a more robust and very striking variant with blood-red veining is known as var. *pearcei* (or *rubra*). *F. argyroneura* (now

Acanthus spinosus

considered a variant of *F. verschaffeltii*) is distinguished by the paler green, white-veined leaves. They are all of dwarf, somewhat trailing habit and come from South America.

Jacobinias are native to the hotter parts of South America and make good winter-flowering plants for greenhouse decoration. However it is important to rest them (that is, keep the pots fairly dry) for a month or so in late summer, otherwise they flower indifferently. The blooms are showy, sometimes solitary, frequently in clusters of varying types. Outdoors they require a fair amount of moisture and for pot culture should be grown in equal parts of leafmould or peat and loam. Cuttings may be rooted in small pots and they should have the tops pinched out when 10–13 cms (4–5 ins) high to encourage a bushy habit.

The most commonly grown are *Jacobinia carnea* (*Justicia carnea*), an upright shrubby plant with grey-green, oval foliage and clustered heads of flesh-pink (sometimes rosy-purple) flowers; *J. coccinea* (now referred to *Pachystachys coccinea,* although *Odontonema strictum* is quite commonly grown as *J. coccinea*) with terminal heads of scarlet blossoms; and *J. pauciflora*. This last grows 30–60 cms (1–2 ft) high and has numerous yellow and scarlet flowers drooping from the leaf axils or standing out almost horizontally. All these species come from Brazil and are easy to grow, requiring much the same treatment as Fuchsias. *J. suberecta* from Uruguay is often used as a basket plant in Britain; it is small and of spreading habit with grey felted leaves and striking heads of one to ten, bright scarlet, tubular flowers. The Mexican *J. ghiesbrechtiana,* also scarlet, is more erect (30–45 cms; 1–1½ ft).

Strobilanthes and *Sanchezia* are striking foliage plants, often with fine leaf markings. *Strobilanthes dyeranus* from Burma is particularly fine, making a branched shrub 60–150 cms (2–5 ft) high with hairy stems and large, opposite, serrated leaves. These are prominently veined and have a bluish-violet metallic sheen on the upper surface but are purplish-red beneath. The lilac flowers come in inflorescences of 15–20 cms (6–8 ins).

Sanchezia nobilis is South American. It makes a branching shrub 60 cms–3.5 m (2–12 ft) high with square, robust stems and opposite, toothed leaves with pale yellow veins.

Crossandras are small evergreen shrubs of 30–90 cms (1–3 ft) which make splendid bedding plants in tropical lowlands. They do not do as well in high altitudes but below 90 m (300 ft) can

Aphelandra squarrosa 'Louisae'

Fittonia verschaffeltii

Pachystachys coccinea

Eranthemum nervosum

be superb, with brilliant four-sided terminal spikes crowded with broad, coloured bracts and flowers. In *Crossandra infundibuliformis* (*C. undulifolia*) these are salmon-pink or orange-yellow—or rosy-orange in the cultivar 'Mona Wallhead'. This Indian species is used lavishly in Asian gardens but can be grown under glass in temperate climates, planted in a peat/loam compost. Three African species *C. nilotica* (brick-red and orange), *C. flava* (bright yellow) and *C. guineensis* (pale lilac, with red furry stems and pink striations on the green leaves) are others suitable for a warm greenhouse.

Ruellias comprise a large genus of annuals and perennials which may be herbaceous or shrubby. The majority are American but representatives are found in Africa, Asia and Australia. They are pretty, free-flowering plants needing a warm climate or treatment under glass and light soil. Several make handsome pot plants, notably *Ruellia portellae* from Brazil, with tubular pink flowers and purplish foliage veined in silver, also R. *paniculata,* the Christmas Pride of Jamaica, with stems up to 90 cms (3 ft) furnished with bluish-purple flowers. The Brazilian R. *rosea* is a profuse bloomer with flowers 2 cms (1 in.) wide of a splendid rosy-red and long tubes of the same shade.

Both this and the dark blue (occasionally white) tuberous-rooted R. *tuberosa* need moist soil and shade from strong sunlight. When moistened the ripe seed pods spring open with a sharp click and eject the seeds over a wide area.

Barleria cristata needs similar cultural conditions. This Indian shrub can be grown as a specimen bush or makes an attractive low hedge, with entire leaves and dense spikes of funnel-shaped violet flowers, with paler blue spots. *B. involucrata,* also from India, is lavender-blue and the Arabian species *B. flava* is yellow.

Beloperone (*Drejerella*) *guttata,* the Shrimp Plant from Mexico, was introduced to Europe in the 1920s. It makes a small twiggy shrub with soft oval leaves and axillary and terminal spikes of brownish-red bracts. From these the white and red flowers emerge like pale tongues and for once the common name suits the species—these do indeed look like pink shrimps. Pot-grown plants usually reach about 30 cms (1 ft) in height and are fairly compact, but when grown outside in frost-free gardens a height of 60–90 cms (2–3 ft) is not unusual and the growth is more straggling. Beloperones are of easy culture provided they have plenty of light and are not waterlogged at the roots.

Eranthemum nervosum (*E. pulchellum*) from India makes a splendid winter-flowering greenhouse plant for cool climates. It has spikes 30–45 cms (1–1½ ft) long of rich blue flowers and very dark green leaves. *Asystasia gangetica* (*A. coromandeliana*) with narrow, wavy-edged leaves, is usually purple-flowered but varies from pink to yellowish-cream. A species of a related genus *Justicia gendarussa,* white-flowered with purple spots, has been used in India for the relief of rheumatism and its leaves introduced amongst clothing will ward off insects. *J. pectoralis* is employed for lung troubles and *Hygrophila spinosa* as a purgative.

Agavaceae

20 genera and 670 species

Many fine ornamental foliage plants belong to this moncotyledonous family, composed of genera once accredited to Amaryllidaceae (like *Agave*) and Liliaceae (like *Dracaena* and *Yuçca*). This family contains many fibrous-leafed plants, often from dry regions, the flowers frequently borne in panicles.

Most Agavaceae have rhizomatous rootstocks, with or without woody stems, and long narrow leaves—often swordlike—crowded together at ground level or on a short trunk. Several are important tropical and subtropical bedding plants and a few are sufficiently hardy to withstand the winters of southern Europe (including southern England) and parts of North America.

The genus *Agave* comprises some 300 species, mostly Mexican or tropical American. The flowers are usually greenish or brown and carried on a tall, pole-like panicle. Their production usually terminates a plant's life, as the surrounding leaf rosette withers and dies when the seed ripens. But a new life springs naturally either from seed or vegetatively through suckers which appear near the base of the old stem.

For garden use the most outstanding species is the American Aloe or Century Plant—*Agave americana.* It is very showy, making a rosette of stiff, fleshy leaves, 90 cms (3 ft) or more in length, and each terminating in a stout spine. There are several variegated cultivars which are popular for pot work or used outside during the summer in raised containers. It is naturalized along the Mediterranean.

The name Century Plant is misleading, for although it was once thought to flower only once a century, there is in fact no certainty when flowers will appear; depending on soil, site and climate it could take five years or sixty. Meantime the plant builds up enormous food reserves in its leaves and when the moment is right a huge terminal inflorescence grows very rapidly, reaching 6 m (20 ft) or more and carrying many greenish blossoms.

The Mexicans make a national drink by cutting off the young flower head and catching the rising sap in a vessel, as much as 1000 litres (220 gallons) reputedly being obtained from a single plant. The fermented juice is called pulque and from this the Mexicans further distil a spirit called mescal.

Another important *Agave* is *A. sisalana,* source of sisal hemp. Before synthetics this was an important Mexican and East African crop, furnishing cord, rope, matting and sacking. Although the industry is declining a certain amount is still grown by native farmers in Kenya.

Furcraea is like *Agave* but with even taller and larger heads of flowers. In the tropics it is not uncommon to see variegated forms of *F. selloa* with huge rosettes of cream and green leaves. They are extremely drought-resistant and native to tropical America. *F. gigantea* (*F. foetida*) is the source of Mauritius hemp.

Dracaena is widely distributed over the tropical Old World. The trees and shrubs of this genus have long, narrow, often

Dracaena draco

Agave americana 'Marginata'

arching leaves and dense panicles of white, yellow or red flowers. Young plants make decorative 'dot' plants amongst summer bedding and are frequently used in jardinieres and conservatories or as pot plants indoors.

D. surculosa (*D. godseffiana*) from the Congo is low growing, with a slender habit and really beautiful leaves. These are oblong, usually arranged in whorls of three and are bright green heavily peppered with cream spots. A cultivar known as 'Florida Beauty' is more heavily patterned.

D. deremensis, also African, is another with beautiful leaves. These are long and narrow, grey-green in the centre with two silver stripes surrounding this and then a border of dark green. Growing 3–4 m (9–12 ft) high, this species carries large bunches of dark red flowers but unfortunately these have a most unpleasant odour—unlike those of *D. fragrans* which are really sweet smelling. This last comes from Guinea and has broad leaves and yellow blossoms. In the West Indies it is often used for hedges. Striped cultivars are popular for tropical bedding and greenhouses.

An interesting *Dracaena* is *D. draco*, the Dragon Tree from the Canary Islands. It has the reputation of being exceedingly long lived, although Alexander von Humboldt was probably a long way out when he quoted the age of a giant plant he saw in 1799 in Tenerife as 5000 to 6000 years. This particular specimen was reputed to be 20 m (70 ft) high and 14 m (45 ft) in girth before it fell in 1867.

A red resinous juice known as 'dragon's blood' exudes from cracks in the trunk of the tree; it has neither taste nor smell and is very brittle, but burns with a bright flame and emits a slight fragrance. It is used in medicine and also in staining marble; in the 18th century Italian violin makers used 'dragon's blood' for varnishing their instruments.

The flowers are like bunches of Lilies and the stiff, rigid leaves are all clustered in tufts at the top of the tree, leaving the rest of the branches bare. Small specimens are often used for subtropical bedding.

Cordyline is often confused with *Dracaena,* as they are similar in appearance although the flowers are tubular in *Cordyline*.

The most commonly cultivated for ornamental purposes is the South Sea Island Ti, *C. terminalis,* a shrub 1.5–3 m (5–10 ft) high with ribbon-like leaves and panicles up to 30 cms (1 ft) across of white or reddish flowers. Several fine foliage forms exist with

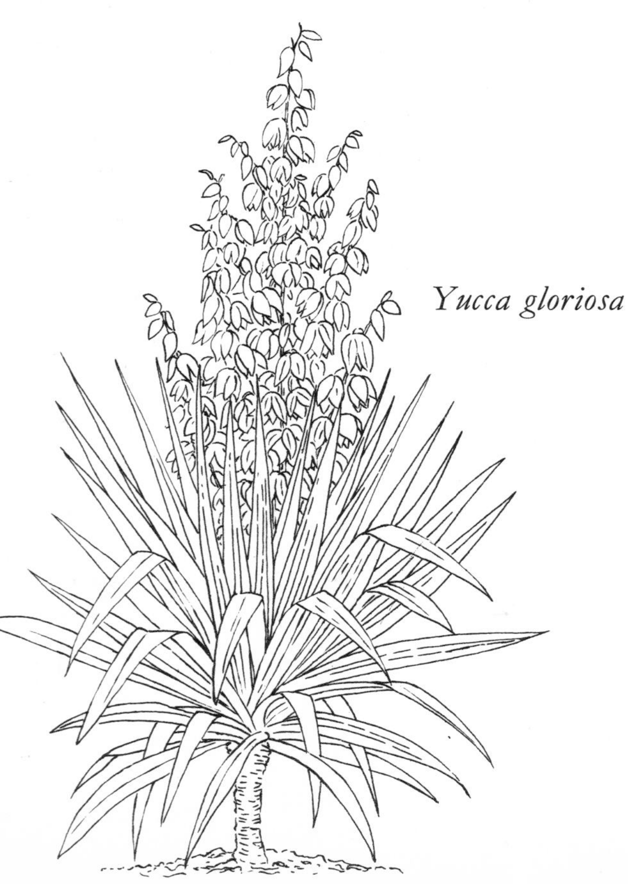

Yucca gloriosa

deep crimson leaves or pink and cream variegations. In Polynesia the leaves are fed to cattle and used in thatching, and also for wrapping fish prior to baking. A rough kind of cloth can be woven from the fibres and is made into fringed skirts by the maidens of Samoa.

C. indivisa from New Zealand is a small tree, frequently under 3 m (10 ft) high with very large pointed leaves 1–2 m (3–6 ft) long and 10–15 cms (4–6 ins) wide which have their mid-ribs picked out in red or yellow. The white flowers are carried in huge, pendulous racemes.

C. australis, also from New Zealand, makes a striking tree 6–12 m (20–40 ft) tall, with the trunks topped by dense, arching, narrow leaves 60–90 cms (2–3 ft) long. The inflorescence is very large, fragrant and creamy-white.

A strong fibre made from the leaves can be used instead of twine or for paper making. It is also known as the Cabbage Tree because early New Zealand settlers ate the crowns as cabbages.

Phormium tenax is the source of New Zealand flax, a tough fibre used for twine and rope and by the Maoris for fishing nets and cloth. This comes from the leaves which are among the toughest in the world. It is unlikely for a man to be able to tear a leaf across horizontally with his bare hands. Fully grown the leaves may be 2.5–3 m (8–10 ft) long and 7–10 cms (3–4 ins) wide. They are green in the type although there are bronze-leafed and variegated sorts.

These are good plants for moist soil in mild climates and were at one time grown in large tubs or pots for standing outside in summer. The flowers are carried on long 'beanstalk' scapes 1–2 m (3–6 ft) tall, individual blossoms being dull red or sometimes yellow, with protruding stamens. They secrete much nectar—it can sometimes be seen dripping from the blossoms—which attracts parrots in their native New Zealand which thus cross-pollinate them. A pulp made from the roasted, macerated roots is sometimes applied as a poultice for abscesses.

Sansevieria (*Sanseverinia*) also yields fibre; indeed *S. trifasciata* is commonly known as Bowstring Hemp. There are about sixty species, all tropical and all xerophytes with fleshy leaves. Extremely resistant to drought and tolerant of shade, they make ideal pot plants for forgetful people; in fact overwatering rots the roots. The leaves are arranged in rosettes and are thick, pointed at the tips, frequently long, fibrous within and usually variegated or mottled with white or cream. The flowers are sometimes fragrant, greenish or cream-coloured and borne in racemes. *S. trifasciata* 'Laurentii' has golden leaf margins; *S. hahnii* (of uncertain origin, sometimes considered a clone of *S. trifasciata*) makes a squat rosette 5 cms (2 ins) high; and *S. cylindrica* has cylindrical rush-like leaves, in two shades of green, and about 90–120 cms (3–4 ft) long.

Yuccas from the US, Mexico and the West Indies have bold tufts of stiff pointed leaves rising directly from the ground or from short woody trunks. The flowers are very spectacular, creamy-white, in large panicles. They have a remarkable method of pollination for at night they emit a perfume which attracts certain insects (the Yucca moths) and so are fertilized. The main garden species are *Yucca filamentosa* or Adam's Needle, a nearly stemless sort with long curly threads on the leaf margins; *Y. gloriosa,* the Spanish Dagger, which has a short trunk and grows to 2.5 m (8 ft); and *Y. recurvifolia* with recurving leaves and branching trunk. The flower buds of *Yucca* species are eaten by the Indians, and the fruits and the roots have saponifying properties so may be used for washing clothes.

Polianthes tuberosa is a fine bulbous plant which may be found in flower throughout the year through successive planting. The blooms are white and extremely showy, in large racemes about 90–120 cms (3–4 ft) high. They are powerfully scented and the source of Tuberose flower oil, used in high-grade perfumery. 'The Pearl' is an improved and shorter cultivar. No wild form similar to the *P. tuberosa* of commerce is known but the plant grown was described by Linnaeus and is probably Mexican. The Chinese in Java use the flowers in vegetable soup.

Dracaena deremensis

Polianthes tuberosa

Dracaena surculosa

Aizoaceae

130 genera and 1200 species

This is a large family of mainly succulent dicotyledonous plants, chiefly from South Africa but also from California, Australia, South America and tropical Asia. The family is composed of xerophytic herbs or undershrubs, with opposite or alternate, usually fleshy leaves. There are four or five (sometimes up to eight) sepals, three, five or many stamens and a two- to five-loculed capsule containing the seeds. The numerous petals are usually very fine, the flowers suggestive of daisies, although there is no close relationship with Compositae. In warm dry climates the mat-forming types produce a mass of brilliantly coloured flowers.

To survive under difficult desert conditions the plants adopt various methods of storing water. Some have fleshy leaves covered with bladder-shaped hairs to ward off sunlight and reduce transpiration. Others reduce their leaves to two, which shrivel to nothing when a new pair are formed in the rainy season; others again bury their leaves in the ground so that only the tips are exposed to the drying effects of sun and wind.

In spite of their frequent low stature and insignificant foliage the flowers are extraordinarily large and showy. They resemble daisies and come in an infinite range of bright colours so that, seen en masse, the impact is perhaps greater than with any other group. This is probably the reason they are so frequently planted on roadside verges and similar open situations in hot countries.

They are ideal for poor soil conditions in full sun, blooming for months on end in garden beds, parks or rock-garden pockets. In a wet season they can be disappointing, for when the weather is dull the flowers remain shut. Those of a spreading habit, like *Mesembryanthemum,* propagate easily from slips or cuttings but the delightful *Drosanthemum* (whose stems are thin and wiry) or the annual *Dorotheanthus* are best raised from seed.

A few species can stand a touch of frost if the weather is dry but cold and wet together turn the plants into a soggy mass. In cool climates rooted cuttings must either be overwintered in a warm greenhouse or seed should be sown afresh each spring.

In South Africa the common name for many of these plants is Vygie, an Afrikaans word meaning 'little Fig' referring to the edible but acid Fig-like fruits. In England, where they have been cultivated for more than 200 years, Fig Marigold is a collective name for all Mesembryanthemums and Hottentot Fig for those which have edible fruits. Another name sometimes used is Ice Plant, presumably because some, like *Gasoul* (*Mesembryanthemum*) *crystallinum,* have cold, glass-like papillae on the leaves. The edible fruits are usually consumed raw, while the leaves of some species are eaten when baked, or chewed for their slightly narcotic content or fed to stock. The crushed fruits of *M. forskahlei* from East Africa are used by the Bedouins in the making of bread; the ash of another species provides a source of soda for washing purposes for the Hottentots.

Dorotheanthus bellidiformis, the Bokbaai Vygie of South Africa, is often listed in catalogues as *Mesembryanthemum criniflorum.* It is a dwarf annual forming a dense mat of narrow, succulent green leaves thickly studded in summer with buff, apricot, orange, pink, purple, crimson or white flowers. The seed is very fine so it is advisable to mix it with silver sand to facilitate sowing—either outdoors, in warm areas, or in colder places in boxes of sandy soil, transferring the seedlings outside when danger of frost is over.

Drosanthemums or Dew Flowers have dewy dots on the leaves like icing sugar and brilliant orange, yellow, white or crimson flowers. These low-spreading plants do best in hot dry areas.

Lampranthus are all perennial and vary from low-growing spreaders to round bushes 45 cms ($1\frac{1}{2}$ ft) tall and 120 cms (4 ft) across. There are countless species including *L. spectabilis,* cerise-pink with a bright yellow centre, a sprawling subject ideal for draping over a wall or rockery; *L. blandus* is pale pink, *L. aureus,* glowing orange, and *L. zeyheri* is purple.

Another important genus in Aizoaceae is *Ruschia,* the species of which form small hard cushions of foliage completely covered when in flower with mauve, pink or white daisies; the species of *Carpobrotus* are perennial succulents with fleshy leaves and juicy fruits which are sometimes obtainable in South African shops and good for jam. *C. edulis* is hardy on the W coast of England.

Lithops is a genus of small South African succulents having two united leaves with a fissure across the top. Their smooth appearance, markings and colours resemble the pebbles among which they grow. The stemless flowers are white or yellow.

Conophytums are succulents from South Africa with very thick glaucous leaves fused together except for a small slit on top. The flowers may be white, yellow or purple and of various sizes. *Conophytum praecox* has white flowers.

Lampranthus spectabilis

Dorotheanthus bellidiformis

Lithops marmorata · *Lithops schwantesii* var. *kunjasensis* · *Lithops mennellii* · *Conophytum praecox*

Alliaceae

30 genera and 600 species

These bulbous or rhizomatous monocotyledonous plants are found all over the world (with the possible exception of Australia) and are intermediate between Amaryllidaceae and Liliaceae. All have flowers carried in umbels like the former but with superior ovaries as in the latter.

Allium is the largest genus with some 450 species distributed throughout the Northern Hemisphere. They include such important economic plants as Onion, Leek, Chives, Garlic and Shallots and some fine ornamentals. Most, but not all, have bulbous rootstocks, the narrow leaves being smooth and hollow in some species, emitting a characteristic Garlic scent when bruised.

Their place in the garden is in mixed borders, naturalized in grass or woodland or in rock-garden pockets. A few (like *A. vineale*) reproduce themselves excessively, not only by seed but from small bulbils produced in the inflorescences (amongst the flowers) and also from underground bulblets. They then become troublesome weeds. Most Alliums like well-drained soil; some are shade-lovers but the majority favour an open sunny position.

Among the most desirable of the smaller species are *A. cyaneum*, 22–30 cms (9–12 ins) high, from China, with close heads of slightly pendent, deep blue flowers; *A. moly*, 15–25 cms (6–10 ins) high, from S and SW Europe, with broad, flat, blue-green leaves and umbels of bright yellow blossoms; and *A. caeruleum* (*A. azureum*), a Siberian plant, slightly taller, 45–60 cms ($1\frac{1}{2}$–2 ft), with rich sky-blue flowers which have deeper blue markings. In midsummer a well-grown group of the latter makes a striking feature in the herbaceous border.

A. karataviense from Turkestan has large round balls of white flowers which may be nearly 30 cms (1 ft) across and flat, blue-green, ground-hugging leaves that are often variegated.

A. bulgaricum (*A. siculum* var. *dioscoridis*) grows up to 90 cms (3 ft) with umbels of drooping, white, green and dark red flowers; these are pretty but have a strange smell like a gas leak or worn rubber so are not for cutting.

A. schubertii from Israel bears up to 200 pink flowers in a single umbel and grows 30–60 cms (1–2 ft) tall; *A. giganteum*, a Himalayan species, reaches 90–120 cms (3–4 ft) but with smaller flower heads and lilac florets.

A. narcissiflorum has bright rose, pendent bells in umbels of three to eight flowers in midsummer. It comes from the Alps and Caucasus and grows about 30 cms (1 ft) tall. *A. neapolitanum* from S Europe has showy umbels of pure white flowers in spring. These are good for cutting and often sold by florists dyed red and greenish-blue. Outdoors the plants like moist soil and grow about 30 cms (1 ft) high. They also make good pot plants.

Tulbaghias are rhizomatous plants from tropical and South Africa, resembling *Allium* and having the same characteristic Garlic smell. Strangely the flowers are often very fragrant, particularly those of *Tulbaghia pulchella* (possibly identical with *T. fragrans*); they are violet and carried in umbels of twenty to thirty on stems of 38 cms (15 ins). *T. violacea,* with purplish-violet flowers, growing to 30 cms (1 ft) in height and *T. alliacea,* with greenish-purple blooms, are perhaps less striking but sometimes grown in warm borders or greenhouses. Happy in full sun or half-shade, the plants need well-drained soil and are propagated by seeds or offsets.

Many plants masquerading under the name *Brodiaea* have now been referred to other genera, although usually still sold as *Brodiaea*. The bulbs are mostly native to western North America,

Agapanthus
hybrid

particularly California, where they stud the meadows with umbels of blue, white and reddish funnel-shaped flowers. The corms require sandy loam soil and are generally hardy in temperate climates if planted in sheltered spots and protected from excessive winter cold. They also make good pot plants.

Among the most satisfactory for garden use are *Dichelostemma pulchella* (*Brodiaea capitata*)—called Blue Dick or Wild Hyacinth in California—a pretty plant with tight umbels of deep violet flowers on stems of 30 cms (1 ft) whose small sweet 'bulbs' are good to eat. Also edible are the corms of *Triteleia* (*Brodiaea*) *laxa* or Grass Nut, which is like a small *Agapanthus* and very variable, with flowers ranging from deep violet, mauve and light blue to white, in large umbels up to 30 cms (1 ft) across on stems 75 cms ($2\frac{1}{2}$ ft) tall. *Triteleia* (*Brodiaea*) *crocea* is yellow and rather small (under 30 cms; 1 ft) and *T. californica* is deep blue and about the same height.

Dichelostemma (*Brodiaea*) *ida-maia,* the Californian Firecracker, is a charming plant with drooping cigar-shaped flowers arranged in tight umbels. These are blood-red with bright yellow and green margins and a white throat.

Agapanthus are beautiful South Africans, freely used in tropical gardens or in cooler climates in conservatories and warm borders. They are also grown in tubs. Non-hardiness in temperate zones seems to be due to excessive winter wet, but covering the crowns of established plants in autumn with glass propped on bricks keeps off rain and yet allows the air to circulate underneath. The beauty of *Agapanthus* lies in their magnificent umbels of flowers, which in most cases are bright blue, and the long, smooth, strap-shaped leaves.

All the species require light well-drained soil, with some well-rotted manure or compost used as top dressing in summer. During the growing season plenty of water should be given but little—if any—in the dormant season.

The genus is much confused, many garden varieties being grouped under the names *A. africanus* and *A. orientalis.* They include forms such as 'Maximus Albus' and 'Giganteus', once referred to *A. umbellatus,* a name no longer valid as the epithet *africanus* was the earliest given to this species and so holds preference. The flowers are borne on immense umbels with up to 200 blooms on a single head and smooth stems of 60–120 cms (2–4 ft). Colours vary from light blue to deep violet; there are white varieties and others with double flowers or variegated leaves. The plants hybridize easily which is one reason for the nomenclatural confusion; some plants play such tricks as producing additional flower spikes from the sides of the umbel (called hen and chickens) or growing out from the top (pagoda). These forms sometimes come true from seed. The hardiest group is the Headbourne Hybrids, a free-flowering race with tight compact heads in rich shades of blue.

A. africanus was introduced to Europe in the 17th century and was flowered first in 1692 at Hampton Court.

Alstroemeriaceae

4 genera and 200 species
all Central and South American

This is a small family of monocotyledonous herbaceous plants, having rhizomatous roots and frequently a climbing habit. The flowers are very showy, occasionally solitary but generally in racemose or capitate heads. Individually these are irregularly shaped, with six perianth segments, six stamens and three-parted capsules containing many seeds. The leaves are smooth, simple and alternate, those of *Alstroemeria* having a twist in the leaf stalks, a unique characteristic which has the peculiar effect of bringing the undersides of the leaves uppermost.

Alstroemeria is named for Clas Alstroemer of Sweden, a pupil of Linnaeus who sent several roots of *A. pelegrina* to his tutor. The plants delighted Linnaeus so much that he kept the fleshy roots in his bedroom for fear they might take cold. This species, commonly called the Lily of the Incas, grows about 30 cms (1 ft) tall and has terminal heads of rosy-lilac or white funnel-shaped flowers with yellow throats and purple blotches and spots.

A. aurantiaca, the Peruvian Lily from Chile, is the hardiest. Growing about 90 cms (3 ft) tall it has many long-stalked orange flowers, individually about 4 cms ($1\frac{1}{2}$ ins) in length and frequently spotted with chocolate. Improved cultivars include 'Dover Orange', 'Moerheim Orange' and 'Lutea' which is yellow. In well-drained soil and a sheltered situation *Alstroemeria* species grow splendidly and make pleasing associates for Scabious, Erigerons and Delphiniums. The trick of establishment is to plant the fleshy roots 30 cms (1 ft) down in previously prepared soil, barely covering the roots at first, but waiting until they shoot and then little by little bringing the soil to ground level.

The same process helps with *A. ligtu* and its hybrids although this species is not quite as hardy and must never be in water-logged soil. *A. ligtu* var. *angustifolia* was introduced to England from Chile in 1933 by Mr Harold Comber and later crossed with *A. haemantha* to produce the Ligtu hybrids. These come in a wide range of colours—pink, orange, flame, yellow and rose. *A. pulchella* (*A. psittacina*) from N Brazil is a striking plant with red and green flowers on stems 90 cms (3 ft) tall.

Bomareas are exceedingly handsome climbers distinguished

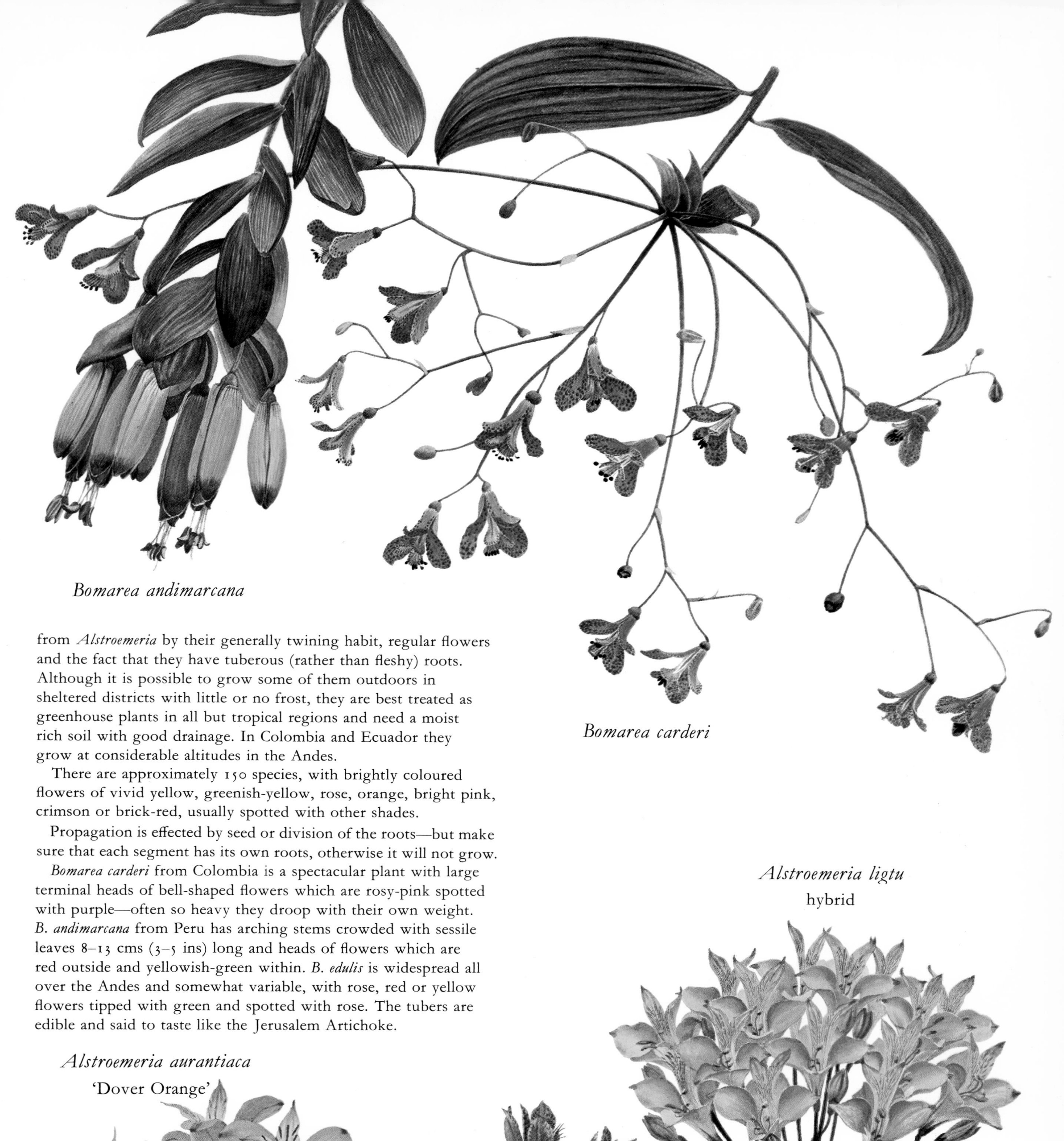

Bomarea andimarcana

Bomarea carderi

from *Alstroemeria* by their generally twining habit, regular flowers and the fact that they have tuberous (rather than fleshy) roots. Although it is possible to grow some of them outdoors in sheltered districts with little or no frost, they are best treated as greenhouse plants in all but tropical regions and need a moist rich soil with good drainage. In Colombia and Ecuador they grow at considerable altitudes in the Andes.

There are approximately 150 species, with brightly coloured flowers of vivid yellow, greenish-yellow, rose, orange, bright pink, crimson or brick-red, usually spotted with other shades.

Propagation is effected by seed or division of the roots—but make sure that each segment has its own roots, otherwise it will not grow.

Bomarea carderi from Colombia is a spectacular plant with large terminal heads of bell-shaped flowers which are rosy-pink spotted with purple—often so heavy they droop with their own weight. *B. andimarcana* from Peru has arching stems crowded with sessile leaves 8–13 cms (3–5 ins) long and heads of flowers which are red outside and yellowish-green within. *B. edulis* is widespread all over the Andes and somewhat variable, with rose, red or yellow flowers tipped with green and spotted with rose. The tubers are edible and said to taste like the Jerusalem Artichoke.

Alstroemeria aurantiaca 'Dover Orange'

Alstroemeria ligtu hybrid

Alstroemeria pulchella

Amaranthaceae

65 genera and 850 species
tropical and temperate

These are dicotyledonous herbaceous plants or shrubs with entire, opposite or alternate leaves. The flowers may be solitary or in axillary cymes; the blooms are regular with four to five perianth segments, a superior ovary and one to five free stamens.

Most members of this large family come from tropical Africa and America, where at one time several had considerable economic importance. *Amaranthus caudatus* was the Achita, Jataco or Quihuicha of the Aztecs, who used its seed for food and for ceremonial purposes. Known as Inca Wheat, eighteen of the Aztec Empire's granaries were once kept filled with its tiny seeds.

Gardeners esteem this annual, 60–90 cms (2–3 ft) tall, for its showy, trailing tassels of red flowers. It is called Love Lies Bleeding and flowers the summer long, often reseeding itself after preliminary establishment. There is a green-flowered form 'Viridis' and a taller species, 120–150 cms (4–5 ft), with erect, deep crimson flower spikes usually called *A. hypochondriacus,* or Prince's Feather.

Celosias are often used for subtropical bedding or as pot plants. In *Celosia argentea* the inflorescences are whitish, stiffly erect, and shaped like small inverted spades, whereas garden forms of the group known as var. *cristata* show a tendency to abnormal flattening or fasciation, so that the blooms are much larger and usually have crimped margins. Their common name is Cockscomb and there are red, yellow and purplish forms. In India the leaves of *C. argentea* are eaten as a vegetable and other species are used medicinally.

Alternanthera and *Iresine* from tropical and subtropical America are mainly grown in gardens for their ornamental leaves. *Alternanthera bettzickiana,* or Joy Weed, has yellow, purple, plain red or red and green leaves and *A. amoena* has distinctive yellow and red blotches on its plain green foliage. The leaves of *Alternanthera* are rather succulent, the foliage of some species being eaten with fish in the Congo or with rice in Indonesia. Iresines are mainly crimson, both in leaf and stem, although gold-foliaged forms are known. *Iresine lindenii* and *I. herbstii* are the kinds most commonly grown.

Ptilotus (*Trichinium*) *manglesii* is a striking Australian plant with narrow leaves and white and pink flower heads, 7 cms (3 ins) across, pink at the tips and silvery at the bases. They have a fluffy appearance for the blooms are covered with delicate hairs.

Amaranthus caudatus

Ptilotus manglesii

Amaryllidaceae

85 genera and 1100 species

This family contains monocotyledonous and frequently bulbous plants which remain dormant for long periods and then leaf in spring or after rains. The bisexual flowers may be solitary or in heads, frequently with a bract below the ovary. They may be regular or zygomorphic, usually with six perianth segments in two whorls of three with six stamens and an inferior ovary.

It is a large family containing many fine ornamental plants which are much cultivated in homes and gardens. The most important group is *Narcissus*—a vast genus which includes Daffodils, perhaps the most widely grown of all bulbs and among the easiest. These are ideal for naturalizing, spring bedding or forcing for home decoration.

Many centuries ago the Ancient Egyptians used the flowers of *N. tazetta* for funeral wreaths, and even after 3000 years the preserved remains are recognizable. Pliny says it was called Narcissus because of the narcotic quality of its scent, but some believe the name refers to Ovid's celebrated youth.

All Narcissi require a long season of growth so they should be planted out early, preferably in late summer (especially the tiny kinds) either in full sun or light shade such as is found in a woodland glade. In wet heavy soils the bulbs should be placed on a layer of sand; they require plenty of moisture after flowering in order to make good bulbs for the following season but cannot tolerate waterlogging. The small kinds should be planted 5–7 cms (2–3 ins) deep, the larger sorts 7–10 cms (3–4 ins). The former are ideal for rock-garden pockets, window boxes, pots in the home or alpine house, or for naturalizing in short grass.

The genus has been divided horticulturally into eleven divisions and subdivided again according to colour and the relative measurements of the perianth and corona.

In the Bulbocodium group the little Hoop Petticoats are distinguished by large funnel-shaped coronas, backed by small, narrow, strap-like perianth segments. They are rarely taller than 15 cms (6 ins) and where they grow wild in the Pyrenees and Spain seem to favour places which are damp in spring. This is brought about naturally by the melting snows and is succeeded by some baking as the moisture drains away and the sun beats down on the mountain slopes.

The same conditions suit Cyclamineus, the second group; indeed these *must* be grown in damp places. The bright yellow flowers have reflexed perianth segments behind long tubular coronas, after the style of a Cyclamen flower. They flower early in spring, almost a fortnight before the first group, and are 10–20 cms (4–8 ins) tall.

The Jonquils in the next group are noted for their rich scent and bunches of short-cupped, small golden flowers on stems 20–30 cms (8–12 ins) high. The leaves are rush-like and deep green in colour. An essential oil obtained from the flowers is used in heavy types of perfumery, such as oriental and floral bouquets. *N. jonquilla* has probably the sweetest scent of any *Narcissus*; *N. juncifolius* has rush-like stems and bunches of clear yellow flowers on stems 20 cms (8 ins) tall; and *N. watieri* is a white-flowered species from the Atlas Mountains. There are also some pretty little cultivars like 'Pencrebar', double gold, and 'Orange Queen', deep orange.

The Poeticus group is also sweetly scented. *N. poeticus* is the Poet's Narcissus, a widely distributed S European plant, with starry, short-cupped flowers with spreading, white perianth segments. It blooms exceptionally late in the season and has a form called *recurvus* with backward-curving perianth segments, which is known as the Pheasant's Eye Narcissus. Both sorts naturalize well in damp meadows and there are many garden varieties of which 'Actaea' and 'Queen of Narcissi' are typical.

The Tazettas have Polyanthus-like bunches of flowers which are sweetly fragrant and are held in high regard for forcing and early cut flowers. They have a wide distribution in S Europe, N Africa and Asia Minor to China and Japan. Well-known garden kinds include the golden 'Soleil d'Or', 'Cheerfulness', 'Geranium', 'Compressa' and the Paper White (*N. tazetta* var. *papyraceus*). All the kinds mentioned can be grown on pebbles standing in a saucer of water.

The Triandrus group includes some dainty little sorts for the rock garden and alpine house. The flowers are generally pendulous, one or more in a loose umbel, on stems 15–30 cms (6–12 ins) long. The corona is cup-shaped, white in *N. triandrus albus,* often called Angel's Tears and corn yellow in 'April Tears'. 'Silver Chimes' and 'Thalia'—both creamy white—are other desirable cultivars.

The larger Narcissi are split up according to the size of the corona (cup) in the centre of the flower. In the true Daffodil this is a real trumpet, as long as or longer than the perianth segments; others are divided into Large-Cupped (corona more than one third but less than equal to the length of the perianth segments), Small-Cupped (less than one third the length of the perianth segments) and Double Narcissi. These again are individually separated into colours or combinations of shades.

There are literally hundreds of cultivars in these sections including such well-known Daffodils as the giant 'King Alfred', white-trumpeted sorts like 'Mount Hood' and 'Beersheba', and also pink varieties such as 'Romance' and 'Mrs R. O. Backhouse'. The doubles can be all white as in 'Snowball', white and orange-red as in 'Anne Frank' or soft yellow as in 'Golden Ducat'. There are also many Large-Cupped and Small-Cupped varieties.

N. asturiensis, with its perfectly proportioned miniature trumpets, is the smallest Daffodil, for it only grows 5–10 cms (2–4 ins) high. Wordsworth's famous *Ode to Daffodils* referred to the little *N. pseudo-narcissus* which still grows freely in meadows in the Lake District.

Other important early bulbs are the Snowdrops (*Galanthus*) which, while mostly native to the eastern Mediterranean area, are widespread in woods all over Europe. Since they are almost always found naturalized near the sites of old abbeys or monas-

1 Leucojum vernum
2 Galanthus nivalis
single and double forms
3 Narcissus cyclamineus
4 Narcissus bulbocodium
5 Narcissus triandrus

6 Narcissus poeticus
7 Narcissus 'King Alfred'

teries it is possible that Italian monks first introduced them in the 15th century.

These much-loved little plants have many names such as Fair Maids of February, Candlemas Bells, Mary's Tapers and Bulbous Violet (this last by the old herbalists Gerard and Parkinson). The bulbs are not difficult to cultivate in cool climates and seem impervious to frost and cold; they require moist but good soil and full sun or light shade. The bulbs soon deteriorate if left lying about so they should be planted as soon as possible after lifting and an occasional top dressing benefits them considerably. The pure white flowers, usually solitary and always nodding, have three long and three short perianth segments, the inner ones bearing green or yellow markings.

Galanthus nivalis is the best known but a very variable species. Some of the choicest varieties or hybrids are 'Flore Pleno' with double flowers; 'Flavescens' which has yellow markings instead of green; 'Straffan' which is very vigorous and late flowering; 'Atkinsii', a vigorous form with long-segmented blossoms; and *reginae-olgae*, an autumn-flowering sort from Greece which produces its flowers before the leaves. *G. plicatus* 'Warham' has very large flowers and broader leaves than *G. nivalis.*

The closely related Leucojums, commonly called Snowflakes, have pendulous bell-shaped flowers and narrow linear leaves. *Leucojum aestivum,* the Summer Snowflake, has a wide distribution in Europe (including Britain), Asia Minor and the Balearic Islands. The largest species, it has sturdy stems 60 cms (2 ft) tall which carry several flowers in terminal umbels, pure white with green markings. It will grow in the open or light shade but likes a moist place, as by a stream. *L. vernum*, the Spring Snowflake, blooms in late winter or early spring, carrying large, usually solitary blossoms on stems of 20 cms (8 ins). It shows the same liking for a damp locality and is native to S Europe.

Crinums come from warm parts of the world, as S and central Africa and parts of Asia where they are usually found in marshy places or by streams. The bulb of a full-grown specimen may be as large as a football at the base and tapering off to a long thick neck. At planting time this neck—which may be 30 cms (1 ft)

Amaryllis belladonna

Hippeastrum hybrid

Clivia miniata

Crinum x *powellii*

in length—should be left above soil level. Crinums are handsome plants with broad strap-shaped leaves and umbels of large—frequently 10 × 7 cms (4 × 3 ins)—flowers on thick succulent stems.

In temperate climates they should have a warm sunny position, rich soil, and protection in winter. Established specimens should not be disturbed as they flower best when crowded, the blooms opening in succession over a period of several weeks. In tropical gardens they are suitable for key positions, near to water.

Crinum asiaticum from China and the Himalayas is the most outstanding as far as size is concerned, with up to fifty white flowers, flushed with pink on stems 90–180 cms (3–6 ft) high. These are sweetly scented and well set off by an enormous bulb and bright green shining leaves. Because the bulbs were once mashed up and used for poisons this species is known as the Asiatic Poison Bulb. *C. moorei* from Natal has light rose flowers on stems of 45–60 cms (1½–2 ft), but more garden worthy is its hybrid *C.* x *powellii* and its cultivars. This has large pink flowers on stems of 60–90 cms (2–3 ft), while 'Album' is white, 'Harlemense' shell-pink and 'Krelagei' deep pink.

Hippeastrums are often erroneously described as *Amaryllis,* a genus with only one species, the South African Belladonna Lily (*A. belladonna*). Blooming in early autumn the latter has fragrant, rosy-red to pink flowers on stems of 45–75 cms (1½–2½ ft) and strap-like leaves which appear when the blooms are over.

The blooms of *Hippeastrum* are much more showy, with large funnel-shaped flowers divided into six segments, in brilliant shades of scarlet, red, pink and white, frequently striped with other colours. *H. equestre,* the Barbados Lily from Central and South America, is widespread in tropical gardens and has bright scarlet flowers. It has given rise to some splendid hybrids, many of which are grown in temperate countries as pot plants. For this purpose the large bulbs should be kept in fairly small containers, with their tops just above the soil. A temperature of 13–16°C (55–60°F) is necessary to induce winter flowers and after the foliage has died down the pots should be kept dry for two to three months to rest the bulbs. *H.* x *johnsonii* is a fine hybrid with bright scarlet flowers which have a greenish-white streak running halfway up each petal. It is named after the English watchmaker who raised it.

Brunsvigia josephinae from South Africa carries twenty-five to thirty scarlet flowers in umbels on stems 60 cms (2 ft) tall. The individual blooms are irregular with a long tube backed by recurved segments. *B. orientalis,* the Candelabra Flower, is similar but flushed with yellowish-green towards the base of the blooms. In their native habitat the bulbs are subjected to long spells of drought, when the umbels of seeds become detached and are blown long distances across the veldt by the wind.

Haemanthus is another South African genus and like so many bulbs these need a rest after flowering. *H. coccineus* is extremely striking; the flowers appear without the leaves and are coral-red, packed closely together and wrapped around by thick, fleshy, scarlet bracts. The whole inflorescence, which is about 7 cms (3 ins) across, is amusingly reminiscent of a house painter's brush. The two leaves produced by each bulb are thick and leathery, up to 60 cms (2 ft) in length and 15 cms (6 ins) wide, and lie flat on the ground. *H. albiflos* is similar but with white blossoms and *H. katherinae,* the Blood Flower, has large (15 cms; 6 ins), round, football-like heads of bright scarlet, stalked flowers with prominent stamens.

Clivias (also known as Imantophyllums) make good shade plants in moist tropical gardens and also dependable pot plants. They were much esteemed by the Victorians on account of the evergreen nature of their broad, strap-shaped foliage and large umbels of red or orange-yellow flowers. They require good soil, plenty of moisture during the growing season, followed by comparatively dry conditions while the plant rests. *Clivia miniata* is the best known and has up to twenty bright scarlet, yellow-throated, tubular flowers on a single stem 60 cms (2 ft) tall. This colour is variable in seedlings and many garden varieties are available. *C. gardenii* has reddish-orange or yellow flowers but fewer in number. All the species are South African.

Zephyr Lilies, from the warmer parts of America, are small bulbous plants with grassy leaves and white, yellow, pink or red funnel-shaped flowers. These are always solitary with six perianth segments united into a tube at the base. The majority flower in summer and are used for naturalizing or edging borders in tropical areas, or grown in sheltered places or pots in temperate climates. *Zephyranthes atamasco,* the Atamasco Lily, has white flowers with pale pink stripes; the bulbs were once eaten by Creek Indians in times of scarcity. *Z. candida,* the hardiest and best known, has pure white Crocus-like flowers on stems of 20 cms (8 ins), and *Z. grandiflora* (*Z. carinata*) has splendid rosy-pink flowers, 7 cm (3 ins) long on stems of 30 cms (1 ft).

Sternbergia lutea

Haemanthus coccineus

Cooperias are similar to *Zephyranthes* except that the flowers are more tubular; *Cooperia drummondii* from Texas is nocturnal and Primrose-scented, with flowers 10–13 cms (4–5 ins) long and 5 cms (2 ins) across on stems 15–22 cms (6–9 ins) tall. The flowers, white when first open, become reddish with age.

Also from South America is *Pamianthe*, a small genus discovered in the Peruvian Andes by Major Albert Pam. *P. peruviana* is a beautiful plant with narrow evergreen leaves and stems up to 120 cms (4 ft) long carrying small umbels of strongly scented white and green flowers. These have a bell-shaped corona—like *Narcissus*—and bloom in early spring. The plant is a prized exotic for British and American gardens and greenhouses.

Hymenocallis narcissiflora (*H. calathina*) is the Chalice-crowned Sea Daffodil, a somewhat similar plant with umbels of pure white, fragrant flowers, about 7 cms (3 ins) across and several on a stem of 45 cms (1½ ft). *H. amancaes* is bright yellow, streaked with green towards the centre. Both plants come from the Andes where the bulbs are used for medicinal purposes. 'Sulphur Queen' is a good cultivar.

Although called Amazon Lily, we are indebted to Colombia for the fragrant *Eucharis,* with their umbels of pure white flowers. In *E. grandiflora* (*E. amazonica*) these are individually up to 13 cms (5 ins) across, the corona tinged with green but the rest of the flower a pure Chinese white. The stems grow about 60 cms (2 ft) tall and the bulbs need plenty of water while growing but should be completely dried off after flowering.

Nerines are South African, autumn-flowering bulbs which are able to withstand a few degrees of frost if planted in a warm well-drained spot, but otherwise must be grown in pots. The most beautiful is *Nerine sarniensis*, the Guernsey Lily, with large umbels of deep rose-pink flowers on sturdy stems of 60 cms (2 ft). Numerous hybrids have been raised from this species with white, pale pink, orange-scarlet and purplish-magenta blooms and, in recent years, polyploids (with even larger flower heads) have become available. The English name denotes its connection with the Channel Islands where it is perfectly hardy and much cultivated for the cut-flower trade. It also commemorates its strange introduction to Guernsey where many years ago a ship was wrecked and bulbs from it were washed ashore and started to grow in the sand. The vessel was believed to have come from Japan and so the bulbs were falsely assumed to be Japanese. They are in fact native only to South Africa and found on Table Mountain where they bloom in March.

N. bowdenii is the hardiest with umbels of eight or more pink tubular flowers on stems 45 cms (1½ ft) tall and strap-shaped leaves. Cultivars include 'Pink Beauty', which is deep pink, 'Hera', cherry red, and 'Fenwick's Variety', a fine and vigorous sort with larger flowers and taller stems.

Vallota speciosa (*V. purpurea*), the Scarborough Lily, received its English name for much the same reason as the Guernsey Lily —through bulbs being washed ashore from a wrecked ship. In Cape Province where the plant grows wild they are known as George Lilies. The funnel-shaped flowers are scarlet and 6 cms (2½ ins) across in umbels on hollow stems 90 cms (3 ft) tall. They bloom in summer and autumn and should be given a sandy compost and repotted and rested after flowering. There is a white form *alba*.

Sprekelia formosissima from Mexico and Guatemala has large, curiously shaped flowers up to 13 cms (5 ins) across and deep. The crimson-scarlet segments are arranged three arching in the rear, the other three thrust forward folding around the stamens at their base to form a sheath. Known as the Jacobean Lily, it makes a striking subject for a warm situation or cool greenhouse.

Pancratium maritimum is the Sea Lily of the Greeks, a plant with white, sweetly scented flowers which is found in sandy situations near the Mediterranean sea shore. The blooms have green stripes on the outside segments. *P. illyricum*, from the same area, carries large umbels of starry white flowers and is also fragrant.

Sternbergia lutea from Greece where it spangles the hillsides in the autumn, is thought by some authorities to be the 'Lilies of the Field' of the Bible. It flowers extremely late in the year, with rich golden Crocus-like blooms and narrow green leaves. In well-drained soil and a sunny situation it makes an attractive rock-garden or edging plant; in Greece it is found among rocks in short grass in the mountains. The variety *angustifolia* has narrower leaves and is freer flowering.

Cyrtanthus are related to *Crinum* and are all native to South and East Africa. The flowers are carried in umbels and bend down in nodding fashion; individually they are funnel-shaped and between 5–10 cms (2–4 ins) in length narrowing to about 2.5 cms (1 in.) at the throat. *C. sanguineus* is bright red, *C. ochroleucus* pale yellow, *C. mackenii* white and fragrant—it also likes boggy conditions—and *C. smithiae* white or pale pink striped with red.

Other good garden plants in Amaryllidaceae include *Habranthus versicolor*, the Changeable Habranthus, so called because while the buds are red, the opening flower fades to white, with a faint pink staining at the petal tips and a deep pink throat. *H. andersonii* grows 15 cms (6 ins) tall and is usually golden-yellow with deep red streaks towards the base and *H. robustus* is pale pink with a green throat and red veining. All *Habranthus* come from tropical and South America.

Lycoris aurea is the Golden Spider Lily, a Chinese species which grows about 30 cms (1 ft) high with an umbel of stalked upright-looking flowers which have narrow, wavy perianth segments. *L. sanguinea* is red and *L. squamigera* pink flushed yellow. Both species come from Japan and require shallow planting, with the necks of the bulbs above the soil.

Eustephia pamiana with umbels of red and green flowers comes from Peru and the Argentine and *Ammocharis falcata* from South Africa. Both are grown as pot plants or planted outdoors in tropical borders. The latter has many-flowered umbels of fragrant bright red blossoms. It grows to about 60 cms (2 ft) and the *Eustephia* to 30 cms (1 ft).

Ixiolirions have regular flowers in loose racemes on stems 30–45 cms (1–1½ ft) tall. They are usually deep blue or violet and like well-drained soil in a fairly dry, sunny situation. *Ixiolirion montanum* has lavender or deep blue flowers 5 cms (2 ins) across and narrow grassy leaves. It is widely distributed in central Asia.

Anacardiaceae

60 genera and 600 species

This family of dicotyledonous trees and shrubs with resinous bark is mostly tropical but represented in temperate regions. The leaves are simple or compound, usually alternate, the flowers small and regular with three to five sepals, three to five petals, three to ten stamens (occasionally many), and the fruit commonly a drupe.

The species best known in Europe is the E North American *Rhus typhina*, the Stagshorn Sumach, so called on account of its branch formation. This has large pinnate leaves which turn red and orange at leaf fall and dense terminal panicles of five-petalled flowers. Male and female flowers are borne on separate trees, the latter having clusters of crimson berries which hang all winter. These are very acid but when soaked in water make a pleasant drink called Indian Lemonade by Americans.

Another much cultivated ornamental is the Smoke Tree, *Cotinus coggygria* (*Rhus cotinus*), a native of the Mediterranean area extending to China. Its smooth oval leaves later assume exquisite autumnal tints and there are also purple-leafed forms. The inflorescences are myriads of tiny flowers on cobwebby branches. At a distance these look like plumes of smoke.

The leaves of most Sumachs contain tannin, that from the S European *R. coriaria* being particularly valuable for tanning pale leathers. Other species provide lacquer and vegetable wax. The flowers of some are used for making a gargle and also an astringent, while the bark of the African *R. albida* is used for tanning.

However, some of the world's most irritating plants are to be found in this family. *R. toxicodendron*, *R. vernix* and *R. radicans*, the Poison Ivy, all cause painful skin allergies and should *never* be planted in gardens. So toxic is the active principle in Poison Ivy (urioshol, an oil in the plant's juice) that even smoke wafted from burning trees can affect a sensitive skin. The blisters it causes are painful and persist for a short period and in some people the symptoms are known to be recurrent.

Other members of Anacardiaceae are the Mango, *Mangifera indica*, and the Cashew Apple, *Anacardium occidentale*. The latter produces large fleshy, red and yellow fruits tasting like apples. The fleshy portion is derived from the swollen receptacle and the bean-shaped fruits attached to the base are the source of Cashew Nuts. Pistachio Nuts come from *Pistacia vera*, now placed in Pistaciaceae. A subtropical and tropical shade tree, *Schinus molle* or Pepper Tree (the seeds are used to adulterate pepper), is also a member of this family, yielding American mastic, a resin.

Cotinus coggygria 'Foliis Purpureis'

Apocynaceae

180 genera and 1500 species

This dicotyledonous family contains many spectacular tropical shrubs, trees and climbers as well as a few modest plants from temperate regions. The twiners, often very vigorous, with large showy blossoms, are much planted in tropical gardens for training over houses, festooning trees and draping fences. Given warmth, light and moisture they are not difficult to establish and in cooler places in the world may be grown under glass.

The leaves are almost invariably simple and entire, often arranged in opposite pairs but also alternate or in whorls of three; the flowers are regular with four or five joined sepals and four to five joined petals—the latter forming a funnel or salver-shaped corolla supporting five adhering stamens. There are often two ovaries which are mostly superior and contain many seeds. The fruit may be follicular, berry-like, indehiscent or in nutlets. The seeds are usually flat, sometimes crowned with hairs which makes them prickly.

Some genera came into prominence during the Second World War as a source of caoutchou (raw rubber). When supplies of natural rubber from *Hevea brasiliensis* were cut off to the Allied powers substitutes had to be found; *Hancornia speciosa,* a Brazilian tree known as the Mangabeira rubber, the African *Landolphia comorensis* and *L. kirkii,* known as African or Madagascar Rubber, and other Apocynaceous vines were all used as a source of latex. Several *Landolphia* species grow in the Congo and terrible tales are told of events which occurred during the height of the extraction of rubber. Apparently King Leopold III's government gave procurement concessions to certain ruthless factions who then demanded the latex from native tribes as a form of tax. As the quota was constantly rising this could not always be met, especially as the vines were invariably cut to ground level to extract the latex, and the unfortunate natives were then maltreated and put into chain gangs to hasten their endeavour; as a consequence many died and thousands were mutilated.

A number of very poisonous plants belong to Apocynaceae, notably *Cerbera tanghin,* the Ordeal Tree of Madagascar. *Strophanthus sarmentosus* from tropical Africa is a source of commercial cortisone but others (*S. gratus* and *S. hispidus*) can be used to produce heart poisons and are used by natives to tip game arrows. The seeds of *S. hispidus* provide the drug strophanthin. Death has been caused by the use of Oleander (*Nerium oleander*) wood for meat skewers.

Some species, as *Wrightia tomentosa* and *W. tinctoria*, furnish yellow and indigo dyes; *Rauvolfia serpentina* from India is used as a tranquillizer and lowers blood pressure without side effects; and the *Aspidosperma* species from South America possess such hard and durable timber that they are used for bridge building and railroad ties as well as for simple objects like tool handles.

Among the climbing ornamentals Allemandas (Allamandas) carry attractive terminal bunches of large, yellow or purple, funnel-shaped blossoms. The leaves are evergreen and there are about fifteen species, all South American and West Indian. *Allemanda cathartica* 'Hendersonii', the Golden Trumpet, is deepest in colour with waxen blooms of deep orange-yellow with five white spots at the throat. In Kenya it is sometimes trained as a bush, constant pinching back producing a branching habit. Buds gathered in the cool part of the day will open up indoors but must be floated on water to prevent wilting. Allemandas are easily increased from cuttings but weak sorts like the rosy-purple *A. violacea* (*A. blanchetii*) are usually grafted on stronger kinds such as *A. cathartica* 'Hendersonii'.

Mandevilla laxa (*M. suaveolens*), the Chilean Jasmine, is another delightful creeper, with very fragrant white or creamy flowers 5 cms (2 ins) across. This Argentinian and Bolivian species does not often grow well in pots and where it cannot be planted outside may be installed in greenhouse borders and trained up the rafters. The same treatment suits the Confederate Jasmine, *Trachelospermum jasminoides,* a plant with fragrant white flowers in showy bunches, which has been a favourite in old southern gardens of the USA since colonial days. Yet it is not indigenous to America but native to China and Malaya.

Another striking climber is the South American *Mandevilla splendens* (*Dipladenia splendens*), a splendid plant indeed with deep rose tubular flowers flaring to wide-open chalices 12 cms (5 ins) or so across. *M. acuminata* is similar but streaked with red at the throat and slightly smaller; *M.* x *amabilis* has its rosy-crimson flowers in clusters.

Beaumontia grandiflora comes from India and climbs by means of its twining young stems. It is a robust plant with broad, oblong-ovate leaves which are downy underneath, and many large,

Mandevilla splendens

Vinca minor
'Bowles' Variety'

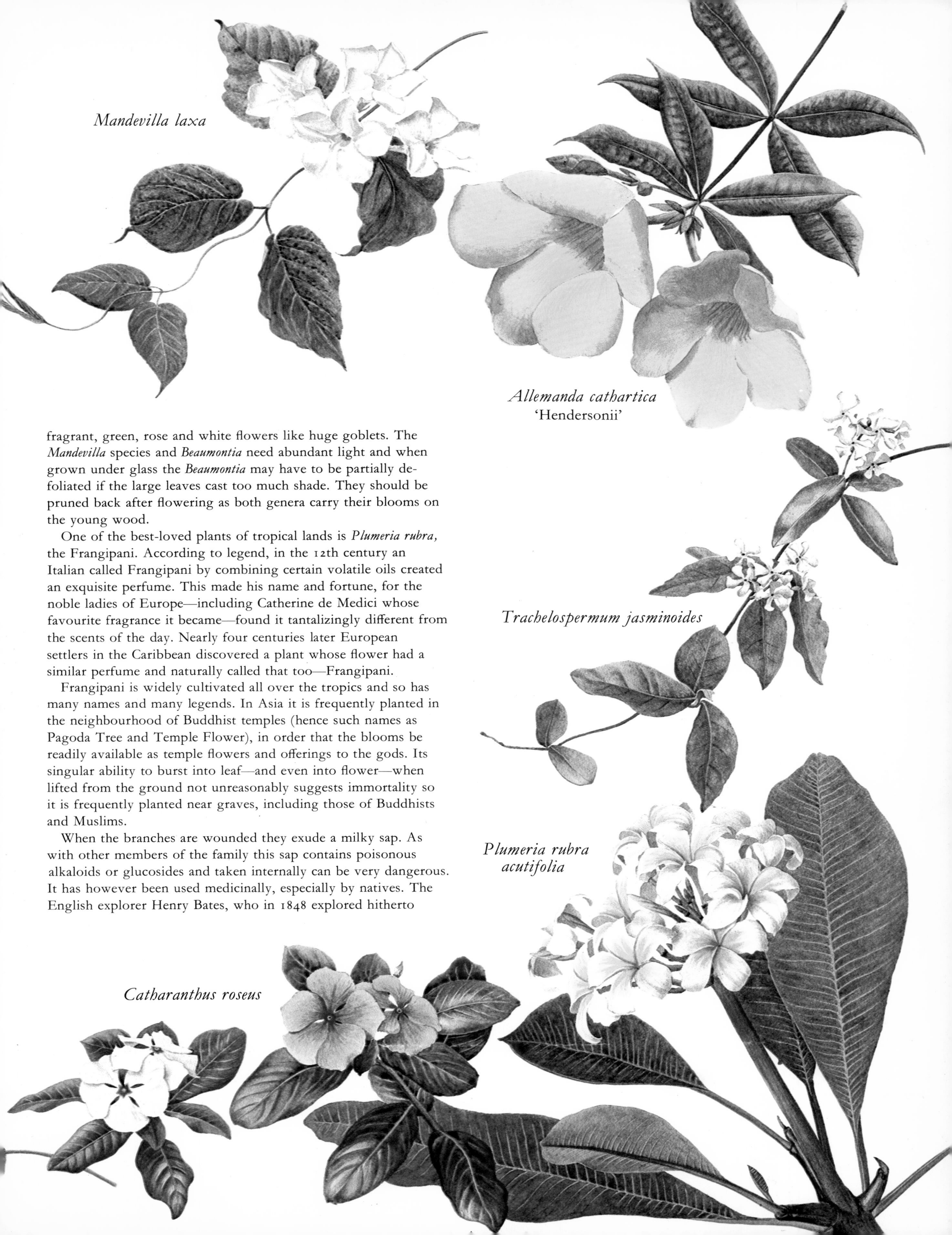

fragrant, green, rose and white flowers like huge goblets. The *Mandevilla* species and *Beaumontia* need abundant light and when grown under glass the *Beaumontia* may have to be partially defoliated if the large leaves cast too much shade. They should be pruned back after flowering as both genera carry their blooms on the young wood.

One of the best-loved plants of tropical lands is *Plumeria rubra,* the Frangipani. According to legend, in the 12th century an Italian called Frangipani by combining certain volatile oils created an exquisite perfume. This made his name and fortune, for the noble ladies of Europe—including Catherine de Medici whose favourite fragrance it became—found it tantalizingly different from the scents of the day. Nearly four centuries later European settlers in the Caribbean discovered a plant whose flower had a similar perfume and naturally called that too—Frangipani.

Frangipani is widely cultivated all over the tropics and so has many names and many legends. In Asia it is frequently planted in the neighbourhood of Buddhist temples (hence such names as Pagoda Tree and Temple Flower), in order that the blooms be readily available as temple flowers and offerings to the gods. Its singular ability to burst into leaf—and even into flower—when lifted from the ground not unreasonably suggests immortality so it is frequently planted near graves, including those of Buddhists and Muslims.

When the branches are wounded they exude a milky sap. As with other members of the family this sap contains poisonous alkaloids or glucosides and taken internally can be very dangerous. It has however been used medicinally, especially by natives. The English explorer Henry Bates, who in 1848 explored hitherto unknown reaches of the Amazon, wrote in *A Naturalist on the Amazon*, 'one of the most singular of the vegetable production of the campas is the Sucu-u-ba tree (Frangipani) . . . . The bark and leaf stalks yield a copious supply of milky sap, which the natives use very generally as plaister in local inflammation, laying the liquid on the skin with a brush and covering the place with cotton. I have known it to work a cure in many cases.'

The Javanese make sweetmeats from the flowers, but strangely the plant rarely sets seed. In cooler parts of the world these plants cannot be grown without a warm greenhouse and are then propagated from cuttings, which must be left after being prepared until the leaves wither and the latex hardens, otherwise they will not root. In the tropics, however, it is only necessary to break off a branch and push it into the soil. Nature does the rest.

Plumerias are small trees with thick stubby branches, smooth, tapering, oblong and long-stalked leaves and clusters, 22–25 cms (9–10 ins) across, of white, yellowish, pink or reddish-purple flowers. These have five petals which overlap at the centre and are curled at the edges. The trees bloom most freely before the rainy season although a healthy plant often shows some blossom throughout the year. *P. rubra acutifolia* (*P. acutifolia*) has white, yellow-eyed blooms.

Nerium oleander, from Mediterranean regions, is one of the most popular evergreen shrubs for warm temperate and subtropical countries. Its ornamental uses extend to roadside and screen plantings, establishment in pots or tubs for patios and terraces and also in the ordinary garden border. Growing 2–3.5 m (7–12 ft) tall (sometimes more), it makes a dominant feature in a key position, with masses of pink, rose, white, crimson or purple, funnel-shaped flowers. There are also double and variegated forms. Oleanders need plenty of sun and air or they do not flower freely, and also plenty of water while growing. *N. odorum,* with sweet-smelling blooms, has a double rose-pink variety.

Strophanthus grandiflorus, the Corkscrew Flower from tropical Africa, is a climber with red-striated yellow flowers which have long (15 cms; 6 ins), twisted tails straggling behind. Alstonias are tropical trees used in California and Florida for outdoor planting, particularly *Alstonia scholaris,* with long leaves arranged in whorls and terminal branches of small white flowers.

double form

Nerium oleander

single form

Beaumontia grandiflora

This is the Devil Tree of India and has medicinal bark which is sold in many oriental drug shops.

Amsonia tabernaemontana from North America is hardy in temperate regions where it is sometimes used in herbaceous borders. Growing 45–60 cms ($1\frac{1}{2}$–2 ft) high it has pale blue flowers in terminal cymes.

Carissa grandiflora, the Natal Plum, is an evergreen shrub with glossy foliage, white star-like flowers and scarlet berries which make good jelly. Its spiny nature commends it for hedge making and it does particularly well in coastal areas.

The most useful members of the Apocynaceae for temperate climates are the Periwinkles (*Vinca*), dwarf sub-shrubs with straggling shoots which thrive in shade and so make useful ground cover between shrubs and under trees. The two most important species, *V. major* and *V. minor*, are both found wild in Europe as well as being well established in Britain, although there is considerable doubt amongst botanists as to whether they should be considered true natives. Nevertheless, *Vinca* was a familiar plant to Chaucer who referred to it as 'the fresh Pervinke rich of hew'. It has had many popular names, as for example Cockles in Gloucestershire, Pucellage or Virgin Flower in France, also Sorcerer's Violet, presumably because witches and sorcerers used it for making love philtres and other potions. In the US it is commonly known as Running Myrtle or Common Periwinkle.

In olden days it was customary to place garlands of *V. minor* on the biers of dead children, which probably accounts for the Italians knowing it as Flower of Death.

But *Vinca* also has medicinal uses. Culpepper, the English seventeenth-century herbalist, tells us it 'stays bleeding at the mouth and nose, if it be chewed. . . .' Lord Bacon on his own testimony affirmed that a limb suffering from cramp would be cured by tying bands of green periwinkle round it. It has also been used for such diverse complaints as nightmare, hysteria, diarrhoea, haemorrhages, scurvy, sore throat and inflamed tonsils. There is, too, the curious statement in *The Boke of Secretes of Albartus*

Magnus of the vertues of Herbs, Stones and Certaine Beastes that 'Perwynke when it is beate unto powder with wormes of ye earth wrapped about it and with an herbe called houslyke, it induceth love between man and wyfe if it bee used in their meales'. Albertus Magnus was the teacher of Thomas Aquinas.

The best form of *V. major* is 'Variegata' ('Elegantissima') with cream-edged leaves which never seem to revert. *V. minor* is blue-purple but there are other forms with white, double blue and double purple flowers and also silver or yellow variegated-foliaged forms. For garden use one of the best forms of *V. minor* is that found by the late E. A. Bowles (one of Britain's best-known horticulturists) at La Grave in France. While waiting for a train one day he chanced to see a fine large blue Periwinkle growing over a grave in a cemetery adjoining the station. Leaning over the railings, he detached a small piece with his umbrella—in later years he used to say it was the only time he had knowingly robbed the dead—and brought it back to his garden. The plant is now widely distributed as 'Bowles' Variety' or 'La Grave' and has blossoms several centimetres wide and a very compact habit.

V. rosea, more properly *Catharanthus roseus* but also known as *Lochnera rosea,* is a tropical Periwinkle with large white or rose-pink flowers on stems of 30–60 cms (1–2 ft). It will grow in very dry places and is, for example, one of the few flowers which will flourish on the volcanic rocks in gardens in the Galapagos Islands.

Strophanthus grandiflorus

Aponogetonaceae

1 genus and approximately 30 species

This family consists of a small genus of monocotyledonous water plants, mainly South African but one or two hardy enough to grow outside in the warmer parts of the British Isles and similar N temperate regions. The leaves are basal, from a submerged tuberous rootstock, either floating or immersed, linear to oval in shape with parallel veining. The flowers, borne on single, forked or occasionally triple spikes, are bisexual, with one, two or three perianth segments, six stamens (occasionally more) and three carpels.

Aponogeton distachyos, the Water Hawthorn or Cape Pondweed from the Cape of Good Hope, is a splendid plant which blooms all the summer. It has floating oblong leaves and surface-riding flowers in forked inflorescences. These flowers are pure white, rather fleshy, with purple anthers and a strong scent of Hawthorn. The egg-like tubers should be planted in deep mud with 15–45 cms (6–18 ins) of water above (the depth does not greatly matter as the plants are surprisingly tolerant) in full sun and still water.

Less hardy is *A. kraussianum,* also South African, with flower spikes borne 15–20 cms (6–8 ins) above the water. These too are forked and sulphur-yellow in colour with pale anthers; they have an almond fragrance. *A. leptostachyum* from Kenya needs very shallow water or wet mud. It grows only about 5 cms (2 ins) high, with narrow lanceolate leaves and twin spikes of soft lavender flowers.

A. angustifolium from Darling, near Cape Town, is like a miniature *A. distachyos*. *A. junceum* has narrow, quill-like foliage and *A. crispum* single spikes of white flowers and crisped lanceolate, completely submerged leaves.

One of the most interesting species, but only suitable for an aquarium, is *A. fenestralis* (*Ouvirandra fenestrale*), the Madagascar Lace Plant. This must have soft, lime-free water and will then produce rosettes of dark green leaves with a lacy appearance because they are full of holes or fenestrations. Sometimes they are 30 cms (1 ft) or more in length and 7–10 cms (3–4 ins) wide. The flower spikes come above the water and are twin or triple spiked, tightly studded with fragrant yellow flowers.

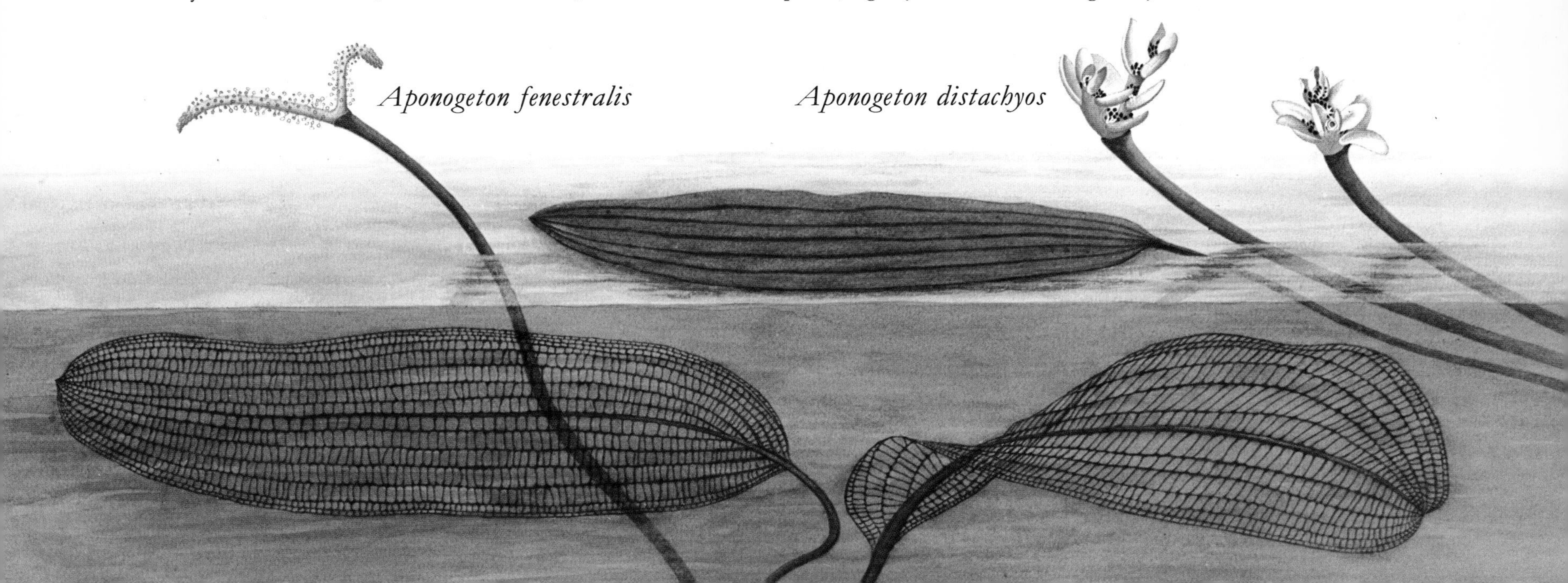

Aponogeton fenestralis *Aponogeton distachyos*

Araceae

115 genera and 2000 species
92 per cent tropical

This is a widely distributed family of monocotyledons, often with strange reptilian-like flowers which yet have a quaint beauty. Many come from the depths of damp tropical forests and have often proved successful as house plants. The northern species are frequently bog plants or aquatics.

The foliage varies greatly in the genus, pinnately and palmately divided leaves are common, while others are simple and entire or develop holes in the leaves, like *Monstera*. Aerial roots are common especially in the larger species. These are used for climbing or they are absorbent and grow down into the soil and take up nourishment.

The flowers have no bracts and are usually massed in a cylindrical spadix, wrapped in an enclosing spathe. The blooms are usually unisexual but may also be bisexual. They may be with or without perianth segments and have one to six stamens. The fruit is usually a berry.

Many Aroids contain latex which is often very poisonous, although it may be dispelled by heat. The rhizomes of others contain starch and are used as food.

Often the flowers have an offensive smell and attract insects which act as pollinating agents; *Arum maculatum,* the Wild Arum or Cuckoo Pint of Britain, has a particularly interesting pollinating mechanism. Male and female flowers are borne on a long poker-like inflorescence called a spadix. The top of this is brown or violet, quite smooth and sterile and protrudes from a pale green or greenish-white spathe. The latter is waisted two-thirds of the way down, forming a pitcher enclosing the lower part of the spadix. Attached to the latter at the base are the female flowers, with ovaries and stigmas. Above these are the male blooms with anthers and above these again, just at the nipped-in area, there are a few rows of down-pointing thick bristles (sterile flowers). As the spathe unfolds an offensive odour is emitted attracting flying insects which, because they cannot find easy foothold, slip down into the cavity.

Only small flies can pass into the spathe and, once inside, they are held prisoner—sometimes for several days. However, they find warmth, shelter and food—the last provided by a special excretion from the stigma.

Eventually the flowers ripen, pollen dusts off on the insects which pollinate the female flowers. The bristles then shrivel and the flies escape, the lower part of the spadix going on to produce bright red berries.

The starchy tubers were used in Tudor times to stiffen courtiers' ruffs, which probably accounts for the name Lords and Ladies. Although very poisonous the tubers (of this and other species) can be eaten after the acrid principles have been destroyed or made harmless by long cooking.

There are many ornamental Aroids, practically all easy to grow if they have sufficient moisture, shade from hot sun and a friable soil. In the water garden *Acorus calamus* 'Variegatus', the variegated Sweet Flag, is highly desirable because of its cream and green striped fans of Iris-like leaves. The type species with plain green leaves is a common stream-side plant in northern and central Europe, temperate Asia and North America. It grows to 60–90 cms (2–3 ft) and has a curious brown spadix.

The *Acorus* found in Europe is said not to fruit, which may be because it has all originated from the same form or clone, nor does it do so in the older settled parts of the United States, to which it has probably been introduced from Europe. In the northern interior of the US, however, to the east of the Rockies (for example in Minnesota) it has the appearance of being native and fruits freely.

A. gramineus is much used as an indoor pot plant, especially in its variegated form. This grows to about 30 cms (1 ft) and has narrow, grassy leaves. *Calla palustris,* the Bog Arum, native to

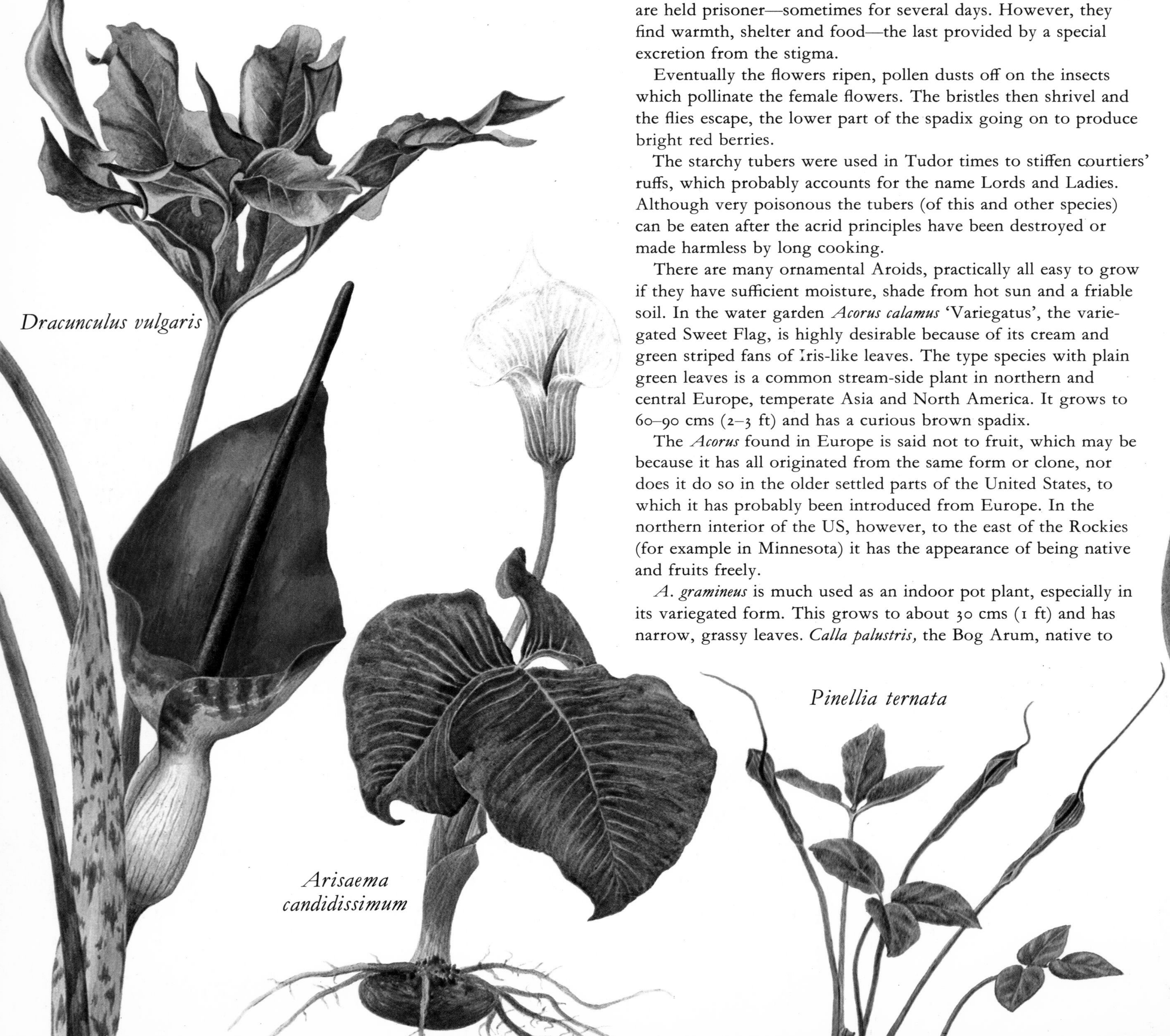

Dracunculus vulgaris

Arisaema candidissimum

Pinellia ternata

North America but widely naturalized in Europe, is one of those rambling, scrambling, rampageous plants that gardeners like to introduce at the edge of a pool, for it is equally happy in shallow water or wet mud. It has glossy, heart-shaped leaves, a white spathe and stubby spadix which are pollinated by water snails (attracted by an offensive odour), then going on to produce bright scarlet berries.

Orontium aquaticum, from the eastern US, is a true aquatic suitable for either deep or shallow water. In deep water the narrow leaves float on the surface and are water repellent, throwing off the errant raindrops like mercury globules. The flowers are bright yellow, terminal to a long white stem which stands well above the water. Under bog conditions the foliage becomes broader and stands upright, often hiding the flowers.

Two splendid bog plants are the Lysichitums, the yellow-flowered *Lysichitum americanum* from W North America and its white counterpart *L. camtschatcense* from Japan, Kamchatka, the Sakhalin Isles and E Siberia. *L. americanum* is perhaps the more showy, having bold spathes of a deep butter shade over 30 cms (1 ft) in length and proportionately broad. These appear in early spring followed by very large leaves which some people (rashly, perhaps) say may be eaten like Spinach. Both plants grow in shallow water or wet mud.

The Mouse Plant, *Arisarum proboscideum,* from southern Europe, is a shade-lover which in damp soil forms a mat of weed-suppressing, glossy green leaves. These are arrow-shaped and when parted hundreds of curious purplish-brown and white flowers can be glimpsed which, with their long tail-like appendages, look like the hind-quarters of numerous burrowing mice.

Pinellia ternata, from China and Japan, likes a hot dry situation and flowers throughout the summer. It has dainty, light green spathes on slender stems of 10–13 cms (4–5 ins). The foliage is ternate and at the base of each leaf stalk will be found a small brown bulbil which can be detached to grow as a new plant.

Among the true Arums is *Arum italicum,* a southern European species somewhat similar to the British Lords and Ladies (*A. maculatum*) but with cream-veined foliage, particularly in certain forms, misnamed in horticulture *pictum* and *marmoratum*. The former name is correctly applied to the Sardinian and Corsican *A. pictum* which has violet (not creamy-white) spathes and blooms in autumn instead of spring and *Arum marmoratum* is a synonym of *A. italicum*.

Closely related are the Arisaemas, notably the Chinese *Arisaema candidissimum* with its large white and rose-flushed spathes and three-parted leaves. *A. sikokianum,* from Japan, has purple spathes with narrow green and white bands, spots and stripes; *A. pradhanii,* another Chinese species, has one or two leaves to each tuber, each leaf divided into threes. They are green with purple veinings and the spathes are deep chocolate-purple netted and striped in green. The American green to purple *A. triphyllum,* while occasionally bearing both male and female flowers on the same inflorescence, changes according to its physiological state. On poor ground only male flowers will be found but on rich soil they will all be female.

Other close relations are the Dragon Arums, *Dracunculus vulgaris* and *Sauromatum venosum* (*S. guttatum*), the former a spectacular plant which attains a height of 60–90 cms (2–3 ft). The large leaves are divided into thirteen to fifteen segments, the stalks and flower stem flesh-coloured but deeply mottled with black. The blooms are large, 20–30 cms (8–12 ins), with a deep purple-red spathe-limb and black-purple shiny spadix with an extremely offensive odour which attracts carrion flies.

Sauromatum venosum (*S. guttatum*), the Monarch of the East from India, is sometimes sold as a plant curiosity, the large tubers sending up odoriferous flowers—dark red-purple outside and greenish-yellow within with purple leopard markings—without

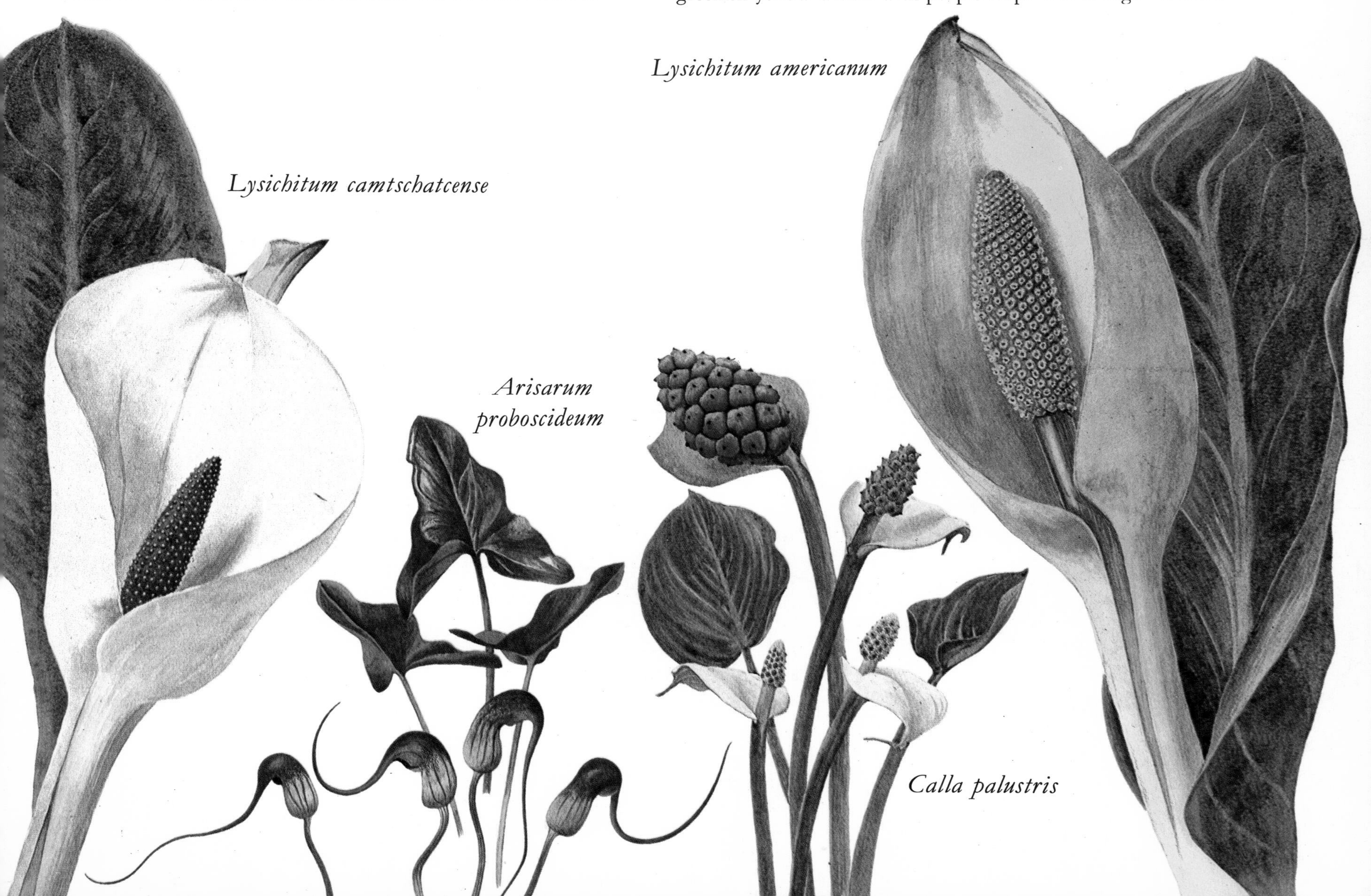

being planted in soil. As soon as the leaves appear the tuber should be potted up or, in frost-free areas, planted outside in a shady corner of the garden.

One of the world's largest flowers is *Amorphophallus titanum,* from W Sumatra. A tuber of this plant, discovered by the Italian botanist Beccari, was 1.5 m (5 ft) in circumference, the leaves were on stalks 5 m (15 ft) tall, much divided and covering an area of 14 m (45 ft) in circumference. The spadix was nearly 1.75 m (6 ft) in height, surrounded by a sheath of beautiful mottled green with purple lining and frilled edge. Plants have flowered at Kew Gardens, London, but the stench is terrible. Another species, *A. rivieri,* from tropical Asia is used as a house plant for its striking, large, pale green and white spotted spathe rose-marbled leaf stalks and large tripartite leaves. In China and Japan *A. konjac* (*A. rivieri konjac*), native to Indo-China, is the source of konjaku flour (konnyaku or mannan), yet *Amorphophallus* sap has been used in the East as a poison.

A newcomer travelling in the outskirts of Cape Town will be thrilled by the sight of the vast quantities of Pig Lilies. They are everywhere, fringing the streams and ditches, in great clusters in wet meadows, and grouped by the roadsides; their large arrow-shaped, tender, glossy leaves and white trumpet flowers are an everyday, year-round sight. Yet in Europe and America they are treasured plants, variously known as Trumpet Lilies, Callas or Arum Lilies—although correctly speaking they are neither Arums nor Lilies. Their proper botanical name is *Zantedeschia* and the term Pig Lily refers to the fact that early settlers found that porcupines (which they called pigs) liked the fleshy roots.

There are about ten species endemic to Africa, seven of which occur in the Transvaal, although many of the cultivated sorts show a combination of characters which make it difficult to determine accurately their original source. The spathes differ considerably in colour, from the pure white of *Z. aethiopica,* bright gold of *Z. elliottiana,* lemon-chrome of *Z. pentlandii,* the greenish-yellow *Z. oculata* which has a prominent black-purple base, and the various shades—ranging through white, pink, wine-red and almost black—of *Z. rehmannii,* sometimes known as Richardia. Some of the yellow-spathed sorts have spotted leaves.

Z. aethiopica can be naturalized in moist, frost-free situations and then remains in character most of the year. In colder climates the tubers may be grown in large flower pots, covered with 5–7 cms (2–3 ins) of soil, a temperature of 10–16°C (50–60°F) being necessary for good results. There are several comparatively hardy sorts for temperate regions; particularly one called 'Crowborough', which is said to have originated in a garden in Crowborough, England, and is hardy in many parts of the British Isles.

Among the most colourful plants in this family are *Anthurium* and *Caladium* species. The latter are grown for their leaves, which are shaped like elephants' ears and extremely variable in colour. An American, H. Nehrling, grew some 1500 named cultivars, with foliage marmorated and blotched in every shade of green, cream, red and pink. *C. bicolor* is native to Central America and northern South America and needs temperatures of 16–27°C (60–80°F), copious waterings and plenty of feeding during its active growth. Later the plants have a rest period when the tubers may be dug and stored, or the pots turned on their sides and left until the season for restarting them into growth (usually spring).

There are about 550 species of *Anthurium* of which *A. andreanum* from Colombia is one of the most important. It has masses of orange-scarlet or vermilion spathes on long, slender stems. The spathes are flat and shiny with many puckers or blisters and in the centre of each arises a cream-coloured spadix. There are countless forms with white, rose, salmon-pink, dark red and yellowish spathes, modern cultivars having smooth rather than puckered spathes.

The soil in which Anthuriums grow must be very porous and contain plenty of organic material such as peat or sphagnum. The plants need a warm, moist (humid) atmosphere, so in the greenhouse aim at 16–21°C (60–70°F) for most members of this family.

A. scherzeranum from Guatemala and Costa Rica is another variable species, the spathes white to yellow, or pink to scarlet, purple or blood-red, often peppered or striated with another colour and the spadix spirally twisted in white, yellow, orange, or red. The leaves are simple, oblong-lanceolate in shape, very dark green and on wiry stems.

Other notable Anthurium species are *A. veitchii* with fine white leathery spathes which have a rosy spadix, *A. acaule* which has fragrant brownish-purple spathes and *A. cristallinum* grown for

Zantedeschia elliottiana

Zantedeschia aethiopica

Anthurium scherzeranum

its fine foliage. The thick leaves reach 30–35 cms (12–14 ins) in length, and are violet when young but develop to deep green with prominent white veins.

Spathiphyllum wallisii from Colombia is used in tropical gardens for its glossy, long-stalked, lanceolate leaves and showy 'Arum' flowers. These are green at first but gradually change to pure white, later reverting to green. Tolerant of shade, the plant can be grown as a house plant in countries which experience frost; it attains a height of about 38 cms (15 ins).

Grown purely for foliage effects are species of *Aglaonema, Alocasia, Philodendron, Monstera, Dieffenbachia, Scindapsus* and *Xanthosoma*.

Dieffenbachias are native to South and Central America with one representative in the West Indies. They need moist but warm conditions and will not thrive in temperatures below 10°C (50°F). They have fine large leaves, usually oblong-oval, in various shades of green—due to markings and blotches of cream or different greens. Years ago in Britain young gardeners (journeymen) were often subjected to painful practical jokes by fellow journeymen who would persuade them that the green and white stem of *Dieffenbachia seguine* was a form of Sugar Cane. But all these plants are extremely poisonous and biting the stem caused agonizing throat swellings so that the victim could not speak for several days. Hence the name Dumb Cane.

Monstera deliciosa from Mexico is sometimes known as the Swiss Cheese Plant, because the leaf blades are perforated with slits or holes, no two leaves being cut in identical fashion. In nature it climbs up the stems of trees, fastening itself by means of aerial roots and sending down long pendent roots which, when they reach the forest floor, take root and become taut like lianas.

In tropical gardens Monsteras are grown to cover the bare trunks of large trees and run up into the branches. In more temperate zones they are used as pot plants in halls or stairways. The squat, creamy flowers give place to long, green fruits which look like slim Pine cones and remain for a year on the plant before they ripen, when they are very succulent with a taste like Pear and Pineapple and the aroma of Banana. Only the pulp is edible but around this are small needle-like crystals (raphids) which can enter the tongue if one is not careful.

Philodendrons, mostly from South America, are used in the same way as Monsteras and have magnificent leaves, 60 cms (2 ft) or more in length, oval-oblong to heart-shaped with striking veining or deeply cut lobes. The juvenile form of *P. oxycardium* (often misnamed *cordatum*) is the most popular US house-plant.

Xanthosoma violaceum is an aquatic plant from Puerto Rico and Jamaica with bright yellow 'Arum' flowers and deep green 'elephant ear' leaves with purple veins and stems.

Peltandra sagittifolia, the White Arrow Arum, and its green counterpart, *P. virginica,* are bog plants with glossy arrow-shaped leaves and white or green 'Arum' flowers which give place to scarlet and green berries respectively. About 45 cms (1½ ft) tall, they come from North America. *P. virginica,* from the eastern states, is hardy in water gardens over most of Britain and the US.

Symplocarpus foetidus, the Skunk Cabbage of North America, has hooded spathes of 7–15 cms (3–6 ins) in a mixture of colours—crimson, purple, green and yellow—and with a skunk-like aroma. These appear before the veined and roughly heart-shaped leaves.

Pistia stratiotes, the Water Lettuce, looks like a green rose as it floats on the water. The leaves, flat on the surface of the water during the day, assume more or less vertical 'sleep' positions at night. The plant is attractive in small ponds and aquaria.

Anthurium andreanum

Spathiphyllum wallisii

Aristolochiaceae

7 genera and 400 species

This is a family of dicotyledonous herbaceous plants or shrubs, the latter frequently twining lianas. The leaves are stalked, usually simple, alternate and often heart-shaped. The flowers are regular or zygomorphic, bisexual, with three petal-like calyx segments; they have between six and three dozen stamens which may be free, or united with the style to form a floral structure, the gynostemium (as happens in Orchids); and they have an inferior ovary split into four to six compartments and a capsular fruit.

The Asarums are mostly shade-loving plants, forming dense ground cover in summer, particularly in leafy soil underneath tall trees and shrubs. Their aromatic rootstock when dried or cooked with sugar makes a spicy substitute for Ginger (*Zingiber officinale*). The leaves are long-stalked, heart-, kidney- or almost spear-shaped, the rhizomes spreading and frequently bunched up above the soil. The flowers are more curious than beautiful, being bell-shaped, with three segments, each terminating in a lobe, or sometimes long whippy tails.

These 'tails' are most marked in *Asarum acuminatum* (now considered a variant of *A. canadense*) from North America and in one which E. A. Bowles called *A. shuttleworthii* and described as a 'buttonhole suitable for the devil'. The plant usually grown under this name in gardens is *A. caudatum* from the Pacific coast of North America which has flowers up to 5 cms (2 ins) long but is shorter-tailed than the kind grown by Mr Bowles and featured in this work. *A. canadense* is the American Wild Ginger and has medium-sized tails and a pungent oil in the rhizome which has a number of medicinal uses. *A. europaeum*, the European Wild Ginger, is used as an ingredient of snuff and bears dull brown flowers with stumpy appendages.

Although the genus contains some shrubs most of the 350 *Aristolochia* species are climbers. A few make splendid plants for the cool greenhouse or may be grown over trellises and fences in frost-free districts. *A. macrophylla* (*A. sipho*) from east to

Araliaceae

55 genera and 700 species

These dicotyledonous plants, chiefly tropical in distribution, are mostly trees and shrubs, often prickly or climbing. The leaves are usually alternate (occasionally compound), the foliage of young seedlings being much simpler than that of adult plants. The small flowers massed into heads or umbels are bisexual, regular, normally with five sepals, usually five petals and five stamens (rarely more), an inferior ovary and drupe fruits. A few have economic uses.

Hedera helix, English Ivy, is a European plant now distributed throughout the world. Some of its many cultivars are prized house plants, others are used as ground cover or climbers. Only the arborescent types bloom, the creamy flowers giving place to black, brown or golden berries. *H. helix* 'Lutzii', one of countless forms, has small leaves mottled in shades of green. *H. canariensis* 'Variegata' has cream variegations and leaf edges.

Fatsia japonica, a Japanese evergreen 2.5–3.5 m (8–12 ft) high, makes a good late-flowering shrub for a shady spot. The large shiny leaves are deeply lobed and it has showy panicles of milky-white flowers. It is hardy in the southern USA.

The pith of *Tetrapanax papyriferum* from Formosa is the source of Rice Paper. *Kalopanax pictus* (*Acanthopanax ricinifolius*) is a small tree with numerous spines and greenish flowers. The forked roots of both *Panax quinquefolium* and *P. schinseng* (*P. gingseng*) are the source of Gingseng, a Chinese medicine which was believed to be a panacea for countless ills; human-shaped specimens were considered to be particularly valuable and, quite literally, worth their weight in gold.

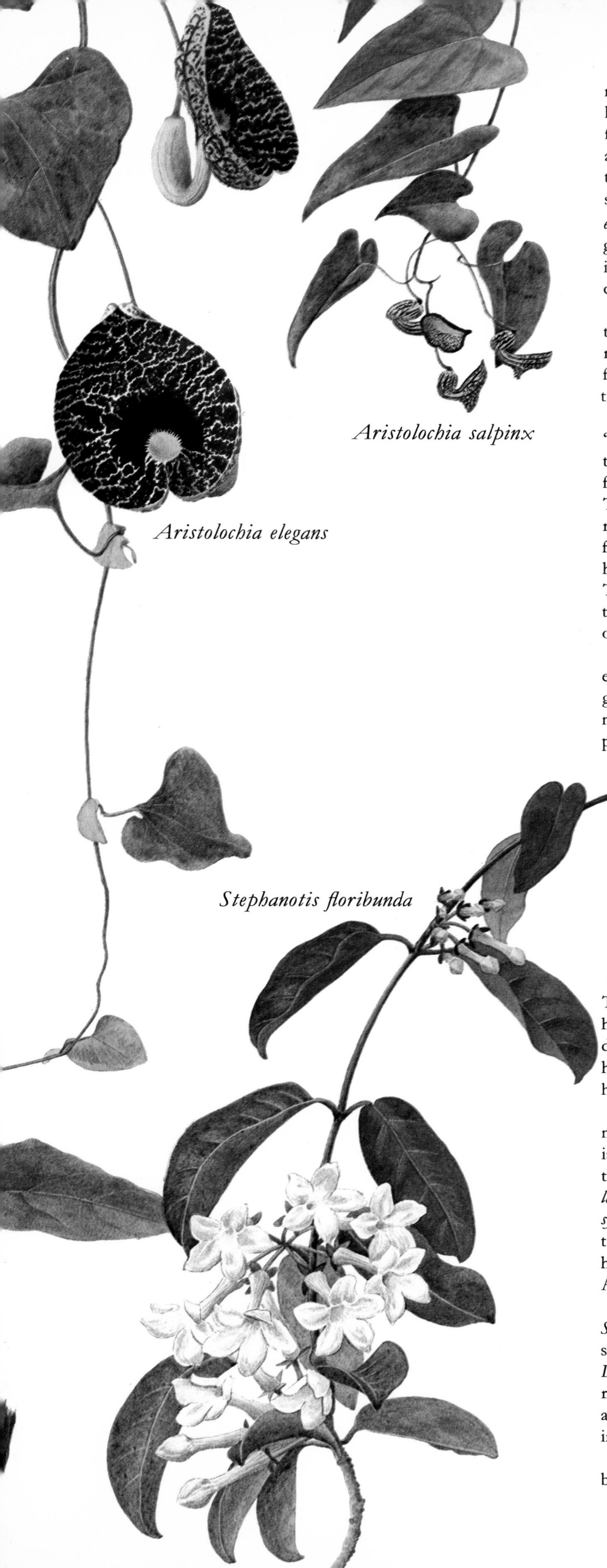
Aristolochia salpinx

Aristolochia elegans

Stephanotis floribunda

midwest US reaches 9 m (30 ft) under good conditions and has kidney- to heart-shaped leaves and brown and black flat-faced flowers attached to swollen, greenish-yellow tubes. The latter are quite long and bent round in several directions, so that from the side they resemble flat-bowled pipes—which gives some significance to their common name of Dutchman's Pipe. *A. elegans,* the warm-weather Calico Flower from Brazil, is graceful with large leaves and flowers. The latter possess rather inflated, pale yellowish-green tubes which expand into shallow cups of rich purplish-brown overlaid with white veinings.

A. saccata (*A. cathcartii*) from India is evergreen, growing up to 6 m (20 ft) with large ovate-cordate leaves and small purplish-red flowers which have large, almost square pouches. *A. salpinx* from Paraguay is smaller in leaf and flower, the latter compressed trumpet-shaped, cream or yellow marked with purple.

A. grandiflora is the largest of the genus, especially the clone 'Sturtevantii' with flowers often up to 50 cms (20 ins) long and tails of 90 cms (3 ft). Alexander Humboldt, the famous explorer, found Indian children in South America wearing them as caps. This species is known as the Pelican Flower because of the resemblance of the unopened bud to a sleeping pelican. Unfortunately its odour when in bloom beggars description, even hogs avoid it, but this does attract flies and ensures pollination. The roots of many species are put to countless medicinal uses, the name Birth-wort (applied to *A. clematitis*) referring to its one-time use by midwives.

All *Aristolochia* species grow best in good loam soil, although efficient drainage is essential. Since they make a great deal of growth in the course of a year an annual mulch of well-decayed manure improves the quality of the flowers. They tolerate hard pruning should this become necessary.

Asclepiadaceae

130 genera and 2000 species

This dicotyledonous family is closely related to Apocynaceae, having similar toxic principles although not to such a marked degree. It contains a number of shrubs—erect and twining—also herbaceous perennials which occasionally are very fleshy and have reduced or non-functional leaves like some Cacti.

Most species have a milky juice, which in the case of *Gonolobus* makes an effective arrow poison and in the East African *Cynanchum* is employed to poison fish. Yet there are contradictions, for while the Singalese use the sap of the Celanese Cow Plant (*Gymnema lactiferum*) as a substitute for milk, to taste the leaves of *Gymnema sylvestre* would be to destroy for a time the power of the tongue to distinguish between sweet and sour. *Oxystelma esculentum* also has drinkable milk and the latex of many Asclepiads is used by American Indians as a source of chewing gum.

The foliage shows great variation, plants like *Huernia* and *Stapelia* having a fleshy Cactus-like appearance; in some *Periploca* species the leaves are like leafless whip lashes and in the epiphytic *Dischidia rafflesiana* fashioned into pouches capable of storing rain-water. This is later made available to the plant by means of adventitious roots which grow into each pitcher, frequently inhabited by ants which rear their young inside.

Normally the leaves are opposite or whorled, usually entire but occasionally lobed or with toothed margins. The rootstock can be tuberous and fleshy or woody, or even completely absent, roots of annual duration growing from the fleshy stems.

Stephanotis floribunda from Madagascar is variously known as Clustered Wax Flower, Chaplet Flower and Madagascar Jasmine. It has many umbels of large, tubular, waxy-white flowers which are very fragrant and smooth fleshy leaves of simple oval shape. Climbing to 3–4 m (10–12 ft) it needs rich soil, plenty of water while growing and a warm temperature—not below 16°C (60°F).

Hoya carnosa was a favourite buttonhole flower of Victorians who greatly admired its pendent clusters of fragrant wax-like flowers. These are creamy-white with beautiful red centres and produce nectar in such profusion that it often drips from the blooms. The plant climbs naturally by means of aerial roots and has fleshy oval-oblong leaves; these are plain green in the type but splashed and margined with cream in the form 'Variegata'. *Hoya carnosa* comes from Australia and is often known as Wax Flower or Porcelain Flower.

H. bella from India is smaller and of drooping habit so it should be grown where it is possible to look up into the flowers, as in a hanging basket or in the fork of a tree. The flowers are similar to *H. carnosa* but smaller and with prominent centres. *H. australis* has pink-tinged flowers and a Honeysuckle fragrance.

Oxypetalum caeruleum (*Tweedia caerulea*) from Argentina is a woody twiner or shrub with five-petalled flowers of a striking and unusual shade of pale blue.

Ceropegia woodii, a curious plant from Natal, has a large root-stock from which emanate slender trailing stems, with opposite pairs of small heart-shaped succulent leaves. These are dark green variegated with white. The lantern-shaped flowers are black and lilac above and light purple below. All the *Ceropegia* species are traps for insects which, attracted by their carrion smell, try to alight on the corolla. Because some have five petals fused into a dome at the top (thus resembling small windows), the flies slip and later creep into the tube which is lined with downward-pointing hairs. There they remain, meantime inadvertently collecting pollinia, until the hairs wither and the flowers assume a horizontal position when the flies escape and carry pollen to other flowers.

Stapelia, Huernia and *Hoodia,* all genera from dry areas, share this habit of attracting insects by means of smell. In some cases this odour is most revolting—like rotting flesh—yet the flowers have a quaint beauty. In *Stapelia variegata* they look like starfish, 5–7 cms (2–3 ins) across, and are pale yellow with dark purple spots and lines. *Stapelia gigantea* has the largest flowers, in some cases up to 40 cms (16 ins) across, ochre-yellow in colour barred with numerous fine crimson lines. Most of this group are native to South Africa.

Huernias are very similar to Stapelias except that the corolla has ten points instead of five and the flowers are usually more bell-shaped. These succulents are frequently grown as rock garden plants in tropical countries or in pots in greenhouses where the climate makes this impossible.

Asclepias species are known as Milkweeds since many contain a milky latex in the stems, roots and leaves. The seeds are carried in bellied follicles which split down one side to expose and release the seeds and their attachment of long silky hair.

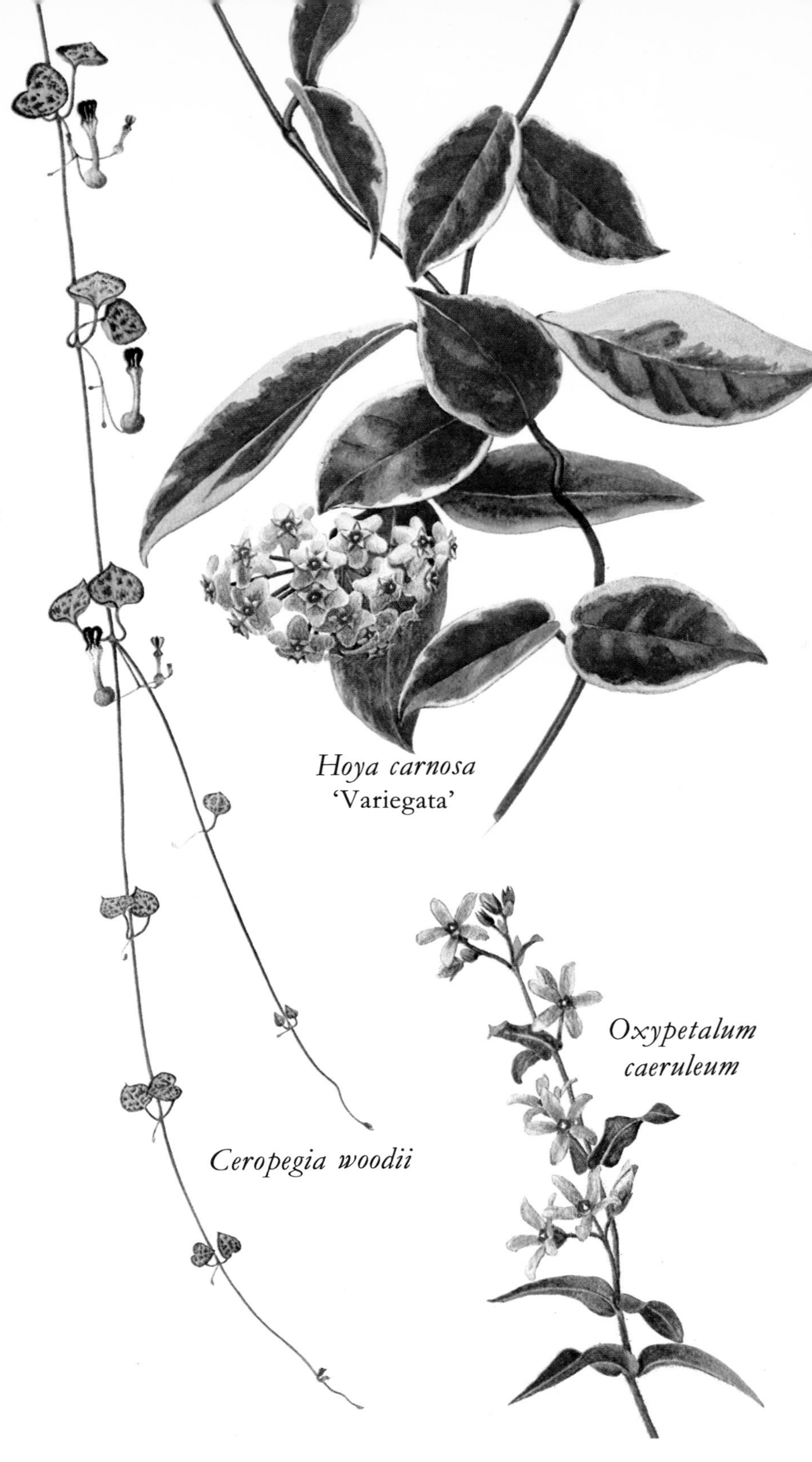

Hoya carnosa 'Variegata'

Oxypetalum caeruleum

Ceropegia woodii

A. curassavica is the Blood Flower, a South American perennial growing up to 90 cms (3 ft) with terminal and axillary umbels of small purplish-red flowers. It is widely cultivated as an ornamental in the tropics and subtropics and as a pot plant in other places.

Other good sorts are *A. tuberosa,* the Butterfly Weed from eastern North America, which is bright orange; *A. rubra,* from the eastern US, which is purplish-red; and *A. incarnata,* the North American Swamp Milkweed, 60–120 cms (2–4 ft) tall, and flesh-pink or rose in colour.

Asclepias species have been used for various purposes by primitive peoples, particularly North American Indians. The seed pods of *A. tuberosa* were boiled and consumed with buffalo meat; the roots were boiled and eaten as were the tender young shoots, while the latex of other species has been recommended as a substitute for rubber in times of emergency.

Stapelia variegata

Balsaminaceae

4 genera and 500 species

This family consists of herbaceous plants with watery translucent stems, alternate leaves, zygomorphic, bisexual flowers possessing three or five sepals (one forming a spur at the back), five petals, five stamens and a superior, many-seeded ovary. They are dicotyledonous and native to Africa, North America, Europe and Asia.

Impatiens, the most important genus, contains some ornamental species esteemed as pot plants or for bedding out in damp places. The flowers contain nectar, which is also found in enlarged glands on the leaf stalks of some kinds, as *I. glandulifera,* probably to attract ants which then protect them from other insects.

The most ornamental are the forms of *I. holstii, I. sultanii* and *I. petersiana* from tropical Africa which are now included under the name *I. wallerana*. *I. sultanii* is usually pink but *I. holstii* or Busy Lizzie comes in brilliant shades of red, crimson, cerise, scarlet, purple and white and flowers for weeks. There is also a fine variegated variety. *I. petersiana* from tropical Africa has deep crimson leaves and scarlet flowers.

The annual *I. balsamina* from India and China has long been cultivated in gardens. Numerous cultivars exist varying in height from 15 cms (6 ins) to 60 cms (2 ft) and with double or single flowers in red, pink or white. *I. congolensis* var. *longicalcarata* (*I. niamniamensis*) from central Africa, called the Congo Cockatoo, has inflated red spurs and yellowish-white petals, the flowers so poised as to resemble perching birds. *I. glandulifera* (*I. roylei*) from India is a waterside plant, up to 2 m (6 ft) tall and varying in colour from white and pale pink to deep rose.

The sap of several species of *Impatiens* yields a red or yellow dye. The native American *Impatiens* is called Touch-me-not because of the propulsion of seeds when ripe capsules are handled.

Impatiens wallerana

Impatiens congolensis var. *longicalcarata*

Impatiens double-flowered cultivar

Begoniaceae

5 genera and 920 species

This family consists of herbaceous dicotyledonous perennials with watery stems and mainly asymmetrical leaves which have one side larger than the other. These are radical (arising directly from the rootstock) or arranged alternately on the flower stems. The flower stems appear in the leaf axils, often having much-branched inflorescences with the lower flowers male and the top ones female. Male flowers often have two or four perianth segments and many stamens; the females frequently have two to five segments and an inferior ovary.

Begonia rex forms

Begonia masoniana

The most important genus is *Begonia* which has some 900 species widely distributed throughout the tropics and subtropics, particularly South America (in the area of the Andes northwards to Mexico), but also in Africa and the eastern Himalayas (southwards to Malaya).

The greater number have tuberous or rhizomatous rootstocks although several climb by means of adventitious roots. They also have succulent stems, which in many species contain oxalic acid. In Brazil the roots of such astringent species as *B. cucullata, B. sanguinea* and *B. bahiensis* have medicinal uses. The leaves of *B. tuberosa* are eaten by natives in the Moluccas as a remedy for scurvy; they are also cooked and eaten with fish. The foliage of the Brazilian *B. luxurians* makes a decoction which will lower the temperature of patients with jungle fevers.

Begonia is an important group horticulturally with many fine foliage plants which can be used for house decoration. There are also splendid and long-blooming bedding plants for gardens and greenhouses, and by careful selection one can have Begonias in flower throughout the year.

Most species require a lime-free, organic soil which is always moist. In made-up composts it is usual to include plenty of peat to retain moisture and give the plants a cool, friable root run.

Among the decorative types *B. rex* and its cultivars hold a distinguished place. They have multicoloured leaves in a variety of shades and patternings, for example 'Silver Queen' and 'Hoar Frost' both with a predominance of silver; 'Curly Carnot' which has a spirally twisted mid-rib; 'La Pasqual', black, purple and silver; and 'Heligoland', red, green, silver and plum-purple.

The flowers of *B. rex* are of secondary importance, being rather small and usually pink. Propagation is effected vegetatively from the leaves which, when cut up into small pieces and 'sown' on pans of peaty soil, develop adventitious buds. Sometimes the whole leaf is pegged down on compost (right side uppermost), after the principal veins have been partially cut through to stimulate rooting. In a warm moist propagating frame, little plantlets develop at the cut areas and can be removed and potted on when they are large enough to handle.

B. masoniana, the Iron Cross Begonia, has purplish marking on its bristly grey-green leaves. It was introduced in 1952 by Mr Maurice Mason who brought it from Singapore.

Other good foliage species are *B. rajah* from Malaya, a plant with rounded smooth leaves, green with red blotches and a highly polished finish and *B. hispida cucullifera* (*B. vitichotoma*), a curious plant which has fibrous roots, tall stems with clusters of showy white flowers and green hairy leaves with many leaf-like appendices attached at irregular intervals along the leaf veins. These cannot be used for propagating purposes for they are not known to root, so the plant must be increased from stem cuttings.

B. luxurians from Brazil has nearly circular leaves split up into many lanceolate leaflets. The whole plant is covered with fine hair and the flowers are cream in a short inflorescence. *B. metallica* from Bahia also has pinkish-white flowers but in larger clusters, and bright green leaves with a metallic sheen.

B. coccinea from Brazil is a tall, fibrous rooted species with roughly kidney-shaped leaves, 10–15 cms (4–6 ins) long, which have reddish undersides; it has smooth fleshy stems and pendulous racemes of coral-red flowers. 'President Carnot' is a magnificent hybrid of it which will reach 2.5 m (8 ft) and has scarlet flowers.

B. x *maculata* has cream-spotted leaves with red undersides and *B. manicata* from Mexico is fibrous rooted with tufts of red bristles along the tops of the leaf stalks and alongside the veins on the undersides of the leaves. It blooms in winter with pink flowers. *B. heracleifolia* is a white-flowered species from Mexico and *B. boweri,* also Mexican, is a small species with creeping rhizomes, white flowers and green leaves zoned in black.

Tuberous rooted Begonias are important bedding plants with large double flowers in a wide range of brilliant shades. The many cultivars available include varieties with frilled and ruffled petals, and sorts with flowers resembling Hollyhocks, Roses, Camellias and Carnations.

Begonia 'Harlequin'

Begonia 'Gloire de Lorraine'

Begonia 'Midas'

Begonia 'Rhapsody'

The plants are usually started each spring from dormant tubers and either put directly out into the garden (in warm districts) or started in a greenhouse and planted outside when all risk of frost is past. At the end of the season they are dried off and stored in peat until the following year.

The original tuberous Begonias were single and came from a number of species introduced between the years 1865 and 1868 by James Veitch and Son of Chelsea, England. The first hybrid raised by this firm was *B.* 'Sedenii' (named for one of their employees) and derived by crossing *B. boliviensis* with an unknown Andean species. During the next twenty years the Chelsea firm continued hybridizing and raising new varieties and then Continental growers like Lemoine and Crousse, both Frenchmen, continued the work, as did the American firm of Vetterle and Reinelt of California. In Britain in recent years, Messrs Blackmore and Langdon of Bath have done more than anyone else to produce quality blooms and outstanding cultivars.

Today double Begonias are available in clear self colours of pink, rose, red, cerise, vermilion, salmon-orange, bronze, clear yellow, cream and white; also in picotees showing a combination of shades, for example apricot with a rose edging, 'Kismet'; white-edged with deep pink, 'Frou-Frou'; and yellow with a pink border, 'Fantasy'.

When plants are grown for display work, as in pots, it is important to remove the two small female flowers which flank each male bloom. These insignificant blossoms not only detract from the male flower, which is much larger and finer in every way, but waste the plant's energy. They should be taken out when quite tiny, except in instances where seed is required.

'Red Cascade' is typical of a type of Begonia usually known as *B.* x *pendula* or the Pendula group introduced in the late 19th century. They make ideal hanging-basket plants with large pendulous blooms in a wide range of colours and are derived from various tuberous species.

The fibrous rooted Begonias come into two main categories, the summer-flowering Semperflorens group and the winter-blooming hybrids (derived originally from *B. socotrana*) of which 'Gloire de Lorraine' is perhaps the best known. The latter makes a small compact plant about 30 cms (1 ft) high which is literally smothered with small pink flowers. It comes to bloom about Christmas which makes it a favourite gift plant and continues in character for many weeks. A peculiarity of the plant is the fact that all the terminal flowers are female, but the side ones (laterals) are all male. It is reproduced by cuttings taken from cut down specimens which have sprouted.

B. socotrana from the Island of Socotra was introduced to England in 1880 and discovered to have the unusual characters of herbaceous stems with clusters of resting buds at their base. These can be detached and grown on. Veitch crossed the species with a tuberous variety called 'Viscountess Doneraile' to produce a hybrid called 'John Head'. This was the first of the winter-flowering types but today there are many named sorts, mostly with pink and red flowers. 'Gloire de Lorraine' is an old plant for it was brought out by the French firm of Victor Lemoine and Son of Nancy in 1892. It makes a good hanging-basket subject in addition to being used as a pot plant, but sudden changes of temperature cause it to drop its flowers and leaves. Among a host of good winter-blooming sorts 'Mrs Peterson' is distinctive by reason of its dark bronzed leaves, which make a pleasing contrast with the deep rose flowers.

Begonia coccinea

Begonia semperflorens

'Whisky'

'Lucifer'

'Pink Comet'

Begonia x *pendula*
'Red Cascade'

B. semperflorens, a Brazilian species (now correctly known as *B. cucullata* var. *hookeri*), is scarcely grown today although its name has been transferred to a group of long-blooming varieties originally derived from this species. These include forms with green, dark red, bronze and yellow foliage and with white, salmon-pink, scarlet, carmine and light rose flowers. All may be raised from seed and make first class bedding plants which seem impervious to wind and weather. They need temperate to warm conditions, damp soil and shelter from the hottest rays of the sun. There are also F1 hybrids available which have larger and better blooms. The general height of the cultivars in this section is around 15 cms (6 ins).

B. semperflorens has also been hybridized with some of the large-flowered tuberous Begonias to produce a race known as Multiflora Begonias. These have many small double flowers in various colours and remain in flower for months.

Berberidaceae

4 genera and 575 species; from N temperate zones and tropical mountains of South America

This is a dicotyledonous family composed in the main of shrubs, many shade-tolerant. The most important are *Berberis* and *Mahonia,* once grouped in a single genus but now differentiated and easily told apart by the foliage.

Most *Berberis* have simple spiny leaves and prickly stems, while *Mahonia* leaves are pinnately divided into leaflets and the stems thornless. Both carry bunches of small yellow or orange flowers which are of regular shape with six sepals and six petals making four whorls, the two inner secreting nectar at their bases. The ovary is superior and there are three plus three stamens which react visibly when touched, as for example by insects seeking nectar, showering pollen all over the intruders which are thus forced to act as unofficial pollinating agents. This device probably accounts for the dense berrying habit of many species.

Berberis vulgaris, the European Barberry, was once cultivated for its fruits which were believed to have antiseptic properties (even the Ancient Egyptians thought this) and up to the 19th century they were frequently made into jellies and preserves. The leaves were also eaten in a sour sauce which the old herbalist Parkinson described as beneficial for 'a fainting hot stomacke and liver' to repress 'soure belchings of choller'. The wood is yellow and very hard but fine grained, so suitable for such objects as toothpicks, mosaic pieces and turnery. A yellow dye lies under the bark (berberin) which has been used to stain wool and leather. Unfortunately *B. vulgaris* has one objectionable trait; it acts as an alternative host during the life cycle of wheat rust so its cultivation has largely declined. In the 15th century, however, it was much favoured for hedging purposes. There are many ornamental *Berberis* species suitable for the front of shrubberies or planting in island beds in lawns or rough grass. Some assume brilliant autumnal colourings when the sealing-wax scarlet of the branch-bending berries looks most effective against the russet, green and flame-red leaf tints. Amongst the most popular are *B.* x *rubrostilla*; the *B. aggregata* hybrid 'Buccaneer' with coral berries; *B. thunbergii* and its purple-leafed form *atropurpurea*. These are tolerant of most soils including those containing lime.

Evergreen *Berberis* include *B.* x *stenophylla*, which makes a dense prickly 'dog- and boy-proof' hedge up to 2.5 m (8 ft) tall, brilliant in spring with masses of small yellow flowers, and *B. darwinii*, one of its parents. This rather straggling shrub was first found in an island off Chile by Charles Darwin in his famous voyage on the 'Beagle'. It has small spiny leaves and orange flowers followed by bluish-purple berries.

Mahonia aquifolium is the Oregon Grape, an evergreen with shiny, glossy leaflets, each terminating in a sharp point. It was first brought to England by the collector David Douglas (circa 1825–27), when it caused great excitement, early layered plants fetching as much as ten guineas apiece. It has yellow flowers followed by blue-black berries—which were once marketed in North America—and will grow in shade.

In order to flower well in a cool climate *M. japonica* must be given a sunny sheltered situation. However most gardeners will find its striking habit and winter blossoms merit such favoured treatment, although some shade is advisable in warmer parts of the world. There seems to be some mystery about its origin, for although the original plants in Britain came from cultivated sources in Japan, the Chinese *M. bealei* (similar but not as good) is often confused with and sold under the name *M. japonica*. The distinctions appear in the flowers which are in long pendulous racemes, 15–20 cms (6–8 ins) long in *M. japonica*, but shorter and more erect in *M. bealei*. In both cases they are primrose-coloured with a strong Lily of the Valley scent. The leathery leaves, glossy

Berberis darwinii

Mahonia japonica

above and pinnately divided, are somewhat spiny and frequently grow 30 cms (1 ft) in length. Ultimately both shrubs attain a height of around 2 m (6 ft).

The species *M. lomariifolia* has the most elegant foliage of any *Mahonia*, with ten to twenty pairs of leaflets. It bears erect racemes of flowers in midwinter, of a deeper yellow than the species described above. It is Chinese in origin.

Nandina domestica was first brought from Canton to England in 1804 by William Kerr and later collected by E. H. Wilson in western China. It requires a deep rich soil and will stand a certain amount of frost and cold, but in very cold climates is best protected by glass. The habit suggests that of a Bamboo, with long 'canes' growing 2–2.5 m (6–8 ft) high terminating in very elegant pinnate foliage. This fact, together with the plant's use in China for altar and temple decorations, probably accounts for its common name of Heavenly Bamboo. The large erect trusses of white flowers are followed by masses of scarlet berries which the Chinese use in the same way as we do Holly. In Japan it is frequently planted close to a door leading into the house from a garden or courtyard, for the convenience of any member of the household who may suffer a bad dream or nightmare. If he confides details of this to the Heavenly Bamboo it is believed that he will suffer no harmful after effects. The aromatic close-grained wood is used for toothpicks.

Vancouveria and *Epimedium* are woodland plants, the former from the western United States (Washington to California), the latter from Asia, southern Europe and North Africa, which form dense ground cover that is effectively weed-smothering. *Epimedium grandiflorum* is often grown in shrubberies where it produces masses of green leaves each with two to three oval- to heart-shaped leaflets on wiry stems 30 cms (1 ft) tall. In autumn these take on a rich bronze hue and remain standing all winter (even when dead and dried) unless cut back by the gardener. The flowers come in early spring on slender racemes which carry six to sixteen blossoms, individually shaped like small bishops' hats. They are variously coloured, white, pale yellow, violet, and carmine in the cultivar 'Rose Queen'.

E. pinnatum from the east and west Caucasus has bright yellow flowers; the hybrid *E.* x *rubrum* has crimson blooms and red blotched leaves.

Barrenwort is a common name for *Epimedium,* which for centuries has been used medicinally in China, and also as an aphrodisiac for sheep. The dried leaves of *E. grandiflorum* constitute a tonic called Fang-chang tsao (literally 'give up stick') presumably because it so strengthens an elderly patient that he can discard his walking stick.

Vancouveria takes the place of *Epimedium* in the New World and is frequently found in the shade of Redwoods (*Sequoia sempervirens*). It has leafless stems carrying small white or yellow flowers and two- to five-parted leaves.

Tecoma stans
Kigelia africana
(K. pinnata)
Jacaranda mimosifolia
Campsis radicans

Bignoniaceae

120 genera and 650 species; a few temperate but mostly tropical

This is a family of dicotyledonous woody plants, generally trees and shrubs but including a number of vigorous climbers. These may be simple twiners, or possess elaborate hooked tendrils or have aerial roots (*Campsis radicans*), with which they clamber over other trees. The leaves are generally compound and paired, but occasionally alternately arranged on the stems. The large ornamental flowers, which are more or less regular and usually tubular, flare out into gaping mouths. The calyx is of five joined sepals, the corolla of five joined petals, with four (rarely two) stamens and a superior ovary. The seeds are usually flattened with a large membranous wing.

The trees are great standbys in tropical countries especially *Jacaranda*. These are used in Africa, South America, the West Indies, India and the warmer parts of the USA and in European countries (like Portugal) for public park and street planting. Others make effective flowering hedges, while in cool countries several climbers are grown in greenhouse borders, or in sheltered places outside during the summer months.

Visitors to equatorial Africa are invariably impressed by the African Tulip Tree (*Spathodea campanulata*) which can be seen as a brilliant patch of scarlet against the nondescript browns and greens of the surrounding jungle. The flowers, borne in dense terminal racemes, are large, individually 7–13 cms (3–5 ins), widely trumpet-shaped and flame-red edged with yellow. The brown buds curve to a point and are inflated with water. The pinnate leaves are large and showy, arranged in opposite pairs or sometimes in threes. The tree grows to a height of about 18 m (60 ft) and has a soft wood which smells of garlic when cut; this is sometimes used for tribal drums and witch doctors' wands.

In Kenya and Tanzania *Spathodea* is known as Nandi Flame or Flame of the Forest. In Gabon it is associated with witchcraft, the blossoms being placed before the huts of those who break tribal laws. When the inmate dies (or is put to death) the flowers are buried with him to prevent his spirit returning.

Kigelia africana (*K. pinnata*) from tropical Africa, is called the German Sausage Tree because of the large grey-green, fibrous fruits (frequently a metre in length) which dangle under the branches. These have very long stalks and succeed trumpet-shaped, maroon-purple flowers, unpleasantly scented, which are pollinated by bats. Slices of the baked fruit are used by Africans to flavour beer and placed in drinking troughs to guard cattle against leeches. The tree grows to 12–15 m (40–50 ft) and has spreading branches and pinnately divided leaves.

Another strange tree in this family is *Crescentia cujete,* the Calabash Tree. This is native to tropical America and carries the flowers directly on the old stems, where they are fertilized by bats. Later the blooms are succeeded by gourd-like fruits with woody exteriors which, when the pulp is removed, are made into drinking vessels (calabashes). The young fruits are also pulled and pickled like walnuts, while the pulp is used medicinally and the cooked seeds eaten as vegetables. They have a more recent use as maracas—the seed-filled gourds with handles, so popular with South American orchestras. Mature fruits are sometimes 45–50 cms (18–20 ins) in diameter.

The tree makes a dense, rather dumpy evergreen about 9 m (30 ft) tall, with clusters of glossy green, broadly lanceolate leaves and bell-shaped flowers.

Jacaranda species from Central and South America and the West Indies have finely divided bipinnate leaves and showy inflorescences of violet-blue foxglove-like flowers. These remain four to five weeks in character and then carpet the ground with the dropped blossoms. *J. filicifolia,* the Fern Tree, grows 18 m (60 ft) or more when conditions are to its liking, bearing dense clusters of flowers which are deep violet-blue and individually 4 cms ($1\frac{1}{2}$ ins) long and 1 cm. ($\frac{1}{2}$ in.) wide. The fragrance from these has been compared to that of wild clover honey. *J. mimosifolia* is probably the most commonly cultivated species. Jacaranda wood is durable and used for carpentry work.

Hardier trees are the Catalpas, particularly the North American *Catalpa bignonioides* which makes a good town plant. It is used for lining streets in many northern capitals. The species makes a small but much-branched tree, 8–9 m (25–30 ft) tall, with large heart-shaped leaves which have a disagreeable odour when

Spathodea campanulata

Catalpa bignonioides

Eccremocarpus scaber

crushed. The common name of Indian Bean or Cigar Tree refers to the capsules and also to the fact that Indians are supposed to have smoked these.

The fragrant flowers come in summer in broad panicles; individually they are bell-shaped with frilled margins and are white with yellow and purple spots. A golden-leafed form called 'Aurea' makes a splendid foil for dark green shrubs or conifers, when planted nearby.

Other Catalpas include the Chinese *C. fargesii* with pinkish flowers, and *C. speciosa* from the US, similar to and possibly conspecific with *C. bignonioides*. *C. ovata* is popular in Japan.

Campsis radicans, from eastern United States, is the Trumpet Creeper, a vigorous deciduous climber with aerial roots—like Ivy—which will scramble 9–12 m (30–40 ft) over buildings, trellises or other trees. It needs a little support at first but then romps away, flowering in the late summer with clusters of vivid orange-scarlet, trumpet-shaped flowers. These are about 6–7 cms (2½–3 ins) in length with flared five-lobed mouths.

C. grandiflora (*C. chinensis*), from China and Japan, is a spectacular species with bunches of flame-red and orange flowers. It is less hardy, however, than *C. radicans* and not so vigorous so it must be given a sheltered site or protected in cold climates. Both species, lacking support, will creep over the ground.

Pyrostegia venusta, from Brazil, is another splendid climber with terminal racemes of many pendulous, trumpet-shaped, orange flowers. These have stamens and pistils protruding from rather shiny corolla tubes and the leaves are opposite and compound. The plant grows quickly in rich soil and full sunshine, attaching itself to supports by means of tripartite tendrils. Plants can be trained up tall trees, balconies and buildings or employed to drape hedges and fences. In Nairobi it is used as a ground cover at the roadsides, achieving really brilliant effects. It is also known as the Orange-flowered Stephanotis.

Incarvillea delavayi

Tecomaria capensis, from South Africa, often called the Cape Honeysuckle, makes a smooth climber, up to 5 m (15 ft) tall, and has opposite, pinnate leaves about 10 cms (4 ins) long which are toothed at the tops of the leaflets. The orange-scarlet flowers are in terminal racemes, narrowly funnel-shaped with protruding stamens and there is a lemon-yellow form called 'Primrose'. In full sun and under fairly dry conditions this makes a good flowering shrub for months on end or it may be used for hedging. *Tecoma* (*Stenolobium*) *stans* is native to Central and tropical America including the West Indies and much planted in tropical gardens, up to 900 m (3000 ft), or occasionally in the cool greenhouse. It is a profusely flowering branching shrub, up to 4 m (12 ft) tall, with bunches of yellow tubular flowers and smooth pinnate leaves which have serrated leaflets. Commonly called Yellow Elder, it is easily reproduced from seed or cuttings and can stand dry conditions. *T.* 'Smithii', a cultivar of unknown parentage which originated in Australia, makes a good pot plant and comes true from seed. The flowers are bright yellow, tinged with orange.

Tabebuias, from tropical and South America and especially the West Indies, are ornamental flowering trees of some stature, 30 m (100 ft) or more in height with great clusters of tubular flowers which are however of short duration. In *Tabebuia serratifolia* they are clear yellow, so that the tree is locally known as the Yellow Poui. The wood is the source of green ebony, a hard heavy timber. The Pink Poui, *T. rosea* (*T. pentaphylla*), is smaller but faster growing with white to deep rose blossoms. These may burst out several times in a season so giving substance to a saying of the Tobago islanders that rain will not fall until the Poui blooms three times in succession.

Eccremocarpus scaber, from Chile, is a semi-woody evergreen climber with shrubby stems, which are ribbed, and the general habit of Clematis. The leaves are in opposite pairs, doubly pinnate, the leaflets unequally lobed or coarsely toothed. Up to twelve tubular flowers, 2–5 cms (1 in.) long and bright orange-red in colour, are carried on the racemes 10–15 cms (4–6 ins) long. The main stem twists round any available object to a height of 3–4 m (10–12 ft). Suitable for growing in pots or borders in the greenhouse, or outdoors in the tropics, it may also be treated as a half-hardy annual where the winters are too cold for it to remain outdoors. Varieties exist with golden flowers, *aurea*, and red-orange flowers, *ruber*.

Incarvilleas are native to central and eastern Asia including the Himalayas. They are showy perennials with pinnate leaves and terminal clusters of large Gloxinia-like flowers which may be up to 7 cms (3 ins) in length.

Suitable for a variety of situations in temperate gardens such as borders, rock-garden pockets or growing in pots, they are easily propagated by seed or division; seedlings take about three years to come to flower. Light soil is best as the roots are inclined to rot off in stagnant moisture. *Incarvillea delavayi* which comes from Yunnan is one of the most popular, with bright rosy-red flowers on stems 30–45 cms (1–1½ ft) tall; 'Bee's Pink' is paler. *I. mairei* var. *grandiflora* (*I. brevipes*) from W China, fuchsia-red, and *I. olgae* from Turkestan, pale pink, are frequently grown.

Other garden plants belonging to this family are *Pandorea pandorana,* the Wonga-wonga Vine from Australia, which has bunches of yellow or pinkish-white flowers; *Doxantha* or *Bignonia capreolata*, a south-eastern US climber with reddish-yellow funnel-shaped blooms; and *Parmentiera cerifera,* from tropical America, sometimes called the Candle Tree on account of its round hanging fruits which look like yellow wax candles. The flowers are large, yellowish-white and often occur on the trunk (as in *Crescentia*) or branch wood. It makes a tree 3–6 m (10–20 ft) high, suitable for tropical gardens or the greenhouse.

Bombax ceiba

Pachira macrocarpa

Bombacaceae

20 genera and 180 species
all tropical especially American

This is a family of dicotyledonous trees, often very large and sometimes with grotesque trunks containing water storage tissue. The leaves may be simple or palmately divided with deciduous stipules; the flowers are asymmetric with five sepals, five or many stamens (free or united into a tube) and a superior ovary. The seeds are smooth, often embedded in hair in large capsules.

Ceiba pentandra (*C. caesaria*) from tropical America (also found in W Africa) is the Silk Cotton or Kapok, a slim tree well buttressed at the roots to support its great height. It has large, whitish and fragrant flowers which are succeeded by pods with many seeds. The latter are enveloped in insect-repelling, water-resistant, silky hairs which are used in many household articles, making an excellent stuffing for cushions, pillows and mattresses.

Another kapok comes from *Bombax ceiba* (*B. malabaricum*), the Red Silk Cotton from India and Burma. This makes a wonderful show when in bloom with tier upon tier of large, scarlet flowers filled with scarlet stamens. As these appear in early spring before the leaves, the effect is even more pronounced; another attraction comes later when the ground is carpeted with the fallen petals. The fleshy calyces are eaten in Burma in curries. The capsules, 15 cms (6 ins) long, disgorge quantities of cotton in which the small seeds are embedded and dug-out canoes are made from the soft wood.

Pachira macrocarpa from Mexico to Costa Rica is a small tree with attractive palmate leaves having seven to eleven nearly stalkless leaflets and white, pink or pale yellow flowers about 22 cms (9 ins) in length that appear above the leathery leaves like aigrettes. These contain quantities of yellowish-red stamens.

P. aquatica, native to tropical America, has even larger flowers (up to 35 cms; 14 ins) with many conspicuous stamens which are half white and half crimson. The petals are white or cream. The roasted seeds can be eaten. This species grows well in wet soil and swampy places. Unlike *Bombax, Pachira* species have no wool on the seeds and flower when the foliage is on the trees.

Durio zibethinus from Malaya produces the evil-smelling but delicately flavoured Durian, the large 5 lb fruits of which fall from the tree when ripe and reek of onions, gas leaks, old cheese and turpentine mixed together. In spite of this they are highly esteemed by connoisseurs and also such animals as tigers, elephants, monkeys and bats. The creamy flowers hang in clusters from the lower branches and trunk.

The balsa raft which Thor Heyerdahl used to cross the Pacific was constructed from wood of *Ochroma pyramidale,* a South American tree noted for its light wood which is difficult to penetrate with nails yet which is so buoyant that it is used in the

Adansonia digitata

manufacture of life-saving apparatus as well as heat insulators and model aeroplanes.

Perhaps the most extraordinary genus is *Adansonia,* especially *A. digitata,* the Baobab. Kenyans say the devil planted this tree upside down and certainly it has a monstrous appearance with its swollen, bottle-shaped trunk and short dumpy branches sticking up in the air like thick roots. One of the largest is the 'Great Tree' of Victoria Falls, Rhodesia, with a trunk 8 m (27 ft) round, but specimens of 12 m (40 ft) are reported. According to Dr E. Swart of University College, Salisbury, Rhodesia, some specimens may be several thousand years old.

The fragrant flowers are white, about 15 cms (6 ins) across and pollinated by bats. They are succeeded by Cucumber-like fruits, 45 cms ($1\frac{1}{2}$ ft) long, on long stalks. These have woody shells, edible seeds and an acid pulp with a cream of tartar taste which makes a pleasant, thirst-quenching drink. The pollen makes a good glue; the young leaves may be eaten as a vegetable when cooked and the old leaves are often fed to stock. The wood is spongy and light so that it may easily be tunnelled out; this fact together with the great size of the trunk probably accounts for the fact that African chieftains were once buried in the heart of a Baobab tree. Formerly the bark fibres were woven into cloth by natives in the Transvaal, and strips of the bark are still used for making ropes and mats and the woody pods for spoons in South Africa.

Adansonia species are also found in Australia and Madagascar; *A. madagascariensis* is commonly known as the Monkey-bread Tree.

Several other genera of Bombacaceae have water-storage stems, as the barrel-shaped *Cavanillesia platanifolia* from Colombia and Panama and the South American *Chorisia*. Often the trunks of the latter are studded with thick thorns that may disappear as the tree ages. *C. speciosa*, the Brazilian Floss Silk Tree, is cultivated in tropical gardens and has spectacular crimson, pink or white five-petalled flowers succeeded by floss-covered seeds in capsules.

Boraginaceae

100 genera and 2000 species
tropical and temperate

This large family consists mostly of herbaceous plants, which are often perennial with fleshy roots or rhizomes, and also a few shrubs and climbers. They are dicotyledons.

Nearly all the plants are roughly hairy especially on the leaves. These are usually alternate but are occasionally opposite and the inflorescences frequently appear as one (sometimes two) coiled cincinnus which look like the twists on a monkey's tail. As the blossoms expand these gradually unfold, so that the newly opened blooms always face the same direction.

Individually the flowers are bisexual, usually regular, with five sepals, five petals (which form a tube or funnel shape) five stamens and a superior ovary with two, four or ten locules. Many flowers have hidden nectaries in the throats, while the blooms of some species, like *Borago* and *Symphytum,* are pendulous.

Favourite garden flowers in this family are the Heliotropes, some 250 species of which exist in tropical and temperate regions. The most important are *Heliotropium arborescens* (*H. peruvianum*) and *H. corymbosum,* both Peruvian and both involved in the parentage of many fine garden forms, which are used all over the world for summer bedding and pot work. They are also grown as standards or large pyramids or trained up walls in warm gardens and greenhouses.

H. arborescens makes a shrub up to 2 m (6 ft) tall with wrinkled, hairy, oblong-ovate leaves and branched spikes crowded with fragrant, violet or lilac flowers. *H. corymbosum* is paler in colour but its blooms are nearly double the size. Named cultivars have to be reproduced from cuttings, but where uniformity is unimportant an interesting miscellany of mauve and purple shades can be obtained by sowing seed.

Heliotropes will not tolerate frost and require a winter temperature of 13°C (55°F) to ripen the wood for the following season. Their powerful scent—the Cherry Pie of gardeners—was once used in perfumery but now the essential source of this (heliotropin) can be synthesized.

Anchusas rival Delphiniums for majesty of bearing and deep blue flower colouring, particularly *Anchusa azurea* (*A. italica*) from southern Europe and the Near East. This grows 1–1.5 m (3–5 ft) tall with rough stems and leaves and large panicles of blue Forget-me-not type flowers which are extremely variable,

Cordia sebestena

so that for garden purposes it is best to grow good cultivars such as 'Loddon Royalist', bright blue; 'Morning Glory', rich deep blue and taller than the preceding, 1.25–1.5 m (4–5 ft); 'Opal', the best light blue; and 'Royal Blue', very intense gentian blue.

Another good species is *A. caespitosa* from Crete, a tufted perennial for the rock garden which never grows more than about 10 cms (4 ins) high, with many sprays of short-stalked bright blue flowers. The leaves come from the base and are irritable of winter wet, so this species is best treated as an alpine house or scree plant. *A. capensis* from South Africa is a biennial with rich blue flowers on stems 45 cms (1½ ft) tall; 'Blue Bird' is a named cultivar, deeper in colour.

Mertensia species are generally low growing and have a smoother texture than most members of Boraginaceae. *M. virginica,* the Virginian Cowslip or Bluebell, is a splendid woodlander from the eastern United States, with smooth, oval, strongly veined leaves and clusters of nodding, trumpet-shaped, blue or lavender flowers which develop from pink buds. *M. sibirica* and its white form *alba* are E Asian representatives, the type plant a deeper blue than *M. virginica*. All the Mertensias need moist soil and light shade.

Cordias are evergreen and deciduous trees and shrubs from tropical or warm temperate regions of Asia, Africa and America. They are often very showy with tubular flowers in terminal and/or axillary clusters. They need greenhouse treatment in cool climates. *Cordia sebestena* from the West Indies grows to 4 m (12 ft) and has scarlet flowers.

Forget-me-nots (*Myosotis*) are much loved, early-flowering plants of annual or perennial nature, widely distributed throughout temperate regions of the world.

All the garden varieties grown today are probably hybrids between *M. sylvatica, M. dissitiflora* and *M. alpestris.* They are easily raised from seed and in the garden provide charming associates for spring bulbs or may be used as pot plants or ground cover between taller plants. The basic colour is deep blue, which may only develop with age from a pinkish juvenile stage, but there are also white and pink forms.

Old English names for *Myosotis,* such as Mouse Ear and Scorpion Grass, refer to the hairy leaves and the curled flower heads which were thought to resemble scorpion's tails and provide an antidote for their sting. The more romantic Forget-me-not is said to be derived from a German legend of a knight who, while gathering flowers of a riverside plant (*M. scorpioides*—originally known as *M. palustris*) for his lady love, fell in and was swept away by the current, crying as he vanished—Forget-me-not!

M. sylvatica is a bushy perennial about 30 cms (1 ft) high and 'Blue Bird', one of its probable cultivars (the nomenclature is much confused), makes a splendid winter-flowering pot plant under glass. Forget-me-nots are easily maintained from year to year if the old plants are shaken over a spare piece of ground before they are destroyed. Enough seedlings will germinate to take care of next season's bedding.

Pulmonarias are coarse perennials with large, bristly, tongue-shaped leaves, which in some species like *Pulmonaria officinalis* are irregularly spotted with white. This characteristic drew the attention of herbalists in the past and according to their *Doctrine of Signatures* they decided that the spotted leaves would cure spots on the lungs and accordingly named the plant Pulmonaria or Lungwort. Other homely names are Soldiers and Sailors (referring to the pink and blue flowers), Spotted Dog and Bethlehem Sage.

All the Lungworts bloom early in the year, before the leaves are much in evidence and so the bunches of tubular drooping flowers can be seen to advantage. In *P. officinalis* they are pink when young but develop purple and then rich blue shades as the

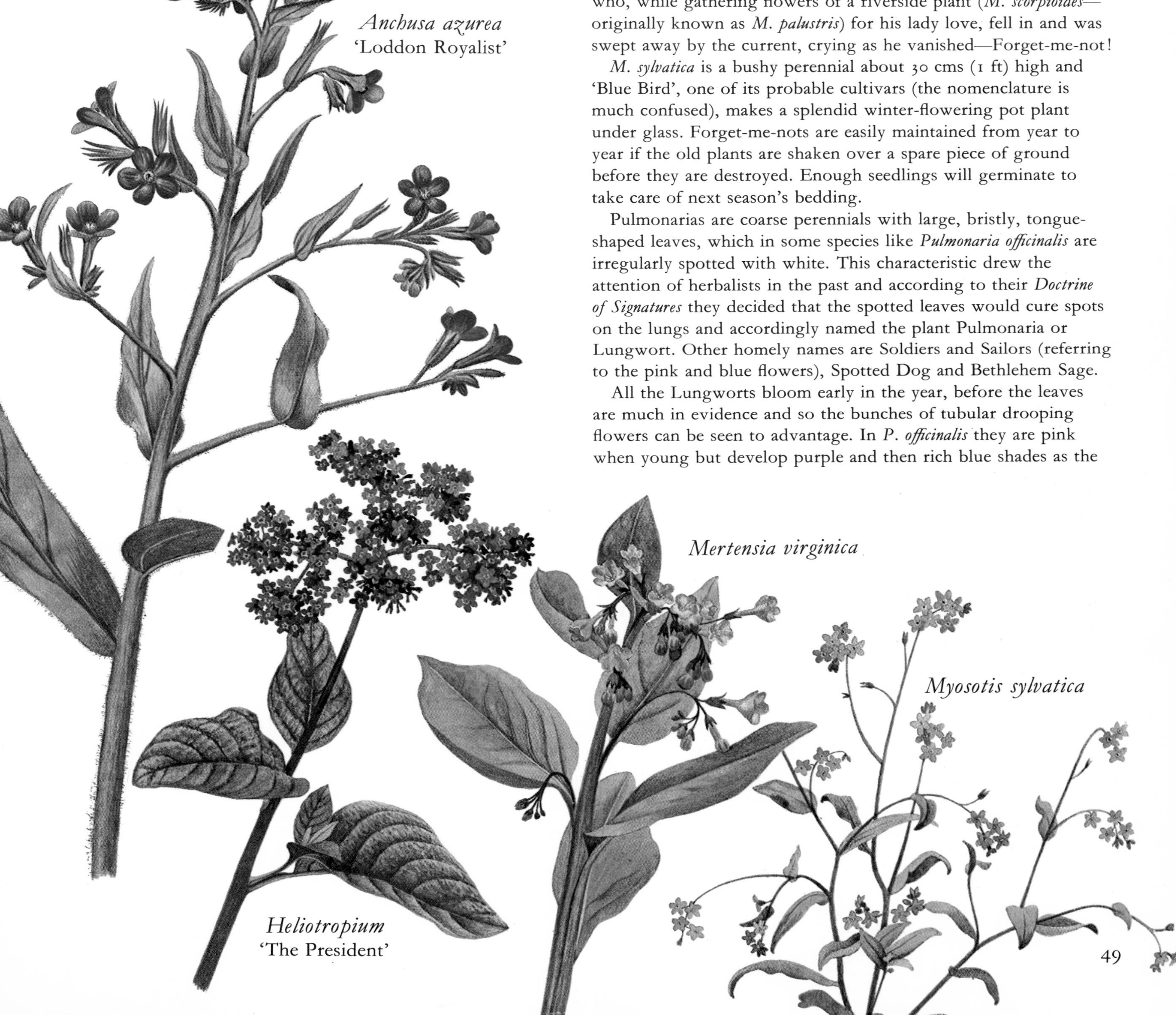

Anchusa azurea 'Loddon Royalist'

Heliotropium 'The President'

Mertensia virginica

Myosotis sylvatica

flowers mature—often all three colours will be seen on the same inflorescence.

P. angustifolia like *P. officinalis* is European and also changes colour from pink to blue. The leaves however are unspotted and the blue more vivid especially in the clone 'Azurea'. There is also a white form. *P. rubra* from south-eastern Europe has plain pale leaves and brick-red flowers and *P. picta* (*P. saccharata* of gardens) has the most sensational foliage, on account of its heavy cream marbling, the flowers pink and blue.

Pulmonarias are not fastidious as to soil or situation but prefer a moist site; their usual height is around 22 cms (9 ins).

Brunnera macrophylla (*Anchusa myosotidiflora*) is a woodland species with sprays of flowers resembling Forget-me-nots, deep blue with yellow throats and rough heart-shaped leaves 7–15 cms (3–6 ins) across. It is a pretty plant for the border or the front of a shrubbery but inclined to be invasive. Growing about 45 cms (1½ ft) tall it was discovered in woods in the Caucasus in 1800. The clone 'Variegata' is slower growing and rather more decorative with its cream-blotched leaves.

There are about sixty species of *Lithospermum,* all from the temperate regions of Europe, Asia and North America, although only a few are worthy of cultivation. General characteristics are rough, hairy, rather narrow leaves alternately arranged on the stems, and white, yellow, blue or violet flowers in terminal cymes and with funnel- or salver-shaped corollas.

The most suitable species for a garden and a highly desirable plant for the rock garden is *L. diffusum* (*L. prostratum*) 'Heavenly Blue', transferred by some botanists to the genus *Lithodora.* This is a shrubby, prostrate evergreen that forms a broad carpet 7–10 cms (3–4 ins) high which is spangled all summer through with deep blue flowers. Old plants tend to disappear in a bad winter, especially in poorly drained soils, but if their leggy habit is kept under control by hard cutting back after flowering they seem to be more durable. Another clone called 'Grace Ward' after the owner of a nursery in Southgate (England), where it originated, has larger flowers and is more tolerant of winter conditions. Both plants should be given well-drained soil and full sunshine but both dislike lime.

L. purpureo-coeruleum (sometimes transferred to the genus *Buglossoides*), a woodland species from Europe (including Britain), thrives in light shade and damp soil—as amongst shrubs. It grows 22–30 cms (9–12 ins) high with sparsely leafy stems terminating in racemes of 1 cm. (½ in.) flowers. These are red at first and then blue; the exact shade seems to depend on locality for they can be both deep blue and intense sapphire. The dark green, narrow leaves are rough in texture. This species is known as the Walking Stick Plant because of a tendency of the long willowy sprays to bend down and root at their tips, a trait which necessitates thinning from time to time.

L. canescens from North America is the spring Puccoon, a soft hairy species with narrow leaves and forked one-sided heads of deep orange-yellow flowers. It likes an open position and well-drained soil and grows to a height of 38 cms (15 ins). The roots yield a red dye which was used by several Indian tribes. *L. officinale,* another European species but also found in West Asia, Caucasia and Iran has yellowish flowers succeeded by pearly-white polished seed pods. The leaves are the source of Croatian tea and its common name is Pearl Plant.

Lindelofia longiflora from the west Himalayas is a beautiful plant with sprays of large, deep sapphire, Anchusa-like flowers uncurling on stems and sprays of 38–45 cms (15–18 ins). The oblong or oblong-lanceolate leaves narrow to a point. Suitable for any open position in rich deep soil, the plant is easily raised from seed and flowers its second season.

Cynoglossum officinale is the common Hound's-tongue, a native of Europe (including Britain) which has a thick tapering tap root with a black rind, rough lanceolate leaves and racemes of dull red flowers. The whole plant smells vaguely of mice which may put some against it and is perhaps the cause of its discontinued use in medicine. It is narcotic and somewhat astringent and amongst physicians of the past was deemed anti-spasmodic. It apparently has some paralysing effect on the central motor system. In spite of all such drawbacks the young leaves were at one time used in Switzerland as a salad vegetable.

C. amabile from China and Tibet is biennial with many blue, white or pink flowers in terminal cymes. The whole plant is very hairy and grows 30–60 cms (1–2 ft) tall.

Borago officinalis, or Borage as it is usually known, is one of the oldest garden herbs. For centuries it was esteemed by herbalists as a cordial and medicine for chest complaints and is still used in the flavouring of beverages, particularly negus and claret cup. Borage is native to the Mediterranean region of southern Europe and very occasionally may be found wild in waste places in Britain, its starry blue flowers with backward-pointing petals (which have a cunning dot of black in the centre) a conspicuous feature of the summer flora. It is an annual plant with rough hairy leaves, easily reproduced from seed, and grows 30–60 cms (1–2 ft) high.

Borage is a good bee plant and the blue flowers are sometimes candied (as with violets) or used in pot-pourri. For the last they should be gathered just before they are fully open as then the brilliant colour is preserved. The young finely shredded foliage has been used as a salad vegetable; indeed the virtues of this plant—which have been known since the Middle Ages—appear to be legion. The leaves, which have a cucumber-like fragrance, impart a faint flavour and coolness to drinks and the flowers and dried shoots were once made into tea and syrups. Borage was also very prominent in Tudor and Stuart needlework and as the embroiderers of those times worked from fresh flowers it could thus have been a widely grown plant. It is said to inspire courage and Gerard the herbalist wrote 'I, Borage, Bring alwaies courage; . . . . The leaves and flowers of Borage put into wine makes men and women glad and merry, driving away all sadnesse, dulness and melancholy.' No-one amongst the ancients it seems credited the wine with this happy state of affairs!

B. laxiflora from Corsica with loose racemes of drooping pale blue flowers and rosettes of oblong, rough hairy leaves is sometimes grown in annual borders.

Arnebia (*Macrotomia*) *echioides* is the Prophet Flower, a handsome perennial 30–45 cms (1–1½ ft) high with bright primrose-

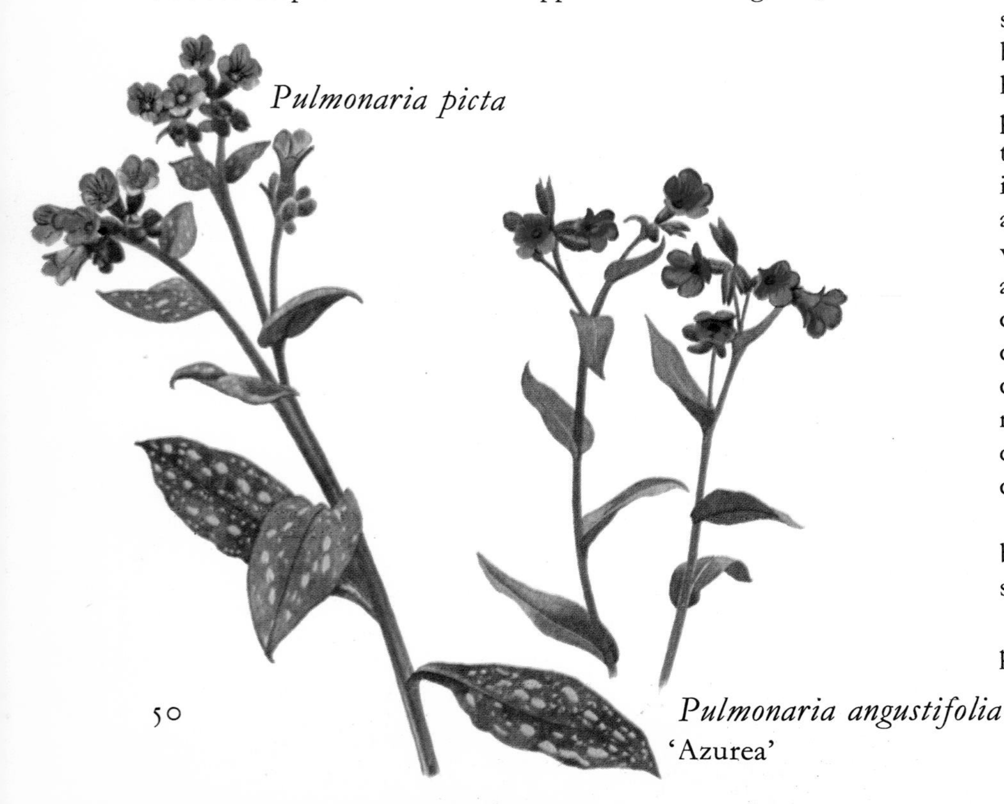

Pulmonaria picta

Pulmonaria angustifolia 'Azurea'

yellow tubular flowers. These have five black spots on the corolla which gradually fade as the bloom ages and finally disappear. By some botanists this species is placed in a separate genus *Aipyanthus* as *A. echioides*. A native of the Caucasus and northern Persia, the plant remains long in flower but deteriorates with age so should be periodically renewed by seed or from heel cuttings taken in autumn. A well-drained situation is necessary for this *Arnebia* which is equally happy in a sunny border, rock garden or growing on a dry wall.

Omphalodes species are light woodland plants favouring cool leafy soil which is yet well drained. Here they often flower for months—*O. verna,* the S European Blue-eyed Mary, first, with plentiful sprays of clear blue flowers on wiry stems 10–13 cms (4–5 ins) long. Later comes *O. cappadocica,* the Dogwood Navelseed from Asia Minor, an attractive plant with flowers like Forget-me-nots and long, mostly basal, heart-shaped leaves coming up from creeping rhizomes. The flowers are a good deep blue, carried in graceful racemes of 30 cms (1 ft) and in the clone 'Anthea Bloom' sky-blue with greyish foliage.

O. luciliae from Greece and Asia Minor is a delightful plant with sprays of flowers, pink in bud but developing to a lovely shade of opalescent blue. Stems and leaves are blue-grey which makes a perfect foil for the blooms. This sun-loving species likes a little mortar-rubble in the moist rich loam. It grows 10–15 cms (4–6 ins) high and flowers on and off all summer.

Onosmas make excellent rock plants with greyish bristly stems and leaves and terminal bunches of drooping tubular flowers, frequently scented. All require full sun and good drainage; in wet situations they often rot or are eaten by slugs. Among the best are *Onosma albo-roseum* from Asia Minor with white flowers which develop with age to deep rose; *O. echioides* from southern Europe, pale yellow; *O. frutescens,* from Greece and Asia Minor, yellow fading to orange; *O. pyramidale,* the Himalayan Comfrey from the west Himalayas, 60 cms (2 ft) tall, bright scarlet; and *O. tauricum,* from south-eastern Europe, deep yellow. Most of those mentioned are roughly 30 cms (1 ft) in height. The roots of *O. echioides* are the source of a red dye (orsanette) sometimes used instead of *Alkanna tinctoria* (another member of Boraginaceae) for dyeing woollens, oil and fat.

Symphytum officinale is the Comfrey, a coarse perennial from Europe (including Britain) suitable for rough places especially near water. It has large, rough, ovate or lanceolate leaves and curled cymes of flowers, which vary considerably in colour—white, yellowish-white, purple, rose and crimson forms being found. *S. caucasicum,* 30–60 cms (1–2 ft) tall from the Caucasus makes a better garden plant, with its deep blue flowers which can be very striking in the right setting. *S.* x *uplandicum* (*S. asperum* x *S. officinale*) is widely naturalized in Britain and has rose-pink flowers which change to blue, both colours being found in the same inflorescence. It is often known as *S. peregrinum.*

Comfrey leaves when boiled make a green vegetable and were once used for flavouring Comfrey Cakes.

Myosotidium hortensia (*M. nobile*) is the Chatham Isle Forget-me-not, a strange plant 30–90 cms (1–3 ft) in height with large, thick, strongly veined leaves and dense heads 5–13 cms (2–5 ins) across of azure-blue, purple-eyed flowers. In the Chatham Isles it is found near the seashore but is much cultivated in New Zealand (where it is known as the Chatham Isle Lily) and also in parts of Scotland and Ireland.

The genus *Echium* contains several good garden plants, some very striking with erect spikes of densely packed flowers. They are mostly annual or biennial but easily raised from seed and do best in mild climates in full sun and any good garden soil. *E. vulgare* is the Viper's Bugloss or Blue Weed of Europe (including Britain), a biennial 60–120 cms (2–4 ft) tall with flowers which have protruding stamens and are purple in the bud but become violet-blue with age. *E. wildpretii,* from the Canary Isles, a 60–90 cms (2–3 ft) biennial, has pale red flowers; *E. rubrum* from Hungary, 30–60 cms (1–2 ft), is reddish-violet; *E. lycopsis* (*E. plantagineum*) from the Mediterranean area, up to 90 cms (3 ft) tall, is purplish-violet; and *E. fastuosum,* from the Canary Isles, 120 cms (4 ft), is deep blue.

E. vulgare dried and powdered was once thought to be a specific against the bites of vipers and mad dogs.

Eritrichium nanum, a gem from the high mountains of the Northern Hemisphere with dwarf tufts of foliage spangled with many brilliant blue flowers, is one of the most desirable plants for any rock garden but extremely difficult to cultivate.

Lobostemon fruticosus from South Africa is the only species in a genus of twenty-five to thirty which is cultivated to any extent. It is evergreen, with numerous branching stems crowded with bell-shaped flowers, pale blue with median stripes. Its English name of Eight-day Healing Bush refers to its uses for medicinal purposes by early settlers. Sun or partial shade, well-drained soil and a moderate climate suit this plant which grows about 90 cms (3 ft) high.

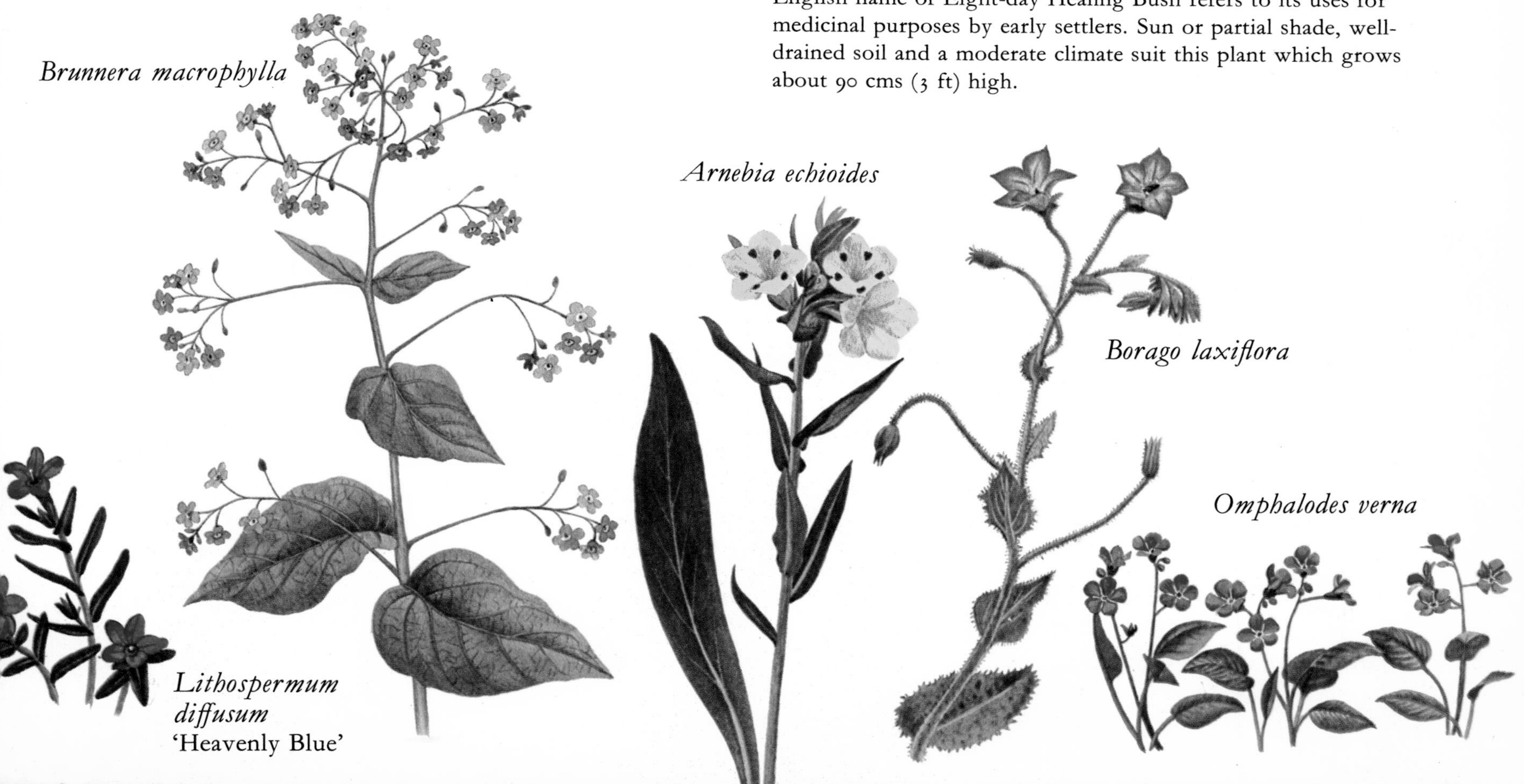

Brunnera macrophylla

Arnebia echioides

Borago laxiflora

Omphalodes verna

Lithospermum diffusum 'Heavenly Blue'

Bromeliaceae

60 genera and 1400 species; all tropical

Bromeliads are monocotyledonous plants, native to south-east US to South America; the greater number are found in Brazil. Some are terrestrial but many live on rocks or are epiphytic on trees and shrubs where, with their brilliant flowers and squat rosette leaf shapes, they look like perching birds.

While shade suits many of the species, a few grow in full sun or survive precariously clinging to telephone wires, or on Cacti or rotting logs; others, like *Puya,* are found in open country high on the Andean plateaus. Most may be discovered in areas of good air circulation which seems essential to their well-being.

Although the inflorescences are in no way similar, individual flowers bear some affinity to Lilies. There are three sepals distinct from three petals, six stamens, one style and a tricarpellate ovary (which may be inferior or superior) and the flowers are regular and bisexual except in *Hechtia*. The size however varies for while single blooms of *Tillandsia lindenii* may be 5–7 cms (2–3 ins) long, those of *Hohenbergia* species are practically microscopic. Often the charm of the plant lies in the brightly coloured bracts which almost obscure the tiny flowers.

Some Bromeliads have a vase- or pitcher-like foliage formation, made up from rosettes of leaves which curve and fit so closely together at their bases that they form a receptacle for food and water. These are particularly well developed in *Guzmania, Aechmea* and *Vriesia* and in their native jungles collect decaying leaves, insects and even small reptiles such as frogs and lizards. It has been shown that these leaves can absorb water and dissolved salts and so take over the function of roots. On the other hand, the sparse anchoring roots of most of the epiphytic Bromeliads are also able to take up food and water. In some of the 'vases' other plants of an insectivorous nature such as *Utricularia* have been discovered.

The largest genus and the most widespread is *Tillandsia* of which there are some 500 species. The best known of these is *T. usneoides,* Spanish Moss or Vegetable Horsehair, a grey slender plant which develops long festoons of thin twisted branches, which hang from forest trees and telegraph wires and look for all the world like giant cobwebs of lichen.

They are covered with tiny scaly hairs which can absorb water like blotting paper, and so the plant dispenses with roots and storage organs and lives on dust, dew and rain. Flowers are inconspicuously coloured so that they can be easily overlooked. However, Spanish Moss grows rapidly and since birds use it for nesting purposes and the wind carries it about, it is widespread all over tropical America and the West Indies. A similar species, *T. insularis,* is found in the moist forest belts of the Galapagos Islands.

The soft stems can be dried and have been used for packing material or in upholstery under the name Louisiana Moss.

Several other Tillandsias are grown primarily for their flowers, either as house plants in temperate countries or attached to trees in tropical gardens. Often the contrast between bracts and flowers is very striking. Among the best are *T. cyanea, T. lindenii* (*T. lindeniana*) and *T. ionantha* (*T. erubescens*). The former is one of the most popular, with graceful narrow leaves which arch out from a central point like *Cordyline*. The flowers open spasmodically on a large flat elliptical panicle which is soft pink and contrasts excitingly with the purplish flower petals. With *T. lindenii* the inflorescence is pink and powder blue and the leaves reddish. Both these species require a temperature around 16°C (60°F) and a certain amount of humidity in order to flower.

T. ionantha is an attractive little perennial scarcely more than 7 cms (3 ins) high, but a sheer delight at blooming time in spring when the tufts of deep green leaves turn reddish and amongst them peep many deep violet flowers. This species can be grown in a light sunny situation, fixed to a log or ball of moss; it is easily cared for—simply spray it over regularly or dunk it in water every other day during hot weather and once a week for the rest of the year.

Vriesea, named for W. H. de Vriese, a Dutch botanist, is similar to *Tillandsia,* but has broader leaves arranged in dense

Tillandsia cyanea

Aechmea tillandsioides

Tillandsia erubescens

rosettes, which are often strikingly blotched with red, brown or a different shade of green. The inflorescences persist for weeks and are often branched with brilliant bracts and contrasting flowers.

Vrieseas like light but abhor bright sunshine so should be grown in a north window indoors, or under trees in the garden in warm climates. They need plenty of air, a temperature around 16°C (60°F) and good humidity. Plant them in small pots, in osmunda fibre—but not too tightly—and keep the vase centre filled with soft water.

The showiest species and one of the easiest to cultivate is *V. splendens*. The leaves are dark green transversely striped with claret-red and the flowers are yellow with bright red bracts. *V. carinata* is the Painted Feather, a plant with rosettes of light green leaves and a flat (7 × 5 cms; 3 × 2 ins) inflorescence of red and yellow. It remains in character for months and its hybrid 'Mariae' (a slightly larger plant) throughout the whole of the winter.

V. fenestralis is a beautiful foliage plant with broad, light green leaves delicately and irregularly banded in darker green, and yellowish-green flowers. *V. hieroglyphica,* with light green rosettes crossbanded with darker green and purplish-black beneath, is another good foliage plant; it has yellow flowers.

Aechmeas are mainly epiphytic with broad leaf rosettes which form an open 'vase'. They vary considerably in the species, being soft or leathery, single coloured or banded, but always with tiny marginal spines. A single flower spike comes from the centre of each rosette, rising well above the foliage; this is usually erect and more or less branched but can be cylindrical or even tufted. The inflorescences are succeeded in some species by showy berries.

One of the commonest species in cultivation is *Aechmea fulgens,* not the type from French Guiana which has plain green leaves, but its Brazilian variety *discolor,* with maroon-purple undersides and a grey bloom on the upper leaf surface. The true flowers are dark purple but ephemeral in nature and only last a couple of days, although the enclosing bright scarlet bracts and rose berries retain their brilliance for two or three months.

A. fasciata (*A. rhodocyanea*) is one of the loveliest Bromeliads with broad (7–10 cms; 3–4 ins) spiny-edged, silver-green leaves, irregularly patterned with horizontal grey bands. A well-grown specimen may be 60 cms (2 ft) across and 30–38 cms (12–15 ins) high. The flower scape (which comes well above the leaves) is rosy-pink with many pink spiny bracts enclosing the lavender-

Aechmea fulgens

Aechmea fasciata (A. rhodocyanea)

blue flowers. The species, a native of Rio de Janeiro, tolerates lower temperatures than most Bromeliads and reproduces easily from offshoots.

A. tillandsioides, a widely distributed plant from Mexico and Brazil, has light green foliage, branched flower scapes with yellow blossoms and red bracts. The ovaries develop to white berries which slowly turn bright blue; the inflorescence meantime retains the red flower bracts.

Cryptanthus species are terrestrial plants which do not have a 'vase' but spread their stiff wavy-edged leaves outwards like the arms of an octopus. Commonly called Earth Stars, they tend to grow on rocks and have a flat habit and translucent white flowers which are not very conspicuous. They need good light and may be grown indoors with other plants in shallow containers, or can be used outside in warm areas to give border colour.

C. fosteranus, named after Mullard Foster of Florida who has introduced more new Bromeliads to cultivation than all the other collectors of these plants put together, has the colours of a pheasant's tail, with zigzag crossbands of tan, copper, green, grey and purplish-red. *C. bivittatus* and *C. zonatus* are both banded with other shades, the former salmon-rose and olive-green, the latter brownish-green and silver.

Guzmanias need higher temperatures, 17°C (62°F) winter minimum, than most Bromeliads and are striking epiphytes (many from the Andean rain forests) with glossy, smooth-edged leaves in typical 'vase' rosettes. In *Guzmania sanguinea* they are blood-red, but a metallic purplish-red in *G. cardinalis* (probably a natural hybrid of *G. sanguinea*). This last has raised cups of bright red, leaf-like bracts and creamy-yellow flowers.

Neoregelias are epiphytic and terrestrial plants from Colombia, Peru and eastern Brazil. Most of them are grown for their striking leaf colour and make excellent house plants. The flowers appear low down in the cup made by the tightly packed foliage bases, the inner leaves of which turn brilliant scarlet at flowering time. Individual blossoms peep above the water in the 'vase' like tiny Water-lilies.

In *Neoregelia carolinae* 'Tricolor' the foliage is particularly fine; the spiny-edged, narrow green leaves are variegated in the centre with cream and pink with the inner leaves becoming bright red when the light blue flowers come to character. Another species commonly in cultivation is the plant known as *N. marechalii,* green with crimson inner leaves and a crimson flush over the rest; *N. meyendorfii* (probably a form of *N. princeps*) is a smaller plant with pink coloration and *N. spectabilis* has blue-tipped reddish flowers and grey-banded leaves tipped with crimson. Neoregelias need shade and more warmth than most Bromeliads.

All the Billbergias are epiphytic, some tall and tubular like *Billbergia zebrina,* others with Aechmea-like rosettes or funnel-shaped forms as in *B. sanderana. B. nutans* is distinct again, with dark green, narrow arching leaves which are almost grassy and not very exciting. Nevertheless this is a popular species probably because it is one of the hardiest and easy to grow in soil compost, or perhaps on account of its flowers. These come in tall inflorescences which droop with the weight of the rose-red bracts and violet-blue, green-tipped flowers with prominent yellow stamens. It is often called Queen's Tears.

B. x *windii*, a hybrid between *B. nutans* and the tenderer *B. decora,* has larger flowers but is less vigorous. Both these plants like sun but more shade-loving is *B. zebrina*, a broad-leafed species with grey-green foliage banded with silver and golden flowers surmounted by showy pink bracts.

Nidularium innocentii is similar to *Neoregelia,* with the flowers emerging from the 'vase' of the leaf bases; the foliage is darker green but purplish beneath and the flowers are white. *Fascicularia pitcairniifolia* from Chile is hardy in sheltered parts of Britain and warm-temperate US and is a striking plant in late summer with numerous, spiny leaves, 90 cms (3 ft) long, which are very narrow and form a rosette, at the base of which there are bright red bracts surrounding a central pincushion head of blue flowers.

Perhaps the most important commercial plant in Bromeliaceae is *Ananas comosus* (*A. sativus*), the Pineapple, which is more extensively preserved than any other tropical fruit, accounting for twenty per cent of the world's canned fruit production. As far as Europeans are concerned the first plants were discovered by Colombus in 1493 on the island of Guadeloupe. Later he saw the fruit on other West Indian islands but it has since been found —apparently wild—in several parts of South America.

A Spanish historian called Oviedo, who was born in Madrid in 1478, wrote about Pineapples in 1539. Apparently the Spaniards saw a resemblance between the fruit and the cones of the Pine Tree (*Pinus*) and called this new fruit *pina-las-India* which was later corrupted by the English to Pineapple. The Latin name *Ananas* is derived from the South American Guarini language, 'a' meaning fruit and 'nana' excellent.

The leaves are the source of strong and durable fibres, used in South America for ropes, cords and packing materials and in Indonesia and the Philippines for such articles as veils. The fibres are extracted by scraping, bleaching and drying the young leaves, when they become moisture-resistant, white and elastic. The cultivar portrayed, 'Variegatus', is a splendid ornamental, the leaves longitudinally striped with green and cream—also pink when grown in a good light. The small fruit is edible.

Puyas are plants of the High Andes, often giants with stems up to 3 m (10 ft), topped by curious greenish or bluish flowers. Essentially adapted to high winds and an arid terrain, they often succeed in rocky exposed gardens of thin soil on ocean cliffs. They can also be seen, towering above succulents and small conifers, in Californian gardens. Their giant proportions, like the Andean Espeletias, may be compared to the great Senecios and Lobelias of the Aberdare Mountains in Kenya. One wonders why these plants grow so tall at such altitudes when all around plants of lower proportions seem to cope so well with the extremes of wind and weather.

The foliage is long and narrow, usually spiny-edged, and sometimes dangerously keen. The upward-pointing leaves form a bristly head, like a porcupine in the juvenile stage, but once the inflorescence extends to its full height it seems to exhaust the plant and the leaves flop downwards. One of the largest is the Peruvian *Puya raimondii* which grows at an altitude of 4000 m (13,000 ft) and reaches 10 m (32 ft) in height, the giant flower spikes tightly packed with greenish-white blooms. More suitable for garden purposes are the following species, all from Chile: *P. alpestris,* 120–150 cms (4–5 ft) tall, the large flowers a metallic blue with orange anthers; *P. caerulea,* deep blue with orange anthers, 120–150 cms (4–5 ft); and *P. chilensis,* greenish-yellow, 60–90 cms (2–3 ft) and a form known as *gigantea* (which may be a species) growing 3–5 m (10–15 ft) high. The leaves of this are used as a source of fibre for fishing nets. *P. berteroniana* from dry, rocky places in central Chile has magnificent spikes 3–4 m (10–12 ft) tall with eighty to 100 blue-green flower spikelets.

All Puyas like plenty of sun and are able to withstand considerable temperature changes, but they dislike winter wet and need a clean atmosphere.

Ananas comosus
'Variegatus'

Guzmania cardinalis

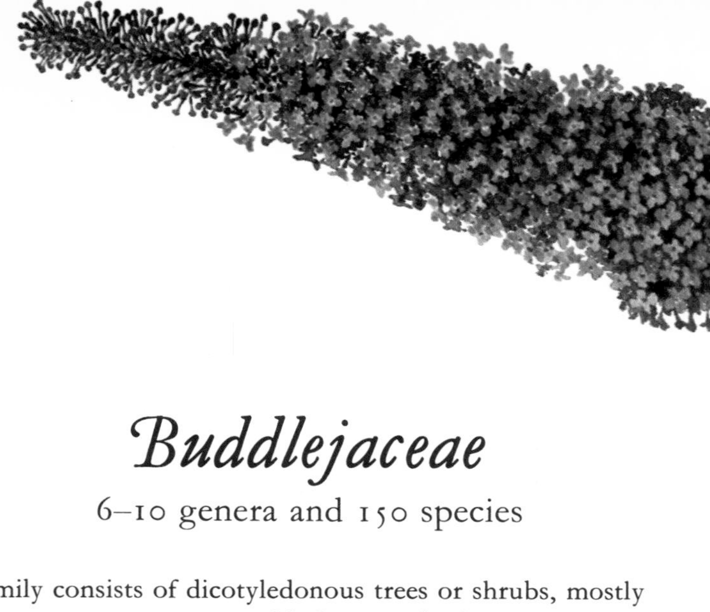

Buddleja davidii 'Île de France'

Buddleja globosa

Buddlejaceae

6–10 genera and 150 species

This family consists of dicotyledonous trees or shrubs, mostly tropical or warm temperate, with four sepals, four petals, four stamens and a superior ovary.

Buddleja (*Buddleia*) *davidii* (*B. variabilis*) is the Butterfly Bush, a vigorous shrub of 4–5 m (12–15 ft), suitable for any reasonably good soil in full sunshine. It is native to China and deciduous with slender panicles 30–38 cms (12–15 ins) in length densely packed with fragrant flowers. These vary from lilac to purple and have a strong fascination for Red Admiral butterflies. For best results cut the previous season's growth hard back in late winter; the bush then produces long stems and flowers in late summer on this young wood. Cultivars include the white-flowered 'Peace'; 'Black Knight', nearly navy-blue; 'Royal Red' red-purple; and 'Île de France', deep rich violet.

B. globosa from Chile and Peru has rounded heads of fragrant flowers like golden cherries and grows 3–5 m (10–15 ft) high. The Chinese *B. alternifolia* produces arching, wand-like, slender stems packed with short clusters of mauve flowers in summer. *B. colvilei* from the Himalayas with crimson flowers, and the Chinese *B. fallowiana,* which is soft lavender and very fragrant, are other good species. *B. madagascariensis* is an evergreen orange-flowered species unusual in producing violet, berry-like fruits. It is widely cultivated in the warmer areas of the world. *B. crispa* from the Himalayas is deciduous with long fragrant panicles of white-eyed, lavender flowers.

Butomus umbellatus

Butomaceae

1 genus and 1 species

This small family has only one representative; a monocotyledonous water plant from temperate Europe and Asia.

Butomus umbellatus is one of Britain's most elegant native aquatics. It is locally distributed in Britain, being more plentiful in the north but rare in Ireland, and often (improperly) called a rush, possibly because of its long, smooth stem.

The rootstock is thickened, fleshy and extensively creeping, producing erect, linear, triquetrous leaves which are purple-bronze when young but become green with maturity. These have enlarged bases which enwrap the rhizome and are about 60–90 cms (2–3 ft) in length. The inflorescence is a many-flowered umbel with an involucre of bracts; individual blooms are regular with six pale rose perianth segments in a saucer-like shape, nine dull red anthers and six follicles containing many seeds.

Variously known as Flowering Rush, Water Gladiole or Grassie Rush (Gerard), it makes a useful ornament at the side of a pond when planted in shallow water 5–10 cms (2–4 ins) deep.

Buxaceae

4 genera and 100 species; tropical and temperate

This family consists of dicotyledonous, mainly evergreen shrubs with the odd tree or herbaceous plant. General characteristics are small simple leaves, entire or toothed, alternately arranged or opposite on the stems. The flowers are individually so small as to be insignificant, but become showy when massed in heads; they are regular in shape, dioecious or monoecious—rarely hermaphrodite—with four to six perianth segments, four to six

Sarcococca humilis

Pachysandra terminalis 'Variegata'

stamens and shiny black seeds in a fleshy drupe or capsule.

The most important plant in the family, both from a garden and economic standpoint, is *Buxus sempervirens,* the common Box. A most accommodating evergreen with small glossy leaves, it is widely used in Europe and North America especially for hedging, topiary, screens, edgings or, when grown to tree proportions, as a lawn specimen.

Box trees are found 'wild' at Box Hill, Surrey, in England, although they may not be indigenous. Some of these trees grow to 9 m (30 ft) and were once so plentiful that in one year alone (1815) £10,000 of box timber was taken from this site. The species generally is native to southern Europe, North Africa and the Orient, where the wood has long been used for making such articles as scientific instruments, flutes, clarinets, tobacco pipes and dagger handles.

Over the years clones have been selected with differing habit. One called 'Suffruticosa' is very dwarf and responds to close clipping; if left uncut (as in many New England gardens) it has a 'billowy' appearance but normally it is used for low neat edgings. It was introduced to the American colonies from England very early in the settlement of that country, particularly in Virginia and Maryland. The flowers are small and insignificant and for some highly fragrant but others can discover no smell at all.

Honey from the flowers is strong and noxious and in Tudor times was believed to cause madness. Box was once used instead of Willow on Palm Sundays and also for funeral sprays. It is thought to be the Ashur Wood of the Scriptures and in the East is often inlaid with ivory. Box grows extremely slowly, but once established may live for several hundreds of years. Varieties with white or golden leaf margins occur, also one with golden foliage.

All the Pachysandras make good ground cover plants for shady places, *Pachysandra terminalis* from Japan being tolerant of very dry sunless situations. It has diamond-shaped leaves and spiky heads of greenish-white flowers; 'Variegata' although slower growing is the most popular for gardens as the foliage is heavily splashed with cream. *P. procumbens* from North America with white or pinkish fragrant flowers grows up to 30 cms (1 ft) high and has the same uses.

Sarcococcas are also shade plants and since they are evergreen with small but sweetly scented blooms, find a place in dark corners or between shrubs where little else will grow. *Sarcococca ruscifolia* and *S. hookerana* bloom during the winter with tiny white blossoms and are succeeded by scarlet and black berries respectively. *S. humilis* has fragrant flowers and black berries. These are all native to China or India and 15–120 cms (6 ins–4 ft) high.

Cactaceae

50–150 genera and perhaps 2000 species

Most Cacti are xerophytic plants from the New World particularly the arid regions of tropical America. A few, however, are native to really cold climates, for example several Opuntias are found in the Southern Hemisphere in Patagonia (50° latitude) and in the Northern Hemisphere in Alberta, Canada (53°). There are also *Neowerdermannia* species 3250 m (10,500 ft) up in the Andes and various *Cereus* grow exposed to the icy winds of the Mexican Cordilleras.

Cacti vary considerably in size and habit, from tree Opuntias such as *Opuntia myriacantha* 11 m (35 ft) tall in the Galapagos and *Carnegiea gigantea,* a branching, candelabra-like Cactus from Arizona, S California and Mexico which grows 9 m (30 ft) high and weighs 6 to 7 tons, to desert dwarfs and epiphytes perched on trees. One of the latter—*Rhipsalis baccifera*—is also found in the Old World, in parts of Africa, Madagascar, Mauritius, Ceylon and the Seychelles. Several species notably Prickly Pear (*Opuntia ficus-indica*) have become troublesome weeds in Australia, India and South Africa.

Most Cacti are characterized by swollen stems which harbour water. The epidermis is considerably thickened; there is a reduction of the pores (stomata) which are involved in transpiration and often a complete absence of the normal leaves. The stems are variously shaped and usually supplied with spines—sometimes barbed—which hinder transpiration, trap morning dews and protect the plants from marauding animals. The root system is generally shallow with elongated slender roots, fleshy in character but sometimes enlarged like the roots of Dahlias. Some, however, have very extensive root systems. A number of Cacti are not completely xerophytic as the epiphytic Cacti and *Pereskia* species which have normally developed leaves and also stalked flowers.

The sap of Cacti is thick and mucilaginous; the water-filled stems are frequently strengthened by means of tough ridges, in the furrows of which lie the stomata. All these characteristics check excessive transpiration and are defences against drought. Spined Cacti can therefore survive in hot and arid places; their

Schlumbergera x *buckleyi*

vitality is great and it would take months or years for them to dry out completely. Such plants in the home and garden must not be overwatered and should always have good drainage at the roots. The majority like plenty of light and sunshine.

The epiphytic kinds on the other hand, like *Rhipsalis* and *Epiphyllum*, are not desert plants at all. They come from deep tropical forests where there is frequent and abundant rain. The medium in which they root consists of fragments of leaves and vegetable matter in branch crevices and tree trunk hollows. This compost is always moist yet not waterlogged, so that epiphytic Cacti grown under garden conditions require less light, but at certain times need more watering than other Cacti and a richer compost containing plenty of leafmould.

Rhipsalis warmingiana is a Brazilian species with pendent stems, narrow, flattened and crenately margined branches, and white flowers.

The flowers of Cactaceae are often very showy, brilliantly coloured, with a satiny sheen and quantities of yellow stamens. They consist of many petals, which show a gradual transition from petaloid sepals to large petals, so that there is no green calyx behind the flowers. They are usually solitary and stalkless (except *Pereskia*), sitting directly on the stems or pads, pollinated by insects or in a few cases by hummingbirds or by bats.

The ovaries of Cacti are inferior and usually develop to pulpy berries. These are often brightly coloured in shades of scarlet, white or lilac, a circumstance which attracts birds which thus disperse the small seeds.

Some have edible fruits, especially certain species of *Opuntia*. These are cultivated in Central America (particularly Mexico) under such names as Indian Fig, Tuna or Prickly Pear. *O. megacantha* has especially large fruits of excellent quality which are reputed to have a high nutritive value. They are eaten fresh or dried and cooked in various ways.

Other byproducts of Opuntias include fruit syrups, a paste made by boiling down the juice and a fermented drink called coloncha. In spring the tender young shoots, flowers and buds are eaten as vegetables and the leaves used as poultices to relieve inflammation. The juice is also employed in the manufacture of candles, gum is extracted from the stem and the plants are fed to stock after the spines have been burnt off with torches. In tropical gardens and also in the Middle East Opuntias are often employed for hedge making.

In the Galapagos Islands the giant Opuntias *O. echios* var. *gigantea* and *O. myriacantha* form the natural diet of the giant tortoises; *O. cochenillifera* was cultivated in tropical America as food for cochineal insects.

Opuntias may be either tree-like or low-growing, often intricately branched and usually with attached ear-like pads. These are covered with spines and in spring carry the brightly coloured yellow, orange, red or purple flowers. Among those frequently cultivated is *O. bergerana* with showy red flowers; this is common on the French Riviera but unknown in the wild state. *O. vulgaris* from South America has golden flowers and there is a variegated-leafed form.

In tropical gardens no plants are more impressive than the night-blooming *Cereus*. With their strange columnar shapes, acute-angled branches and ribbed stems they present a weird and fascinating effect in the garden landscape. The flowers are funnel-shaped, usually white or cream.

Closely related are the giant Carnegieas with candelabra branches and such tenacity of life that some specimens in Arizona are thought to be more than 250 years old. *Cephalocereus senilis,* the Old Man Cactus, is crowned with long mops of white hair when the stems reach maturity, and amongst a number of genera cultivated in subtropical gardens or greenhouses are the *Cleistocactus* (*Cleistocereus*) with very elongated stems and mostly red flowers, and *Borzicactus* (*Oreocereus*) with white hair and yellow spines, both capable of withstanding fairly low temperatures (to minus 9°C; 15°F) and *Heliocereus speciosus,* a tropical species with drooping or climbing stems and magnificent blue-tinted, scarlet flowers. In the Jardin Exotique in Monaco there are impressive specimens of this *Cereus* group growing outdoors.

Mammillaria species are generally small plants of squat or dumpy shape, heavily spined and with bright sessile flowers. They comprise a large group which find favour with collectors as they are undemanding and take up little space. They grow very slowly, require little watering and thrive happily in any light compost.

But perhaps the most popular members of the Cactaceae are the epiphytic types known horticulturally as *Schlumbergera,*

Echinocactus grusonii

Lophophora williamsii

Rhipsalidopsis rosea

Zygocactus and *Epiphyllum* (or *Phyllocactus*). These make excellent pot plants and remain in flower for several weeks.

The commonly grown Christmas Cactus—so called as its blooms often reach perfection at this season—has been the subject of much controversy from the nomenclatural viewpoint. It is usually known as *Schlumbergera truncata* (*S. bridgesii*) or *Zygocactus truncatus,* but most plants in cultivation are now considered to be hybrids of this species with *S. russelliana,* and should be grouped under the name *S.* x *buckleyi*, according to Hunt in a recent revision in the Kew Bulletin, 1969. Both parents are native to Brazil.

The flat leaf-like stems are segmented and terminated at flowering time by many-petalled carmine flowers 5–7 cms (2–3 ins) across. The whole effect is reminiscent of a lobster's claw, which accounts for its other English names, Lobster Cactus and Crab Cactus. The parent species are rarely seen in cultivation but horticultural forms or hybrids abound with pink, red, white, cerise and dark purple flowers.

Once the buds appear the plants should not be moved or these will shrivel and drop before they open. The plants need light but not bright sunshine and the dry heat of central heating must be tempered by some form of humidity, such as a moist gravel tray below the pots.

Britton and Rose in their monumental work *The Cactaceae* published a new genus by the name of *Rhipsalidopsis*, in which they placed the single and only species *R. rosea*. However the genus they dealt with is doubtfully distinct from *Rhipsalis,* the plant they dealt with having small flattish joints which sometimes are three- or four-angled. From one viewpoint *Rhipsalidopsis rosea* can be considered as another *Rhipsalis*; from another, a species of the *Schlumbergera* of Britton and Rose. In both cases flowers grow from the tips of terminal joints, and the plant has a chain-like formation of sometimes flat links. The *Schlumbergera* has coloured flowers while the *Rhipsalis* has whitish flowers.

Epiphyllums have continuous flattened stems (not jointed like *Zygocactus*) which resemble long thick leaves. These grow 60–90 cms (2–3 ft) long but are rather weak so that they trail downwards unless supported by thin canes. The flowers are large, up to 25 cms (10 ins) in diameter in some kinds, and very showy with the

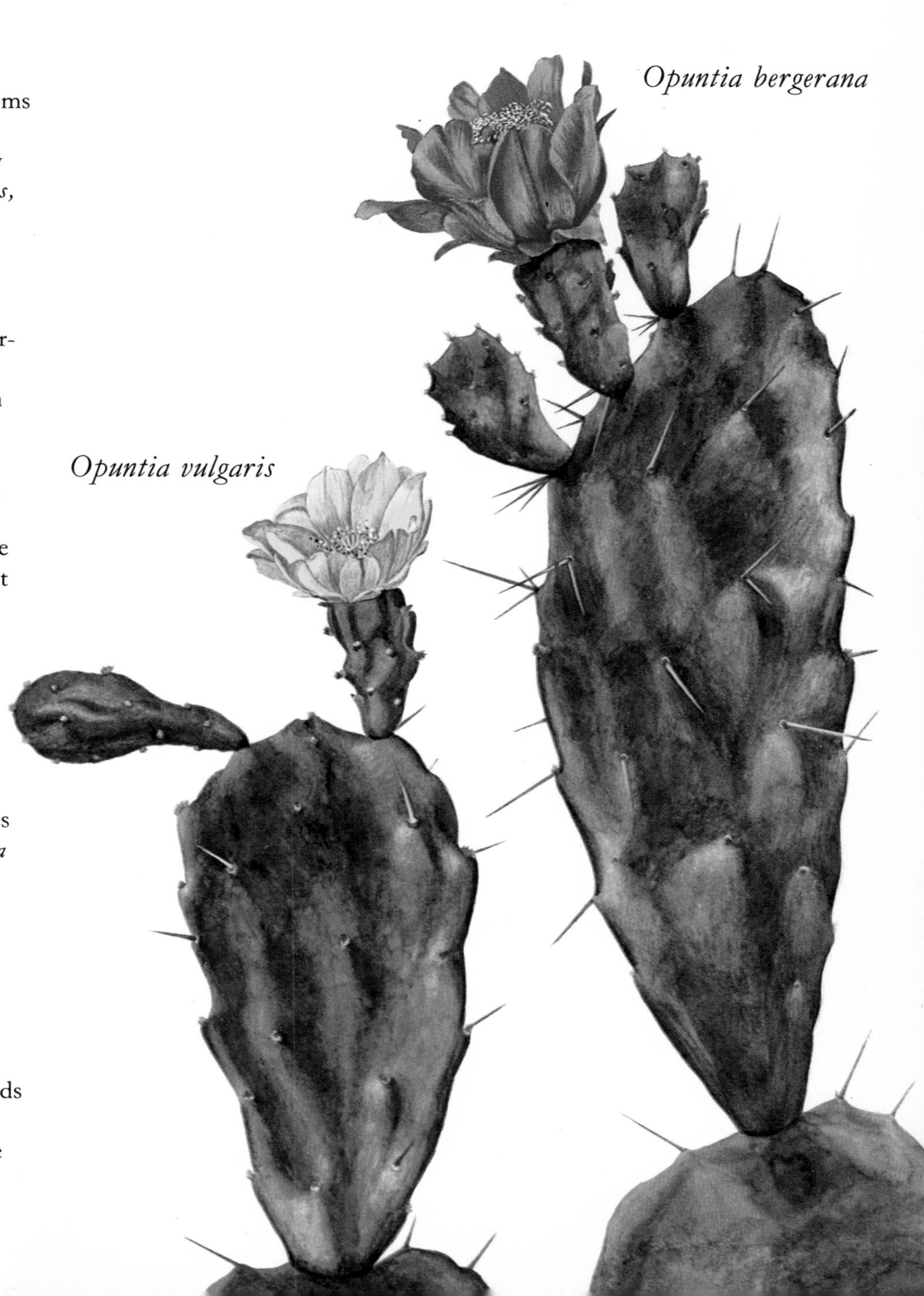

species white and the intergeneric hybrids having yellow, pink, cream, crimson or deep red, many-petalled flowers; the species except for *E. crenatum* are nocturnal. Most garden forms are derived from *Epiphyllum* x *ackermannii* or *E. oxypetalum*. Their common name is Orchid Cactus.

Epiphyllums grown in pots must have good drainage and a friable compost so that water runs through freely. They require a moist atmosphere and must not be exposed to hot sunshine.

Frequently epiphytic Cacti are grafted on *Opuntia* or other Cacti in order to counteract the trailing habit and bring the flowers up to a more convenient height.

Other interesting Cacti for greenhouse or subtropical gardens include the grotesque Barrel Cactus, *Echinocactus grusonii* or—as some unkindly name it—Mother-in-law's Armchair. Even a small specimen can be striking, although it may reach 120 cms (4 ft) or so across and 75 cms ($2\frac{1}{2}$ ft) high. It is a rounded, barrel-like plant with long, tapering, golden spines mathematically arranged all over its surface. Through these formidable thorns the red and yellow blossoms can be glimpsed at flowering time.

Aporocactus flagelliformis, the Rat-tail Cactus, is well named for the long flexible stems, barely 1 cm. ($\frac{1}{2}$ in.) in diameter, hang down all round the plant. They are covered with bristly hairs and carry masses of crimson flowers 7 cms (3 ins) across. Easy to grow and very free, this species again is often grafted on the straight stem of another cactus to give it height.

Selenicereus grandiflorus is the Queen of the Night, a straggling Mexican climber which has little to commend it until it flowers. Then it deserves every complimentary epithet, for at dusk the great trumpet-shaped blooms expand, 30 cms (1 ft) long and 20 cms (8 ins) wide, white on the inside and yellow without, exuding a rich vanilla fragrance which can be discerned at some distance.

Lobivias come from South America (the generic name is an anagram of Bolivia) and although small growing—usually 7–10 cms (3–4 ins) tall with cylindrical or globular stems—have large flowers for their size. These are often 5–7 cms (2–3 ins) across and lemon-yellow in *Lobivia aurea,* scarlet in *L. cinnabarina.*

An interesting Cactus because of its association with drug addiction since pre-Columbian times is the Peyote, *Lophophora williamsii.* The dried crowns, called Mescal Buttons and looking something like toadstools, contain the drug Mescaline (trimethoxyphenephylamine). When chewed these induce hallucinations and a sense of well-being. Aldous Huxley is said to have written a book under the influence of mescaline which is a well-known intoxicant first used for ritual purposes by the Aztecs.

Calycanthaceae

2 genera and 7 species

This is a small family of shrubs usually aromatic, with opposite, simple leaves and terminal flowers with cup-shaped receptacles on short shoots. The floral parts (perianth segments) are spirally arranged and may be numerous, especially in *Calycanthus.* Here the petals and sepals are indistinguishable and the outer ring of stamens are also petal-like and consequently do not produce pollen. When the real anthers ripen these false stamens bend inwards over the stigmas and so prevent self-fertilization.

Calycanthus are peculiar to the SW and E United States and owe their value as garden plants more to fragrance than flowers. *C. floridus,* the Carolina Allspice or Sweet Bush, is a deciduous shrub 2–3 m (6–9 ft) tall, with dark green, oval leaves which are rough on the upper surface. The wood is very aromatic, especially when dried, and the bark has been used as a substitute for cinnamon. The fragrant, many 'petalled' flowers are reddish-purple and leathery in texture. The plant was first introduced to Europe by Mark Catesby in 1726.

Another very fragrant species, *C. occidentalis,* comes from California where it was discovered by David Douglas in 1831. It is commoner in British gardens than *C. floridus* but has the same spicy odour, but larger leaves and purplish-red flowers.

Chimonanthus species are Chinese, the most important being the well-known Winter Sweet, *C. praecox* (*C. fragrans*). This blooms in winter, the small but exceedingly fragrant yellowish-green and chocolate flowers studding the naked stems. As it does not flower well generally in the open it is best grown against a warm wall or fence in a sunny situation. To encourage regular blooming the secondary shoots should be spurred back after flowering to one or two eyes. The species has been cultivated for centuries in China and Japan on account of its perfume but was not introduced to Britain until 1760 where it was found to be hardy and became very popular. Several good forms, probably clones, have arisen including one with larger flowers *grandiflorus,* and another having primrose-yellow flowers called *luteus.*

In China *C. praecox* flowers about the time of the Chinese New Year celebrations, when girls gather its blossoms to deck their hair. The aromatic twigs are also tied in bunches and used to scent linen in the same way that English women use lavender. The shrubs grow around 2–3 m (6–10 ft) tall in the open.

Campanulaceae

60–70 genera and 2000 species; mostly temperate or subtropical, but also representatives in the mountains of the tropics

This large family comprises many good garden subjects, particularly *Campanula* which includes plants suitable for ornamental flower beds, herbaceous borders, greenhouse display and the rock garden.

Most members of Campanulaceae are herbaceous perennials, but there are also annuals and biennials as well as a few trees and shrubs. There are also many alpines.

The leaves are usually alternate and entire, rarely lobed or divided. The inflorescence is generally racemose, ending with a terminal flower in *Campanula* and its near relatives like *Platycodon* and *Wahlenbergia*. In other cases, instead of a single flower, small branching inflorescences occur in the axils of the racemes or the whole flower stem may be cymose as in *Canarina*.

The flowers can be regular or irregular and are bisexual (rarely unisexual) and usually epigynous with five separate sepals, five petals, five stamens, five, three or two stigmas and carpels and an inferior (rarely superior) ovary. The seeds may be enclosed in a fleshy berry but most commonly are in a capsule, which splits in various ways in the different genera.

Campanula contains some 300 species most of which come from N temperate regions, especially in the Mediterranean area. Few genera display greater diversity of form and habit, for while there are many lilliputians for the rock garden from the mountains of Europe and Asia, some of the border perennials may grow 1.5–2 m (5–6 ft) in height. These will grow in practically any garden soil which is cool and yet well drained and while the majority are happier in sun several favour shade conditions. Indeed, when well established in light woodland, species such as *C. persicifolia* and *C. rapunculoides* (a tenacious weed) sometimes become naturalized. For display work in beds and borders Campanulas should be grouped. Propagation is effected by seed, or in the case of cultivars by division of the roots in spring or autumn, or by soft cuttings rooted under glass in early spring.

1 *Campanula lactiflora* 'Loddon Anna'
2 *Campanula latifolia*
3 *Campanula medium*
4 *Campanula persicifolia*
5 *Campanula glomerata*

A number of species (like *C. rapunculus*) have edible roots, but the plants are cultivated primarily for the beauty of their flowers.

Among several good sorts for border work is the *C. glomerata* or Clustered Bellflower of Europe (including Britain). The inflorescence comprises a clustered head at the top of the stem, generally with an interrupted spike beneath it, the terminal flower opening first. Individual blooms are large and showy in a rich violet shade, funnel-shaped and erect; the stems, 45–60 cms ($1\frac{1}{2}$–2 ft) tall, and leaves are roughly hairy. There is a white form *alba* and a more compact but deeper coloured cultivar called 'Dahurica'.

C. latifolia, the Giant Bellflower, is widespread from Europe to Kashmir. Its stout, leafy stem of 120–150 cms (4–5 ft) carries many pale blue flowers. A fine plant for cool moist woodlands or damp borders, it seeds freely and becomes naturalized; it has a white form *alba* and several cultivars which are deeper blue.

C. lactiflora from the Caucasus is the Milky Bellflower, a fine free-flowering sort with 1.5–2 m (5–6 ft) stems bearing loose or dense panicles of milky, blue-tinged flowers. There is a mushroom-pink cultivar called 'Loddon Anna' and a white form *alba*. *C. persicifolia*, again European but also from North Africa and N and W Asia, has smooth stems, 30–90 cms (1–3 ft) tall, narrow, rather leathery leaves and racemes of wide, cup-shaped blue flowers. The cultivars are superior for garden purposes especially 'Telham Beauty' which has large china-blue flowers; 'Misty Morn', a semi-double American sort of frosty blue; 'Alba Flore Pleno', double white; and 'Delft Blue', double china-blue.

C. grandis (*C. latiloba*), the Olympic Bellflower from Siberia, has pale violet blossoms on stems of 30–90 cms (1–3 ft) and is shade-tolerant; *C. rapunculoides,* the Creeping Bellflower, with unbranched stems of 60–120 cms (2–4 ft) carrying narrow, funnel-shaped, bluish-violet flowers naturalizes itself in woodland or moist borders by means of underground creeping stolons. It is a European plant (including Britain) as is *C. rapunculus,* a biennial with a swollen root, once grown in Britain and blanched like celery for eating raw in salads, or boiling like asparagus. It has blue or white flowers on stems of 60–90 cms (2–3 ft).

The best-known biennial, however, is *C. medium,* the Canterbury Bell, a beautiful plant from southern Europe with long, rough, wavy-edged leaves and strong spikes of showy, bell-shaped, blue, white, pink, mauve or rose-coloured flowers. The calyx is frequently large and saucer-shaped, coloured like the flower—the Cup and Saucer varieties of the trade, which are also known as var. *calycanthema*—and there are double sorts in various shades. The name Canterbury Bells is said to honour St Thomas à Becket and is an allusion to the horse bells once used by pilgrims visiting his shrine at Canterbury Cathedral.

Alpine Campanulas are legion, some very easy, others intractable, taxing the patience and skill of the most talented gardener. Among the most accommodating are *C. rotundifolia,* the Harebell of England but the Bluebell of Scotland, a sun-lover growing 13–30 cms (5–12 ins) high with small, rounded, stalkless leaves and many drooping, bell-shaped flowers, several to a raceme on thread-like stems. *C. portenschlagiana* from southern Europe is an easy species with light blue-purple or white flowers.

C. carpatica is another favourite, a variable plant with wide, erect, bell-shaped solitary flowers in various shades of blue in the cultivars 'Loddon Fairy' and 'Riverslea', or white as in 'White Star'. It is native to the Carpathian Mountains.

C. cochleariifolia (*C. pusilla*) from mountainous areas in Europe grows 10–15 cms (4–6 ins) tall with solitary, drooping flowers in great profusion in summer. It is easy to grow and useful to edge beds or plant in rock-garden pockets. Var. *alba* is white and 'Oakington Blue' and 'Miranda' deep blue.

Other noteworthy sorts are *C. arvatica,* 5 cms (2 ins), from Spain, a blue-flowered species for scree or rock crevices; *C. garganica,* of S Europe with white to light blue open stars on stems of 5 cms (2 ins), its deeper coloured hybrid 'W. H. Payne' and *C.* 'Warleyensis' a beauty 15 cms (6 ins) high with double, powder blue flowers.

For hanging baskets in the home or greenhouse, even outside in summer in good climates, few plants surpass *C. isophylla.* The long trails of growth are smothered for months on end with blue or white star-shaped flowers and there is one with cream-splashed leaves.

Perhaps the most easily grown of Campanula's near relations is *Platycodon grandiflorus* from NE Asia. It is commonly known as the Balloon Flower because of the shape of the inflated buds, which are joined at the petal margins and split on opening. Growing about 60 cms (2 ft) tall it carries a number of flowers of a pleasing blue shade on every stem. Forms are available with white flowers, *albus*; semi-double as in 'Plenus'; and deep rich blue in 'Mariesii'

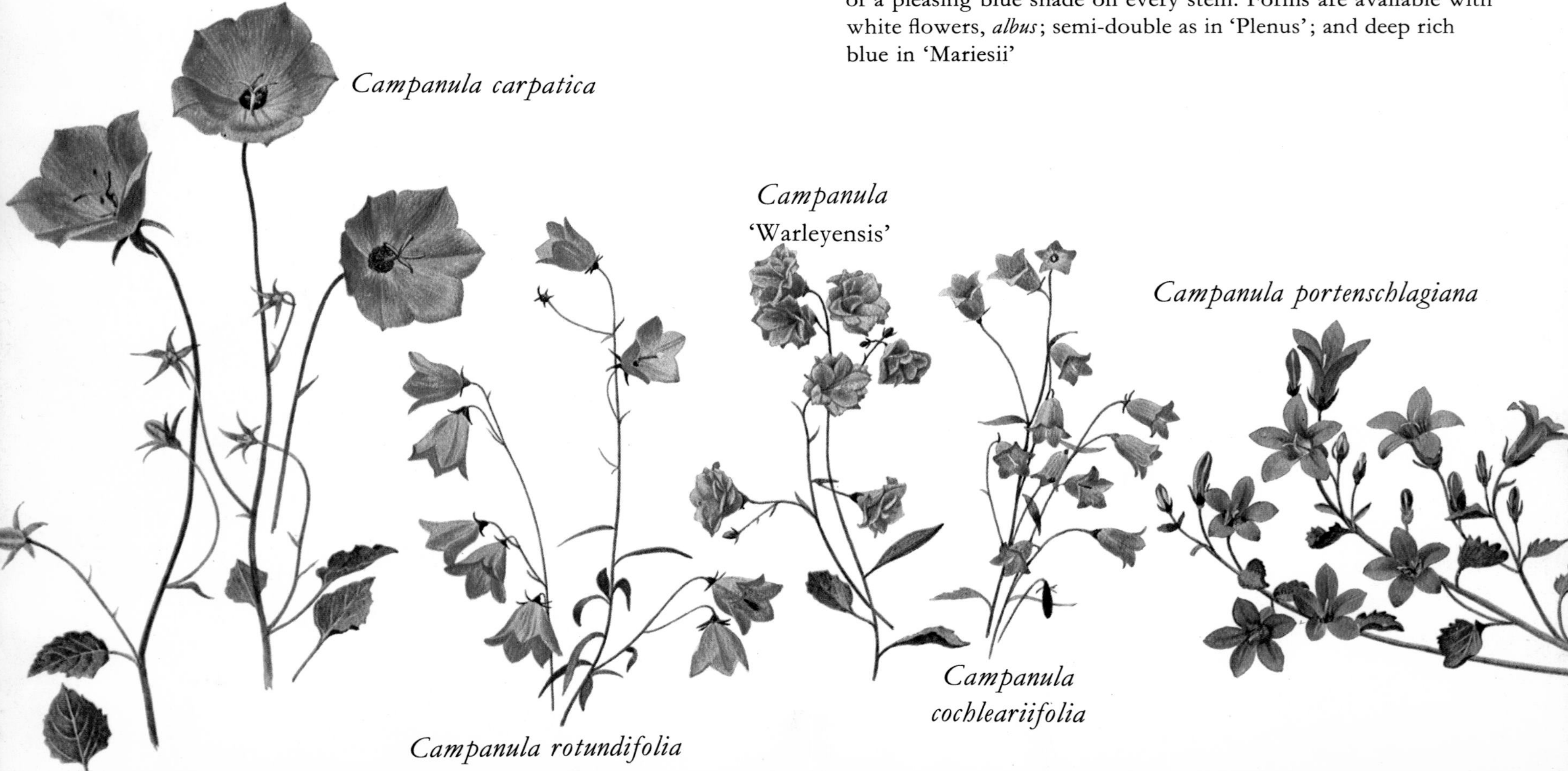

Campanula carpatica

Campanula 'Warleyensis'

Campanula portenschlagiana

Campanula cochleariifolia

Campanula rotundifolia

Wahlenbergia, with more than 150 species, is very like *Campanula* except for slight botanical differences in the capsules.

Edraianthus, a small genus from the Mediterranean regions, differs horticulturally from *Wahlenbergia* for usually its flowers are borne in clusters instead of singly. They are, however, much confused in growers' catalogues. Commonly grown sorts are the beautiful *E. pumilio* from the Balkan Peninsula, only 5 cms (2 ins) tall, suitable for scree or crevice with lavender flowers and linear silvery leaves; also *E. serpyllifolia* from the same area and another mat-forming species of which the best garden form is 'Major', a delightful little plant which when well grown is spangled with bells of Imperial purple. *E. caudatus* (*E. dalmaticus*) has purple flowers and *E. niveus* is white. They grow about 7 cms (3 ins) high.

The Throatworts, *Trachelium rumelianum* and *T. asperuloides* (by some authorities placed in *Diosphaera*), both come from Greece and require full sun and good drainage; they make good alpine house plants in cold climates. The former appreciates lime and grows about 15 cms (6 ins) with rounded heads of small lilac-blue flowers; but *T. asperuloides* is more cushion-like with tiny stalkless leaves and clusters of one to five lilac flowers.

Phyteuma differs from Campanula in the shape of the flowers, which are small and massed into heads with long thin petals and protruding stigmas. *P. spicatum* is the Rampion, a white or pale blue British native, but best for the garden is *P. comosum* from the calcareous regions of the S European Alps. This forms ground-hugging tufts of glossy, rounded, sharply toothed leaves and in summer carries many heads of comparatively large flask-shaped flowers on stems of 7 cms (3 ins). These present an inflated appearance and are lilac with purple beaks.

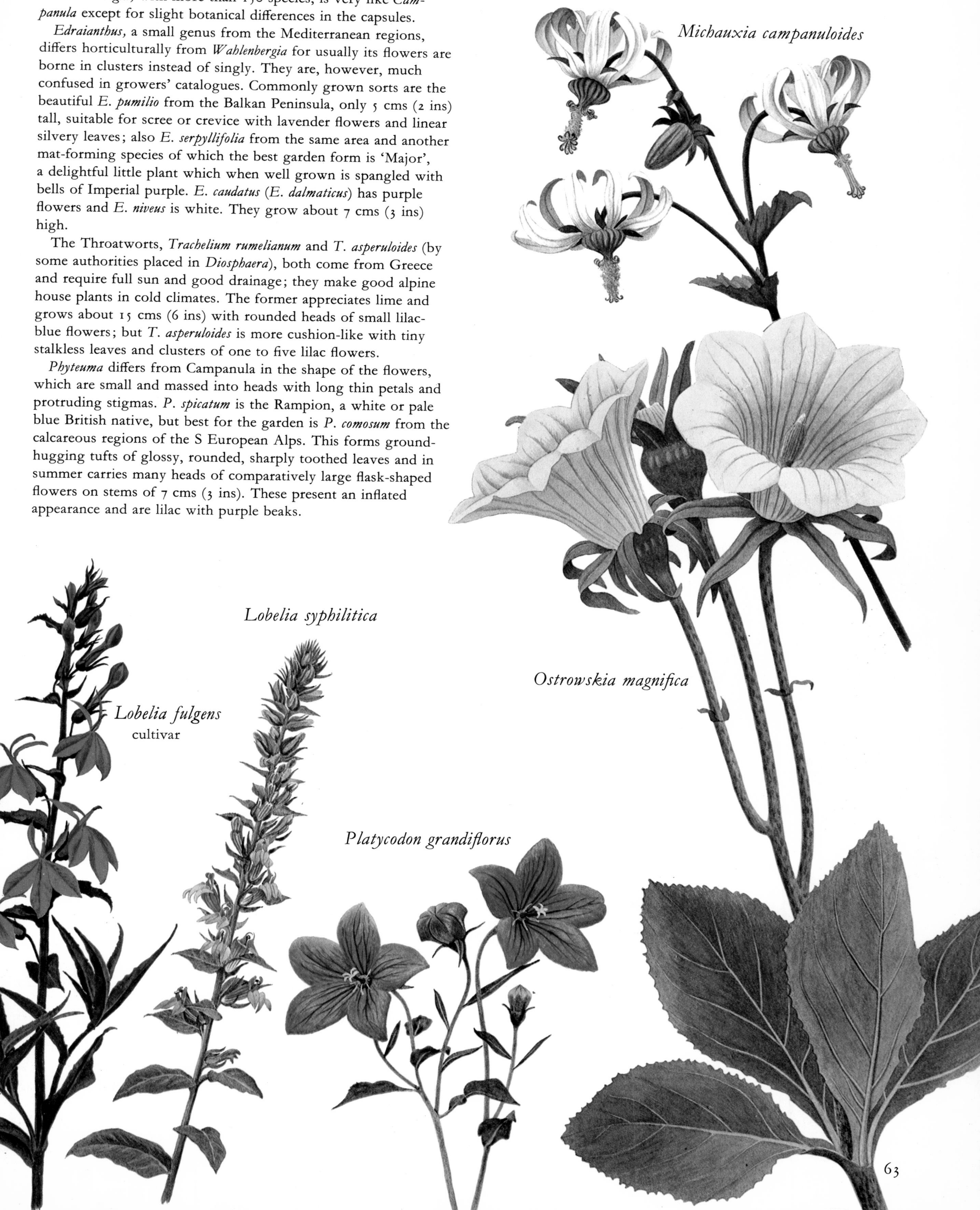

Michauxia campanuloides

Lobelia syphilitica

Lobelia fulgens cultivar

Ostrowskia magnifica

Platycodon grandiflorus

Canarina canariensis

Jasione montana is the Sandbell, a plant which at first sight so resembles Scabious that in England it is called Sheep's Scabious. In parts of Britain it is found in sandy fields and by the roadsides, with many sturdy stems carrying 2.5 cms (1 in.) heads of lilac-blue flowers. In the rock garden *J. jankae* from Hungary and *J. humilis* from the Pyrenees, both bright blue and from 15–25 cms (6–10 ins) tall, are desirable for their summer blossoms.

Adenophora resembles the perennial types of *Campanula* with broad, stalked leaves and racemes of drooping, bell-like flowers. The species like moist but well-drained soil in a warm sunny spot and should be propagated by seed sown as soon as it is ripe. *A. bulleyana* is a Chinese species up to 120 cms (4 ft) with pale blue flowers; *A. potaninii,* the Bush Ladybell, also Chinese, is 60 cms (2 ft) and light blue; and *A. verticillata,* 60–90 cms (2–3 ft), from Siberia and China, pale blue and many flowered.

Symphyandra hofmannii from Bosnia can be used as a ground cover in shady places amongst shrubs. It has nodding, white tubular flowers 2.5–4 cms (1–1½ ins) long and rough oblanceolate leaves. *Michauxia campanuloides* is biennial so needs frequent renewal from seed. It comes from the Levant and is a lovely plant 1–1.5 m (3–5 ft) tall with spikes of drooping white flowers which are suffused with purple outside.

Ostrowskia magnifica, a monotypic genus from Turkestan, is a lovely perennial with large tuberous roots and leafy stems with large leaves carrying several fine cup-shaped flowers. These are of a delicate light purple with deeper veining and marks.

Codonopsis ovata

Lobelias are distinctive and much-prized plants. Several are good waterside plants, as *Lobelia syphilitica,* the Blue Cardinal Flower, and its white counterpart *alba.* The bold spikes of blue or white flowers make pleasing associates for the red Cardinal Flowers *L. cardinalis* and *L. fulgens* which are very similar, except that the foliage of *L. cardinalis* is green and *L. fulgens* deep wine-red. The latter from Mexico and south USA is more tender than the other two but looks very striking against the spikes of sealing-wax scarlet flowers 60–90 cms (2–3 ft) tall. The other Lobelias mentioned are all native to eastern North America.

Interesting forms and also hybrids between the species have evolved through the years, the best of such cultivars being 'Jack MacMasters', violet-blue with purplish foliage; 'Huntsman', brilliant red; 'Queen Victoria', deep purple leaves and crimson flowers; and 'Purple Emperor', purple.

Apparently *L. cardinalis* was introduced to England in 1629 from Canada (then a French colony) and sent to Henrietta Maria, Charles I's queen. She, it is said, laughed excessively on seeing the flowers and said their colour reminded her of the scarlet stockings of a cardinal. Parkinson therefore called it the Cardinal Flower in his *Paradisi in sole Paradisus terrestris* (1629) which he later published and dedicated to Her Majesty.

Linnaeus gave the blue-flowered *L. syphilitica* its specific name because Peter Kalm, who visited America in 1747, reported that the Indians esteemed the roots for their medicinal uses.

The dwarf forms of *L. erinus* are much prized as bedding plants. They make good edging and pot plants, are excellent for winter flowering in cool greenhouses and the trailing forms are in demand for hanging baskets and window boxes. *L. erinus* is a South African perennial with pale blue or violet, white-throated flowers; its various forms are grouped according to growth habit, for example 'Compacta' for all the bushy compact kinds, 'Pumila' includes the exceptionally dwarf varieties and 'Gracilis' the pendulous types. All are readily raised from seed and now come in many shades of blue, and also white and purple.

In equatorial East Africa there are Lobelias of giant proportions; the largest, *L. gibberoa,* is usually 5–6 m (15–20 ft) and has racemes of greenish-white flowers. In *L. wollastonii* they are powder-blue and in *L. bequaertii* deep purple.

For tropical gardens and warm greenhouses Canarinas make unusual wall climbers. *Canarina eminii* and *C. abyssinica* both with hanging, pale orange bells, lined with red, are found growing over old trees or rocky cliffs in damp places in Kenya, often at altitudes of 2000–3000 m (7000–9000 ft). Inside the bells are prominent club-shaped stigmas which look like clappers. *C. canariensis* (*C. campanulata*) from the Canary Isles has deep reddish-orange, pendent bells with reddish veins and toothed, roughly ivy-shaped leaves. The berries are edible.

From Asia (the Himalayas to Japan) comes *Codonopsis*, a group of annual and perennial plants, often with twining stems but chiefly noteworthy on account of the beautiful colour markings inside many of the bell-shaped flowers. These have the usual *Campanula* shape and appearance without and are normally pale blue, but the interior is often intricately patterned in various shades of green, chocolate, orange and blue. Because of this the plants should be grown in raised situations so that the interior as well as the exterior of the flowers can be seen. All require well-drained soil and full sunshine.

The most important horticulturally are *Codonopsis ovata,* 15–30 cms (6–12 ins) tall, the flowers pale blue, speckled inside with gold and green; *C. meleagris,* 30 cms (1 ft), porcelain-blue with zones of chocolate, purple and green; and *C. tangshen,* a twiner up to 3 m (10 ft), the flowers greenish with purple stripes and spots. *C. convolvulacea* is plain blue both inside and out and of climbing habit as is *C. vinciflora*, which is bluish-lilac.

Canna
'Louis Gagueaux'

Canna
'J. Anderson'

Cannaceae

1 genus and 55 species; tropical and subtropical America

This family consists of monocotyledonous plants which at first sight bear some resemblance to members of the Zingiberaceae and Marantaceae. Even out of flower, however, they can be identified by the fact that Zingiberaceae and Marantaceae have outgrowths or small swellings at the leaf bases, neither of which is present in Cannaceae.

The inflorescence is terminal, individual blooms being irregular and bisexual with three petals and three sepals. These are small, the showy part of the flower being provided by the stamens. Some are abortive, but one has half an anther on one edge with several petaloid appendages making the flower look many-petalled. The round black seeds are very hard, which may be the reason one species is known as Indian Shot, *Canna indica* (*C. bidentata*), although it is a Brazilian plant from shady marshy places. In spite of their American ancestry many Cannas have become naturalized in parts of Asia and Africa.

The fleshy roots are used as food by the head-hunting Jivaros and also the Tunebo Indians of Colombia and Ecuador. Indeed, the Tunebos not only use Canna roots as a staple diet but wrap their new-born babies in the leaves and also employ the latter for thatching.

The garden Cannas in their hot and splendid colours are the results of hybridization between several species, notably *C. indica, C. flaccida, C. glauca* and *C. coccinea*. These are invaluable wherever subtropical bedding is practised, either as 'dot' plants amongst lower-growing subjects or as a feature in their own right.

The innumerable varieties show flowers in a wide range of lively colours, brilliant scarlet, warm apricot, vivid orange and deep yellow, and the foliage is frequently colourful as well, purple and bronze leaves being common.

The first attempts to use Cannas as bedding plants in Europe were made in 1846 when M. Année, the French Consular Agent at Valparaiso in Chile, brought a collection he had found while travelling in South America to his garden near Paris. These proved so successful that he took to hybridizing the species and thus laid the foundation of the splendid collections now grown in Europe.

Cannas can be grown from seed, but these are so tough-skinned that it is necessary to soak them in warm water for twenty-four hours before sowing or else file through the seed coats. They need a temperature of 18–27°C (65–80°F) for successful germination.

Named cultivars must be propagated vegetatively by splitting the fleshy rhizomes into sections of convenient size in spring. These are then potted or boxed up in leafy soil or fibrous peat and kept in a temperature around 16°C (60°F). When there are plenty of roots they can be transferred to pots for flowering or be planted outside in a sunny spot when all danger of frost is past. In tropical countries, division and replanting will take place at the same time.

In temperate latitudes the plants can only be left outside if protected by leaves, ashes or glass. Most gardeners prefer to lift them in autumn and store the tubers in moist soil or leaves in a frost-proof building until the following spring.

C. edulis is the source of Queensland Arrowroot, used for babies and invalids, especially in the West Indies and Hawaii. The leaves and tubers provide fodder for dairy cows and the seeds of other species such as *C. indica* are used for such items as necklaces and rosaries and the leaves for wrapping food.

Caprifoliaceae

12 genera and 450 species

This family consists of dicotyledonous trees and shrubs, with a few lianas and a small number of herbaceous plants, the majority native to the Northern Hemisphere.

The leaves of Caprifoliaceae are opposite, usually entire and simple but occasionally lobed or with stipules. The flowers are bisexual, regular or irregular, usually carried in cymose inflorescences. The calyx has five or sometimes four sepals which may be open or imbricate; there are four or five petals, four to five stamens, an inferior ovary with many seeds and a simple style and stigma. The seeds may be enclosed in a fleshy berry, drupe, capsule or occasionally an achene.

Lonicera is the largest genus with 200 species scattered throughout North and Central America (to Mexico), South to North Africa and Eurasia including particularly the Philippines, Himalayas and SW Malaysia. Most of these are erect shrubs, although there are also twining lianas such as *L. periclymenum,* the Honeysuckle, and *L. caprifolium* which are useful for walls and fences.

The flowers are usually irregular with four petals at the top and one below forming, with their jointed bases, a long narrow tube. Nectar is secreted at the base of these tubes but their length is such that it prevents all but insects with a long proboscis from reaching their sweetness. Those with fragrance to the flowers, which is most marked towards evening, attract night-flying hawkmoths which hover near the blooms and insert their proboscis into the interiors. This effects cross-pollination since self-fertilization is prevented because the stamens and stigma ripen at different times. The age of some flowers can be detected by colour, for when white the anthers are ripe, but as these fade the flower changes to yellow, the stigmas then becoming receptive.

Although shrub Loniceras are in the majority they are less popular and showy than the climbers—many of which are sweetly scented. *L. periclymenum,* the common Honeysuckle, is a well-known deciduous plant of the English countryside where it clambers up trees and festoons hedges with its strong clasping stems and creamy fragrant flowers. British or not (and some authorities credit the Romans or William the Conqueror with its introduction from Asia Minor or southern Europe) it has long been grown in English gardens. The Elizabethans probably used it to cover arbours and pleached walks, for Shakespeare gives hints of such culture in *Much Ado About Nothing*, III. i

'And bid her steal into the pleached bower,
Where Honeysuckle, ripened by the sun,
Forbids the sun to enter.'

Steriphoma paradoxa

Capparis spinosa

Capparidaceae

30 genera and 650 species
tropical and warm temperate

This family (also spelt Capparaceae) consists of dicotyledonous small trees or shrubs (some of the latter lianas) and more rarely herbaceous plants.

The leaves are simple or palmate, sometimes with stipules which take the form of glands or thorns. The flowers are perfect and regular, usually in racemes; the calyx and corolla are similar to Cruciferae (two plus two sepals and four diagonal petals), with four to six or more stamens, two carpels and the fruit a berry, nut or drupe.

One of the most important plants in this family is the Caper, *Capparis spinosa*, the unopened flower buds of which provide the capers of commerce. These are gathered and pickled in wine vinegar for use in cooking.

It is native to the Mediterranean area, although now widely grown in other warm territories, and is a thorny shrub with long trailing branches, carrying large cup-shaped, white or violet flowers which may be 7 cms (3 ins) across and brimful of long anthers. There are many substitutes for Capers, particularly Nasturtium (*Tropaeolum majus*) seeds. The bush is not hardy in cool climates and the so-called 'Caper' of English gardens is really a Spurge (*Euphorbia lathyris*), which has poisonous fruits.

Steriphoma paradoxa (*S. cleomoides*) comes from Venezuela and grows to 2 m (6 ft) with long-stalked, oblong-lanceolate leaves and showy orange-brown calyces with yellow petals and stamens.

Viburnum plicatum var. *tomentosum*

Viburnum x *bodnantense* 'Dawn'

florum' is sometimes called the Japanese Snowball. *V. macrocephalum* is the Chinese Snowball, a shrub with spreading branches and large globose heads of white flowers.

The European *V. opulus* 'Sterile' is another Snowball Tree, producing masses of large round heads of sterile white flowers. These often grow as large as cricket balls.

V. opulus is known as the Guelder Rose, a small tree or shrub valuable in autumn for its showy bunches of bright red translucent berries (golden in the variety *xanthocarpum*) and brilliant leaf tints. Its name refers to an erroneous belief that it originated in the Dutch province of Gelderen but it has other names such as Cranberry Tree and in the 16th century was known as Rose Elder because its pithy branches resemble those of the Elder (*Sambucus nigra*). When Parkinson was writing his herbal in 1629 he declared '. . . there is but one kinde of Rose Elder . . . it groweth like an hedge tree, having many knottie branches or shootes coming from the roote, full of pith like the common Elder.'

V. tinus from SE Europe is the evergreen Laurustinus, an admirable winter-flowering plant with clusters of pink and white flowers, which needs sun if it is to bloom well. Nevertheless it tolerates shade and the soot and dust of urban conditions better than most shrubs, although it is important to sweep up any of its fallen leaves close to the house, as they rot down and smell.

Some of the American Viburnums have delightful common names. *V. alnifolium*, for example, with white flowers and magnificent autumnal tints goes by such picturesque names as Hobblebush, Tangle-legs, Down-You-Go, Moosewood or Devil's Shoestring. Their origin lies in the habit of the lower branches which droop to the ground and root at their tips. The bark and roots of several species have medicinal properties and their leaves have been used as a source of tea.

The Chinese *Dipelta floribunda*, an upright shrub, 3–4 m (10–12 ft) tall, has the habit and general appearance of *Diervilla* with nodding, fragrant pink and orange flowers on long (up to 180 cms; 6 ft) downy shoots.

A small but important genus in the Caprifoliaceae is *Linnaea*. When the great Swedish botanist Carl Linné selected a plant to bear his patronymic he chose the Twin Flower, *Linnaea borealis*, a lowly woodlander with fragrant, nodding pinkish-white blossoms carried in pairs. For garden use the American form var. *americana* is much more amenable to cultivation, providing a rapid ground cover with tiny trumpet-shaped flowers. It dislikes lime and favours a shady place but preferably not under trees.

Viburnum opulus 'Sterile'

Viburnum tinus

Caryophyllaceae

70 genera and 1750 species

These are dicotyledonous plants with a world-wide distribution, mostly herbaceous in character but with a few undershrubs. The flower stems often have swollen joints, as in Carnations, and are normally branched; the leaves are opposite, usually entire and often with stipules. The flowers are regular and symmetrical with a distinct calyx and corolla, and are clearly designed to attract insects. Sometimes the five sepals are free right down to their bases thus ensuring that the nectar is freely accessible to small insects such as flies, but with others (like Carnation) the nectar is secreted deep in a tubular calyx of joined sepals, so that only insects with a long proboscis like butterflies and moths are able to reach it. These are probably attracted by the colours (red seems to lure butterflies and white the moths) and sweet scent.

The flowers have five (sometimes more) plain or clawed petals and the same number or twice as many stamens, a superior ovary and a capsule containing many seeds.

Several well-known garden plants belong to this family of which *Gypsophila, Dianthus* and *Lychnis* are the most important. It also contains many common weeds of cultivated ground such as the annual Chickweed (*Stellaria media*).

The Carnation is one of the most widely cultivated flowers under glass and belongs to *Dianthus,* which with some 300 species is the second largest genus in Caryophyllaceae. *Dianthus* are exclusive to the Old World and generally found in dry sunny situations.

D. caryophyllus, parent of the modern Carnation, has a natural distribution in southern Europe but is now widespread on the Continent, having been brought to England it is thought about the time of the Norman Conquest. The name Carnation refers to its use as a coronary or garland flower and was first used by an English author (Henry Lyte) in 1578. According to Chaucer, plants were extensively cultivated in England during the reign of Edward III and we know they were widely grown in the 16th century.

But even before this *D. caryophyllus* was known as the Divine Flower by the Ancients and so described by Theophrastus in the 3rd century B.C. The variations in colour, size and petal fullness of modern Carnations are the result of continuous cultivation under artificial yet favourable circumstances. John Rea's *Flora, Ceres et Pomona* of 1676 lists 360 named varieties, since which time countless thousands have been evolved.

As for the historical and fragrant Old Clove Carnation, it is known that the original stock came from Holland, but was later lost and substituted with another. Its name refers to the similarity of its scent to the cloves of commerce, the unopened flower buds of *Syzygium aromaticum* (see page 191).

Other old names are Gillyflower and Sops in Wine, a reference to the habit of using the petals to impart a pleasant flavour and bouquet to wine and ale. Even the named varieties of Elizabethan days were charmingly labelled, as 'Ruffling Robin', 'Lustie Gallant' and 'Master Bradshawe his Daintie Ladie'.

Besides various garden uses the cut flowers of Pinks and Carnations are in great demand for floral decoration, buttonholes, corsage sprays and all forms of floristry.

The main groups available come under the following headings: PERPETUAL FLOWERING CARNATIONS (Tree Carnations). Of mixed

Border Carnations
1 'Oadby Glory'
2 'Margaret Palmer Clove'
3 'Orange Maid'
4 'Mendip Hills'
5 'Pink Pearl'
6 'Santa Claus'
7 'Leslie Rennison'

Rock Pinks

Dianthus
'Mrs Sinkins'

parentage probably from varieties derived from *D. caryophyllus* and the Indian Pink, *D. chinensis.* These are the Carnations grown commercially under glass and come in a wide range of shades and mixture of colours.

HARDY BORDER CARNATIONS. Hybrids derived from *D. caryophyllus* and Perpetual Flowering Carnations. First raised in 1913, they are hardier, stockier plants suitable for outdoor cultivation.

GARDEN PINKS. Old garden plants probably derived from *D. plumarius* with one period of flowering. Many are sweetly scented. The Laced Pinks of the muslin weavers of Paisley in Renfrewshire (who were greatly devoted to the flowers and raised many new colours) originated from these around 1785. Other groups are known as Herbert's Pinks, London Pinks, Imperial Pinks and so on. Incidentally the double white 'Mrs Sinkins', perhaps the most famous Pink of all times, started life in a workhouse garden. Its name commemorates Catherine Sinkins of Slough, Buckinghamshire, whose husband was master of a workhouse there from 1867 to 1900. It was here that the Pink was raised in the 1870s and later Mr Sinkins disposed of the stock to Charles Turner, a nurseryman of Slough, on condition that he retain the name.

This variety had tremendous success and gained many awards, and when Slough gained Borough status in 1938 the armorial bearings incorporated a Buckinghamshire swan with a Pink 'Mrs Sinkins' in its beak.

ALLWOODII. Perpetual Flowering Carnations crossed with the Fringed White Pink, *D. plumarius.* These have a more branching habit than Garden Pinks.

SWEET WILLIAM. Garden forms of *D. barbatus* and well-known biennials (although some seem to be perennial) with single and double flowers in bunched heads. The main colours are white, pink, scarlet and crimson; many are Auricula-eyed.

SWEET WIVELSFIELD. Sweet Williams crossed with Allwoodii resulting in loose trusses of flowers, which may be single or double and are larger than Sweet Williams. Propagated by seed.

There are also many small *Dianthus* for the rock garden; good forms of such species as *D. alpinus*, *D. gratianopolitanus* (*D. caesius*), *D. deltoides* and *D. pavonius* (*D. neglectus*), or cultivars and hybrids of merit. These must have good drainage conditions and full sun. The family generally does well on calcareous soils and can be propagated by seed or cuttings.

Gypsophila differs from other genera in Caryophyllaceae by having larger panicles of tiny flowers, a circumstance which is noted in the specific name of the E European *G. paniculata.* This is a perennial with long thonged roots and branching stems of 90–120 cms (3–4 ft) with narrow, 5–6 cms (2–2$\frac{1}{2}$ ins), glaucous leaves and myriads of small white flowers. The common name Baby's Breath refers to its light frothy appearance. Cultivars and forms with double and pink flowers exist, of which 'Rosenschleier' ('Rosy Veil') and 'Bristol Fairy', a double pink and white respectively, are the best known.

G. elegans is an attractive annual from the Caucasus, useful in the garden and greenhouse and also for cutting purposes. The plant grows about 45 cms (1$\frac{1}{2}$ ft) high with a profusion of white, pink or rose flowers (considerably larger than *G. paniculata*) on slender branching stems. *G. repens* from the European Alps and Carpathians has a place in the rock garden where it forms broad mats, smothered in summer with tiny white or pink flowers.

All Gypsophilas are sun-lovers and like an alkaline soil. The annuals and species are easily raised from seed but cultivars do not root readily and are usually grafted on *G. paniculata* or similar perennial species.

Several *Lychnis* are useful in the herbaceous border but require careful siting as their vivid colours can clash with other plants.

L. chalcedonica, the Jerusalem Cross, received its name because

Sweet Williams
(*Dianthus barbatus*)

Garden Pinks
1 'Show Charming'
2 'Farnham Rose'
3 'Prudence'
4 'Vanessa'
5 'Gran's Favourite'

it was thought to have been brought to western Europe and Britain from Jerusalem during the time of the Crusades. Other countries, as for example Portugal, call it Maltese Cross for similar reasons, although in point of fact its native home is Russia. Another name, Flower of Bristow, refers to the similarity of the flower colour to a popular bright red dye, manufactured in Bristol in the 16th century under the name Bristol (or Bristow) Red. All this points to the plant's antiquity and its continued cultivation in gardens.

The species produces many stiff, leafy stems, 60–90 cms (2–3 ft) tall, terminating in flat heads of small scarlet flowers. There is a white form and several doubles.

L. coronaria (*Agrostemma coronaria*) is the Rose Campion from southern Europe, a well-known border plant with silvery stems and foliage and a soft furry effect due to myriads of fine silvery hairs. This downy substance was at one time picked from the leaves and used for lamp wicks. Only the vivid cerise flowers escape this plush wrapping and make a striking contrast to the argentous tones of the rest of the plant. There are white forms and a good red-crimson cultivar called 'Abbotswood'.

Richard Folkard Jr (*Plant Lore, Legend and Lyrics*, 1884) says that the scarlet *L. coronaria* is in the Catholic Church dedicated to St John the Baptist and that the text 'a light to them that sit in darkness' refers to a belief that the flame-coloured flower 'was said to be lighted up for his day' and called *Candelabrum ingens*.

L. x *haagena* is a composite race of large and brilliantly coloured *Lychnis* derived from crosses between *L. fulgens* and a form of *L. grandiflora*. The seedlings are variable with flowers of 5 cms (2 ins) in orange, scarlet, crimson and white. The splendid scarlet-flowered 'Arkwrightii' belongs here. These hybrids are less sturdy than most garden *Lychnis* and need constant renewal.

L. viscaria (*Viscaria viscosa*) is the German Catchfly and takes its name from the sticky joints which are thought to protect the plant against insects, particularly ants. The double-flowered 'Splendens Plena' is a showy plant for the rock garden with long narrow leaves on stems of 45 cms (1½ ft), which terminate in groups of startling carmine-pink flowers.

A pleasing annual for the summer border is the Corn Cockle, *Agrostemma githago* (*Lychnis githago*). It is a European (including British) species, with erect, slender stems, 60–90 cms (2–3 ft) tall, bearing narrow, greyish, hairy leaves and large bright magenta flowers which close towards evening.

When Job said 'Let thistles grow instead of wheat, and cockle instead of barley', he was referring to the weediness of this beautiful plant in the cornfields. If the seed is ground with flour the latter will become bitter and poisonous; as John Gerard said in his Herbal of 1636, 'What hurt it doth among the corne, the spoile of bread, as well in colour, taste and unwholesomenesse, is better knowne than desired.'

All these plants appreciate moist but well-drained soil with plenty of sunshine, a circumstance which also suits several small rock plants. The best of these is perhaps *Petrorhagia* (*Tunica*) *saxifraga* 'Rosette', a carpeter with tiny glaucous leaves and many double 'polka dot' rosettes of lilac-pink flowers.

The Moss Campion, *Silene acaulis,* is a good crevice plant, forming a tight wad of 5 cms (2 ins) with tiny leaves and pale pink spring flowers. There is a double white form of *S. alpestris* (*Heliosperma alpestre*), which is showy in spring, and for autumn, *S. schafta* 'Abbotswood' with pink flowers.

Arenaria montana, 10 cms (4 ins) high with white flowers, and *A. purpurascens,* 5 cms (2 ins) and pink (also white), are carpeters for well-drained soil and useful for spring and early summer.

Cerastium tomentosum from SE Europe, often called Snow in Summer because of its many single white flowers and frosted silver leaves, is a rampant spreader which needs careful siting in a rock garden. It has a place nevertheless as a wall drape, to trail over banks or mask unsightly stones or tree stumps.

Saponaria officinalis from temperate Asia and Europe (including Britain) has several English names of which Bouncing Bet and Soapwort are best known. The last refers to a curious property of the leaves which when bruised or boiled in water become saponaceous, producing a lather which has detergent effects and even removes grease. It was used in ancient times (by mendicant friars particularly) as a substitute for soap and has also been given internally for the relief of gout and rheumatism. The 'wilde sope-wort' was also used according to Parkinson for scouring wooden and pewter vessels (china and earthenware were rarities in those days), hence other names such as Fuller's Herb and Latherwort.

The double forms 'Alba Plena' (white), 'Rosea Plena' (pink) and 'Rubra Plena' (red) make useful plants for the wild garden with showy panicles of flowers, 6 cms (2½ ins) across, on stems of 30–90 cms (1–3 ft).

S. ocymoides from the European Alps is an easy trailing rock plant, useful at the corners of steps, to drip over sink gardens and similar aspects. It needs plenty of sun and good drainage and carries many rose-pink Campion-like flowers in early summer.

The Cow Herb, *Vaccaria pyramidata* (*Saponaria vaccaria* or *Vaccaria segetalis*), is so called because it was thought to be a good fodder plant. It is a European annual with graceful sprays of showy, deep rose-pink flowers on stems of 60 cms (2 ft). Often grown in mixed borders it also makes a good cut flower.

Casuarinaceae

2 genera and 65 species

This is a small family of tropical trees and shrubs, mostly from NE Australia, Polynesia and South-east Asia, with long or short, deeply grooved, leafless branches which are of slender habit and frequently pendulous. Whorls of scale leaves surround the stem joints (as in *Equisetum*) and because the green tissue and stomata are tucked into the branch grooves (there are no leaves in the normal sense), the plants are remarkably drought-resistant.

There are two kinds of flowers usually borne on separate trees: the male blossoms in spikes, the female in dense rounded heads.

The Casuarinas or She-oaks of Australia are among the most characteristic trees of that continent, having a thin topped habit with wind-pollinated flowers and slender green branchlets. The female flowers are red or reddish-brown followed by quaint globular cones; the male blooms (which produce the pollen) are golden-brown with rusty brown tips. In some species the last are remarkably beautiful.

Casuarinas do well in both brackish and alkaline soils. *Casuarina equisetifolia* is widely cultivated in S Florida and California, also India, Egypt and tropical Africa, where it remains green when other trees wilt from heat and drought. In favourable climates it grows quickly and up to a height of 45 m (150 ft).

C. cunninghamiana, the River Oak, is a tall erect tree which needs moister soil and has been used to prevent river bank erosion.

C. sumatrana, a Sumatran and Javanese shrub, has a natural close growth of formal habit and is clipped to various shapes in Malaysia. It is slow growing and has been planted outdoors in sheltered parts of the British Isles; the branches are used for floral work.

C. distyla is an Australian shrub 1–3 m (3–10 ft) high with attractive male flowers 2.5–7 cms (1–3 ins) in reddish-brown spikes.

The She-oaks yield a hard timber which in British colonial days was used by settlers in place of the European Oak for construction work. However it is much redder in colour and so called Beefwood in some parts. Various species yield fine-grained wood suitable for shingles, staves, wagons, gates, furniture, and turnery and the bark of some is used for tanning.

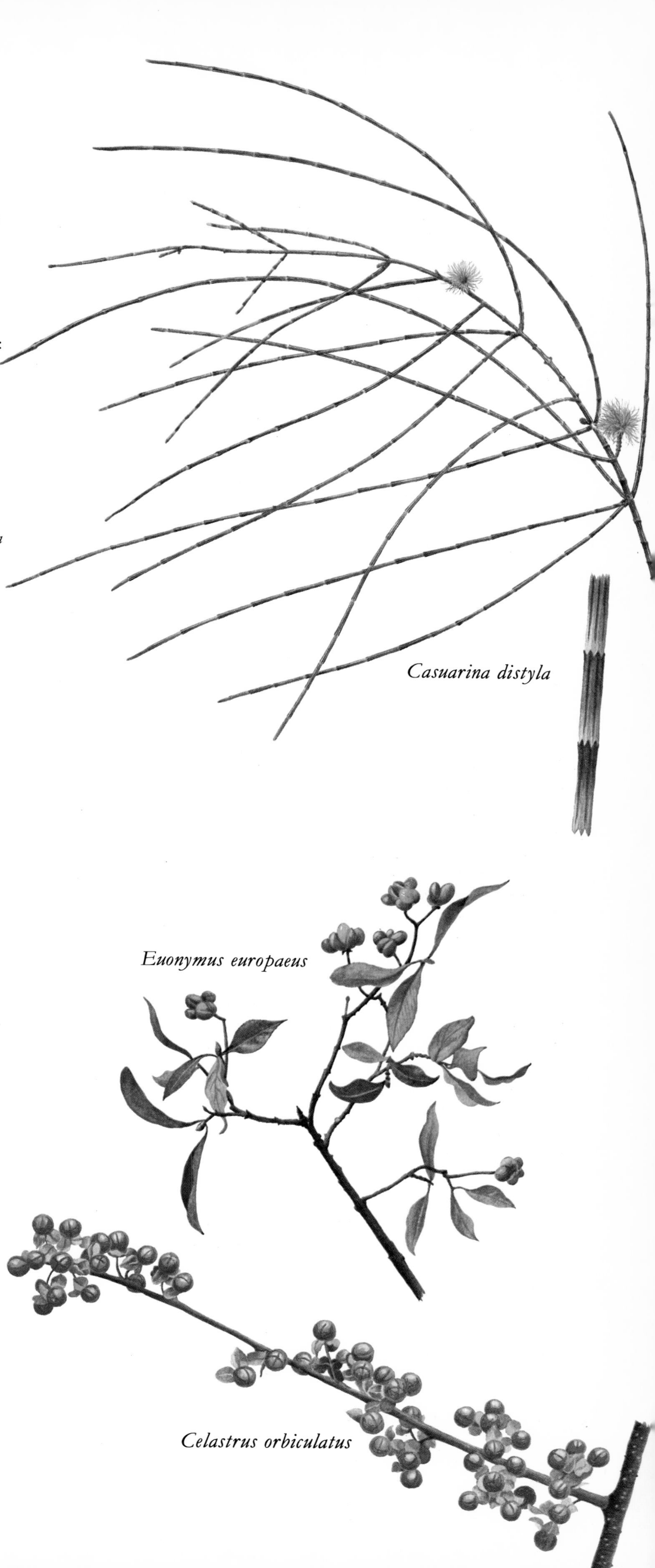

Casuarina distyla

Euonymus europaeus

Celastrus orbiculatus

Celastraceae

55 genera and 850 species; tropical and temperate

This family consists of trees or shrubs with simple, opposite or alternate, often leathery leaves, and small, regular and usually bisexual flowers in cymose inflorescences. The calyx has three to five (free or united) sepals; three to five imbricated petals; three to five stamens and a superior ovary; the capsular fruits contain seeds which frequently show a brightly coloured appendage or aril.

Several useful shrubs belong to this family of which the most important are included in *Celastrus* and *Euonymus*. *Celastrus orbiculatus* from NE Asia is a vigorous deciduous climber, up to 9–12 m (30–40 ft), with simple alternate leaves and clusters of rather dull greenish flowers followed by spectacular yellow fruits. These split to reveal bright yellow interiors cushioning vivid red seeds. However they only fruit when a clone with bisexual flowers is obtained. The foliage turns gold in autumn. *C. scandens,* the Climbing Bittersweet from North America, is similar with yellowish-white flowers and scarlet and orange fruits.

Euonymus is split into two groups: those grown primarily for foliage effects, the others for their autumn colouring and brilliant fruits. The former includes the Japanese *E. japonicus* often grown as a pot plant, but also as a garden evergreen 3–5 m (10–15 ft) tall. The leaves are so densely packed on the stiff upright branches as to resemble Boxwood and there are forms with white leaf borders, gold centres and several kinds of variegation; also dwarf cultivars of compact habit, with tiny narrow leaves which may be green, silver-margined or gold-variegated.

E. fortunei, another Japanese species and particularly its variety *radicans,* makes splendid ground cover in sun or shade—or it can be used as a wall climber. Besides plain green and variegated leafed kinds, one called 'Colorata' has purplish-red leaf suffusions and remains colourful all winter.

E. europaeus is the European Spindle Tree, a shrub of 4–5 m (12–15 ft), which adds glowing scarlet and bronze to the myriad shades of autumn foliage. As the leaves fall the pendent fruit clusters of pink and orange are revealed. Picked just before the capsules split they last weeks in water. There is a white-fruited form and others with purple or variegated leaves. *E. alatus* from China and Japan is called the Winged Spindle Tree because the branches bear two thin corky wings, nearly 1 cm. ($\frac{1}{2}$ in.) wide. The fruits are purplish and the bush smaller, 2–2.75 m (6–9 ft), than the preceding. Both plants like a chalky soil.

The wood of *E. europaeus* is very hard and close-grained and according to John Evelyn (circa 1776) was used to '. . . make bows for viols, and the Inlayer uses it for its colour, and the Instrument makers for the toothing of organs, and virginal keys, toothpicks etc.' Butcher's skewers were also fashioned from the wood and the same hard timber is still employed for making artist's charcoal. Gerard says it received the name Spindle Tree 'because spindles bee made of the wood hereof'.

In the Middle Ages the country folk of Europe would watch the budding Spindle Trees with close concern. If the flowers were scanty all would be well, but great apprehension was felt if there was a profusion of blossom for this meant that plagues would sweep the land. In order to ward off such dire disaster great quantities of the revealing flowers (Nature's own antidote) were consumed, in fearsome concoctions which caused violent nausea and purging.

The fruits are in fact even more poisonous as Theophrastus noted three centuries before Christ, 'This shrub is hurtfull to all things and namely to Goats . . . the fruit hereof killeth; so doth the leaves and fruit destroy Goats especially.'

Cistaceae

8 genera and 200 species

This small family consists of shrubs and herbaceous plants, mostly found in the N temperate zone in dry sunny places, particularly on chalk and sand.

The flowers are bisexual and regular, either solitary or in cymes. The calyx usually has five sepals, the two outer frequently being smaller than the three inner; there are five, three or no petals but many stamens and a superior ovary. The leaves are usually small and opposite, often with glandular hairs (as in *Cistus*) which secrete an ethereal oil.

Several genera have horticultural importance, notably *Cistus, Helianthemum* and *Halimium,* which in the garden prefer open well-drained soil in full light—damp and shade are anathema to them all. They are particularly suitable for bank planting or rock-garden pockets or between the stones of crazy paving. All the shrubby types mentioned can be increased by soft cuttings taken during the growing season.

Helianthemums are the Sun Roses, shrubby carpeting evergreens with a long flowering period, when they display sheets of blossom, the five-petalled flowers in white, yellow, orange, pink and red shades. There are also double cultivars in various colours.

Most of the garden sorts are derived from *Helianthemum nummularium* (*H. chamaecistus*), a yellow-flowered species found wild in many parts of Europe, including Britain. The Sun Roses associate splendidly with Aubrietas and Rock Phlox (*Phlox subulata*), and are great weed suppressors, but need trimming back after flowering to keep them compact and healthy.

Cistus are small evergreen shrubs, mostly from southern Europe, which have a pleasant smell when handled and masses of large simple flowers like Dog Roses (*Rosa canina*), many with coloured blotches at the petal bases. Although these last only a few hours (they drop soon after noon), a plentiful supply of buds extends the pageant of colour for some weeks each summer.

Among the most important species are *C. laurifolius*, 2 m (7 ft), with white flowers which have a yellowish base; *C. ladanifer,* 1.5 m (5 ft), with large crimped white flowers having chocolate basal patches and very gummy leaves; and *C. incanus* (a very variable species now held to include *C. creticus* and *C. villosus*), 90–120 cms (3–4 ft), soft mauve-purple. Good hybrids and cultivars include *C.* x *purpureus,* red-purple with deeper blotches; 'Silver Pink', clear pink; *C.* x *lusitanicus,* white and crimson with downy leaves, and its form *decumbens* with smooth leaves; and

Helianthemum hybrids

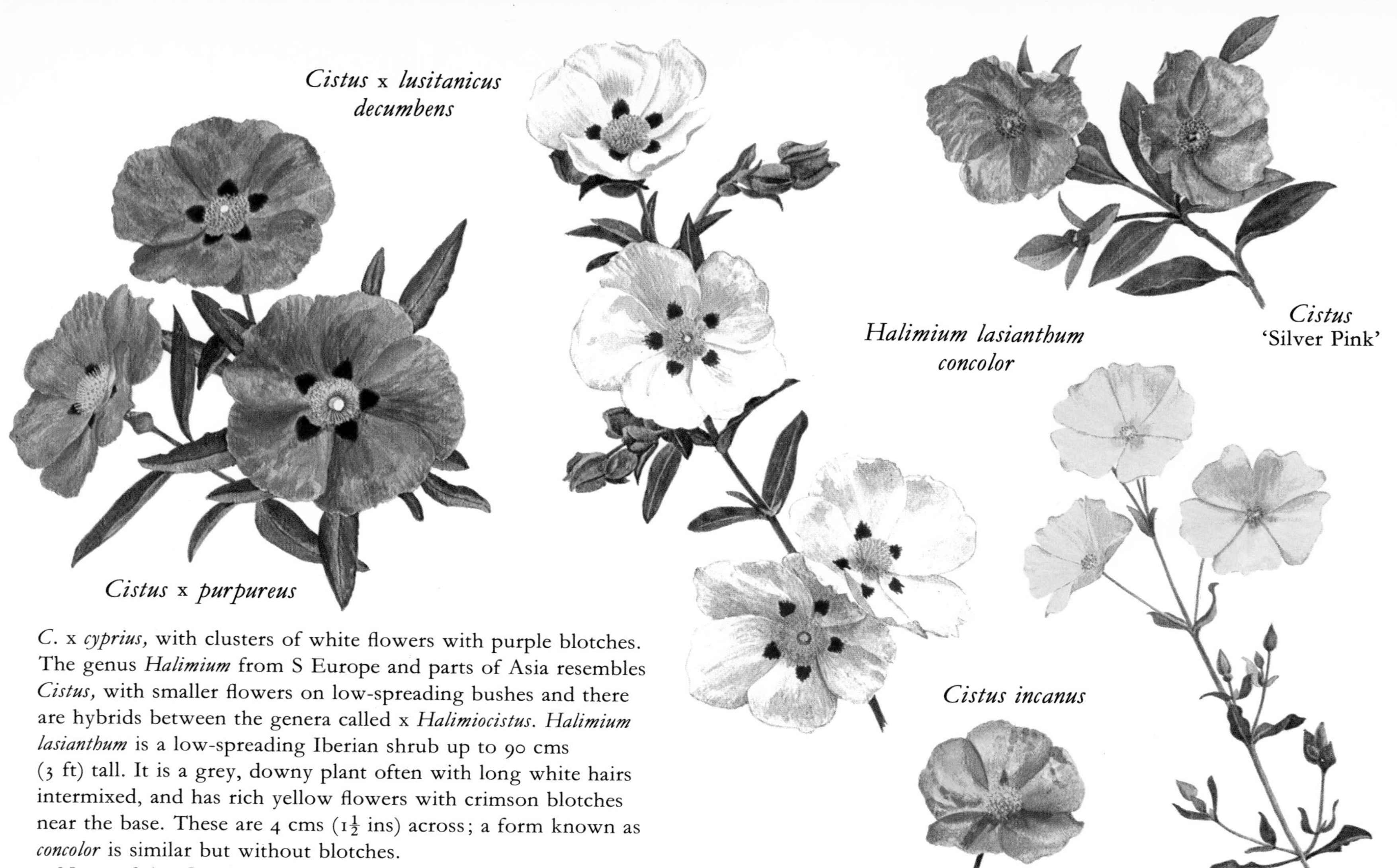

C. x *cyprius,* with clusters of white flowers with purple blotches. The genus *Halimium* from S Europe and parts of Asia resembles *Cistus,* with smaller flowers on low-spreading bushes and there are hybrids between the genera called x *Halimiocistus. Halimium lasianthum* is a low-spreading Iberian shrub up to 90 cms (3 ft) tall. It is a grey, downy plant often with long white hairs intermixed, and has rich yellow flowers with crimson blotches near the base. These are 4 cms ($1\frac{1}{2}$ ins) across; a form known as *concolor* is similar but without blotches.

None of the *Cistus* are long-lived, particularly in cool climates, so need frequent renewal from seeds or cuttings. They also resent disturbance so should be left alone when established.

In some of the smaller Greek Islands, oil from the foliage is used to make a crude form of incense. Collection is made by goats which small boys round up and drive through the *Cistus* thickets. The gum sticks to the goats' hair which at the end of the operation smells considerably sweeter than it did beforehand and then the hair is cut, placed in vats with water and brought to the boil. Besides cleaning the hair, this releases the gum which can be scraped off when the liquid cools.

The substance extracted is known as Ladanum and is most plentiful in the young leaf tops from which it is now removed with the aid of alcohol. Centuries before Christ, however, Herodotus described another method of collection: 'Ladanum . . . is found in a most incongruous place. It is the sweetest of scented substances. It is gathered from the beards of he-goats, where it is found sticking like gum, having come from the bushes on which they browse.'

The main species used for this purpose are *C. salviifolius* and *C. incanus*, which are commonly known as Gum Cistus or, in New Zealand, as Gallipoli Roses because troops from the First World War took back seeds of these species from Gallipoli. Ladanum is also thought to be the Myrrh of the Bible referred to in Genesis 37: 25 '. . . behold a company of Ishmaelites came from Gilead with their camels bearing spicery and balm and myrrh.'

The gum is dark with a heavy odour resembling ambergris and makes a fixative for perfumes; it is used in the manufacture of soap, powders, creams and cosmetics. The leaves are also made into tea by Arabs in Algeria.

Another genus belonging to Cistaceae is *Hudsonia*. These North American plants have Heather-like, sprawling stems with small leaves like needles and many yellow five-petalled flowers at the ends of short branches. They are native to poor sandy soils and beaches, hence the name Beach Heather or Poverty Grass for *H. tomentosa. H. ericoides* is another coastal species.

Cleomaceae

12 genera and 275 species; tropical and subtropical

These are dicotyledonous, generally glandular annuals, although there are also a few trees, shrubs and climbers. The flowers are usually irregular with long protruding stamens.

Cleome hasslerana (usually grown as *C. spinosa* or *C. speciosa*) is the Spider Flower, a South American native now widely naturalized in Africa and other tropical regions. Growing 75–90 cms ($2\frac{1}{2}$–3 ft) tall, it has palmate leaves and racemes of rose, pink or white flowers (also purple in the cultivar 'Purple Queen') with long protruding stamens and four large petals. The plants bear small spines on the leaf stalks and are sticky to the touch. They bloom for weeks on end and so make useful bedding subjects in summer. Other interesting species are *C. rosea* from Rio de Janiero with rosy flowers and *C. gigantea,* a South American shrub, with greenish-yellow flowers with pink filaments and yellow anthers.

Cleome hasslerana

Cobaea scandens

Cleome species are perennial in warm localities but are usually treated as annuals in temperate countries.

Gyndandropsis is similar to *Cleome*, with white or purplish flowers which only have six stamens (there are four to ten in *Cleome*) and three to seven leaflets on the foliage. The only species is the annual *G. gynandra* whose seeds are the source of an essential oil used instead of garlic or mustard oil. The leaves are also eaten in some countries as a vegetable or used for sauce flavourings.

Cobaeaceae

1 genus and 18 species

Cobaea scandens is the Cup and Saucer Plant, a climbing shrub from Central and South America with alternate, pinnate leaves which have their terminal leaflets converted to tendrils. The large and showy flowers have long stalks, five green sepals, five petals, five anthers and a superior ovary. On first opening, the blooms are green and unpleasantly scented but with age they become purple and then have a pleasant honey fragrance.

The plant likes a rich soil in sunshine or partial shade and makes a useful hedge drape or wall climber in the tropics. In cooler climates it is grown in a greenhouse or as an annual.

Combretaceae

19 genera and 600 species

This family consists of trees and shrubs, sometimes climbing, mostly confined to the tropics of both hemispheres.

Family characteristics include opposite or alternate, simple, entire leaves and spikes or racemes of usually bisexual flowers, regular, with four or five sepals; four, five or no petals; four, five, eight or ten stamens alternating with the petals and an inferior ovary. Many make attractive ornamentals for tropical gardens, where they are used as shrubs or shade trees and the climbing sorts to drape other trees or buildings.

Terminalia is a genus with 250 species of trees, some famous for hard, close-grained wood which is brown with dark streaks. It is used for house and boat building, and agricultural implements; tannin from the bark supplies snuff and perfume for Sudanese women; the fruit is used for dyeing and in E India to colour teeth black. It is also preserved as a food. Myrobalan (*T. catappa*) is grown for its nutty seeds and as a shade tree.

Quisqualis indica, the Rangoon Creeper or Drunken Sailor from tropical Asia, is said to have been given its generic name (which literally translated means Who? What?) by Rumphius, in astonishment at the variability of the plants' growth. It starts by growing erect for about 90 cms (3 ft) then producing a new basal shoot which is entirely climbing. After that the original stem dies. It is extensively grown in the tropics for its showy racemes of five-petalled flowers with long perianth tubes. These open white, then change to pink and finally become deep crimson. They smell sweetly at night.

Another exciting climber is *Combretum coccineum* from Natal and Madagascar, a splendid plant which uses other trees for support and drapes them with large inflorescences of rich scarlet flowers. *C. grandiflorum* from tropical Africa has brush-like inflorescences in clusters, each bloom about 1 cm. ($\frac{1}{2}$ in.) across with five dark red petals. The oblong leaves are arranged in opposite pairs on a very hairy stem. It is used for growing over walls and trellis.

Quisqualis indica

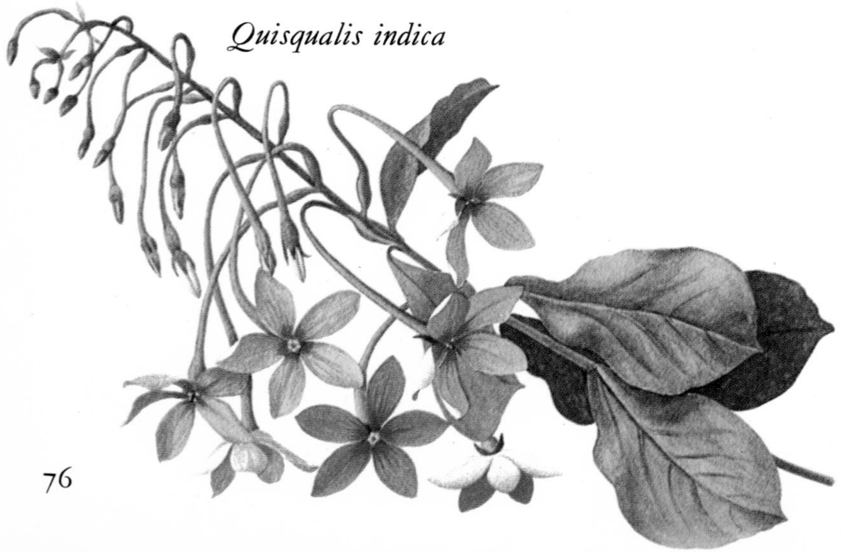

Commelinaceae

38 genera and 500 species

This family of monocotyledonous plants comes mainly from the tropics and subtropics. It is characterized by succulent (rarely twining) stems with prominent joints and simple leaves arranged alternately and partially clasping the stems.

The inflorescence is often a cincinnus (like Boraginaceae) with the individual flowers bisexual, usually regular, but sometimes irregular and commonly blue. They have three sepals, three petals, six stamens and a superior ovary.

Many greenhouse ornamentals and house plants belong to this family, notably *Tradescantia, Zebrina, Rhoeo* and *Cyanotis*. These are commonly used in tropical gardens as ground cover in shady areas or as carpeters between taller plants in display beds. A few are hardy enough to grow outside in temperate climates.

The best known Tradescantias are those long, trailing indestructible plants so often used for home decoration. They come in infinite variety and are more valuable for their silver, gold or even pinkish leaf patternings and easy-going nature than for their flowers. Most are variants of *Tradescantia fluminensis,* the American Wandering Jew, a green-leafed plant which is seldom seen except when the striped kinds revert to type. When this happens the green parts must be ruthlessly cut or in no time at all they will 'take over'. Tradescantias root easily (even in water) but for attractive containers the pots should be well filled so use several cuttings, not one or two, in small pots of sandy soil.

T. blossfeldiana from Argentina has purple stems and fleshy dark green leaves with purplish undersides, the whole covered with a fine coating of hair. It is a more substantial looking plant than *T. fluminensis* but less adaptable and hardy.

T. virginiana is the common species from which the Tradescantias of herbaceous borders (known collectively as *T.* x *andersoniana*) with grassy leaves and showy blue, purple, red or white flowers are derived. The blooms are three-petalled with beautiful fluffy stamens. The plant rejoices in a number of names such as Spiderwort, Moses in the Bulrushes, Devil in the Pulpit and Widow's Tears. The original species was brought to England from North America about 1629 and named for John Tradescant, gardener to King Charles I. His son, John Tradescant the second, visited Virginia in 1637 and brought back many plants such as Virginia Creeper (*Parthenocissus quinquefolia*), Swamp Cypress (*Taxodium distichum*), Cardinal Flower (*Lobelia cardinalis*), Red Maple (*Acer rubrum*) and the first Michaelmas Daisies (Wild Asters in the US).

He also collected natural history curiosities such as a stuffed dodo and 'two feathers of the Phoenix tayle'; he assembled these,

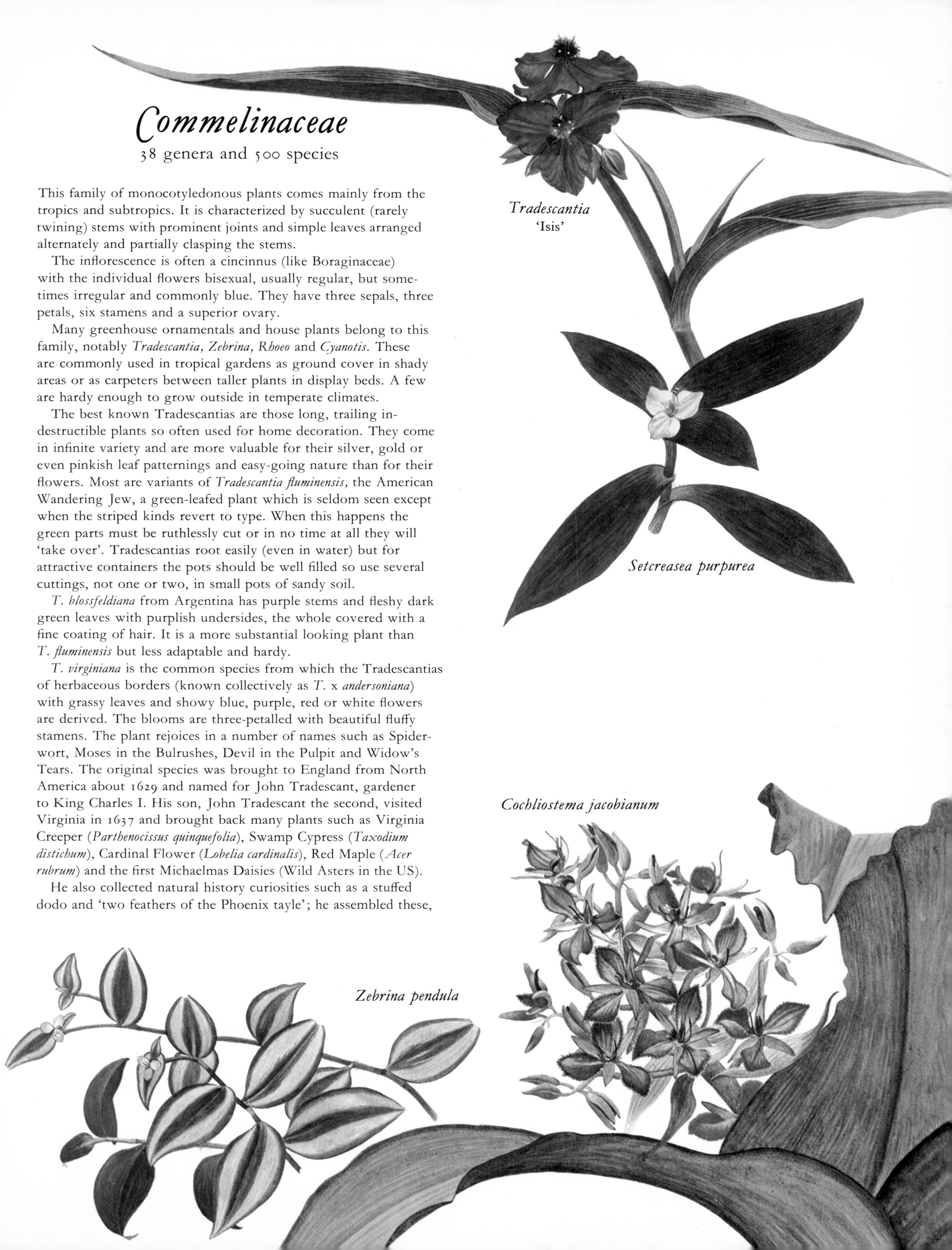

Tradescantia 'Isis'

Setcreasea purpurea

Cochliostema jacobianum

Zebrina pendula

together with articles supplied by 'noblemen, gentlemen and sea commanders with such toyes as they could bring from other parts' to make a 'Museum Tradescantium'—commonly known as Tradescant's Ark. When the King lost his head Tradescant fell on hard times so that the Ark became an important source of revenue. On his death the collection was bequeathed to his friend Elias Ashmole and became the foundation of the famous Ashmolean Museum at Oxford. It is said that Mrs Tradescant contested the will and on losing the action drowned herself.

Zebrina pendula from Mexico is closely related to *Tradescantia* but has thicker and larger leaves with silvery stripes and red undersides. The cultivar 'Quadricolor' is most spectacular in shades of white, purple, dark and silvery green. *Setcreasea purpurea,* the Purple Heart, also Mexican, does not trail. Both stems and foliage are Tyrian purple and the flowers cerise—a ghastly combination! *Cyanotis kewensis* from Malabar is similar to *Tradescantia* but the leaves are smaller and more fleshy, olive-green with purple undersides and wholly covered with woolly brown hair. The flowers are red and violet-blue.

Two Commelinas are suitable for sheltered borders in temperate climates, otherwise the tubers must be lifted and stored in ashes for the winter like Dahlias. *Commelina erecta* from the SE United States has blue and white flowers but the better *C. coelestis* from Mexico has flowers of the most beautiful shade of blue imaginable. It grows about 45 cms (1½ ft) high and blooms all summer. *C. diffusa* from the Argentine, commonly but erroneously known as *C. nudiflora*, is of creeping habit with cobalt-blue flowers.

Cochliostema jacobianum from Peru is a handsome perennial with numerous, sweetly scented blue flowers in crowded cymes and large oblong-lanceolate leaves, 30–90 cms (1–3 ft) long.

Another genus cultivated for its foliage is *Dichorisandra,* all species of which are native to tropical America. They are herbaceous perennials with erect, often branching stems, smooth glossy leaves which are frequently beautifully marked and racemes of variously coloured, attractive flowers. These all require warm growing conditions—such as a tropical garden or stove greenhouse—and being forest plants they appreciate a moist humus type soil and shelter from direct sunlight.

D. thyrsiflora from Brazil is one of the most effective. It grows about 120 cms (4 ft) tall and has broad, dark green, lanceolate leaves with purple reverses and striking electric-blue flowers with yellow anthers. *D. reginae,* a Peruvian species, has white or lavender flowers and the green leaves are banded and spotted with silver. The centres are red-violet and reverses deep purple.

Compositae

900 genera and over 13,000 species

More than a tenth of the world's flowers belong to Compositae, which is one of the largest dicotyledonous families. With global distribution from the Arctic to the tropics, the plants are with very few exceptions all herbaceous in character. The 'daisy' inflorescence is typical but not universal, for the heads may be arranged in racemes, corymbs or compound heads as in *Echinops.* Nevertheless the flower structure in nearly every case is sufficiently distinctive as to be immediately recognizable.

The success of Compositae in colonizing so much of the earth's surface is thought to be due to special characteristics, such as the way the flowers are massed in heads, which creates a greater impact and makes them more conspicuous. It also allows for economy in floral parts, as well as ensuring a high degree of cross-pollination, for a single insect can fertilize several blooms at the same time. Again, the simple flower mechanism (a narrow tube enveloping the nectary and essential sex organs) protects both pollen and honey and also hinders self-pollination—without however preventing it taking place if cross-fertilization fails.

In certain genera such as *Taraxacum* (the Dandelion) the calyces play a part in seed distribution, for after the flowers are fertilized they grow on to a feathery pappus, light and buoyant enough to wing the seeds—as if by parachute—for long distances.

The inflorescence of many members of Compositae is a capitulum (head) with a lot of individual blooms packed so tightly together as to resemble a single flower. This effect is strengthened by the aggregation of green bracts surrounding the head in much the same way as a calyx encloses a normal flower. Often the outer ring of flowers (ray florets) are different in shape and colour from the inner flowers (disc florets), which again heightens the resemblance to a single bloom.

Structurally, disc florets (such as may be found in *Bellis* and *Pyrethrum*) bear some affinity to the flowers of Campanulaceae; both have five petals with five regular lobes joined in a tube or funnel. Some genera have in addition ligulate (ray) florets in the outer ring of flowers, for example *Bellis perennis,* or all ligulate blooms as in Dandelion. These flowers are irregular with a flattened group of united petals attached to a very short tube.

The leaves are usually alternate although in some genera (as *Helianthus* and *Eupatorium*) they may be opposite. Often they

Semi-pompon Chrysanthemum 'Nosegay'

Rayonnante Chrysanthemum 'Soleil Langraise'

Spoon Chrysanthemum 'Honeysuckle'

Reflexed Chrysanthemum 'Tipoff'

spring directly from the ground (radical) as in *Taraxacum* or are whorled as in *Zinnia verticillata*. The roots may be fibrous (*Aster*), tuberous (*Dahlia*) or carrot-like (*Cichorium*).

The family contains many ornamentals of which *Dahlia, Chrysanthemum, Aster, Calendula* and *Zinnia* are perhaps the best known. Comparatively few have economic uses but some *Scalesia* and *Brachylaena* are timber trees and Lettuce, Jerusalem Artichoke, Chicory and Sunflower have edible properties.

Among florists' flowers Chrysanthemums reign supreme. They are widely cultivated for their attractive and long-lasting blossoms which are at their best late in the floral year, when other flowers are scarce. Most garden varieties stem from two plants of Chinese origin—*Chrysanthemum vestitum* (*C. morifolium*) and *C. indicum*—and it would appear that the first flowers were yellow since Chrysanthemum literally translated means 'Golden Flower'.

In the Orient they are held in particular esteem and we know that the Chinese cultivated the flowers in 500 B.C., that the Mikado adopted them as his personal emblem in 797 and that the Japanese held Chrysanthemum Shows in A.D. 900. At one period the cultivation of Chrysanthemums was reserved for the gardens of the Emperor and nobility. The Rising Sun in the Japanese flag represents a Chrysanthemum (not the sun) with a central disc and sixteen flaring petals.

Not until the 18th century did the plants find their way to Europe, where they are now nearly as popular as the Rose. They are also widely grown in America, and to Alex Cumming of Connecticut we owe the hardy free-flowering Koreans so useful for open border work in temperate gardens. His first hybrid, derived from a plant known in gardens as *C. koreanum* and an early outdoor Chrysanthemum called 'Ruth Hatton' came in 1928. The similar, but even more adaptable *C. rubellum* hybrids were introduced by Amos Perry in 1938. Both these groups come in a wide range of colours—from white, yellow and bronze to pink and deep red. They are characterized by many-flowered sprays of 60–75 cms (2–2½ ft) of single or double blooms, which are excellent for cutting.

The Pompons, with small, button-like double flowers were derived from a semi-double, mahogany-red variety, identified by the name Chusan Daisy, which Robert Fortune sent home from China in 1846. Ignored in Britain, it was taken up by the French and the subsequent race is said to have received its name from the similarity of the flowers to the pompons worn on the hats of French sailors.

Among the large-flowered cultivars there are classifications

1 Intermediate Decorative Chrysanthemum 'Oh Boy'
2 Incurved Decorative Chrysanthemum 'Primrose Dorothy Else'
3 Intermediate Decorative Chrysanthemum 'Kelvin Hoskins'

1

2

3

Reflexed Decorative Chrysanthemum 'Crimson Pretty Polly'

Reflexed Decorative Chrysanthemum 'Yvonne Arnaud'

according to the season of flowering, that is, Early-flowering, Mid-season (October in Europe and North America) and Late-flowering, and the way the petals turn. The last point is highly important for shows and exhibition purposes, the three main groups being Incurved, Intermediate and Reflexed.

Incurved varieties have all their florets curved inwards, tightly and firmly in regular formation making a globular-shaped bloom.

Intermediate varieties (cultivars) come between Incurved and Reflexed, that is, they are semi-reflexed or partially incurved.

Reflexed blooms have florets that point outwards, the outside petals very often drooping.

There are also Singles, Spoon Chrysanthemums, Rayonnante, Anemone-flowered types and subdivisions within the groups relating to the size of the blooms.

With such sizeable flowers and so many to sustain, all florists' Chrysanthemums require generous soil treatment. The large-flowered sorts are reproduced from cuttings and grown on in pots, even bigger blooms being obtained by stopping and disbudding. Those grown outdoors in borders need well-cultivated soil and full sun and are propagated by cuttings or division of the roots.

Other species of Chrysanthemum include *C. coronarium,* the Crown Daisy from southern Europe, a variable plant with yellow and white flowers, often double in cultivation, on stems of 90–120 cms (3–4 ft) and the annual *C. carinatum* (*C. tricolor*) from North Africa. This grows around 60 cms (2 ft) tall with large white, purple-centred flowers with yellow bands. Cultivars show striking combinations of colour, for example, 'Pole Star', white, purple and yellow; 'Eclipse', buff, brown and crimson; and 'Burridgeanum', crimson, white and yellow.

C. frutescens, the rather more tender Marguerite from the Canary Isles, is often used for window boxes, hanging baskets and summer bedding. It has a yellow form called 'Comtesse de Chambord'.

C. maximum, the Moon Daisy, is a native of the Pyrenees and the parent of several striking border plants known as Shasta Daisies: 'Esther Reed' (double white), 'Snow Princess' (fringed petals) and 'Wirral Pride' (anemone centred) are all typical cultivars.

Chrysanthemum coccineum (*Pyrethrum roseum*) is the well-known Pyrethrum, a variable plant from the Caucasus with many fine cultivars having large white, pink and red flowers on straight stems of 75 cms (2½ ft). These may be single or double and are excellent for cutting. *C. cinerariifolium* is the source of Pyrethrum powder used as an insecticide. Vast quantities are grown for this purpose in the Kenya Highlands.

Michaelmas Daisies (Asters) are important Composites which grow equally well in town or country gardens and flower in late summer when most other border plants are finished. They are not particular as to soil but like full sun and need plenty of water during the growing season. Since the roots are mat-like and invasive, frequent division is necessary to maintain the quality of the blooms. Asters associate pleasingly with Golden Rod (*Solidago*), Red Hot Pokers (*Kniphofia*) and the White Mugwort (*Artemisia lactiflora*). For best effects the plants should be grouped. Propagate by division or soft cuttings in spring.

Asters are native to many parts of the world, from Europe to the Himalayas, China and Japan. But the greater number are found in Canada and the north US; the first recorded introduction from this continent being *Aster tradescantii,* which John Tradescant brought to Britain around 1633.

The more important horticultural groups such as *amellus, novi-belgii* and *novae-angliae* have received considerable attention from plant breeders, so that today pink, red, white, blue, mauve and purple cultivars are available in heights varying from 1.75 m (6 ft) to 7 cms (3 ins). Credit for considerable pioneer work with the genus belongs to the late Ernest Ballard, a Midland brewer, who found a chance Michaelmas Daisy seedling in his Malvern garden. Thinking this superior to any garden form then in cultivation he took it to the Royal Horticultural Society in London, hoping for an award. Failing in this he took it up again

Perennial Asters (Michaelmas Daisies)

Aster novi-belgii 'Crimson Brocade'

Aster novae-angliae 'Harrington's Pink'

Aster amellus 'King George'

Aster novi-belgii 'Erica'

the following season with similar results and then resubmitted it a year later. This time (1907) he acquired a First Class Certificate for it under the name 'Beauty of Colwall'.

The first pink Michaelmas Daisy, a *novae-angliae* seedling called 'Harrington's Pink', was received from the raiser in North America by Amos Perry in 1938 in exchange for a rare fern and was introduced in 1943.

A. amellus comes from S Europe and western Asia and is characterized by rough leaves and stems, short stature (60 cms; 2 ft) and large, golden-centred flowers. 'King George' is an old but favourite cultivar.

A. cordifolius, the Wood Aster or Beewood of North America, has hundreds of pale blue flowers on arching sprays of 90 cms (3 ft). *A. multiflorus* (better known as *A. ericoides*) is the North American Heath Aster, an apt name because of its fresh Heather-like foliage and slender arching sprays of tiny white, mauve or lilac flowers. *A.* x *frikartii* is a name coined for a group of blue-flowered seedlings from *A. thomsonii* and *A. amellus,* raised in Switzerland by Monsieur Frikart, and *A. tongolensis* (*A. yunnanensis*) 'Napsbury' is an early summer-flowering cultivar with large (6–7 cms; $2\frac{1}{2}$–3 ins) flowers on stems of 38 cms (15 ins).

To Mexico we owe Dahlias, which first reached Europe in 1789 through Vincente Cervantes, Superintendent of the Botanic Gardens of Mexico City who sent seeds to the Abbé Cavanilles of Madrid.

The Abbé named the resultant plants *Dahlia pinnata* in honour of Dr Andreas Dahl, a botanist and pupil of Linnaeus, but looked upon them more as a food crop than a garden ornamental. Although the tubers are still eaten by the Tunebo Indians in Colombia their flavour did not please Europeans, and since these early flowers were insignificant they soon dropped from cultivation. And then the Empress Josephine, whose gardens at Malmaison were world-famous, realized their beauty as garden plants. She was proud of her Dahlias which were reserved for the Royal gardens.

An amusing tale relates how they reached a wider public. One of the Ladies-in-Waiting (some accounts say it was the Countess of Bougainville) wishing to have Dahlias for her own garden persuaded her lover to obtain roots, which he did by bribing the

Pompon Dahlia 'Whale's Rhonda'

Anemone-flowered Dahlia 'Comet'

Semi-cactus Dahlia 'De Ruyters' Yellow'

Miniature Ball Dahlia 'Worton Joy'

Small Decorative Dahlia 'Twiggy'

Collerette Dahlia 'Allegro'

Large Decorative Dahlia 'Perran Garnet'

Empress' gardener. One hundred plants changed hands for 100 golden louis; the lady bragged of her acquisition and the news reached Josephine. She, annoyed, sacked the gardener, banished the lady from court and destroyed her entire Dahlia stock.

As garden plants modern hybrids and cultivars are unequalled for brightness of colour, easy cultivation and long season of flowering. They also make good cut flowers.

In the tropics the roots can be left outside all the year round but in areas where frost is liable it is necessary to lift them in autumn, cut the stems back to 15 cms (6 ins) and store the tubers (after drying) in straw, peat, sand or newspaper in a cool but frost-free shed. They can be restarted in spring and planted outside when the weather improves and there is no risk of frost.

Helianthus 'Monarch'

Helianthus 'Loddon Gold'

Dahlias rival Chrysanthemums in variety of form and size of flower and plant, so that again exhibitors adopt a system of classification. The main types are Singles, Anemone-flowered, Collerettes, Paeony-flowered, Decorative (these are fully double but subdivided again according to the size of the flower), Double Show and Fancy (double but more globular than the preceding), Pompon (very small, rounded doubles), Cactus with quill-petalled flowers with a mysterious origin: apparently a plant reached Holland from Mexico between 1864 and 1872, and was named *D. juarezii*, the Cactus Dahlia, for its resemblance to a Cactus called *Heliocereus speciosus*; and various Dwarf Bedding Dahlias.

Plants can be propagated from seed although the results will be variable or, in the case of cultivars, by soft cuttings taken in spring or by careful division of the tubers.

A number of Composites find favour as border perennials in temperate countries. Among the earliest of these are Doronicums, especially the Leopard's Bane (*Doronicum plantagineum*) from W Europe and its cultivars 'Harpur Crewe' and 'Miss Mason'. These make sturdy plants with rough heart-shaped leaves and large yellow 'daisies' on branching stems of 60–75 cms (2–2½ ft). They are good plants for light shade and flower with the Daffodils—but they resent drought.

Fleabanes (Erigerons) soon follow and remain in character, intermittently for most of the summer. They are generally mauve or violet, although pink forms exist and there is an orange species, *Erigeron aurantiacus,* the so-called Orange Daisy from Turkestan. The latter is a poor perennial, however, and succumbs to cold in a hard winter. A collective group of uncertain origin, but most probably derivations of *E. macranthus* and *E. speciosus,* provides the best garden plants. Such cultivars as 'Wupperthal' (violet), 'Sincerity' (pale mauve), 'Merstham Glory' (deep mauve) and 'Felicity' (clear pink) are typical. These grow 45–60 cms (1½–2 ft) tall, with branching, rather weak stems and many daisy flowers 5–6 cms (2–2½ ins) across. *E. mucronatus*, the Bony-tip Fleabane from Mexico, is a scrambling species 20 cms (8 ins) tall with many small pink and white flowers. It makes a good subject between crazy paving, at the corner of a flight of stone steps or similar situation.

In midsummer *Helianthus, Helenium* and *Heliopsis*—in autumnal shades of yellow, orange and sometimes crimson—take the field. *Helianthus* is the Sunflower, the perennial forms of which, such as *H. salicifolius,* 1.75–2.5 m (6–8 ft), deep gold, and *H. decapetalus,* light yellow, 60–120 cms (2–4 ft)—both North American—and such cultivars of the latter as 'Soleil d'Or', double sulphur-yellow, 'Capenoch Star', single yellow, and 'Loddon Gold', double deep yellow, make good background plants.

Less hardy but a splendid cut flower which lasts nearly five weeks in water is *H. atrorubens* 'Monarch'. This is a branching plant 1.75–2 m (6–7 ft) high, with golden, semi-double flowers the size of a man's hand. The annual *H. annuus* is a native of South America and Peru where it was once an emblem of the Sun God of the Incas. One can imagine the rivalry between gardeners of that time (which has persisted to this day) aiming at the biggest specimens. The plate-like heads and stout stems are invariably large—usually around the height of a man. Gerard (circa 1597) claimed he grew flowers 40 cms (16 ins) across on stems of 4 m (14 ft); capped by Crispin de Pass (1614) who declared that '. . . sown in the Royal Garden at Madrid in Spain, they grew to 24 feet in height; but at Padua in Italy it is written they attained to the height of 40 feet.'

This huge annual has sundry economic properties, the seeds being the source of Sunflower Oil used in many foods and as a butter substitute. Oil cake for cattle is made from the seeds which are also fed to poultry. The flowers are a source of honey and a yellow dye and pith from the stems is used in the preparation of microscope slides. *H. tuberosus* is the Jerusalem Artichoke.

Heleniums are easy perennials with branching stems of 60–120 cms (2–4 ft) carrying a number of daisy flowers, all of which have prominent central discs, a characteristic of the genus. *H. autumnale*, a species of Canada and the E United States, is the parent of most garden cultivars of which 'Moerheim Beauty' (bronze-red), 'Riverton Beauty' (yellow) and 'Copper Spray' (copper-red) are representative. The common name—Sneezeweed—refers to the Indians' use of the plant as errhine (promoting nasal discharge).

Heliopsis scabra from E and S USA resembles a small refined Sunflower, with narrow-petalled flowers which can be dried for winter decoration. All bloom in late summer and grow around 90 cms (3 ft) tall. Cultivars include 'Goldgefieder', double yellow with greenish centres to the blooms; 'Orange King', bright orange; and 'Gold Greenheart', buttercup-yellow with an emerald-green heart to the young flowers.

Rudbeckias are characterized by prominent cones in the flower centres, a circumstance shared by the closely related *Echinacea,* so that the name Coneflower is peculiarly apt. Best known of the former (which are all North American) is the Black-eyed Susan, *Rudbeckia fulgida* var. *speciosa,* a sturdy plant with rough narrow leaves and stems of 60 cms (2 ft) carrying golden flowers with blackish-purple centres; var. *sullivantii* is similar but larger.

R. *laciniata* 'Herbstsonne' is much taller, up to 1.75 m (6 ft), with smooth leaves and stems and large yellow flowers which have the petals flopping back after the habit of Cyclamen, so that the green central cones protrude like thimbles. 'Goldquelle' is a cultivar with deep yellow, semi-double flowers on stems of 90 cms (3 ft). The Gloriosa Daisies are annual (sometimes short-term perennial) Rudbeckias with variously coloured flower heads, ideal for cutting.

Echinacea purpurea is the North American Hedgehog Coneflower, a plant of similar habit but with larger flowers than *Rudbeckia fulgida*. 'The King' is a popular cultivar of 90–120 cms (3–4 ft) with wine-crimson flowers 13–15 cms (5–6 ins) across with mahogany-red centres. 'White Lustre' is an albino form.

Anthemis tinctoria is a European (including British) species known as Ox-eye Chamomile. It is a useful summer-flowering

Rudbeckia laciniata
'Herbstsonne'

Rudbeckia fulgida

Helenium
'Moerheim Beauty'

Echinacea purpurea
'The King'

Leontopodium alpinum

plant for well-drained soil in a sunny situation, blooming profusely with many golden 'full moon' daisies on stems 75 cms (2½ ft) tall. These yield a yellow dye. For garden purposes cultivars such as 'Perry's Variety' (golden yellow), 'Mrs E. C. Buxton' (lemon-yellow) and 'Grallagh Gold' (deep gold) are preferable. Cutting the plants hard back directly after flowering stimulates secondary flushes of bloom.

In the rock garden *Leontopodium alpinum* is treasured for its romantic associations but as an ornamental plant it is in fact less arresting than some species. Growing about 15 cms (6 ins) high, it has woolly leaves and clustered heads of small flowers which are surrounded by a 'collar' of greyish-white, felted bracts. It needs firm, well-drained soil and a sunny situation, and resents winter wet which can cause it to rot. It comes readily from seed.

This is the Edelweiss of the Swiss mountains, which grows in the region of perpetual snow, from whence it used to be brought down as proof of having reached high altitudes. Since in many districts it only grows in nearly inaccessible places, Victorian maidens valued the flowers as a proof of their lover's daring. At one time to prevent its extinction German and Tyrolese Alpine Clubs imposed fines for plucking Edelweiss and it is still protected in many districts. However, in other areas away from the tourist routes, it grows in profusion. There are carpets of it in the Italian Alps.

Most Golden Rods are native to North America, from whence the first species to reach Britain—*Solidago canadensis*—was brought over in 1648 by John Tradescant. Although many can only be described as weedy perennials, certain modern cultivars such as 'Golden Wings', 'Lemore', 'Goldenmosa' and 'Lesdale' with their dense panicles of yellow florets are worth a place in the autumn border. They will grow in any good garden soil, attaining a height around 75 cms (2½ ft). The European *S. virgaurea* was formerly called Wound-weed because of its reputed healing properties.

Achillea millefolium, the White or Common Yarrow, according to Gerard grew in churchyards as a reproach to the dead 'who need never have come there if they had taken their yarrow broth faithfully every day while living' (Marion Cran. *Joy of the Ground*). It has produced several coloured forms, the best 'Fire King' (deep red) and 'Cerise Queen' (rose-cerise). These grow about 60 cms (2 ft) high with ground-hugging leaves and leafy stems bearing flat terminal heads of flowers. *A. filipendulina* from the Caucasus is taller, 120–150 cms (4–5 ft), with plate-like heads of yellow flowers and deeply cut, aromatic foliage. 'Gold Plate' is one of its cultivars. The European *A. ptarmica* is known as Sneezewort because according to Gerard 'the smell of this plant procureth sneesing . . .' It has many small white flowers on branching stems 45 cms (1½ ft) tall; 'Perry's White', 'The Pearl' and 'Snowball' are common garden plants.

Anaphalis are grey foliaged plants with crowded heads of white 'everlasting' flowers often used by florists and in dried arrangements. The most common are the North American *A. margaritacea* and *A. triplinervis* from the Himalayas. Other silver-leafed plants are *Senecio* (*Cineraria*) *maritima, Artemisia stellerana, A. ludoviciana* and the cultivar 'Silver Queen'. All are useful for planting between brightly coloured summer bedding.

Artemisia lactiflora from S Asia is the White Mugwort, a perennial 120–150 cms (4–5 ft) tall with Chrysanthemum-like foliage and many small white flowers in crowded feathery plumes. This species likes moist soil although the others mentioned do better in sunny well-drained situations. The liqueur absinthe comes from the European *A. absinthium.*

A few Composites will grow in very damp places notably *Ligularia, Petasites* and some species of *Eupatorium.* They are all rather coarse so that the proper place for most is in the wild garden or towards the back of a moist border. Ligularias have large leaves and branching inflorescences of yellow or orange flowers. Eupatoriums have flat-topped clusters of purplish or white blooms and *Petasites* are chiefly valuable for their early flowers. *P. fragrans* is the European Winter Heliotrope, a prized pot plant in Victorian times on account of its sweet scent and clusters 15 cms (6 ins) across of white winter flowers; the leaves which follow are large and coarse. It grows rapidly and once established in the garden is almost impossible to eradicate.

Coreopsis and *Centaurea* provide the gardener with valuable cut flowers, particularly *Centaurea cyanus,* the European (including British) Cornflower or Blue Bottle, and the Persian *C. moschata,* the popular Sweet Sultan. Both are annuals and both bear variously coloured flowers, those of *C. moschata* being larger and also sweetly scented. The Cornflower was once a common weed of cornfields where it was apparently known as Hurt-sickle. Painters used to prepare a blue-colour from the petals.

Several perennial Centaureas are valuable in the herbaceous border particularly the Caucasian *C. dealbata* with rosy-purple or pink flowers; *C. montana,* the European Mountain Knapweed, grows 45 cms (1½ ft) tall with white, blue, purple, pale yellow or reddish flowers; *C. glastifolia,* is a pretty yellow species which comes from Asia Minor; and *C. macrocephala,* another Caucasian, has stems 90 cms (3 ft) high carrying yellow thistle-like flowers the size of a clenched fist.

The best of the Tickseeds (*Coreopsis*) are garden forms of *C. grandiflora* of the southern USA such as 'Mayfield Giant', 'Perry's Variety' and 'Badengold'. They are smooth perennials with leafy branching stems and many deep yellow, daisy-like flowers 2.5–6 cms (1–2½ ins) across. Although these only reach about 90 cms (3 ft) the habit is straggling so that staking is advisable. Dwarfer, 60 cms (2 ft), is *C. verticillata* from the E United States, a plant which flowers throughout the summer and has fine, thread-like foliage. 'Golden Shower' has larger and deeper coloured flowers. There are also a number of annual Coreopsis of which *C. tinctoria* (usually known as *Calliopsis*) from the central United States is a most vivid plant, with finely cut foliage and bright gold, maroon and crimson flowers (often with rings or suffusions of other colours) on stems 90 cms (3 ft) tall.

Catananche coerulea, a Cornflower-like plant 30–38 cms (12–15 ins) high from the Mediterranean region, has mauve, semi-double flower heads which are protected behind by silver papery bracts. These rustle when touched and the blooms can be dried. There is a white form. The narrow, hairy leaves are almost ground-hugging. Good drainage is essential for Catananches, which are reproduced from seed or root cuttings. The common name, Cupid's Love Dart, refers to their one-time use in love potions and philtres.

Echinops is the Globe Thistle, a coarse perennial of easy culture with pinnate, usually spiny foliage (often white and woolly beneath) and large, round, Thistle-like flowers. They are usually blue although there is a red species (*E. hispidus*) in Kenya. The most frequently cultivated are the steel-blue *E. ritro* from S Europe and its metallic blue hybrid 'Taplow Blue'. These reach a height of 90–150 cms (3–5 ft) and associate pleasingly with pink Phlox and golden Achilleas. *E. viscosus* from Greece is the source of a gum (Angado Mastiche) used in some parts as chewing gum.

Liatris are North American perennials which bloom in late summer with clusters of magenta, stalkless flowers—like little shaving brushes—on leafy spikes of 60–90 cms (2–3 ft). *L. spicata* is known as the Gayfeather, *L. pycnostachya* as the Kansas Feather and *L. scariosa* as the Button Snakeroot. The root of the latter was once prized by Indians as a stimulant for their horses. The globular tubers were dried and pulverized and then blown into the horse's nostrils when a long hard race lay ahead.

Stokesia laevis or Stoke's Aster is a monotypic genus from the SE United States. It resembles a China Aster (*Callistephus*) with large mauve flowers set off by a 'collar' of green leaves. Growing 30–45 cms (1–1½ ft) tall it has several good forms, notably

Solidago 'Golden Wings'

Echinops ritro

Achillea filipendulina 'Gold Plate'

Catananche coerulea

Cynara scolymus

'Blue Star', *alba*, white, and *rosea*, pink. Easily propagated by seed or root cuttings it is perennial and thrives in well-drained soil.

Several Composites are food crops, for example Lettuce (*Lactuca sativa*), Endive (*Cichorium endivia*), Chicory (*C. intybus*), Black Salsify (*Scorzonera hispanica*), several Artichokes and Dandelion (*Taraxacum officinale*). Some make attractive border plants, particularly the Globe Artichoke, *Cynara scolymus* and Cardoon, *C. cardunculus*. These have large, handsome, deeply cut, silvery leaves and large blue or purple Thistle-like flower heads on stems 90–180 cms (3–6 ft) tall. In parts of Spain an extract of dried Cardoon flowers is used to curdle milk for cheese-making.

Chicory occurs wild in Europe, growing 60–120 cms (2–4 ft) tall with wand-like leafy branches carrying many-petalled azure-blue flowers 4 cms (1½ ins) across. White and pink forms exist but unfortunately all the blooms close in dull weather. The roots of this plant have long been used as an adulterant for coffee, and the practice at one time reached such extremes in England that a law was passed requiring vendors to print the amount of chicory blended with their coffee on every packet. Indignant housewives (to whom this was news) accordingly rejected ground coffee and went back to purchasing the beans. One firm then roasted its chicory to coffee-brown, pulverized the roots and made them into moulds which looked like coffee beans—and so the merry deception went on.

Many species of Thistle come in Compositae, most of which do not make satisfactory garden plants. Exceptions include the Scotch or Cotton Thistle, *Onopordum acanthium,* a beautiful biennial growing up to 1.5 m (5 ft), with striking, sculptured leaves, covered with white cobwebby hairs and with silvery winged stems. Young plants are almost entirely white. The flowers are rich purple and constitute the badge of the Stuarts and are the national emblem of Scotland. This honour was accorded it following an invasion of the Danes. In the early days of Scottish history armies only marched during the day; but the Danes thinking to catch the enemy unawares, moved at night and so that they should not be heard went barefoot. One warrior, stepping on a particularly fine specimen of Thistle cried out in pain, thus arousing the Scots who drove them off with great slaughter.

Another Thistle, which is sometimes used as a border plant or grown with dark-leafed shrubs is the Blessed Mary Thistle, *Silybum marianum*. Growing 90–120 cms (3–4 ft) high, this beautiful plant has deep glossy green leaves with milk-white veins and blotches and also has yellowish spines. The stalks may be eaten.

Many annuals belong to Compositae including *Callistephus chinensis,* the China Aster, a plant showing a tremendous colour range from white and soft yellow to pink, deep red, crimson and various shades of blue and mauve. Easily raised from seed, there are single and double types, with large flowers in such strains as Ostrich Plume and Chrysanthemum-Flowered or small in the Pompons. All the China Asters require rich moist soil in sun or light shade.

Zinnias, sometimes called Youth and Age, are Mexican plants with stiff-stemmed, showy flowers which are splendid for cutting. Most of the present-day strains—Burpee Hybrids, Giant Double Dahlia and the like—have been derived from *Zinnia elegans* and have large double flowers in shades of rose, pink, lilac, red, crimson, scarlet, orange, yellow and white. A cultivar called 'Green Envy' is an unusual shade of chartreuse-green. Most grow 45–75 cms (1½–2½ ft) tall and have rough, oval or heart-shaped leaves, but dwarf strains exist with small double flowers on stems of 23–30 cms (9–12 ins), like Double Lilliput and Mexican; also a pygmy race only 15 cms (6 ins) tall called Thumbeline.

Early-flowering varieties of the Mexican species *Cosmos bipinnatus* sown in spring will provide cut flowers all through the summer. The flowers are rose-pink, crimson or white in colour on stems 75–90 cms (2½–3 ft) tall, and they bear some resemblance to *Coreopsis*.

Gaillardia is the Blanket Flower, a useful North American plant with single and double flowers in autumnal colours. These

Onopordum acanthium

Helichrysum bracteatum

cut well and have a long flowering period and also withstand hot dry weather conditions better than most plants. *G. pulchella,* an annual, and *G. aristata*, a perennial, are the parents of most of the garden hybrids.

Ageratums are annuals and perennials from Central and tropical America, grown for their long-flowering qualities. In temperate countries they are half-hardy and much esteemed for border edging or as summer bedding; they also make good pot plants for winter flowering under glass. Commonly called Floss Flower, they have bunches of small, pleasantly scented, mauve or blue flowers; their value depends on a compact and dwarf habit—10–15 cms (4–6 ins)—and free-flowering propensities. Although white and pink forms exist it is the blue shades which are most popular. These are mostly forms of *Ageratum houstonianum* (*A. mexicanum*). Varieties such as 'Blue Mink' and 'Blue Chip' are easily raised from seed and come true to type.

Calendulas are known as Marigolds, or sometimes more simply as Golds. *Calendula officinalis* is also called Pot Marigold, presumably because the plant has slight culinary properties. At one time the dried petals were in demand as a cheap substitute for Saffron and according to Gerard were used in stews and broths and to impart a rich flavour to cheese. Fresh petals can be eaten in salads or cooked in Marigold Pudding, made from chopped petals, sugar, lemon juice, peel, breadcrumbs and cream.

Although native to S Europe, Pot Marigolds are fairly hardy and make so much seed that after frost cuts the foliage, continuity is assured. In milder climates blooms persist all the year round. The flowers come in shades of orange or yellow and are single or double in the cultivars.

In spite of its name the African Marigold, *Tagetes erecta,* is native to Mexico but apparently received its misleading appellation because, although brought to Spain early in the 16th century, it was rediscovered in 1535 naturalized (from earlier introductions) in North Africa. It is a vigorous annual, 60–90 cms (2–3 ft) in height, with thick green, branching stems, dark green, evil-smelling leaves and large showy heads of citron-yellow flowers. Modern cultivars have huge double blossoms as large as a man's clenched fist, in various shades from pale lemon to deep orange.

T. patula, the French Marigold, is another misnamed annual which is also native to Mexico. It is a more compact plant growing 15–30 cms (6–12 ins) high and has deeply cut, dark green leaves and small red or yellow flowers. There are numerous varieties; some tall, others dome-like or compact, with double or semi-double flowers in yellow or orange, sometimes striped or suffused with red.

The plants like full sun and good soil and drainage. Another species, *T. minuta,* contains an essential oil which is reputed to repel flies and vermin and is sometimes planted between food crops to ward off pests.

The unpleasant smell and other properties possessed by Tagetes is noted by many authors, including Gerard. In his *Historie of Plants* he quotes Dodonaeus (circa 1552) who '. . . did see a boy whose lippes and mouth when hee began to chew the floures of (*T. erecta*) did swell extreamely . . . likewise . . . we gave to a cat the floures . . . tempered with fresh cheese, she forthwith mightily swelled and a little while after died; also mice that have eaten of the seed thereof have been found dead.'

Gazanias are low-growing South African perennials with long narrow leaves with silver undersides. The flowers are unusually large and brilliant in shades of white, yellow, orange, pink, red, ruby, light and dark brown, frequently with rings of other shades such as green, white or pink. No bedding plant is more beautiful on a sunny day with hundreds of flowers out together; nothing more disappointing in a wet summer for then the blooms refuse to open. They like well-drained soil and full sunshine and in regions which experience much frost should be lifted and over-wintered under glass. Gazanias are sometimes called Treasure Flowers.

Several species of Compositae are grown for the 'everlasting' texture of the flowers. These when they have been dried retain their colour and thus provide suitable material for arrangements in winter. The most important are *Helipterum* (the double varieties of which are often sold under the name *Acroclinium* or *Rhodanthe* and mostly come from South Africa and Australia) and *Helichrysum* or Straw Flower, whose range extends from Europe, Asia and Africa to Australia.

The most effective of these is *Helichrysum bracteatum,* an Australian perennial treated as annual in gardens. It is one of several plants sometimes referred to as Immortelle.

Some of the lesser-known annuals are worth growing for their large and showy flowers. *Arctotis* are brilliant South Africans with colourful blooms up to 9 cms ($3\frac{1}{2}$ ins) in diameter, which have glistening metallic centres. A warm sunny position is essential, for the flowers will only open in full sunlight. Normally this is in early spring but in temperate countries the seed is sown in spring to produce summer flowers. The nomenclature of *Arctotis* (and *Venidium*), however, is much confused and probably many plants grown as *A. acaulis*, *A. grandis* and *A. stoechadifolia* are not the true species but *A. venusta*. Nevertheless the most usual name in catalogues is *A. acaulis* which provides an array of golden-yellow flowers backed with wine-purple and dark blackish centres. This has produced numerous colour forms in shades of

Gazania hybrids

'Freddie' 'Sunbeam' 'Renée'

crimson, brick-burgundy, pink, purple, mauve, buff and cream. The leaves have white woolly undersides.

A. stoechadifolia var. *grandis* (*A. grandis*) is known in South Africa as the Trailing Arctotis because its thick evergreen stems form great mats. In less-favoured climates it is treated as an annual, the original species being white with lavender undersides and a brilliant steel-blue central disc. A miscellany of hybrids in other shades have come with the passing of time, many of which are grouped in seedsmen's catalogues, under the name *A. hybrida.* These range through cream, yellow and orange to red, crimson and purple. Even the heights vary—from dwarfs a mere 22 cms (9 ins) to some at 120 cms (4 ft).

Venidium fastuosum (which is also commonly known as *Arctotis aurea*) is another South African annual; the flowers are brilliant orange with deep purple-black shining discs. Both leaves and flower buds are densely covered with white cobwebby hairs which provide a striking background to the blooms. The flowers appear to have two rows of 'petals', due to the placement of the disc florets. If cows graze on the plant their milk is said to become bitter.

A bi-generic cross between *Venidium fastuosum, A. speciosa* and *Arctotis stoechadifolia* var. *grandis* was made in 1950 by Suttons of Reading, England. The resultant hybrids, called *Venidio-arctotis,* can only be reproduced from cuttings as they do not set seed. Colours range from white, pale yellow and chestnut to salmon, deep pink, wine, mahogany and rosy-mauve. They flower all summer if the sun shines and make good bedding plants.

Dimorphothecas are other South Africans which are variously known as African Daisies, Cape Marigolds and Namaqua Daisies. In light sandy soil and a sunny situation they flower for several months, with hundreds of smooth, shining daisy flowers in orange, white, buff, yellow or salmon. In warm climates many are perennial but temperate gardens can only use them under glass or treat them as half-hardy annuals.

Dimorphotheca pluvialis is the Rain Daisy, a charming annual 22–30 cms (9–12 ins) tall, with golden-centred, glistening white flowers which are mauve-backed. They have a deep violet ring round the yellow centre. A closely related species *D. sinuata*, with an orange centre to the yellow or orange flowers 7 cms (3 ins) across, grows 30–35 cms (12–14 ins) high.

The related Osteospermums are annuals or shrubby perennials, mostly with pink, white, orange or golden flowers. Some are evergreen like the popular blue and white flowered species *Osteospermum ecklonis*.

Cotulas, which resemble Daisies with the ray florets removed, are showy plants for a moist situation. In South Africa they are known as Button Flowers or Duck's Eyes, the bright yellow flowers being freely produced for months on end. *Cotula coronopifolia* is a bog plant often grown outdoors in England. *C. barbata* associates very pleasingly with blue-flowered annuals such as the cobalt-blue Kingfisher Daisy, *Felicia bergerana,* another South African Composite.

The homely Daisy, *Bellis perennis,* is a common plant of the English countryside and one of the worst lawn weeds. It grows to a height of around 7 cms (3 ins), with basal clumps of ground-hugging, spoon-shaped foliage and slender, leafless stems terminating in white, yellow-centred flowers. The species has produced both single and double varieties with pink, red, white and rose flowers. These are frequently used between spring bedding or as ground cover amongst tall Tulips.

Although perennial the plants are usually treated as annuals or biennials—a wise precaution as the seedlings may fall and germinate on nearby lawns. In the variant known as Hen and Chickens, little flower buds are formed in the axils of the bracts; sometimes as many as ten or twelve of these minute Daisies surrounding the parent flower. Gerard called this 'the childing Daisie' and claimed great virtues for Daisies which '. . . do mitigate all kind of paines, but especially in the joints, and gout, if they be stamped with new butter unsalted and applied upon the pained place. The juice of the leaves and roots . . . given to little dogs with milke, keepeth them from growing great.'

Chamomile, *Chamaemelum nobile* (*Anthemis nobilis*), is a plant with a double capacity for it makes a good drought-resistant lawn and the flowers have medicinal uses. Between the 17th and 19th centuries Chamomile lawns, paths and even turfed seats were common, for prior to that time there was considerable difficulty in obtaining good grass seed. It is pleasantly aromatic when walked on and as Spenser remarked '. . . the more it is trodden on the better it grows'. There is a Chamomile lawn at Buckingham Palace and when Drake played his famous pre-Armada game of bowls it is thought to have been on a Chamomile lawn. The flowers provide an excellent oil used in perfumery and for flavouring fine liqueurs. The double variety with small white flowers is the most attractive for garden purposes. *Chrysanthemum parthenium* which is the *Matricaria* of seedsmen has the same strong smell as Chamomile; the garden forms have round double heads of white or yellow flowers on stems of 30 cms (1 ft). These are often used for summer bedding.

Cinerarias
(*Senecio cruentus* hybrids)

Cineraria hybrid

One splendid Composite used as a greenhouse perennial in cool climates (outdoors in warm) is the South African *Gerbera jamesonii*, the Barberton Daisy. This is the best of some seventy species native to Africa, Madagascar, Asia and Indonesia and produces tufts of large, deeply toothed basal leaves. Between these come the leafless flower stems, 60 cms (2 ft) high, with flowers 7–13 cms (3–5 ins) across. In *G. jamesonii* they are red but hybrids show a wide range from white, cream, buff and gold to orange, pink, flame, rust and pink with red or yellow centres.

Barberton Daisies make splendid cut flowers and will keep on blooming if not allowed to seed. Propagation of colour forms is by division, but since the plants do not like being disturbed this should not be practised more often than necessary. They like rich but well-drained soil and full sunshine.

Another favourite greenhouse plant for cool climates is the florists' Cineraria hybrids of *Senecio cruentus,* from the Canary Isles. These make valuable winter-flowering pot plants in cool climates, producing large massed heads of white, pink, red, purple, deep and light blue daisy flowers. Some variants have rings of other shades embodied in the blooms. There are a number of strains, which come true from seed, namely: Large Flowered; Double Duplex, containing a large percentage of double flowers; Stellata, star-shaped blossoms; and Multiflora Nana, dwarf plants.

Although perennial the florists' Cinerarias are almost invariably treated as annuals because they are at their best the first year from seed. They require fairly cool conditions, germinating in a temperature of 10–13°C (50–55°F) and grown on in 7–10°C (45–50°F). The plants need plenty of room as the leaves easily damage—which spoils their appearance and also affects flower production.

Other species of the large genus of *Senecio* worth cultivating are the shrubby, grey-felted foliage evergreen *S. laxifolius* from New Zealand, which makes a good seaside plant and has rich golden daisies, and several bog plants. The best of these are *S. tanguticus,* a Chinese perennial with beautifully deep-cut leaves and stems 180 cms (6 ft) tall terminating in panicles of small golden flowers, and *S. smithii* from the Falklands with white, yellow-centred flowers and large oval to heart-shaped foliage.

Olearias are evergreen Australian and New Zealand trees and shrubs, all of which do well near the sea. They are often known as Daisy Bushes. The hardiest are *Olearia macrodonta,* up to 6 m (20 ft) and white flowered; *O. haastii* (probably a hybrid) a bush 1.25–3.25 m (4–8 ft) high with clusters of white fragrant flowers; and *O. phlogopappa* (*O. gunniana*), 1.5 m (5 ft), white-flowered in the type, but with pink, mauve, blue and purple forms.

Mutisias are South American climbers with simple or pinnate leaves which often terminate in tendrils. These help the plants to climb over trees or buildings—or up greenhouse walls in colder climates. The large terminal flowers are very showy, those of *Mutisia clematis* being bright orange-scarlet. *M. linearifolia* is scarlet, *M. ilicifolia* pale mauve-pink, *M. decurrens* bright orange and *M. oligodon* satiny-pink.

Other annual Composites include *Tithonia rotundifolia* (*T. speciosa*), the Mexican Sunflower, 1.5–1.75 m (5–6 ft), which looks like an orange-red single Dahlia; *Ursinia anethoides* from South Africa with vivid orange flowers on stems 30 cms (1 ft) tall; and *Layia elegans* from California which is often called Tidy Tips, a plant 30 cms (1 ft) tall with yellow Daisy flowers tipped with white.

Celmisia species are Australian and New Zealand perennials with long, narrow radical leaves and large solitary flower heads on stems 15–90 cms (6–36 ins) tall, usually white with yellow centres. All the leaves, stems and flower bracts are densely covered with white or buff hairs. They require sun and rich but well-drained soil but resent winter wet, so in cool climates should be grown in pots and wintered under glass.

Mutisia decurrens

Convolvulaceae

55 genera and 1650 species; tropical and temperate

This is a family of dicotyledonous plants, shrubs and trees which includes many tropical and subtropical climbers.

The rootstocks may be tuberous or rhizomatous; the leaves are alternate and usually stalked and the flowers regular and usually bisexual. Most parts of the blooms are in fives, the ovary is superior and the petals variously shaped—sometimes joined and funnel-shaped.

The genus *Convolvulus* contains some 250 species, the majority too weedy for the garden. One exception is *C. cneorum,* a small Mediterranean shrub, 30–90 cms (1–3 ft) tall, with white flowers and beautiful little silver leaves covered with plush-like silky hairs. It needs sun and good drainage or, in cold climates, makes a pretty pot plant.

An attractive annual for summer bedding is *C. tricolor* from Spain and Portugal. This grows about 30 cms (1 ft) high with many deep blue flowers which have yellow throats and white tubes. In the 18th century the species was known as the Life of Man, because according to the Reverend William Hanbury (*A Complete Book of Planting and Gardening,* 1770) 'it has flower buds in the morning, which will be full blown by noon, and withered up before night'. Some plants grown as *C. tricolor* are in fact the related *C. gharbensis.*

Convolvulus cneorum

Convolvulus tricolor
'Blue Ensign'

Among the perennials *C. mauritanicus* from North Africa makes a trailing basket plant for the greenhouse or may be used as a scrambling rock plant in sheltered gardens. It has blue flowers, 2.5 cms (1 in.) wide, with white throats. *C. althaeoides,* a trailer from S Europe, carries quantities of pale red or silvery-pink flowers and possesses two kinds of foliage; the basal leaves are heart-shaped with scalloped edges and those on the flower stems deeply cut—as in some Aconitums.

A purgative medicine called Scammony is obtained from the resinous root exudation of *C. scammonia,* a perennial vine native to the Mediterranean region. The Lesser Bindweed, *C. arvensis* from Europe and Asia, has pretty little fragrant pink flowers. It has sometimes been used in hanging baskets but only a foolhardy gardener would let it loose in the garden for this is one of our most pernicious weeds. The roots have been known to penetrate 180 cms (6 ft) into the ground, so that country gardeners call it Devil's Guts. Once introduced it is extremely difficult to eradicate.

The hitherto separate genera of *Batatas, Calonyction, Pharbitis* and *Quamoclit* are now usually included under *Ipomoea* which thus becomes a large group of some 500 species, the majority evergreen or deciduous climbing plants or trailers.

One of the most popular is the annual Morning Glory, *I. purpurea* (*Pharbitis purpurea*) from tropical America. This twists its way upwards for approximately 3 m (10 ft), the slender stem carrying smooth, heart-shaped leaves and dark purple funnel-shaped blooms in several-flowered peduncles. Another from the same area, *I. tricolor* (*Pharbitis rubro-coerulea*), is even better, the blooms white in bud with reddish lines but opening a rich china-blue or purple. Garden varieties exist with azure-blue flowers ('Heavenly Blue'), wine-red flowers ('Scarlett O'Hara') and white flowers ('Pearly Gates'). All these Morning Glories make fine greenhouse climbers or in the tropics can be used to drape hedges, mask buildings or cover fences. They should be planted where they may be seen in the morning as the blooms close at midday.

I. learii (*Pharbitis learii*), another tropical American, is called the Blue Dawn Flower because the ephemeral flowers open so early, and have intensely blue flowers like a tropical sky. These turn magenta as the blooms age and are produced in lush profusion all summer. The species has been introduced to the Old World tropics and become particularly widespread in India, where it is commonly known as Railway Creeper or Porter's Joy. The evergreen, small, deeply cleft leaves are divided into seven lobes of varying sizes. The plant is easily propagated and will clamber over any obstacle, producing—in such warm climatic conditions—flowers every morning of the year. *I. learii* is now considered to be a form of the variable *I. acuminata*.

I. alba (*I. bona-nox* or *Calonyction aculeatum*) is the Moonflower or Moonvine or the tropics. It is a strong twining species, often prickly, with milky sap and thin, lobed, heart-shaped leaves. The flowers are pure white and fragrant with greenish ridges; they have a long narrow tube which flares out to a wide

Ipomoea
'Heavenly Blue'

Convolvulus arvensis

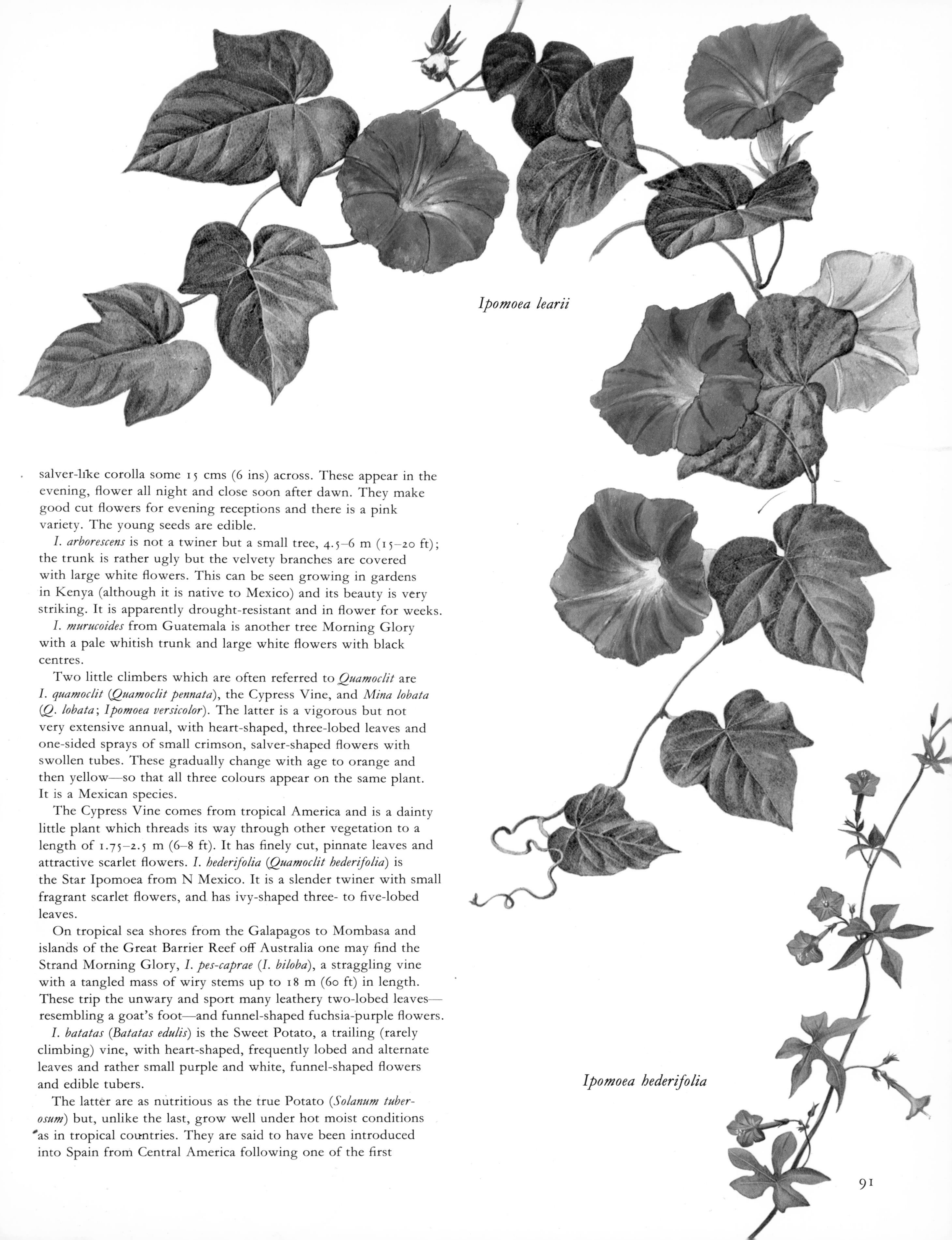

Ipomoea learii

salver-like corolla some 15 cms (6 ins) across. These appear in the evening, flower all night and close soon after dawn. They make good cut flowers for evening receptions and there is a pink variety. The young seeds are edible.

I. arborescens is not a twiner but a small tree, 4.5–6 m (15–20 ft); the trunk is rather ugly but the velvety branches are covered with large white flowers. This can be seen growing in gardens in Kenya (although it is native to Mexico) and its beauty is very striking. It is apparently drought-resistant and in flower for weeks.

I. murucoides from Guatemala is another tree Morning Glory with a pale whitish trunk and large white flowers with black centres.

Two little climbers which are often referred to *Quamoclit* are *I. quamoclit* (*Quamoclit pennata*), the Cypress Vine, and *Mina lobata* (*Q. lobata*; *Ipomoea versicolor*). The latter is a vigorous but not very extensive annual, with heart-shaped, three-lobed leaves and one-sided sprays of small crimson, salver-shaped flowers with swollen tubes. These gradually change with age to orange and then yellow—so that all three colours appear on the same plant. It is a Mexican species.

The Cypress Vine comes from tropical America and is a dainty little plant which threads its way through other vegetation to a length of 1.75–2.5 m (6–8 ft). It has finely cut, pinnate leaves and attractive scarlet flowers. *I. hederifolia* (*Quamoclit hederifolia*) is the Star Ipomoea from N Mexico. It is a slender twiner with small fragrant scarlet flowers, and has ivy-shaped three- to five-lobed leaves.

On tropical sea shores from the Galapagos to Mombasa and islands of the Great Barrier Reef off Australia one may find the Strand Morning Glory, *I. pes-caprae* (*I. biloba*), a straggling vine with a tangled mass of wiry stems up to 18 m (60 ft) in length. These trip the unwary and sport many leathery two-lobed leaves—resembling a goat's foot—and funnel-shaped fuchsia-purple flowers.

I. batatas (*Batatas edulis*) is the Sweet Potato, a trailing (rarely climbing) vine, with heart-shaped, frequently lobed and alternate leaves and rather small purple and white, funnel-shaped flowers and edible tubers.

The latter are as nutritious as the true Potato (*Solanum tuberosum*) but, unlike the last, grow well under hot moist conditions as in tropical countries. They are said to have been introduced into Spain from Central America following one of the first

Ipomoea hederifolia

Spanish voyages to the New World, although some people believe them to be of African origin. Whatever the facts it is certain they spread quickly and unobtrusively through S Europe to the Middle and Far East. In the Pacific Islands, China, India and the East Indies they are today an important item of food, both for stock and humans. The Japanese ferment alcohol from the tubers and also make from them a dried, pulverized meal.

A. W. Anderson (*The Coming of the Flowers*) tells a fascinating story of their introduction to New Zealand. Apparently Sweet Potatoes were taken there from the warmer islands of the Pacific about the end of the 14th century by a man called Taukata. Taukata explained to the Maoris the need to protect the stock tubers against cold in winter. This they understood but felt that only a powerful guardian would ensure their safety. Accordingly, since Taukata seemed a natural for this position he was slain, his blood sprinkled over the door frame of the storehouse and his skull preserved. Each year at planting time, the skull dressed with feathers and small potatoes was taken to the fields and left to guard the crop, and after harvest returned to the storehouse to continue the vigil throughout the winter.

Several other Ipomoeas have slight economic importance. *I. purga* is the Jalap, a purgative derived from the dried tuberous roots. Ashes from *I. murucoides* are used in Guatemala instead of soap for washing; the root of *I. digitata* makes a root tonic in India and the leaves and shoots of *I. aquatica* (*I. reptans*) are cooked and eaten as 'Chinese Cabbage' in tropical Asia.

Jacquemontia, a genus of about 120 species, mostly warm American, is represented in tropical gardens by *J. martii,* a perennial climber from Brazil. This has very slender stems but will climb to 6 m (20 ft) under good conditions. It has small heart-shaped leaves and clusters of light blue, long-stalked, bell-shaped flowers like miniature Morning Glories and can be propagated by seed or cuttings. Some of the African Jacquemontias are used as vegetables.

Argyreia species mostly come from India and Malaya. They are handsome, large-leafed, silvery climbers with large and showy funnel-shaped flowers of violet, red or white. They are suitable for running up trellis or covering buildings in tropical gardens or may be used as greenhouse climbers in colder climates. *A. splendens* has light red flowers, *A. cuneata* deep bright purple, *A. speciosa* deep rose and *A. wallichii* white tinged with pink. The leaves of *A. speciosa* are used by Malayans for poulticing.

Falkia repens is a small South African evergreen creeper with little red flowers which have a paler throat and oval to heart-shaped leaves. It makes a pretty greenhouse plant and needs light soil. Propagation is best undertaken by means of cuttings.

Evolvulus purpureo-coeruleus is widespread in tropical America and is a perennial, 45 cms ($1\frac{1}{2}$ ft) tall, with small, entire, lanceolate leaves. The flowers are borne at the ends of long branches and are ultramarine-blue with white centres and have a purple line running up the middle of each lobe.

Hewittia sublobata (*H. bicolor*), a species from tropical Africa, has entire, often lobed, broadly cordate leaves and white or pale yellow, bell-shaped flowers with dark purple centres.

Porana racemosa, the Snow Creeper from Burma and India, has many small funnel-shaped white flowers in a loose leafy panicle. It is an annual with cordate, slender, pointed leaves which are often downy and 7–10 cms (3–4 ins) in length. It makes a pretty pot plant twined round sticks or can be trained up rafters and similar supports.

Rivea hypocrateriformis, the Midnapore Creeper from W India, is also white but a nocturnal species, the large fragrant flowers smelling of cloves and flaring wide to a flat salver shape at the top of the corolla tube. The leaves are rounded heart-shaped and sometimes very hairy on the undersides.

Cornaceae

12 genera and 100 species; N and S temperate regions and also the tropics

Most representatives of this family are trees and shrubs, although there are also a few herbaceous plants. They are dicotyledonous with simple opposite or alternate leaves, usually entire, and inflorescences in heads, umbels or corymbs. The flowers have four, five or no sepals, four, five or no petals, four to five stamens and an inferior ovary developing into a drupe or berry.

Cornus is the most important genus, although certain of its species are now referred (by some authorities) to other genera such as *Chamaepericlymenum, Cynoxylon, Thelycrania* and *Dendrobenthamia.*

Cornus alba, a wide-spreading Siberian shrub with brilliant red shoots in winter (particularly in the clone 'Sibirica' or 'Westonbirt'), is one of the most popular for garden work, especially in its cultivars 'Sibirica Variegata' (which has silver variegated foliage), 'Spaethii' (golden variegated) and 'Kesselringii' (with dark purple stems). The small white flowers are insignificant.

C. sanguinea, the Cornel or Dogwood, grows wild in England on chalky soils. It is a much-branched shrub, 1.25–2.5 m (4–8 ft) tall, with grey bark and clusters of small creamy-white four-petalled flowers which are succeeded by black berries. The young shoots are bright red in autumn and winter, a characteristic exploited by gardeners who cut the shrubs almost to ground level

Cornus kousa var. *chinensis*

Aucuba japonica 'Crotonoides'

Cornus mas

in spring for the sake of the winter colour of the new growths.

The wood is extremely hard and when bows and arrows were the national means of defence, arrows were made from the timber. It has also been used for ramrods of fowling pieces (especially in France) and for skewers and pipes. According to Mrs Lankester (*Sowerby's English Botany*), the best charcoal for gunpowder is obtained from the wood of *C. sanguinea*. Such names as Dogberry Tree, Dogwood and Hound Tree recall a one-time use of the bark for washing mangy dogs, although Parkinson says that the name Dogberry refers to the bitter-tasting berries 'which are not fit to be given to a dogge'. These fruits however contain good-quality oil which has been employed in soap making.

The Dogwood of America, *C. florida* (*Cynoxylon floridum*) is the State Flower of Virginia, a beautiful small tree 3–6 m (10–20 ft) tall which bears large, white, purple-tinged bracted flowers in spring. The bracts are the distinctive feature, the flowers they enclose being small and yellow. Probably nowhere else do they bloom to such perfection as in the eastern United States where, with Redbuds (*Cercis canadensis*) and Maples, they make an unforgettable picture each spring. Pink forms (var. *rubra*) are often grown in gardens. The leaves of *C. florida* assume brilliant red tints in autumn.

Another fine tree is *C. nuttallii* (*Cynoxylon nuttallii*), the Pacific Dogwood. This grows 4.5–6 m (15–20 ft) tall or more with four to eight (usually six) large, rounded, white petal-like bracts surrounding a button-like inflorescence of small flowers.

The wood of both species is hard, heavy and close-grained and used for such things as tool handles, cabinet work and turnery. The bark of *C. nuttallii* makes a quinine substitute and was much prized by the Confederate army during the Civil War when true quinine was unobtainable because of the blockades.

Their Asiatic counterpart *C. kousa* (*Dendrobenthamia japonica*) from Japan is very similar except that the bracts are pointed instead of rounded. The best garden forms are the large bracted sorts known as var. *chinensis*. This species has beautiful round Strawberry-like fruits and is hardy in Europe and widely grown in gardens.

C. mas is the Cornelian Cherry, a small tree of spreading habit with dense clusters of small yellow flowers in spring, before the leaves. These are succeeded by large, bright red, edible but acid fruits which may be eaten fresh or in preserves and are also the source of an alcoholic beverage, known in France as Vin de Cornoulle. The tree is a native of southern Europe and has tough wood which is said to have been employed for the making of the Trojan Horse. There are several foliage forms of Cornelian Cherry, with yellow and variegated leaves, and also a yellow fruited variety.

Chamaepericlymenum canadense (or *Cornus canadensis*) is a small woodlander with creeping rootstocks from which come leafy stems of 10 cms (4 ins) with about six leaves crowded in a whorl. The stems terminate in tiny umbels of greenish-yellow flowers, encircled by four white bracts. It is a pleasant little plant for spring blooming, known in Canada as Squaw Berry because the small round scarlet berries which follow the flowers were eaten by Indians. The plant is almost circumpolar in distribution and is represented in North Cape, Norway, as well as the borders of the Arctic Zone in North America.

C. suecicum (or *Cornus suecica*) is very similar with purplish flowers set off by four large white bracts and has much the same geographical distribution. The fruits are bluish-black.

Aucubas are Asiatic shrubs noteworthy for their tolerance of shade and poor soil conditions, which are often grown in town gardens as hedging plants or to mask unsightly buildings.

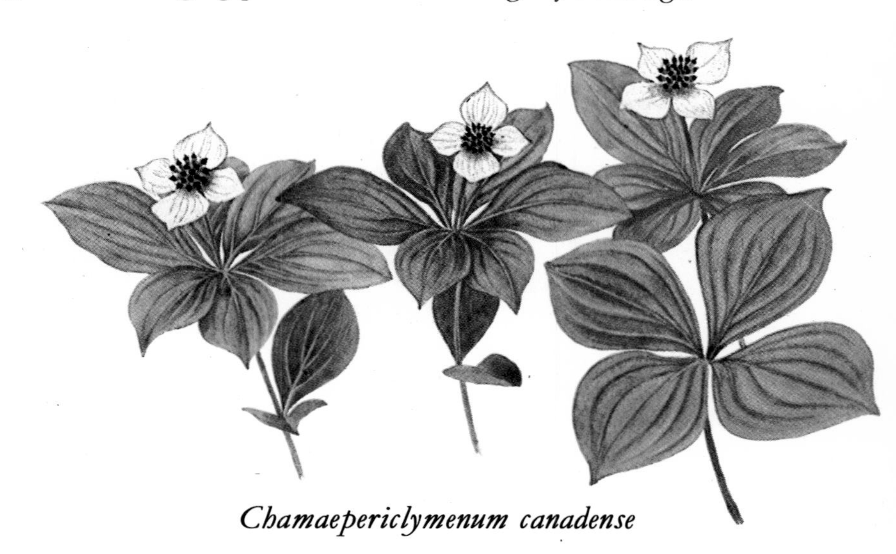

Chamaepericlymenum canadense

They are frequently (but erroneously) called Laurel, a name which rightly belongs to the Bay (*Laurus nobilis*).

The most commonly cultivated species is *Aucuba japonica,* an evergreen with large, shiny, deep green, simple leaves. Garden forms with toothed margins or cream variegated leaves exist, but it is necessary to grow both male and female plants in order to obtain the showy scarlet berries (sometimes yellow), which are the plants' chief charm. These persist for months.

A. japonica was introduced to Britain from Japan in 1783, but only a female plant. Naturally this never produced fruit, until Robert Fortune in 1861 procured a male plant in Yokohama. When berried specimens first appeared they were very expensive and the few males available were hired out for cross-fertilization. The spotted-leafed form is often used today as a house plant and artificially pollinated to obtain plenty of berries.

Griselinia littoralis is a New Zealand shrub with leathery, oval-oblong leaves which have a silvery lustre. They are evergreen and do particularly well near the sea. *G. lucida,* commonly called Shining Broadleaf, makes a small tree with glossy green, simple leaves which have a varnished appearance. Male and female flowers are on separate trees, the former yellow and attractive, the latter green. The fruits are fleshy and purple. This species is often epiphytic with descending roots resembling huge lianas.

Costaceae

4 genera and 200 species

These tropical monocotyledonous plants are closely related to Zingiberaceae but lacking the aromatic foliage of that family.

The largest genus is *Costus* with some 150 species, whose flowers possess striking enlarged lips and projecting petal-like stamens. One of the showiest is the Brazilian *C. igneus* with orange and yellow flowers on leafy stems. This foliage is oblong-lanceolate and spirally arranged, so that the blooms peep out above a mass of herbage. Warm conditions are required for cultivation with rich soil, good drainage and protection from direct sunlight. *C. spectabilis* from West and East Africa forms a rosette of ground-hugging, fleshy leaves in the centre of which sit the stemless yellow or orange flowers.

Tapeinocheilos (*Costus*) *ananassae* is a curious tropical plant from the Moluccas with scattered leaves and an ovoid cone-shaped inflorescence—something like a Pineapple without its crown. This has many deep scarlet to crimson, recurving bracts enclosing sunken, yellow flowers. A well-grown plant is most impressive and can reach 180 cms (6 ft) in height.

Costus igneus

Tapeinocheilos ananassae

Crassulaceae

35 genera and 1500 species

This is a large family of dicotyledonous and often xerophytic plants mostly from South Africa but also Mexican. Tufted growth, fleshy stems and leaves, waxy surfaces and tightly packed flowers are common characteristics. The blooms are usually arranged in cymes and are mostly bisexual, with a superior ovary and anything from three to thirty sepals, petals and stamens.

Members of the genus *Bryophyllum* form adventitious buds on the leaves which drop off to produce new plants; other genera develop bulbils or may be propagated vegetatively by division of the rhizomes or removal of offsets and detached leaves.

Crassula is a large genus of some 300 succulent shrubs and herbs, mostly South African. The leaves are always opposite, sometimes packed into basal rosettes, but variable in form and texture. Usually the flowers have their parts in fives, the number of the stamens equalling that of the petals. They are small and white, yellow or red in colour. Propagation is by means of seed or stem or leaf cuttings.

Among the most important horticultural species is *C. lactea,* a shrubby plant with thick stems and dark green leaves with white dots and racemes of starry white flowers. *C.* (*Rochea*) *falcata* is a useful decorative plant called Red Treasure in South Africa, with leafy stems 60 cms (2 ft) tall terminating in inflorescences of brilliant red flowers. These last three to four weeks in bloom. The foliage is very fleshy and stands edgewise on the stems—looking like sickles. It makes a delightful pot plant.

C. sarcocaulis is hardy enough to plant outside in sheltered spots in temperate climates. It is a small shrubby species with pink flowers. *C. acinaciformis,* one of the most striking species, has rosettes of fleshy leaves up to 30 cms (1 ft) in diameter and flower spikes of 90–120 cms (3–4 ft) terminating in dense inflorescences of creamy-white flowers. All the species mentioned are South African.

Cotyledon species are also mostly South African, needing sun, well-drained soil and a temperature which does not fall below 5–10°C (40–50°F). The thick succulent leaves are frequently pale green with a mealy covering. In poor soil and dry conditions these often colour up red, although prolonged shortage of water causes them to drop.

Among the most ornamental are *C. undulata,* which has wavy-edged, mealy leaves and cream and red flowers; *C. paniculata* with bell-shaped flowers 2.5 cms (1 in.) across bordered with

Kalanchoe 'Brilliant Star'

Echeveria gibbiflora

Echeveria glauca

green on stems of 120–150 cms (4–5 ft); also *C. orbiculata*, the Pig's Ear, with tubular, orange-red flowers and oval, waxy-bloomed leaves which are outlined in red.

The genus *Aeonium* includes a number of popular greenhouse plants for cool or temperate countries. From the Canaries comes *A. tabuliforme* with flat rosettes of pale green leaves and flower stems of 30–60 cms (1–2 ft) with starry inflorescences of pale yellow flowers. *A. undulatum,* also from the Canaries, is bright yellow and 90–120 cms (3–4 ft) in height and *A. arboreum* from Morocco, another with bright golden flowers, makes an attractive small tree with dense rosettes of leaves at the ends of the branches.

Echeverias are other rosette-forming succulents, often with waxy or glaucous leaves and urn-shaped flowers of orange, red, yellow or white on long stems. Two attractive species are the scarlet-flowered *Echeveria gibbiflora* with grey-green, fleshy leaves flushed with red or purple and *E. setosa,* the Mexican Firecracker, which has green foliage closely covered with white bristles and red flowers margined with yellow.

E. atropurpurea has its broad leaves crowded in rosettes at the tops of the stems. These are not very thick but an attractive dark purple colour with a glaucous bloom. The flowers are red. *E. glauca* has a compact habit and glaucous blue foliage which is red margined. The flowers are bright red with yellow tips. All the species mentioned are Mexican and need sun and light soil, with protection from wet and hard frost during the winter.

Kalanchoe blossfeldiana from Madagascar has vivid scarlet flowers in large clusters at the top of wiry stems 30 cms (1 ft) tall. It is easily raised from seed and has succulent leaves with notched edges. *K. flammea* from Somaliland, also scarlet, and *K. teretifolia,* a South African species with white flowers and grey-green leaves, are other good ornamentals for the greenhouse or warm tropical

Sempervivum tectorum

gardens. *K. tomentosa* from Madagascar has a plush-like quality and curious markings which have earned it the name Panda Plant. The oval leaves are densely covered with silvery hair, which at the tips are reddish at first but develop to deep brown. Although possible, it seldom flowers under cultivation.

Bryophyllums are more bizarre than beautiful on account of the little plantlets which develop along the edges of the leaves. These fall or may be detached to form new plants. The flowers are often quite showy, being borne in considerable numbers at the tops of the stems. They are pendent, with a tubular (often inflated) calyx. Many botanists now include *Bryophyllum* under *Kalanchoe* and the species are sold under either generic name. Mostly commonly grown are *B. daigremontianum,* with yellow and violet-pink flowers on stems 60 cms (2 ft) tall, and *B. tubiflorum* which has red blooms and snake-like, purple and pink mottled stems and leaves. The latter are spike-like with small plants at their tips. Both species come from Madagascar. *B. pinnatum* (*B. calycinum*) with fleshy pink and green flowers on leafy stems of 60–90 cms (2–3 ft) is widely cultivated in the tropics and subtropics.

The genus *Sedum* comprises some 600 species, mostly N temperate plants except for one Peruvian representative and a number of Mexicans. They have succulent leaves which help them to withstand dry periods and inflorescences made up of many small flowers. These flowering shoots die after blooming, although the leaf-bearing stems persist. Many exist under the poorest soil conditions, often clinging precariously to rocks and walls, but also clambering accommodatingly over stones in the rock garden.

The best of the border kinds are the European Ice Plant, *S. maximum* (now considered a sub-species of the variable *S. telephium*), and the Japanese *S. sieboldii* and *S. spectabile.* The last is commonly grown for its showy, plate-like heads of pink flowers on stems of 30–45 cms (1–1½ ft) in early autumn. Cultivars such as 'Carmen' and 'Brilliant' are much more vivid in colour; *album* is white and 'Meteor' glowing red. The glaucous leaves are slightly toothed. *S. sieboldii* has pink flowers and nearly round, fleshy leaves which become rosy in autumn. *S. maximum* var. *atropurpureum* (a superior plant to the type) has carrot-like roots, thick fleshy leaves which are green at first but then become deep claret, and cream and mahogany flowers. There is also a green and cream leafed form known as 'Versicolor'.

Countless dwarf species with white, yellow, pink or red flowers are available for the rock garden. *S. acre* which grows wild in Britain has tufted stems, small fleshy leaves and yellow flowers. It is commonly called Stonecrop or Wall Pepper although other names exist such as Jack of the Butterie. Pliny recommended wrapping the plant in a black cloth, secreting this (unbeknowingly) beneath the pillows of sufferers to cure insomnia.

A circumpolar species, *Rhodiola rosea* (*Sedum rhodiola* or *S. rosea*), is known as Rose-root because the fleshy broken root has a delightful perfume, which persists for several weeks.

The flowers of *Sempervivum* may be white, yellow, red or purplish, often in dense cymes, with fleshy leaves in crowded rosettes. The rosette dies after flowering but many young ones appear round the base and ensure continuity. Countless species and garden varieties exist and may be grown in well-drained soil in the rock garden, scree or alpine house.

S. tectorum is the Houseleek, so called because it was once planted on the roofs of houses as a protection against lightning. Because of its supposed efficacy against lightning Charlemagne ordered every house roof to be planted with it. It was considered unlucky to uproot Houseleeks or to allow them to flower. Another name is Welcome-home-husband-though-never-so-drunk. The plant needs very little moisture and was once used threaded into a framework of wire or wood to make living firescreens.

Cruciferae

375 genera and 3200 species; cosmopolitan distribution but mainly in N temperate regions

This is a very large dicotyledonous family containing several important economic plants and also some well-known ornamentals. The species are chiefly herbaceous, although a few are shrubby with alternate and usually simple leaves, which however show various modifications in some cultivated plants, as *Brassica*. A raceme or corymbose inflorescence is usual, individual flowers being sub-regular (rarely zygomorphic), bisexual, with four sepals, four petals, six stamens and a superior ovary. The unusual placement of the four petals—to form a cross—is one of the most striking characteristics of Cruciferae. The ovary consists of two carpels and generally takes the form of an elongated pod (siliqua) or a short rounded pod (silicula). Members of the family are rich in sulphur (detected in the smell of boiling cabbage) and other trace elements and vitamins and thus useful in deficiency diseases such as scurvy.

The most important garden ornamentals are Wallflowers (*Cheiranthus*), Stocks (*Matthiola*), *Aubrieta* and *Alyssum*.

Cheiranthus species are found indigenously from Madeira and the Canary Isles to the Himalayas. The commonest species, *C. cheiri,* is naturalized in Britain and the parent of a wide selection of garden varieties with white, different shades of yellow, purple, pink, red and golden-brown flowers. These come relatively true from seed and are delightfully scented. Although perennial it is usual to treat Wallflowers as biennials, sowing the seed after harvesting to flower the following spring. They associate particularly well with spring bulbs, Forget-me-nots and Violas.

In Elizabethan days Wallflowers were known as Gilliflowers or Yellow Flowers, in allusion to their colour. The poet Robert Herrick ascribed their origin and name to the spirit of a young fourteenth-century maiden who fell and died while attempting to escape from imprisonment and to elope with her lover.

'Up she got upon a wall,
'Tempting down to slide withall,
But the silken twist untied,
So she fell and bruised and died.

Love, in pity of the dead,
And her loving luckless speed
Turned her to this plant we call
Now the flower of the wall.'

The shorter-growing Siberian Wallflower, 30 cms (1 ft), usually known as *C.* x *allionii*, is a brilliant orange plant which flowers for weeks; it is now referred to *Erysimum* x *marshallii* although its origin is obscure and there is dispute as to the correct name. It is a hybrid coming true from seed, best treated as a biennial.

Most garden stocks are derived from *Matthiola incana* a S European biennial with a woody base, downy, lanceolate leaves and racemes, 30–60 cms (1–2 ft) long, or purple flowers.

For garden purposes the races are divided into two main groups—the Summer Stocks (often called Ten Week because they bloom ten to twelve weeks after the seed has been sown) and Winter Stocks, which are sown one year and flower the next. The last are subdivided again into Intermediate, East Lothian, Brompton, Emperor and Perpetual Stocks.

There is much variation within the groups but all are free flowering and have a delightful fragrance and sturdy habit.

Both single and double forms appear in every batch of seedlings from these groups, but the doubles are the more popular because they are more compact and showy and also last longer in flower. Since the doubles are not fertile, seed must be saved from the singles, the aim being to grow a strain which throws a high percentage of doubles. The latter can be detected in the seedling stage; dark-leafed plants produce single flowers and light green double flowers. It has also been shown that old seed produces a higher proportion of doubles than fresh seed, presumably because seed of the single-flowered plants has a shorter viability.

Stocks are excellent for cutting purposes, most kinds growing around 45 cms ($1\frac{1}{2}$ ft) tall, but they also make good border plants as well as subjects for pot work in the cool greenhouse.

Matthiola incana

Erysimum x *marshallii*

Cheiranthus cheiri

The Night Scented Stock is *M. bicornis* (now considered a subspecies of *M. longipetala*), a small somewhat straggling annual 22–30 cms (9–12 ins) high with lilac flowers. Insignificant during the day these come into their own at nightfall, when their sweet fragrance is most marked. Pinches of seed raked into the soil provides an easy method of sowing and if this is near windows and doorways the scent is appreciated at its fullest.

Malcolmia maritima is the Virginian Stock (although from Greece and Albania), a variable annual, 15–30 cms (6–12 ins) tall, with racemes of lilac, rose, red or white flowers. There is also a yellow variant. By successive sowings flowers may be had from spring to autumn. Several well-known rock plants belong to Cruciferae. Among the most important are two evergreens *Aubrieta* and *Aurinia* (better known as *Alyssum*), both spring-flowering and mat-forming, producing vivid patches of colour—the former in various shades of mauve, purple and crimson, the latter in gold and yellow. They have many uses, such as draping walls or banks, planting between crazy paving, as a ground cover between spring bulbs or planted in rock-garden pockets and as border edgings.

Aubrieta (*Aubrietia*) *deltoidea* has a wide distribution from Sicily to Asia Minor. A variable plant, it is the parent of many named hybrids, including several with variegated foliage. After flowering the clumps should be cut back to maintain their shape and vigour and prevent them smothering weaker neighbours. Seed gives variable results so that named forms are usually propagated from summer cuttings struck in sandy soil.

Aurinia saxatilis (*Alyssum saxatile*) or Gold Dust is an eastern European species, with silvery oval-oblong leaves, densely covered with downy hairs and numerous golden-yellow flowers in crowded corymbs. A double form 'Plenum' and a light yellow cultivar called 'Dudley Neville' are worth growing. The overall height is around 15–20 cms (6–8 ins) and cultivation and propagation are similar to Aubrieta.

Sweet Alyssum, *Lobularia maritima* (*Alyssum maritimum*), is a charming free-flowering annual about 7 cms (3 ins) high, which throughout the summer is smothered with clusters of honey-scented white or lilac flowers. It is represented in cultivation by a wide range of small compact varieties which are used for border edgings, carpet bedding, window-box planting and similar purposes. These have white, pink, rose or violet flowers.

Candytuft is a name given to several species of *Iberis*. *I. gibraltarica,* a sub-shrubby evergreen from Gibraltar, with rounded heads of white or pink-tinged flowers and wedge-shaped leaves is less hardy than the S European perennial *I. sempervirens* or annual *I. amara* from W Europe. Garden forms of the latter are frequently sold as *I. gibraltarica*. The plants bloom in early summer and form broad carpets of leaves and white flowers.

Aethionemas from the Mediterranean region are sub-shrubby perennials (or rarely annuals) similar to *Iberis* but with reddish or purplish flowers. They require full sun and well-drained soil for successful establishment and appreciate a little lime. One of the most popular is the hybrid 'Warley Rose', 15–20 cms (6–8 ins) high, with deep pink flowers.

Aubrieta deltoidea

Draba is a large genus of some 300 dwarf, tufted species from cold Arctic regions and the mountains of Europe, Asia and America. Their use in the garden is normally confined to the rock garden, moraine or alpine house where they form cushions of compact foliage spangled with yellow, white, lilac, violet, orange, or reddish four-petalled flowers. As with many alpine plants, snow creates no problems but many require protection from winter wet, which rots the crowns. A sheet of glass or polythene shelters them from cold rains, while a 'collar' of granite chippings round the plants draws off standing moisture.

Cardamine (now considered to include *Dentaria*) is a genus of about 160 species of annual or perennial smooth, leafy herbs with a cosmopolitan distribution. Garden representatives are usually bog or woodland plants, requiring damp soil and sometimes shade.

C. pratensis, found wild in many parts of the Northern Hemisphere (including Britain and the US), is the Cuckoo Flower, named according to Mrs Loudon (*The Ladies' Flower Garden*) 'from its blossoming when the cuckoo sings'. Another name, Lady's Smock, is derived says the same author 'from its flowers being produced in such abundance in the meadows as to give them the appearance of a bleaching ground, or of being covered with clothes from a wash, laid on the grass to dry.' The plant has pinnately divided leaves on the flower stems but rounded and angulate foliage from the base. The flowers are pale lilac to mauve on stems of 30–60 cms (1–2 ft) and there is a double variety. In damp seasons the stems frequently bear bulbils round their bases and adventitious buds on the radical leaves; these propagate the plant.

The flowers at one time were esteemed in Britain for their properties in alleviating hysteria and epilepsy, the blooms being roasted on pewter dishes over a fire and the resultant powder stored in bottles stoppered with leather—never with a cork for some reason.

C. trifolia, a species only 15 cms (6 ins) tall with white flowers, and *C. macrophylla,* pale purple and 30–45 cms (1–1½ ft) tall, are others sometimes cultivated.

The early flowering species once included in the genus *Dentaria* have scaly rhizomes and thrive best in shady, woodland soils. They have smooth, deeply cut leaves and racemes of white or pinkish flowers. *C. bulbifera,* a rare British native, has foliage with five to seven leaflets and racemes of white or lilac flowers which become rosy with age. Viable seed is rarely produced, but bulbils appearing in the leaf axils fall off when ripe to produce new plants. The English name of this species is Coral Root, referring to the fleshy creeping rootstocks which resemble strands of coral.

C. diphylla, the North American Pepper Root, has long, crisp, succulent roots tasting like Watercress and white flowers which are pinkish outside. *C. laciniata,* the North American Cut Toothwort, has whorls of leaves with deeply lobed leaflets and large white flowers in early spring.

Isatis tinctoria is a biennial with branched leafy stems 60–120 cms (2–4 ft) tall, carrying many small yellow flowers. It is not unattractive in a mixed flower border but is chiefly noteworthy as the source of Woad, an indelible juice with which the early inhabitants of Britain stained their bodies. The old Celtic name for paint was Brith and Brithon signified a stained man, so that some historians believe that Britain or Britannia was derived from this noun. It has frequently been used as a substitute for indigo.

The leaves are the important part and are richest in dye when gathered from the flowering stems. These are partially sun dried and then ground into paste and left in the air (protected from rain) to ferment. At this stage the smell is most offensive, so that in Tudor times it was an offence to dry Woad within the vicinity of the royal palaces. Eventually the material is fashioned into cakes, then moistened and again fermented, the colour finally being brought out by infusing the Woad with lime water. The plant grows wild throughout central Europe and in England.

I. glauca from Asia Minor is the more attractive for garden purposes with oblong greyish leaves and bright yellow flowers.

Hesperis matronalis from S Europe is the Sweet Rocket or Dames' Violet, a plant long esteemed for its sweet perfume. This emits a violet-like fragrance towards evening whereas, as Parkinson noted 300 years ago, this 'pretty sweet scent' is almost entirely absent during the day.

Growing 60–90 cms (2–3 ft) tall the plant has smooth, oblong toothed leaves and long spikes of four-petalled white, mauve or purple flowers in early summer. Except for their scent these bear some resemblance to the flowers of Honesty (*Lunaria annua*) and are good for cutting. Seedlings vary considerably so that good colour forms should be vegetatively increased by division. Double forms exist both in white and purple shades, but being difficult to propagate are scarce. Rockets grow most satisfactorily when left undisturbed and can often be found in English cottage gardens. Escapes from American gardens are now widely naturalized in eastern USA and Canada.

Lunaria annua (*L. biennis*) is the Honesty, a biennial with scentless violet flowers, which later give place to large, flat seed pods. When these are ripe the outer valves can be rubbed off, when the seeds drop and reveal the satiny inner septums looking like mother-of-pearl pennies. These are popular for dried winter arrangements. There is a variegated-foliaged form which comes almost true from seed but only colours up to full advantage prior to flowering.

Lunaria, a Mediterranean genus, was once credited with such supernatural powers as possessing the ability to cast the shoes of horses which walked on it. It was believed to be efficacious in the cure of madness and there is a popular superstition that wherever Honesty flourishes, those who grow it are exceptionally honest. The roots are said to be edible.

Among a number of important food crops included in Cruciferae are the various forms of Cabbage—Savoys, Brussels Sprouts, Cauliflowers, Red Pickling Cabbage, Kale, Broccoli and the like—all of which have evolved from a single species. *Brassica oleracea* is indigenous to the coastal cliffs of S and W Europe (including Britain) and through the centuries has mutated to the various types we know and grow today.

Turnips and Kohlrabi, Radish, Rape and Swede (Rutabaga) are other food plants in the family; also Garden Cress, *Lepidium sativum,* Watercress, *Rorippa nasturtium-aquaticum* (*Nasturtium officinale*), Horse Radish, *Armoracia rusticana* (*A. lapathifolia*) and Black and Yellow Mustard *Brassica nigra* and *Sinapis alba.* Oil derived from the seeds of Turnip and Rape are amongst the ingredients used in the manufacture of margarine and Rape oil is used in the making of lubricants and soap.

Crambe maritima is the Seakale, a coastal plant with fleshy, waxy leaves and white flowers, the young growths of which when blanched make a succulent vegetable. Other species, such as *C. cordifolia* from the Caucasus and *C. orientalis* (which grows on the shores of the E Mediterranean), make handsome foliage plants for key positions near the back of the border. They reach 120–180 cms (4–6 ft) at flowering time and have much-branched panicles of white flowers. All the species appreciate sharp drainage.

One curious crucifer is *Anastatica hierochuntica,* the Rose of Jericho, which was brought back to western Europe by the Crusaders as a symbol of resurrection. It is found growing from Morocco to S Persia, an annual 15 cms (6 ins) tall with thick branches, silvery leaves and numerous white flowers. As the seeds ripen during the dry season the leaves fall off and the branches curve inwards to make a round lattice-type ball. In this state it is blown out of the soil and rolls about the desert until it reaches a moist spot or the rainy season begins. The name Resurrection Plant as it is sometimes called, refers to this trait, for if the dried plants (sometimes sold in florists' shops) are brought near water, moisture is absorbed and they appear to grow.

The Rose of Jericho is regarded with some reverence in Israel as the plant alluded to in Ecclesiasticus 'I was . . . as a Rose-plant in Jericho'. It is supposed to have first blossomed at the Saviour's birth, closed at His crucifixion and blossomed again at Easter—hence the name Resurrection Flower.

Lunaria annua

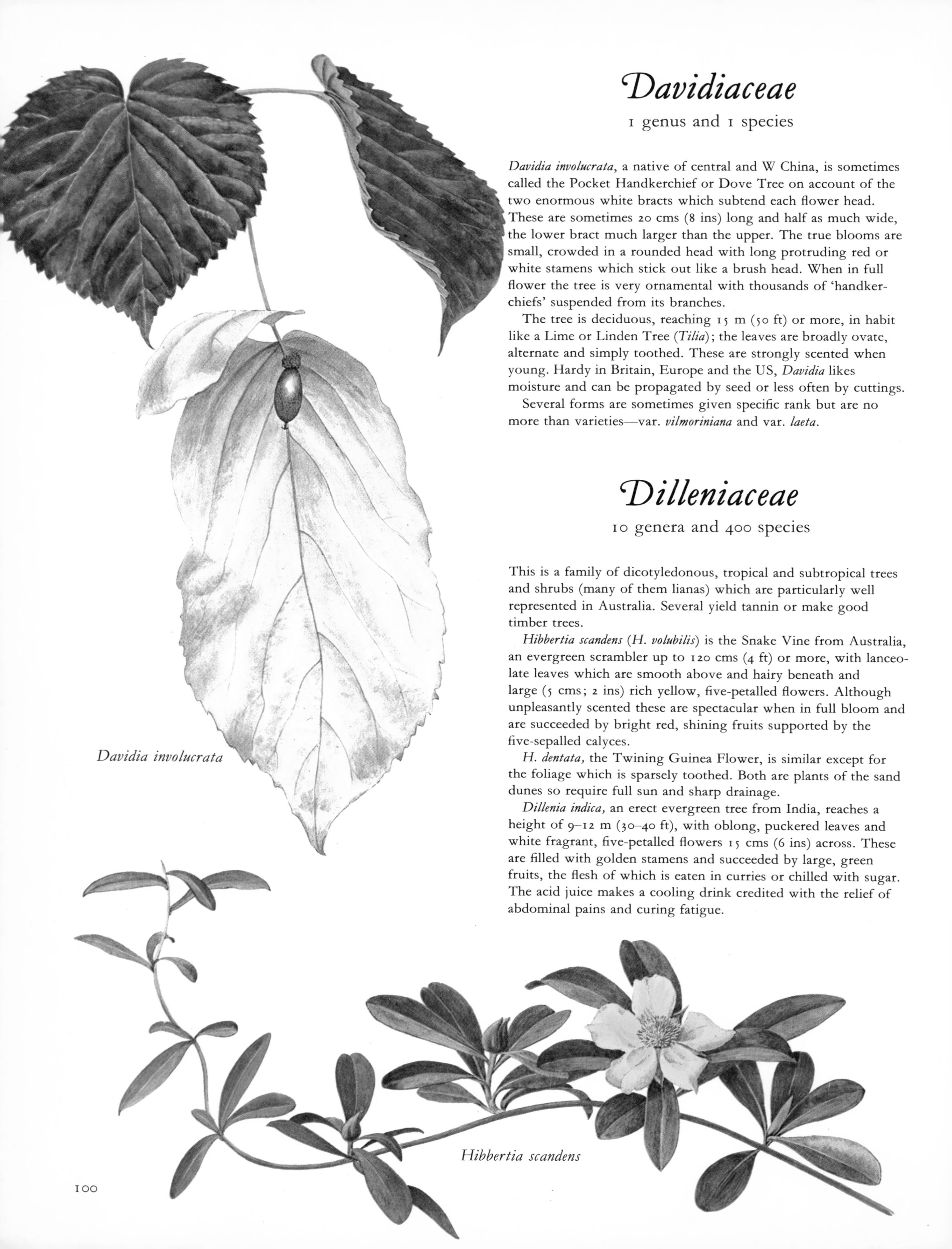

Davidiaceae

1 genus and 1 species

Davidia involucrata, a native of central and W China, is sometimes called the Pocket Handkerchief or Dove Tree on account of the two enormous white bracts which subtend each flower head. These are sometimes 20 cms (8 ins) long and half as much wide, the lower bract much larger than the upper. The true blooms are small, crowded in a rounded head with long protruding red or white stamens which stick out like a brush head. When in full flower the tree is very ornamental with thousands of 'handkerchiefs' suspended from its branches.

The tree is deciduous, reaching 15 m (50 ft) or more, in habit like a Lime or Linden Tree (*Tilia*); the leaves are broadly ovate, alternate and simply toothed. These are strongly scented when young. Hardy in Britain, Europe and the US, *Davidia* likes moisture and can be propagated by seed or less often by cuttings.

Several forms are sometimes given specific rank but are no more than varieties—var. *vilmoriniana* and var. *laeta*.

Dilleniaceae

10 genera and 400 species

This is a family of dicotyledonous, tropical and subtropical trees and shrubs (many of them lianas) which are particularly well represented in Australia. Several yield tannin or make good timber trees.

Hibbertia scandens (*H. volubilis*) is the Snake Vine from Australia, an evergreen scrambler up to 120 cms (4 ft) or more, with lanceolate leaves which are smooth above and hairy beneath and large (5 cms; 2 ins) rich yellow, five-petalled flowers. Although unpleasantly scented these are spectacular when in full bloom and are succeeded by bright red, shining fruits supported by the five-sepalled calyces.

H. dentata, the Twining Guinea Flower, is similar except for the foliage which is sparsely toothed. Both are plants of the sand dunes so require full sun and sharp drainage.

Dillenia indica, an erect evergreen tree from India, reaches a height of 9–12 m (30–40 ft), with oblong, puckered leaves and white fragrant, five-petalled flowers 15 cms (6 ins) across. These are filled with golden stamens and succeeded by large, green fruits, the flesh of which is eaten in curries or chilled with sugar. The acid juice makes a cooling drink credited with the relief of abdominal pains and curing fatigue.

Davidia involucrata

Hibbertia scandens

Dillenia indica

Dipsacaceae

8 genera and 150 species

These are dicotyledonous plants, mostly herbaceous, with opposite leaves and dense heads of flowers; the individual blooms have five petals and five sepals (occasionally four), four stamens and achenial fruits. The majority are native to the N temperate regions of Europe and Asia with a few in tropical and southern Africa.

Scabiosa is the largest genus, the most important garden plants being the perennial *S. caucasica*—which blooms for months—and the annual *S. atropurpurea*. *S. caucasica,* the Pincushion Flower, was introduced to Britain in 1591 and has become a favourite for border and cut-flower work. The round flower heads on stems of 30–45 cms (1–$1\frac{1}{2}$ ft) are pale mauve but cultivars exist in other colours, such as rich mauve, white, cream and deep violet blue. New varieties come and go for the results are variable from garden to garden; it is advisable for those with sickly plants to abandon them and switch to another cultivar.

This species should be grown in full sun and well-drained soil; it likes lime but resents being moved which, if it becomes necessary, is best undertaken in spring.

S. atropurpurea from SW Europe, often called Sweet Scabious or Mournful Widow, is a free-flowering and long-lasting species with stems of 60–90 cms (2–3 ft) terminating in fragrant deep crimson flowers. The blooms are ideal for cutting and varieties exist with pale blue, dark blue, purple, cherry-red, scarlet, salmon, white, violet, rose and double flowers.

The name Scabious was bestowed on the genus because of the supposed efficacy of certain species to cure skin diseases, such as scabies. It was cultivated for that purpose in the 15th century. In the language of flowers Scabious signifies 'I have lost all' so that it was considered 'an appropriate bouquet for those who mourn for their deceased husbands' (*Flowers and Flower Lore* by the Reverend Hilderic Friend). *S. atropurpurea* was at one time in great demand for funeral wreaths in Portugal and Brazil.

Cephalaria gigantea (*C. tatarica*) is the Giant Scabious, a Caucasian plant of branching habit reaching 150–180 cms (5–6 ft) in summer with many soft yellow, Scabious-like flowers. It is the most garden-worthy of the genus and will grow in most soils, in sun or partial shade.

Dipsacus fullonum ssp. *fullonum* (*D. sylvestris*) is the common Teasel, a plant found in moist places in Europe, including Britain. It has a prickly branching stem of 120–180 cms (4–6 ft), and pale lilac flowers with prickly bracts in cone-shaped heads. The leaves are opposite, have jagged edges and are usually connate at the base so that they form a cup which traps water—some of which is probably absorbed by the plant. This water was at one time considered a remedy for poor eyesight.

D. fullonum ssp. *sativus* is the Fuller's Teasel and similar to the preceding except that the bracts on the flower heads are hooked. This circumstance accounts for their use in the textile industry where they are still employed for raising the nap on woollen cloth. They do this better than any machine. The Fuller's Teasel is known only in cultivation, but both plants can be raised from seed and are suitable for the rougher parts of the garden. The dried flower heads are frequently employed for winter bouquets.

Succisa pratensis is the Devil's Bit Scabious and earned its name according to the *Grete Herbal* of 1526 'Bycause the rote is blacke and semeth that it is iagged with bytynge, and some say that the devyll had envy at the vertue thereof and bete the rote so far to have destroyed it.' The root is certainly astringent and has been used as the source of green and yellow dyes. Growing 30–90 cms (1–3 ft) tall with blue-purple or white flowers, the plant is native to Europe and NW Africa. It too is suitable for the wild garden, especially on chalky soils.

Knautia macedonica is an ornamental from E Europe with red Scabious-like flower heads on stems 60 cms (2 ft) tall. It is sometimes grown as a cut flower, the blooms lasting several weeks.

Dipsacus fullonum

Scabiosa caucasica

Droseraceae

4 genera and 105 species

This is an interesting family of dicotyledonous insectivorous plants, all of which are herbaceous and usually found in acid bogs. *Aldrovanda* is a submerged aquatic.

The leaves of Droseraceae frequently grow in rosettes from small perennial rhizomes and exhibit various devices to trap their prey. The flowers may be solitary or in cincinni (coiled spikes) of racemes, and are regular in shape and bisexual with five sepals, five petals, usually five stamens and a superior ovary.

The need to trap insects and later digest their bodies is linked with their waterlogged environment. All plants need nitrogen and usually obtain this via the roots in the form of soluble nitrates. Organic matter (plant and animal remains) breaks down to simple components such as nitrates, through a complicated cycle, in part due to the action of nitrifying bacteria. The latter are naturally present in well-drained soil but since they require oxygen are usually absent or sparse in marshland and peat bogs.

The plants growing in such situations adopt various methods of survival, some forming associations with fungi from which they derive additional organic material. Others, like Ericaceae, have the leathery leaves normally found in such plants as Firs (*Abies*), a characteristic often exhibited by plants inhabiting an environment where nitrogen is in short supply.

Insectivorous plants have found a way round such difficulties by finding their own source of nitrogen (protein) and are thus capable of surviving in situations which would be anathema to the majority of plants.

Drosera, with 100 species both tropical and temperate, is especially plentiful in Australia and New Zealand. It is the largest genus in the family, the foliage varying greatly in the different species—from a rounded spoon shape to long, narrow, elongated or forked leaves. All exhibit trapping devices, however, in the form of reddish glandular hairs thickly studding the upper leaf surface, each of which terminates in a globule of sticky glistening fluid.

As these tentacles sparkle in the sunlight their common name—Sundew—seems singularly appropriate. Their appearance attracts insects which alight on the leaves. The lightest touch affects the extremely sensitive tentacles and by some form of communication this is relayed to the other tentacles. Aware of their prey these bend over the unfortunate victim, gumming it fast and suffocating it by blocking the breathing pores. The glandular heads of the tentacles now secrete an acid peptonizing ferment which, acting upon the proteins, dissolves all the soft parts of the insect's body. Brought into solution this 'fly broth' is absorbed by the leaves and so nourishes the plant. Afterwards the tentacles expand, the indigestible parts (wings and chitinous remains) blow away and the leaves manufacture and secrete more sticky fluid.

The cultivation of Droseras is dependent upon a source of salt-free soft water. Salt is anathema to them, even the slight amounts found in tap water proving fatal to many, particularly the British species, *D. rotundifolia, D. anglica* and *D. intermedia* (*D. longifolia*). Less sensitive are certain species from the Cape and Australia, such as *D. capensis, D. spathulata* and *D. binata*; in consequence these are the kinds most frequently seen in cultivation.

The easiest way to grow them is in pots or pans half full of crocks and then filled with a compost of one third chopped sphagnum moss, one third sifted loam mixed with a little salt-free silver sand and one third fern fibre, with the dust shaken out. Set the plants firmly in this, but do not ram them too hard and insert small pieces of live sphagnum (as cuttings) in the spaces around. Plunge the pans up to their rims in damp sphagnum moss, with the spaces between heaped up and planted with live sphagnum. Keep the plants in a cool, moist, even temperature such as a cold greenhouse or, in subtropical countries, in a spot where the soil is unlikely to dry out.

Those without such facilities can stand each pot in a saucer, keeping a little water always at the bottom and cover the whole with a polythene or glass cover. This should stand several centimetres above the plants, so that it maintains an airy but a moist atmosphere inside—most important in centrally heated establishments or hot, dry positions.

Watering should be carefully carried out, a light overhead spray proving most beneficial. If the sphagnum moss can be kept green and healthy the Droseras will probably be all right. Only use soft water or, if this is unobtainable, the liquid obtained after defrosting a refrigerator can (if brought to room temperature) be used instead. Propagation is normally by division or by seeds sown in live sphagnum moss.

Several have slight economic importance, the roots of *D. whittakeri,* an Australian species, being the source of a red dye suitable for dyeing silk. In Madagascar *D. ramentacea* is employed as a remedy for coughs and dyspepsia and is supposed to preserve the teeth. The European Sundew (*D. rotundifolia*) was once dried and used medicinally as an antispasmodic, for whooping cough and bronchitis. In Britain years ago it was also valued as a cosmetic and the leaves used to curdle milk in the same way as rennet (see also *Pinguicula*).

Some of the most interesting or attractive species include *D. binata* with slender, forked leaves 30–38 cms (12–15 ins) long, arising from an erect stem carrying pretty white papery flowers in loose inflorescences; *D. capensis* with rosettes of purplish, linear, oblong leaves and purple-red flowers; *D. cistiflora,* a South African species with white to magenta and also scarlet flowers 5 cms (2 ins) across; and *D. spathulata* which has white flowers and rosettes of spoon-shaped leaves.

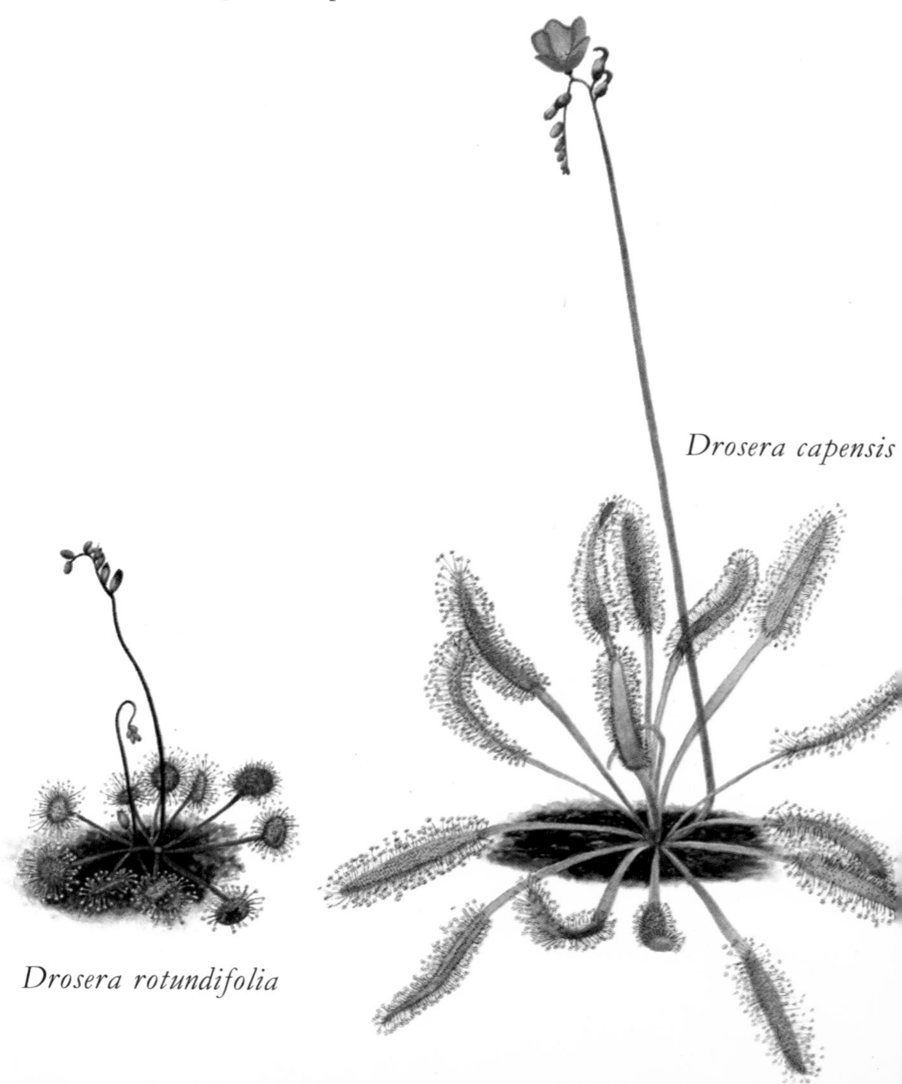

Drosera capensis

Drosera rotundifolia

Drosophyllum lusitanicum, a monotypic genus endemic to Portugal and N Morocco, grows in rock crevices and in both countries is hung up indoors as a fly catcher. From a woody stem bearing the accumulated remains of old leaf bases, rises a leafy stem 30 cms (1 ft) tall carrying an inflorescence of yellow flowers each about 3 cms ($1\frac{1}{2}$ ins) across. The leaves are long and narrow, 10–20 cms × 3 mms (4–8 × $\frac{1}{8}$ ins), and thickly covered with purplish viscid hairs, which however are not mobile as in *Drosera.*

Aldrovanda vesiculosa is another monotypic genus with a wide distribution (although never very abundant) from central Europe and the Caucasus, eastern and South-east Asia to Timor and Queensland, Australia. It is a rootless water plant which, but for its flower, remains entirely submerged.

It has whorls of leaves, individually almost round, with bilobed leaf blades which carry long sensitive hairs near the mid-rib. The touching of these by any passing creature results in the closure of the lobes, thus imprisoning the prey which is later digested through the action of gland ferments. This process closely resembles that of *Dionaea.*

Winter buds, about the size of peas, form in cold climates and normally sink to the bottom of the water in autumn. This however does not occur in warmer countries where the plant continues to grow. These interesting carnivores will only thrive in acid pools, where they should be placed amongst such plants as Water-lilies (*Nymphaea* species) or reeds, to provide shade.

Dionaea muscipula, the Venus Flytrap, which Darwin once called 'the most wonderful plant in the world' has a restricted range, being found only in wet pinelands and sandy bays in North and South Carolina, USA.

It is nevertheless a much sought after plant and frequently offered for sale in the dormant 'bulbil' stage, although sadly most of the resultant plants die of wrong treatment. If these tubers are planted in pans of sphagnum moss as recommended for *Drosera* species, are always watered with soft water (not over the leaves as this causes them to blacken), and are kept in a cool but frost-free situation they will grow and multiply and also flower well during the summer. Seed is sometimes obtainable and can be raised—sown in similar compost—in a closed propagating case.

The plant produces a short stubby rhizome carrying a rosette of leaves which lie close to the ground. These form a most unusual catching device, each leaf having a basal part (like a winged petiole) which functions as a normal leaf, also an upper portion, quadrangular in shape, with a hinged mid-rib. The margins of this upper part are edged with long spiky teeth and have three spines—trigger hairs—on the inner surface of each blade. These hairs are jointed at the base so that they can fold down when the leaf blades close together.

While the edges and the backs of the leaves are green, the inner blades are covered with reddish dots—glands which secrete the digestive juices which prepare a trapped insect's body for absorption.

The sensitive parts are the trigger hairs which, when lightly touched several times or if vigorously stimulated, cause an impulse to be transmitted to the mid-rib portion. Then the leaf halves close together and the interlocking marginal teeth prevent the escape of small insects. A discharge of acid, digestive fluid from the glands (dry until such a moment) now appears and acting upon the proteins of the prey, gradually dissolves them. The soluble parts are then reabsorbed by the same glands that discharged the acid ferment, so that when the trap reopens the glands are dry. The unwanted debris is taken by the wind and the leaf is ready for more victims.

Darwin's great work (*Insectivorous Plants*) led the author to experiment with *Dionaea* some years back. It was found that a non-nitrogenous object such as a bead or match head caused closure of the leaves but only for a short while. They were not long fooled and soon reopened. Hard boiled egg or fragments of raw meat induced the same reactions as an insect and if a plant had been long deprived of 'meat', it exhibited marked symptoms of excitement—even to the extent of dribbling out an overflow of the colourless, slimy gland secretions.

On the other hand, too much food forced on the leaves induced a form of indigestion for the leaves sickened and turned yellow. Passing a pad soaked in ether over their surface held up proceedings until the effects wore off—surely a form of anaesthesia!

The flowers of *Dionaea muscipula* are quite pretty although growers often nip them off in order to encourage sturdier foliage. They are white and borne in clusters, each bloom having a calyx with five sepals and a corolla with five separate petals. The genus is monotypic.

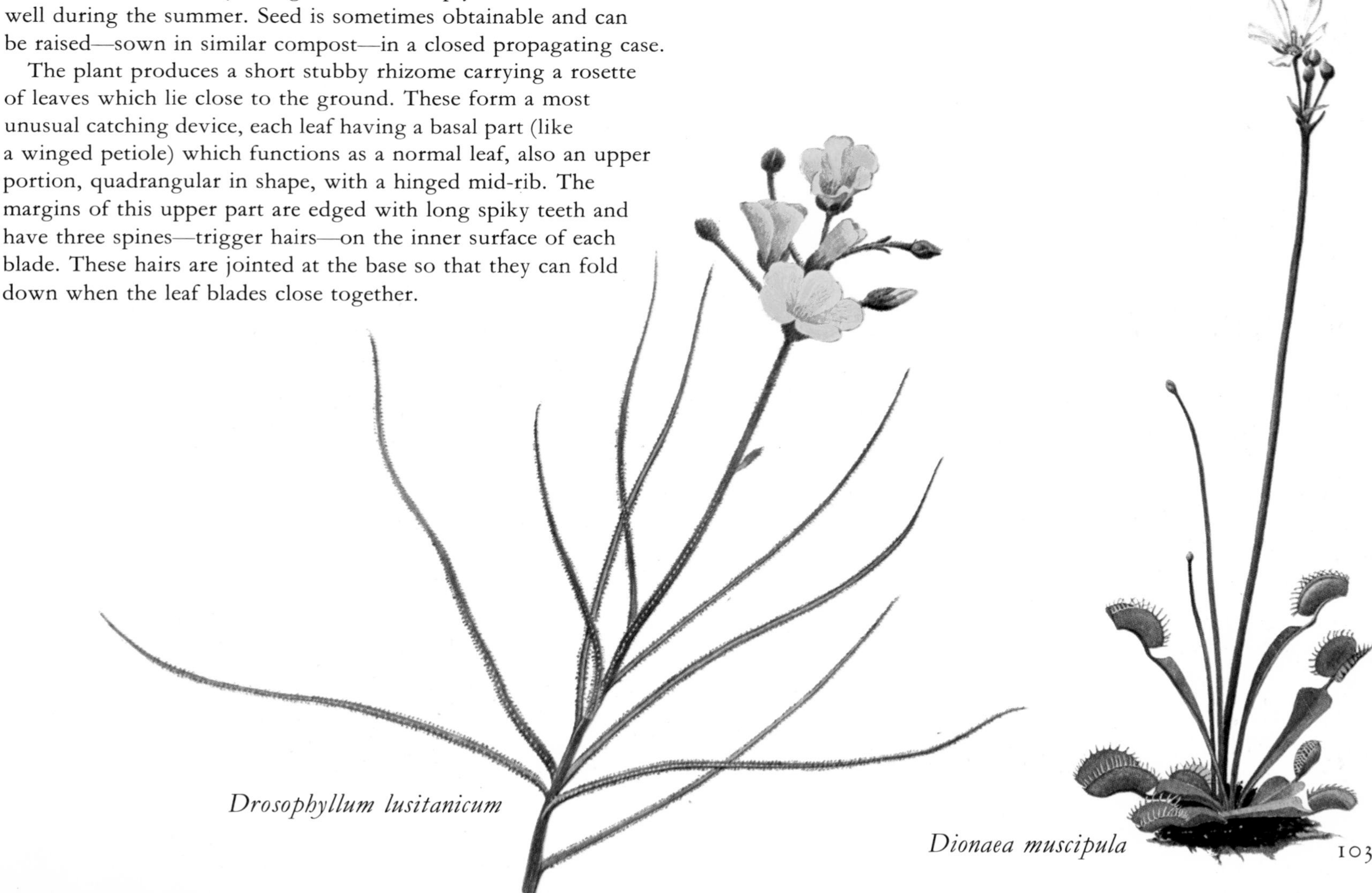

Drosophyllum lusitanicum

Dionaea muscipula

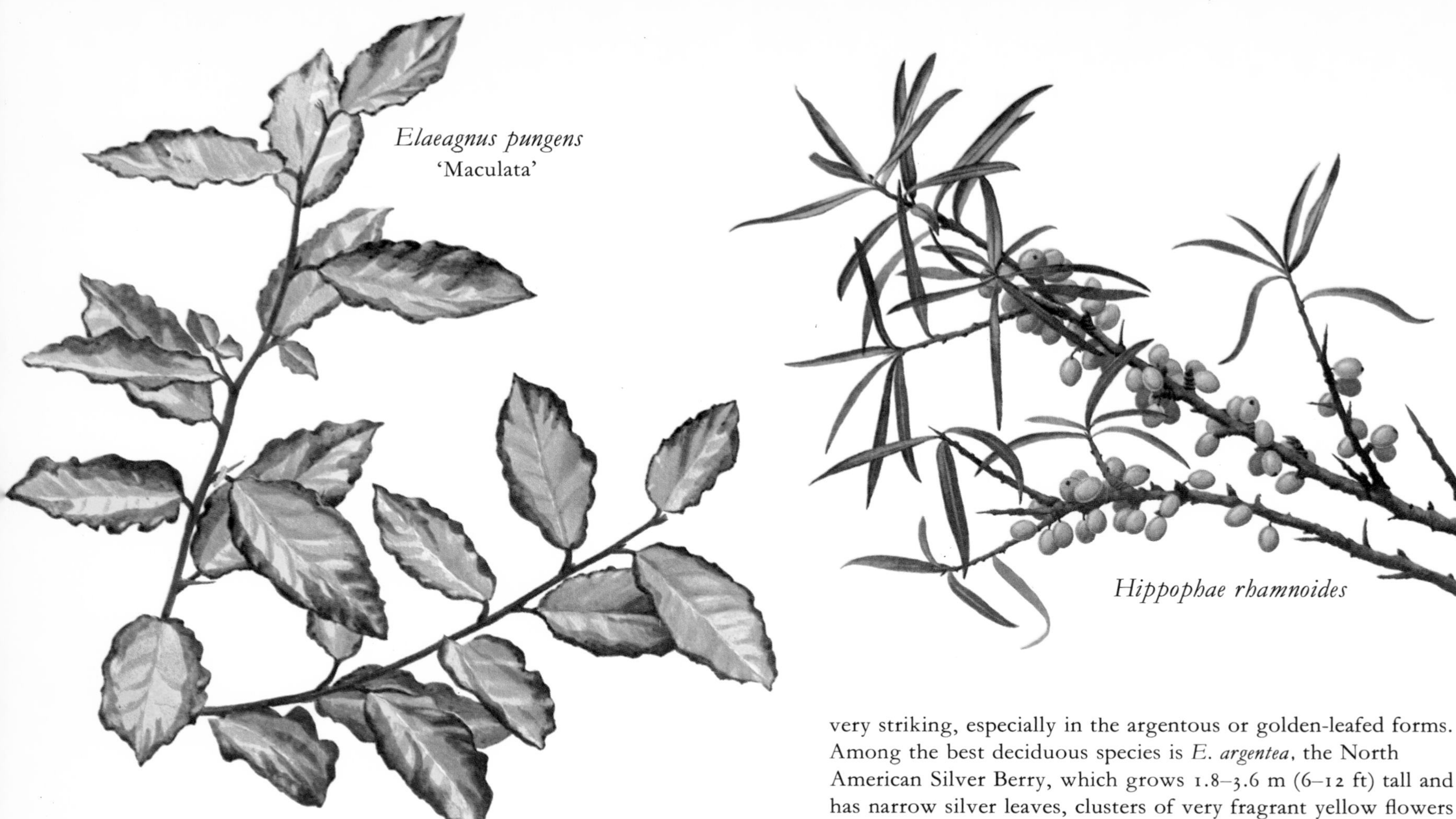

Elaeagnus pungens 'Maculata'

Hippophae rhamnoides

Elaeagnaceae

3 genera and 50 species; all N Hemisphere

This family consists of dicotyledonous shrubs or small trees, often thorny and intricately branched, with alternate or opposite leaves. Both these and the stems are thickly covered with scaly hairs, while the inflorescence is often racemose with unisexual or bisexual flowers.

The genus *Elaeagnus*—commonly known as Oleaster—consists of about forty-five species of evergreen or deciduous trees and shrubs with mostly fragrant, perfect or unisexual flowers in axillary clusters. They are primarily sea shore or steppe plants with entire, rather leathery leaves which are tolerant to exposed conditions, and are usually grown for this foliage which is often very striking, especially in the argentous or golden-leafed forms. Among the best deciduous species is *E. argentea*, the North American Silver Berry, which grows 1.8–3.6 m (6–12 ft) tall and has narrow silver leaves, clusters of very fragrant yellow flowers and small egg-shaped silvery fruits said to be edible.

E. pungens from Japan is evergreen and grows 3–3.5 m (10–12 ft) high. The young growth is densely covered with brown scales and it has oval leaves, thick and leathery, green above and white beneath. The silvery fragrant flowers hang in clusters in early winter. Garden forms include 'Aurea' with yellow-margined leaves and 'Maculata' or 'Aureo-variegata', a splendid plant, especially in midwinter, the leaves heavily splashed with gold. Any tendency to revert to green must be checked by prompt removal of the offending branches. The evergreen species of *Elaeagnus* should be increased by cuttings; the deciduous kinds by seed. A poor soil brings out the silvery colouring of the latter better than good loam.

Hippophae rhamnoides is the Sea Buckthorn, a European (including British) native. A deciduous shrub or small tree with male and female flowers on separate plants, it has stiff twigs, narrow, tapering, silvery leaves and clusters of small flowers. The bright orange berries hanging in clusters on the female plants are very beautiful and since birds—unless desperately pressed—seem to dislike their acid flavour, remain attractive throughout the winter. They are sometimes made into jelly or a sauce eaten with fish and meat. The hard wood has been used for turnery.

Crinodendron hookeranum

Elaeocarpaceae

12 genera and 350 species
tropical and subtropical

This family contains trees and shrubs with alternate or opposite leaves and racemes or panicles of small flowers with four to five sepals, four or five petals (often fringed), many stamens and a superior ovary. The seeds of *Elaeocarpus ganitrus* are made into sacred beads used by Brahmins. Each chain contains 101 beads—representing the 101 eyes of Shiv.

The Lantern Tree, *Crinodendron hookeranum* (*Tricuspidaria lanceolata*), is a handsome Chilean evergreen suitable for a sheltered spot in lime-free but warm peaty or loamy soil. It grows 3–9 m (10–30 ft) tall with green, broad-lanceolate leaves and many crimson, urn-shaped flowers swinging from long stalks. Where it thrives it makes a handsome shrub or small tree for open or light woodland. In colder climates it must be grown in a greenhouse but since it responds to clipping can be kept fairly small. *C. patagua,* another Chilean, has white flowers with fringed petals.

Elaeocarpus are evergreens for subtropical gardens or else must be grown under glass. They have simple, toothed leaves and small, frequently fragrant, white or soft yellow flowers in racemes. One of the most decorative is *E. grandiflorus,* a shrub from Java about 2 m (7 ft) tall with many white, drooping flowers. The Bead Tree from India and Malaya, *E. ganitrus,* makes a large tree; its creamy flowers are succeeded by brown warty seeds used for beads, rosaries, hatpins and buttons. *E. dentatus* from New Zealand has tough, leathery, lanceolate leaves with toothed margins and masses of pretty, white, drooping flowers. The latter are succeeded by Damson-like fruits which are the source of a blue-black dye used by the Maoris. The bark is also rich in tannin. This species is widely planted in parks and gardens in New Zealand and makes a medium-sized tree—maximum height 12 m (40 ft). *E. reticulatus,* the Blueberry Ash of Australia, grows 6 m (20 ft) tall with small white fringed flowers in spring, followed by blue berries.

The fruits of *E. serratus* from the East Indies are picked and used like Olives and also used in curries. This makes a tree 15 m (50 ft) high with fragrant white flowers.

Aristotelia peduncularis from Tasmania, an evergreen shrub 90–180 cms (3–6 ft) high, may be grown under the same conditions as *Crinodendron.* It has solitary, axillary, white and orange flowers, succeeded by edible, juicy black fruits.

Epacridaceae

30 genera and 400 species
chiefly Australia and Tasmania

This is a family of dicotyledonous small shrubs, or rarely small trees, which bear a close resemblance to Ericaceae. The leaves are narrow, rigid and entire, usually alternate, stalkless and often sheathing the stem.

The types of inflorescence vary but frequently take the form of spikes or racemes; individual blooms are perfect, regular and often fragrant with usually five free sepals, five petals joined to form a campanulate or cylindrical tube, five stamens and a five-celled superior ovary.

The forty species of *Epacris* are confined to Australia and Tasmania, New Caledonia and New Zealand. They are small-leafed evergreens with tubular flowers which last well when cut. As a general rule the plants thrive in a sand-peat compost but are not hardy outside in Britain, possibly because there is insufficient sun to ripen their wood.

E. impressa is from Australia where it is frequently used as a rock plant and known as Common Heath. It makes a small erect shrub of 90–120 cms (3–4 ft) with slender, downy shoots bearing many narrow pointed leaves and five-lobed, tubular, delicate pink, white or red flowers. Surprisingly it grows well both in sandy soils and swamps. A variety with oval leaves is known as *ovata* and there is a double-flowered form. This is the State Flower of Victoria.

E. longiflora from New South Wales is known in Australia as the Native Fuchsia or Fuchsia Heath. It is a straggling shrub of 120–150 cms (4–5 ft) with long spectacular spikes closely set with sessile, round to oval-lanceolate leaves and many red flowers, tipped with white. It makes a good plant for a tub or large flower pot and, commencing in late spring, continues in flower for weeks. *E. obtusifolia* from Tasmania and Australia has long heads of fragrant, cream, bell-shaped flowers in spring and summer.

Dracophyllum secundum, another Australian and Tasmanian native, is characterized by rigid leaves, 5–10 cms (2–4 ins) long, and many short, tubular cream or pink flowers in long, one-sided inflorescences.

Styphelia triflora is the best of a genus of eleven species, with closely set, broad, stiff leaves (partially stem-clasping) and clusters of pale pink flowers with long protruding stamens. It comes from Australia where it is known as Pink Fivecorners.

Astroloma and *Lysinema* (*Woollsia*) have tubular, five-lobed flowers of various colours on low-growing shrubs. The *Leucopogon* species with smaller white flowers are shrubs or small trees.

Ericaceae

50 genera and 1350 species

This is an important family of dicotyledonous plants which, excepting in deserts, has an almost cosmopolitan distribution. In the tropics the members are essentially plants of high altitudes, although certain genera often form a distinctive part of the vegetation of peat soils in swamps and moorlands. The family has sparse representation in Australasia, where the closely related Epacridaceae takes its place.

While all the species are woody, they vary in size from dwarf undershrubs to large shrubs or even small trees, with either broad or needle-like and xerophytic foliage. The majority are evergreen although some are deciduous and there are many epiphytes, especially in Asia.

The inflorescences may be terminal or axillary, with the individual flowers bisexual (usually) and regular, each with a four- or five-lobed corolla, four or five sepals, five to ten stamens and a superior (occasionally inferior) ovary with two to five cells. The fruit is usually a capsule, more rarely a drupe or berry.

The family contains many highly decorative and well-known plants, which are often seen in gardens, although a very large number of these require acid conditions and fail completely in calcareous soils.

Ericas have a wide distribution, particularly in South Africa where there are said to be 605 distinct species.

In Europe *Erica cinerea* and *E. tetralix* carpet vast areas of moorland with their springy low growth, which in early autumn—when the flowers are in character—turn to sheets of purple.

Epacris longiflora

Epacris impressa var. *ovata*

In stature members of the genus vary from shrubby dwarfs 7 cms (3 ins) high to bushes or small trees 6 m (20 ft) tall. All have small, narrow, folded leaves and flowers with three bracts, four sepals and usually urn-shaped corollas.

The European dwarf Heaths play an important part in modern planting techniques, especially in Britain. They have a place in both large and small gardens as well as parks and roadside verges. Besides being accommodating and easy to grow (providing their soil requirements are met) they make splendid ground cover and consequently suppress weeds. By careful selection it is possible to have varieties in bloom during the whole twelve months, the winter sorts being particularly valuable.

These Heaths, usually forms of *E. carnea* (*E. herbacea*), *tetralix, cinerea, vagans* and *ciliaris*, are best planted in an informal way, in drifts, sloping the ground if possible for the most telling effects. The majority favour acid conditions so work plenty of granulated peat into the existing soil; moorland peat is often excessively acid and if used must be broken down with plenty of sand. Avoid planting Heaths in lime-impregnated soil or using fresh manure or artificial fertilizers for their cultivation. Ground which has been dressed with lime over the years (as distinct from natural calcareous soil) can be induced to grow Heaths if plenty of peat is used during the initial planting. The most lime tolerant and likely to succeed are variants of *E. carnea* (which contains many winter-flowering sorts), *E. mediterranea* (now *E. erigena*), *E. terminalis* (*E. stricta*) the Corsican Heath and *E.* x *darleyensis*.

The Tree Heaths are valuable to give height in a mixed bed of Heaths. The hardiest are *E. arborea* from North Africa and S Europe which grows 3.5–4.5 m (12–15 ft) and sometimes up to 6 m (20 ft), with tiny, evergreen leaves in whorls of three and white, honey-scented flowers; *E. terminalis,* the Corsican Heath, which is rosy-pink and 120–180 cms (4–6 ft) high; and *E. lusitanica* from Portugal, a species 3 m (10 ft) tall with fragrant, pink-budded flowers which expand to white. *E.* x *darleyensis* (*E. mediterranea hybrida*) is an elegant hybrid, about 60 cms (2 ft) high, with long-lasting trusses of rosy-purple flowers.

The hard, knobby roots of *E. arborea* furnish the wood used for making 'briar' (bruyère) tobacco pipes. Two stories account for the discovery that they were useful for this purpose. One relates how a French pipe maker, while holidaying in Corsica lost his cherished meerschaum and begged a native craftsman to fashion him some sort of substitute. Using the knobby root of Tree Heath the native produced such a handsome and satisfying pipe that when the Frenchman journeyed home a bagful of roots accompanied him. He became the first manufacturer of briar pipes.

The other tale concerns the town of Saint-Claude in eastern France—now a great centre for the manufacture of pipes, but at one time noted for its wood carvings made of boxwood, particularly articles such as rosaries, snuffboxes and beads. The boxwood was gathered in the Pyrenees by peasants, some of whom to fill a bag more quickly slipped in a few chunks of Tree Heath root. These proved unsuitable for snuff boxes so the thrifty French turned them to other uses—first beads, which polished so beautifully that someone thought of making pipes from them and so an industry was born.

Another Heath used for this purpose is the Besom Heath (*E. scoparia*), the twigs of which are also fashioned into besom brooms in France.

The Cape Heaths need greenhouse conditions in colder climates but make beautiful pot plants for winter flowering, with elegant wands of closely packed, white, cream, yellow, pink, red or crimson flowers.

Erica pillansii is a showy distinct form which, according to H. A. Baker (*Ericas in Southern Africa*), was originally found in a Cape flower seller's bucket by a Mr Pillans. It was some years after that before it was discovered in the wild. It has brilliant bright red flowers in terminal racemes on bushes up to 1.25 m (4 ft) high. *E. grandiflora* has very large flowers of orange-red on bushes 1.5 m (5 ft) or so high but its variety *exsurgens* is smaller and more diffuse. The colour of this varies according to soil conditions but is usually deep yellow.

Calluna vulgaris is the Scotch Heather or Ling, the only species in the genus, with purple-pink flowers in autumn. There are numerous named varieties with white (*alba*), dark crimson, double white, double pink flowers, and many gold- and silver-foliaged forms. The flowers provide a rich source of honey, and at one time this Heather was used to thatch cottages and also (together with earth and straw) for making their walls. The crofters in the Scottish Highlands dyed their wool with a yellow dye extracted from the bark and also made beer and tea from the leaves and stems.

Daboecia cantabrica, St Daboec's Heath, found in heaths in S Ireland is an attractive shrub of upright habit with rosy-purple flowers. It has a white form *alba* but 'Bicolor' is an oddity with both purple and white flowers.

Several ericaceous plants have edible fruits, notably *Vaccinium uliginosum,* the Bog Whortleberry, a plant from high mountain areas in the Northern Hemisphere and *V. myrtillus,* the Blueberry, Whortleberry or Bilberry, a low, deciduous shrub from N European mountain slopes and moors. Both have blue fruits and

Pieris japonica

assume handsome rose-red autumnal leaf tints. Besides being consumed fresh, the berries (of the Blueberry particularly) are cooked in soups, preserves and pastry, used to give wine a good red colour and made into a wine (Heidelbeerwein) on their own account. *V. vitis-idaea,* variously known as Cowberry, Red Whortleberry and Red Huckleberry, is an evergreen with bitter red berries which are used in sauces and jellies; *V. oxycoccus* from mountain bogs in the Northern Hemisphere is the Small Cranberry. In the USA the native Blueberry (*V. angustifolium* and hybrids) is grown as an important fruit crop.

A bog plant of circumpolar distribution, particularly in Lapland, is the sweetly scented *Ledum palustre.* It is a small evergreen shrub 60–120 cms (2–4 ft) high, with dense, terminal clusters of white flowers. The leaves have been used as a tea substitute and can be employed to produce an aromatic oil with medicinal and insect-repellent properties. When any of these plants are used in the garden they must have cool conditions and lime-free soil. Plenty of peat should be worked into the planting sites and in some cases light shade provided.

Andromeda polifolia from arctic and temperate regions of the Northern Hemisphere (including Britain) is a straggling evergreen with umbels of white (or pale pink), bell-shaped flowers, rather like Lily of the Valley, and narrow, oblong leaves. The foliage and twigs are used in Russia for tanning purposes.

Pieris floribunda from the Allegheny Mountains, USA, and *P. japonica* from Japan are attractive evergreen ornamentals for lime-free, well-drained soil. Both have long sprays of white urn-shaped flowers at the tops of the branches and oblong tapering leaves. The flower sprays droop in *P. japonica*, 3–4.5 m (10–15 ft) tall, but are upright in *P. floribunda,* 1.25–1.75 m (4–6 ft).

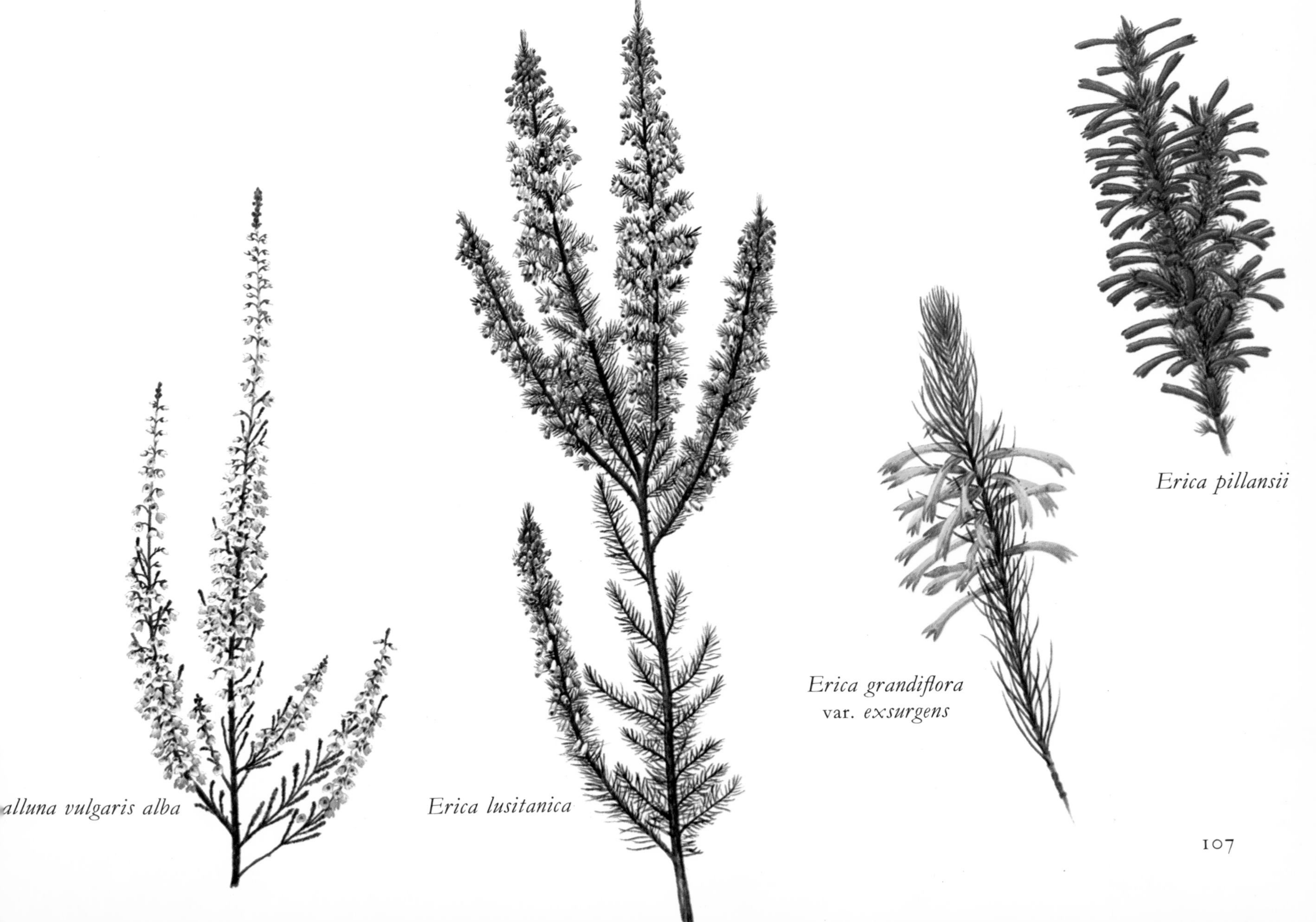

Erica pillansii

Erica grandiflora
var. *exsurgens*

alluna vulgaris alba

Erica lusitanica

Arbutus unedo

Enkianthus campanulatus

But, beautiful as these are, they in turn are surpassed by the Himalayan *P. formosa,* especially in a form of this collected and sent home from the Yunnan by George Forrest in 1910—and named appropriately enough var. *forrestii.* In sheltered gardens, as in the south of England, this grows up to 6 m (20 ft) tall and, apart from the beauty of its smooth, dark green foliage and long, white flower sprays has an added charm in the brilliance of its new young leaves, which are glowing scarlet. From a distance these look like flowers.

Arctostaphylos are shrubs or small trees requiring similar growing conditions to Rhododendrons. They are useful for shady situations and have fragrant, white or pinkish, urn-shaped flowers and leathery, evergreen leaves. The Californian *A. manzanita* possesses peeling bark and Red Currant-like fruits which are made into drinks and jellies. Growing 1.25–3.5 m (4–8 ft) in Europe, but up to 6 m (20 ft) in California, it blooms in spring. Widespread in N temperate regions the Bearberry, *A. uva-ursi,* is a prostrate species which makes a useful ground cover and has pinkish flowers and scarlet fruits. The leaves are used for tanning in parts of Sweden and Russia and also in the preparation of 'Russian Leather'.

An ericaceous tree that is more tolerant than most of chalky soils is the Strawberry Tree, *Arbutus unedo.* This native of SW Ireland and S Europe grows to a height of 4.5–6 m (15–20 ft). It has evergreen, oblong glossy leaves, white, urn-shaped flowers in clusters and round, crimson fruits which look like Strawberries. Frequently flowers and fruits are in character at the same time, the latter often used for making preserves and alcoholic drinks.

A. menziesii, the Californian Madrona, is a bark-peeling species, the smooth, greenish outer coat flaking off in strips to reveal the light red inner bark beneath. It has white flowers and orange-red fruits. The wood is close-grained and used for furniture and the bark contains tannin.

Another popular ornamental suitable for lime-free soils in partial shade is the Japanese *Enkianthus campanulatus,* a fine deciduous shrub 1.25–3 m (4–10 ft) tall, whose fiery leaf brilliance in the fall matches its beauty in spring. Then the branches bear bunches of drooping, urn-shaped flowers which resemble fat Lily of the Valley bells. They are white or yellowish in colour with crimson veining.

Kalmia latifolia is the Mountain Laurel and State Flower of Pennsylvania, a showy shrub forming dense thickets which in spring light New England Highways with terminal clusters of bright pink flowers. At close quarters these resemble ten-ribbed parasols and possess an explosive device which releases pollen on insects foraging for honey. The stamens are bent like bows in the open flowers and when touched shower pollen over the intruder. The long, narrow leaves are shiny, leathery and evergreen. The Sheep Laurel, *K. angustifolia,* has narrower leaves and flowers which are *not* terminal but surmounted by foliage. Kalmias need moist but well-drained and lime-free soil in light shade. The roots are sometimes made into tobacco pipes and the wood into such articles as spoons; hence the Indian name of Spoon Wood for this species.

Gaultherias are evergreens for light shade, the North American *Gaultheria procumbens* making good ground cover—it is only 10–13 cms (4–5 ins) tall—at the fringe of tall trees. It has oval, glossy, aromatic leaves which yield Oil of Wintergreen and are used to make a fragrant drink called Mountain Tea. The bell-shaped flowers are white tinged with pink, maturing to bright

Cassiope
'Muirhead'

Gaultheria procumbens

Kalmia latifolia

Form of *Rhododendron simsii*

red edible berries like Holly. Common names for this species are Spicy Wintergreen, Winter Berry and Checkerberry. *G. shallon,* also North American, is 60–120 cms (2–4 ft) tall with white occasionally pinkish flowers and black fruits. The foliage is popular with American florists.

Cassiopes are plants with an alpine-arctic distribution in the Northern Hemisphere, from the Himalayas to the Alaskan sea shore. They are dwarf, frequently prostrate shrubs for arid soils, needing cool constantly damp positions—so are not suitable for hot climates or situations. The foliage is adpressed—like some of the Heaths—to prevent moisture losses.

One of the most attractive for the garden is the nearly prostrate *Cassiope lycopodioides,* which appears to be leafless because of the tight pressing of the foliage against the stems. It has many white, pendent, bell-shaped flowers—like Lily of the Valley bells—on thread-like stems and is native to NE Asia and NW America.

'Edinburgh', a cultivar under 30 cms (1 ft) tall, is a large-flowered white form and the Himalayan *C. wardii* has white flowers which are suffused inside with rose and unpleasantly scented. *C.* 'Muirhead' is a hybrid between this species and *C. lycopodioides.*

Epigaea repens, the Trailing Arbutus, a native of Canada and the northern United States is a delightful woodlander for cool, dry, sandy positions—as under Pine Trees—but extremely difficult to establish. This may be due to some mycorrhizal association. It has trailing hairy stems, oval leathery leaves which remain green all winter and clusters of tubular pink or white, sweetly scented flowers. *E. asiatica* from Japan is similar but easier to grow.

Among the showiest dwarf berrying evergreens for moist, lime-free soil are Pernettyas, especially the Magellan *Pernettya mucronata,* 60–90 cms (2–3 ft). It is important to remember that male and female flowers are sometimes borne on separate plants and that both should be grown to ensure fruiting. Blooming in early summer with white Heath-like flowers, the female plants in autumn and winter become truly magnificent with clusters of large round berries the size of marbles and variously coloured in white, pink, lilac, crimson, purple, magenta and almost black. No regular pruning is necessary beyond cutting back over-long straggling shoots. Pernettyas have a wide distribution in the Antarctic Islands.

P. pumila was first collected by Charles Darwin in the Falklands but is lower growing than *P. mucronata.* It has white, bell-shaped flowers on prostrate shrubby shoots which are often only a few centimetres high. These are followed by round, white or pink berries which are eaten like Huckleberries in their native habitat.

Although most members of the genus are predominantly South American and Mexican there is a Tasmanian species called *P. tasmanica* which according to Hooker forms large green cushions a few centimetres high on the granite slopes of its native mountains. The fruits are normally red but sometimes cream or yellow. It can be grown in a rock garden—preferably in peaty soil. If the berries of the Chilean *P. furens* (sometimes grown in rock gardens) are eaten they cause wild excitement, dementia and sometimes death. The genus commemorates Antoine Pernetty, the historian who accompanied Bougainville on his voyage to the Falklands in 1763–64.

From a spectacular point of view few shrubs—with the possible exception of *Rosa*—rival the Rhododendrons. Their brilliantly coloured flowers in rainbow hues appear over a long season, with the greatest representation in late spring. With 500 to 600 species

Form of *Rhododendron simsii*

Pernettya mucronata

and countless forms and cultivars the genus (which includes Azalea) is a large one and also extremely variable, from dwarf sub-shrubs to trees 12 m (40 ft) tall.

With a few exceptions like *Rhododendron lochae* which comes from tropical Australia, the species are native to N temperate regions, with the greatest representation in eastern Asia (from S China to Japan and the Himalayas) and also in New Guinea and to a lesser degree in North America. All the species require a considerable amount of atmospheric moisture, which is the reason they do so well in high altitudes or in woodlands and are seldom, if ever, found in arid spots.

The best possible soil for their cultivation is one of a peaty nature, although light soils can be improved by the addition of decayed leaves, moist peat or well-rotted manure. Rhododendrons as a whole dislike lime (R. *hirsutum* from the limestone formations of the European Alps is an exception) and are also moisture-loving, so any exposed roots should be mulched with 10–15 cms (4–6 ins) of moist peat or decayed leaves to prevent them drying out. Most of the broad-leafed evergreen species and hybrids and some of the deciduous kinds do well in light shade; the narrow-leafed evergreens succeed, like Heathers, in more open situations.

After flowering the old blooms of both Rhododendrons and Azaleas should be removed as seed production has an adverse effect on next season's blossom.

Most of the large-leafed evergreen cultivars are grafted on species, generally R. *ponticum,* although other types are commonly increased by seed sowing, layers or cuttings.

Hybrid Rhododendrons are the most popular for garden purposes, particularly in northern countries. In the main these have been derived from seven species—R. *ponticum* from Asia Minor, R. *caucasicum* from the Caucasus, R. *maximum* and R. *catawbiense* from North America and R. *arboreum,* R. *campanulatum* and R. *griffithianum* from the Himalayas.

The results are varied; there are short cultivars and tall ones; thin bushes and squat fat kinds; some varieties bloom early, others very late; and the flowers come in all colours from white, cream, yellow and orange to pink, rose, deep red, blue, purple and violet. In the garden these hybrids have many uses; they can be planted in tubs and containers, employed for hedging, used in mixed woodland plantings, especially with Birches, Heathers and Maples, grown by themselves in island beds on lawns or a single plant may be treated as a specimen feature.

The deciduous Azaleas are almost as well known; their chief assets are brilliantly coloured flowers of flame, orange, apricot and scarlet—many of them richly scented—and the almost as

Hardy hybrid Rhododendrons
1 Rhododendron 'Tessa'
2 Rhododendron 'Sappho'
3 Rhododendron 'Pink Pearl'
4 Rhododendron 'Bo Peep'

vivid autumnal tints assumed by the foliage prior to leaf fall. They need plenty of light, although a little shade during the hottest part of the day prolongs the life of the flowers.

The main groups include the Ghent Azaleas, which were originally raised by a Ghent baker, who was interested in these plants, between 1805 and 1830. Derived from the North American R. *viscosum,* a richly scented, late-flowering species, the yellow R. *luteum* from Asia Minor and the vivid scarlet and orange R. *calendulaceum* from North America, the hybrids are characterized by tall growth (1.5 m; 5 ft), late flowers in many colours (some double), hardiness and a delightful perfume. Mollis Azaleas flower earlier in the year and Knaphill Azaleas combine the best qualities of both groups.

For rock-garden work or the front of shrub borders dwarf evergreen Azaleas are ideal as they spread out fanwise without making much height. The flowers are small but produced in such profusion that the whole plant becomes a splash of flame. The largest group is the Kurume section, but there are others such as the American Glenn Dale Azaleas and the R. *indicum* cultivars—which however have no connection with florists' Azaleas.

Several dwarf Rhododendrons are also useful in the rock garden, particularly the blue hybrids derived from R. *lapponicum* (from Europe, Asia and North America) and the Chinese R. *augustinii.* Well known amongst these are 'Blue Tit' and 'Blue Diamond', both free-flowering and 60–90 cms (2–3 ft) tall.

There are also some large tree Rhododendrons for large gardens, such as R. *arboreum* from the Himalayas which has blood-red flowers and which will grow 9–12 m (30–40 ft) tall. R. *falconeri,* also from the Himalayas, has fragrant, creamy flowers and R. *lacteum* has canary-yellow flowers and large green leaves with fawn or pale red-brown felted undersides. Both of these will grow up to 9 m (30 ft).

The well-known greenhouse 'Indian' Azaleas are evergreens of mixed parentage, mainly derived from the Chinese—not Indian—R. *simsii.* These provide a wonderful colour range of single and double cultivars and need to be grown in a porous compost containing a large proportion of peat with some coarse sand and lime-free loam.

They make valuable pot plants for spring display, needing cool but frost-free conditions. In summer they can be plunged in soil in a cool shady spot in the garden. Ripe cuttings of the current season's growth root readily in a closed propagating case although the trade often graft them on to stocks of a strong growing cultivar.

Rhododendron honey possesses poisonous principles and the late Frank Kingdon Ward used to relate how—during a plant-hunting trip in the Himalayas—his bearers found a wild bees' nest in Rhododendron country. Thinking to enliven a somewhat monotonous diet they partook freely of the fresh honey with the result that they became so intoxicated as to be practically unconscious for several days. This toxic property disappears as the honey ages or can be dispersed by heating.

In spite of this, crystallized Rhododendron flowers are a great luxury in Tibet. Yak saddles are also manufactured from the wood and Tartarian Tea, used in parts of central Asia, is made from the leaves of R. *chrysanthum.* In E America the wood of R. *maximum* is fashioned into tools used for engraving, the roots are made into tobacco pipes and the leaves employed in the treatment of rheumatism.

Phyllodoces are tough, Heath-like shrubs from the alpine arctic areas of the Northern Hemisphere which thrive best in lime-free, humus soils in cool gardens. They have linear, evergreen leaves and stalked, solitary flowers, which are urn-shaped and arranged in racemes or terminal umbels. *Phyllodoce breweri* from California is semi-procumbent and 15–30 cms (6–12 ins) high with racemes of rosy-pink flowers. *P. caerulea,* a widespread species in northern Europe, N America and N Asia, has light to deep purple blossoms and *P. glanduliflora,* from W North America, has fragrant, yellowish-green flowers. Among the best of the garden forms are *P.* x *intermedia,* a natural hybrid between the latter and *P. empetriformis* (a brownish-pink species), with variable blooms from pink to mauve to light and dark purple and 'Fred Stoker', a light purple form.

Menziesia purpurea, the best of a rather small group of seven species, is a charming Japanese deciduous shrub, 90–180 cms (3–6 ft) tall. It makes an attractive subject for well-drained, lime-free soil but since it likes a cool root run it should be afforded some shade during the hottest part of the day. If this is not possible mulching the ground with generous quantities of moist peat fulfils the vital function of keeping the roots moist. The species has many red or purple, bell-shaped flowers, pendent in terminal umbel-like clusters and oblong, bristly leaves.

4

Agapetes serpens

Agapetes serpens (*Pentapterygium serpens*) is a somewhat bristly evergreen shrub from W China. It has a rather large tuberous rootstock from which arise several slender, drooping stems 60–90 cms (2–3 ft) long thickly covered with alternate, small, lanceolate leaves. The lantern-like flowers 2 cms ($\frac{3}{4}$ in.) are suspended along its length and are five-angled and bright red with darker V-shaped marks. The plant will not tolerate frost but in colder climates makes an attractive greenhouse perennial which can be grown in pots in a peat/sand mixture.

A. flava from NE India has yellow flowers margined with red; *A. incurvata* (*Pentapterygium rugosum*) from the Khasia Mountains is white and green with purple or blood-red bands and *A. macrantha,* also from Khasia, has variable flowers of white, yellow, pink or red with darker striations. Like so many ericaceous plants the leaves of some Agapetes (like *A. saligna*) are used as tea substitutes.

Leucothoë is a small genus of evergreen and deciduous shrubs from Japan and North America. They share the family liking for cool, humus type, lime-free soil and must have plenty of moisture throughout the growing period. The largest flowered is *L. keiskei* from Japan, an evergreen rarely exceeding 25–30 cms (10–12 ins) with smooth, prostrate, zigzagged stems that are red when young. These are furnished with thick, oval, tapering and alternate leaves and have racemes of five-lobed, white flowers.

As with most of the dwarf, ericaceous plants propagation is possible by means of half-ripe cuttings, taken with a heel of the old wood, in summer.

Zenobia speciosa (*Z. pulverulenta*) is a monotypic species from eastern North America, a deciduous or semi-evergreen (in mild districts) shrub, 1.25–1.75 m (4–6 ft) high, of somewhat irregular, thin habit with oblong-ovate, intensely glaucous leaves. In var. *nuda,* which grows with the type, the shiny green foliage lacks the glaucous bloom. The pendent flowers, produced in fours or fives in axillary clusters on the terminal shoots of the previous year's growth, are bell-shaped and glistening white with five shallow lobes, a little reminiscent of Lily of the Valley.

Escalloniaceae

7 genera and 150 species

This is a family of woody shrubs with simple, perfect flowers having four to five sepals, four to five petals, four to five stamens, a superior or an inferior ovary and a capsule or berry fruit.

The most important genus represented in gardens is *Escallonia,* the species of which are mainly evergreen (*E. virgata* is a notable exception) and native to South America, especially in Andean regions. Except in very mild areas they are generally too tender for gardens experiencing much frost, although the protection of a wall increases their chances of survival. Nevertheless in warm gardens, especially near the coast, they are invaluable for their freedom of flower and long blooming season.

Escallonias thrive in soils which would prove fatal to many shrubs, even in poor ground polluted by builder's rubble; wind does not worry them unduly and they like lime. Since the habit is inclined to be thin and straggling, it is important to prune the bushes back in spring to keep them compact and encourage the production of young flowering wood.

One of the hardiest is *E. virgata,* an open-habited shrub 2.15–2.5 m (7–8 ft) tall, with arching sprays of small white flowers in summer. Hybrids from this species make the most satisfactory garden plants, especially the Donard hybrids, raised by Slieve Donard Nursery of Ireland, as 'Slieve Donard' (red); 'Donard Beauty' (purplish-pink and very free); 'Donard Brilliance', with its large crimson flowers, a splendid shrub for growing against a whitewashed wall, for example; and 'Donard Seedling' (which is white with pink buds).

Agapetes macrantha

E. macrantha (now considered a variety of *E. rubra*) is commonly planted for hedging in southern Ireland and as windbreaks in the Isles of Scilly. It is evergreen with crimson flowers. Some good kinds of hybrid origin are *E.* 'Langleyensis', an evergreen 2.5 m (8 ft) tall with small glossy leaves and short sprigs of rosy-carmine flowers all along the arching branches, and *E.* 'Exoniensis' which has white or rose-tinted blooms.

Some Escallonias, notably *E. illinita* (which now includes *E. viscosa*), have an unpleasant pigsty smell, which is even retained (after years of drying) in herbarium specimens. It is white-flowered and evergreen.

E. rubra from Chile is a red-flowered evergreen shrub up to 5 m (15 ft) in height with reddish glandular shoots. In the Andes this species is called *siete camisas* (seven shirts) because of the ease with which the bark scales rub off and are replaced by other layers of loose covering.

Escallonia 'Slieve Donard'

Eucryphiaceae

1 genus and 5 species; all S temperate

These are deciduous trees or shrubs from the Southern Hemisphere with evergreen, opposite, simple or pinnate leaves. The flowers are large, white and showy, with four petals, four sepals, many stamens and winged, fleshy seeds.

Eucryphia glutinosa is a beautiful summer-flowering tree from Chile, growing up to 6 m (20 ft) high, with upright but pliant branches carrying masses of glistening white, bowl-shaped blooms of 7 cms (3 ins) filled with long golden stamens. They somewhat resemble the yellow Rose of Sharon (*Hypericum calycinum*). The dark green, divided, rose-like leaves assume orange and red tints in autumn which gives the tree a second attribute. This is the hardiest species but slow to establish, especially in poor ground. It requires a rich acid soil which is mulched each spring and flowers when the plants are about 90 cms (3 ft) tall. It is propagated from seed which unfortunately throws many doubles—the latter generally inferior to the singles and usually discarded. Layering is probably a better if slower method of increase.

E. cordifolia, also Chilean, is less hardy but makes a large tree—24 m (80 ft) in its native haunts, where the timber is used for making railway sleepers, flooring, furniture, telegraph poles, oars and yokes. The bark is the source of a commercial tannin and Indians use the wood for making canoes.

The white flowers 5 cms (2 ins) across, produced singly in the terminal leaf axils, are an excellent source of honey. The leaves are simple and heart-shaped with wavy margins.

In spite of the fact that *E. cordifolia* is not really hardy in Britain, a natural hybrid between this species and *E. glutinosa* occurred about 1915 in the gardens of Colonel Messel at Nymans in Sussex. Now known as *E.* x *nymansensis* this beautiful hybrid flowers earlier than its parents, is hardier than *E. cordifolia* and tolerates a lime soil—which *E. glutinosa* detests.

The dark green, slightly downy, leaves resemble both parents, in that some are simple and others compound. The flowers are 6 cms (2½ ins) across, pure white with many golden anthers.

E. lucida (*E. billardieri*) is the Leatherwood of Tasmania and makes a tree of 8–20 m (25–70 ft)—usually the smaller size—with pure white four-petalled flowers 2.5 cms (1 in.) across and glossy green oblong leaves which are silvery beneath. It likes moist soil and is the source of pinkish timber used for cabinet work and building. *E. moorei,* the Plumwood from New South Wales, is used for the same purposes.

Eucryphia glutinosa

Euphorbiaceae

300 genera and 5000 species
cosmopolitan except in the arctic

This is a variable family of shrubs, trees and herbaceous plants, many xerophytic; also a few lianes, several with stinging hairs.

The leaves are usually alternate although in some instances opposite; in a few genera both kinds are present, the opposite leaves being at the tops of the stems and the alternate ones below. The stalks frequently secrete latex.

The inflorescences are complex since so many kinds are evidenced in the various genera. The regular flowers are always unisexual but may be monoecious (on the same plant) or dioecious (on different plants), with the perianth segments in two whorls but more often in one. The stamens may be one or many—free, united or sometimes branched—and there is a superior ovary. Flowers are often in combined clusters, simulating single flowers.

Several plants of economic importance belong to this family, notably *Manihot esculentus* (*M. utilissimus*) (Cassava and Tapioca); *M. glaziovii* (Ceará Rubber); *Hevea brasiliensis* (Para Rubber); *Croton tiglium* (Croton Oil) and *Ricinus communis* (Castor Oil).

The largest and most interesting genus is *Euphorbia* with 2000 species, the majority from subtropical and warm temperate regions. These differ considerably in form and habit, those from dry places often so like members of Cactaceae that—when out of flower—it is difficult to tell them apart. These have fleshy stems of various shapes, for example squat, cylindrical and spherical, often with short thorns—just like Cacti, but one sure way of identifying the families is by snapping a stem. The Euphorbias have a milky latex; Cacti (except for some Mammillarias) do not.

One of the most popular is the Poinsettia, *Euphorbia pulcherrima,* a plant widely grown for the Christmas trade in America and Europe. It is a Mexican species, the showy part not the flowers (which are grouped terminally and are small and greenish-yellow) but the large, scarlet, petal-like bracts surrounding them. These

Euphorbia pulcherrima
'Paul Mikkelsen'

Euphorbia fulgens

Euphorbia epithymoides (*E. polychroma*)

Euphorbia griffithii

remain in character for weeks. There is also a white form.

In tropical gardens a single Poinsettia makes a striking feature, growing up to 3 m (10 ft) tall and nearly as much across. Hedges of Poinsettias can be seen in the West Indies but for pot work a shorter plant is required and one known as 'Paul Mikkelsen' fulfils all the requirements of a house plant. It has extra large heads of flowers, is easier to grow than the type species and it responds to chemical dwarfing techniques so that flowers are produced on stems of 60 cms (2 ft).

Pot-grown Poinsettias need rich soil and feeding, with slight shade from burning sun; as garden plants they require full sun and good soil and are best cut back after flowering. They are propagated from cuttings—which should be dipped in powdered charcoal to stop the latex 'bleeding'.

E. fulgens, also Mexican, has an entirely different appearance. It is a slender, willowy shrub with oblong, alternate leaves and clusters of small orange-scarlet flowers along the length of each stem. Here again appearances are deceptive for the 'petals' are really bracts and the 'stamens' are flowers. This species is good for cutting or pot work, but when grown in tropical gardens should be given a sheltered position.

E. milii var. *splendens* is a spectacular, sharply thorned shrub from Madagascar, about 90 cms (3 ft) high and irregularly branched with a few oval leaves near the tips of the branches. The flowers come in the leaf axils, in branching cymes; they are long lasting and brilliant scarlet.

Several South African species, some more quaint than beautiful, are commonly met with in greenhouses or on dry rocky slopes in warm climates. Representative are *E. meloniformis,* ribbed and squat like a dumpty floor cushion; *E. caput-medusae,* the Medusa's Head, with radiating branches from a central head; *E. grandicornis,* fiercely spined and *E. obesa,* which has male and female forms and is spherical to sausage shaped and eight-angled.

Hardier Euphorbias, grown in cold temperate gardens, are valued for their early—frequently greenish-yellow—flowers. They are known as Spurges and the effective parts are the bracts which surround the small, usually insignificant true flowers. *E. cyparissias,* the Cypress Spurge, grows about 30 cms (1 ft) high, with very narrow leaves and greenish-yellow 'flower' heads in spring; *E. epithymoides* (*E. polychroma*), also 30 cms (1 ft), is the Cushion Spurge, very bright golden and *E. characias* and its subspecies *wulfenii* grow taller, the stems of 120 cms (4 ft) densely clothed with striking blue-green leaves and surmounted by yellowish-green bracted blooms. These are all European.

E. griffithii is a splendid species from the Himalayas, the leafy stems 75 cms (2½ ft) high topped by rich burnt-orange flowers. This one likes a little shade while the British *E. palustris,* whose lush, leafy stems of 90 cms (3 ft) are topped by sulphur-yellow blooms, likes plenty of moisture. An annual worth noting is the North American *E. marginata* or Snow on the Mountain, a graceful plant up to 60 cms (2 ft) tall, its leaves and bracts heavily margined with white. The latex from this species is used as Chewing Gum by Indians in New Mexico. The European *E. lathyris* is called the Caper Spurge for its fruits resemble the flower buds of the Caper, *Capparis spinosa.* They are, however, poisonous to eat but the stiff, angular, blue-green leaves have a sculptured appearance which is pleasing in some settings. This species is biennial and has the unfounded reputation of driving moles away from any areas where it is planted.

Many Euphorbias have economic properties, being used for such things as purgatives, rubber, as arrow poison or to stupefy fish. The young shoots of *E. antiquorum,* a xerophytic tree from

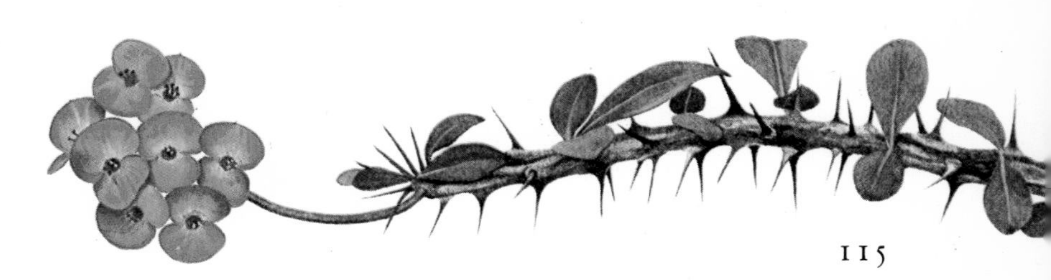

Euphorbia milii var. *splendens*

SW Asia, are boiled and reboiled by the Chinese before being made into a sweetmeat with sugar. *E. antisyphilitica* is the source of Candelilla Wax, which when refined is used for polishes, gramophone records, dental moulds, metal lacquer, sealing wax and similar objects.

Jatropha podagrica, the Guatemala Rhubarb, is a strange Colombian plant with a swollen fleshy stem of 45 cms (1½ ft) which sticks out of the ground like a bottle. This bears large lobed leaves and erect inflorescences carrying numerous, small coral-red flowers on red branches. It is easily raised from seed and may be grown in light soil in deep containers or on rock gardens in warm countries.

J. curcas is the French Physic Nut from tropical America, the seeds of which contain a purgative oil that is nevertheless often used in cooking—as well as during the manufacture of soap and candles. It is a deciduous shrub 3 m (10 ft) tall with many yellowish-green flowers and three- to five-lobed leaves.

Acalyphas are bushy shrubs, many of them having their leaves variegated with red or creamy-white. These are alternate and ovate, more or less toothed and nettle-like while the flowers are small and in drooping catkins. Except for the female *Acalypha hispida* the blooms are inconspicuous; Acalyphas are normally grown for their bright foliage effects and in tropical gardens are frequently used for hedging purposes.

Several make attractive foliage plants for the greenhouse and indoor decoration, the most common being *A. godseffiana,* which has green leaves margined with cream and *A. hispida,* the Chenille Plant or Red Hot Cat's Tail. The last is a shrub 90–180 cms (3–6 ft) high from the Malay Archipelago, with alternate, bronze-green leaves which have red veins and long, pendulous, blood-red inflorescences, 30–50 cms (1–1¾ ft) long. The plants are easily struck from cuttings and grow quickly but require a sunny position and rich but well-drained soil. For pot work in greenhouses the plants should be established in a loam/leafmould compost and kept warm.

A. wilkesiana, a native of Polynesia, is called 'Match Me if you Can' in the tropics, because no two leaves are identically patterned. This shrub, 1–6 m (3–20 ft) tall, is essentially grown for its foliage, which at a distance is as gay as flowers, being brilliantly marked with red, crimson, green, fawn-brown and bright red.

Codiaeum variegatum is a well-known shrub from the Moluccas, cultivated in warm greenhouses or tropical gardens for its brightly coloured leaves. Commonly known as 'Crotons' these are used in hedging, as specimen plants in key situations, where they sometimes run up to 4.5–4.75 m (15–16 ft), for group plantings on lawns or as small pot plants in homes and greenhouses. When they get too tall for the latter purpose—45–60 cms (1½–2 ft) is an ideal size—they should be scrapped in favour of younger plants.

Leaf colours vary from green to yellow and red, variously blotched, spotted or striated with other shades. In some forms the leaves are curled and twisted, and one has two leaf blades (on each leaf) separated by a length of mid-rib. The cultivar 'Van Ostensee' has long, thin, grassy leaves of dark green, spotted with orange.

The true Crotons, with some 750 tropical and subtropical species form a much larger genus with more medicinal than decorative interest. They make small to medium-sized shrubs or sometimes small trees with narrow, lanceolate leaves, slender, pointed and often hairy underneath and they have spikes of insignificant (monoecious or dioecious) flowers.

C. tiglium is the source of the powerful purgative Croton Oil. This is expressed from the seeds which are also used to stupefy fish. Others yield lacquer used in varnish-making (*C. laccifer*), or medicines employed for a number of ailments, such as skin diseases (*C. cortesiana* and *C. humilis*), eye troubles (*C. echinocarpa*), fevers (*C. nivea*) and toothache (*C. alanosana*). Gum exuded from the trunk of *C. xalapensis* is used instead of toothpaste by Mexicans and Dragons-blood resin obtained from *C. draco* is the source of a red dye used in Mexico for hoof ailments of horses.

Ricinus communis is the Castor Bean or Castor Oil Plant, a monotypic genus from tropical Africa and Asia. Although it makes a tree in the tropics—up to 12 m (40 ft) tall—in temperate regions it has to be grown as an herbaceous annual or perennial. Under these conditions the plants grow 90–150 cms (3–5 feet), rarely taller, and make effective 'dot' plants among lower growing summer bedding or they may be used for background planting

Jatropha podagrica

in borders. The red-leafed forms are more ornamental than the green and there are literally hundreds of variants in cultivation.

The seed coats are very hard and take a long time to break down, so that soaking the seed in tepid water for twenty-four hours prior to sowing is advisable to hasten germination. The plants require rich fertile soil in an open sunny position and staking may be necessary in windy situations. The large, peltate leaves are palmately lobed, the flowers appearing in terminal panicles and the fruits quite showy, something like Horse Chestnuts (*Aesculus hippocastanum*)—usually green but red in the crimson-leafed cultivars.

The Castor Oil of commerce, used for medicine and technical purposes and as a lubricant, is derived from the seeds. A secondary or cruder oil is employed in the manufacture of transparent and textile soaps, typewriter inks and imitation leathers and, when dehydrated, in the making of paint, enamel and varnish.

Although a useful shade tree for tropical situations—especially in poor soils near the coast—and notwithstanding the fact that Botanic Gardens sometimes grow it, few will wish to cultivate *Hippomane mancinella*. This is the Manchineel, one of the world's most toxic plants; even dew or rain dripping from the leaves carries with it sufficient poison to cause dermatitis. A drip in the eyes induces temporary blindness and should any of the latex touch the skin it causes a burning sensation, followed by blistering and painful swelling. Even smoke from the burning wood can cause headache and sore eyes while eating the green, small, apple-like fruits is to invite terrible internal torment.

Phyllanthus is a large genus of some 600 species but few have any horticultural value and these can only be grown in greenhouses in cold climates.

An unusual feature of some species is the flattened stems which look like leaves and indeed perform the function of those organs. They are known as phylloclades and when the flowers are borne along their margins it appears to be the leaves which bear the blooms. *P. angustifolius* is an example, a shrub 60 cms (2 ft) tall with red flowers on the phylloclades.

Acalypha hispida

Some *Codiaeum* leaf forms

P. acidus is the Indian Gooseberry, a tree with round, fleshy, ribbed fruits which are used for pickles and preserves—particularly in Ceylon.

P. pulcher from Java is a small shrub with oblong small leaves and yellow and red flowers. *P. grandifolius* is a South American native which appears to have large pinnate leaves of 30–60 cms (1–2 ft), but closer inspection reveals that these are lateral shoots with small flowers in their axils. One species, *P. fluitans*, is a small, free-floating aquatic sometimes used in aquaria.

Undoubtedly the most important plant in the genus however is *Hevea brasiliensis,* the Brazilian Rubber Tree. Before Columbus discovered America, Mexicans and Indian Aztecs were tapping trees in the Amazonian forests to collect the milky sap. From this latex they fashioned waterproof clothing and 'bouncy' ball-like playthings which Spaniards of the 16th century found better than any then known in Europe.

In 1823 Mr Mackintosh of Glasgow discovered that rubber was soluble in naphtha and impregnated fabrics, turning these into waterproofs marketed under his name.

Gradually more uses were found for rubber—particularly tyres and footwear—and the product became extremely valuable. At first the latex was collected exclusively from wild plants, especially in Brazil, and mostly from *Hevea*—although other genera were also used. With increasing demand supplies became more difficult and prices soared and then Sir Henry Wickham, resident in Manaos took seeds (some reports say he smuggled them out) from Brazil to Kew. Some of the resultant seedlings were shipped in 1876 and 1877 to Ceylon and Malaya where they formed the foundation of the vast rubber plantations of Malaya, Java and Sumatra.

The latex is extracted by tapping, that is, by making a V-shaped cut in the tree bark early in the morning and affixing cups to catch the white or yellow fluid.

One of Hong Kong's most spectacular trees, much used in parks and for roadside planting, is the Mu Oil Tree (*Aleurites montana*). It is also planted in Florida where it makes a semi-evergreen with white, star-like flowers in spring, 5 cms (2 ins) across. *A. fordii*, the Tung Oil Tree, flowers in spring when the trees are bare of leaves. These are white and cup-shaped. The nuts contain much oil and in Malaya a dozen of them stuck on the end of a spear are used to make long-burning torches.

Berberidopsis corallina

Azara lanceolata

Flacourtiaceae

93 genera and 1000 species
tropical and subtropical

These are dicotyledonous trees and shrubs, generally with alternate leathery leaves and either cymose or racemose inflorescences or showy, often solitary, axillary flowers. These have many stamens, a superior ovary and anything from two to fifteen sepals and fifteen to no petals. Some species have edible fruits and a few supply oil and timber.

Berberidopsis corallina from the mountain forests of Chile, is a low, scrambling shrub with large pendent racemes of crimson flowers on long stalks. In temperate gardens it requires a warm sheltered situation.

Idesia polycarpa of Japan and China grows to 13 m (40 ft) and is noted for its long drooping panicles of fragrant, yellowish-green flowers. In male plants these may be 13–15 cms (5–6 ins), although longer on female plants. The round, orange fruits are the size of peas.

Azaras are Chilean evergreen shrubs or small trees with small, fragrant, many-stemmed flowers in short clusters. *Azara microphylla,* growing 4–9 m (12–30 ft), with dark, shiny, toothed leaves and tiny, yellow, Vanilla-scented flowers is the hardiest. It blooms in spring and has a yellow leaf-edged variety 'Variegata' which makes a good wall shrub. *A. integrifolia* is a useful street tree or shrub for warm climates and has bright yellow flowers with small purple sepals. *A. lanceolata* is a small Chilean tree needing sheltered conditions in frost-prone climates. It has coarsely-toothed, lanceolate and evergreen leaves and soft yellow flowers in short axillary corymbs.

Oncoba spinosa is a beautiful tree for the tropics with solitary or paired, white, fragrant flowers 5 cms (2 ins) across, full of golden stamens. It is an African native often known as Fried Egg Tree. The large, 6 cms ($2\frac{1}{2}$ ins), spherical fruits have a leathery rind and when pulverized are used as snuff by natives. The seeds are also made into ornaments and Africans hollow out and polish the fruits and use them as snuff boxes or rattles. The tree is widely planted in Florida and when in full bloom, with hundreds of flowers like wild white roses, it is an arresting sight. The leaves are elliptical and the trunk spiny.

Fumariaceae

16 genera and 450 species
mostly N temperate

This family consists of dicotyledonous plants of a herbaceous nature, sometimes bulbous and occasionally climbing, with watery stem juices. The leaves are alternate and usually compound, and the flowers are irregular but bisexual with two sepals, two or four petals, six stamens and a superior ovary.

Dicentras are pretty spring-blooming perennials for cool climates, with fleshy brittle roots and finely cut fern-like foliage. The flowers are pouched and hang in racemes. They are much visited by bees which hang on the pendent blooms and probe for honey—first on one side of the pouch and then the other. Dicentras flourish in cultivation but need rich, moist, humus soil.

An outstanding species is *Dicentra* (*Dielytra*) *spectabilis,* an elegant Japanese plant with soft green, compoundly cut leaves and arching sprays, sometimes 90 cms (3 ft) or so in length. These are hung with large, rosy-pink, heart-shaped flowers with whitish tips—which glisten when the blossoms are fresh. This tearful effect makes appropriate its common name of Bleeding Heart. Another name, Lady in the Bath, refers to the effect produced when a flower is turned upside down and the pouches pulled downwards. It should be planted in rich cool soil in sun or light shade and protected against slugs early in the year. Lifted plants force easily in a warm greenhouse and there is a white form *alba.*

D. cucullaria, the North American Dutchman's Breeches, has deeply cut, fern-like foliage, sappy stems and dainty cream or primrose flowers with sac-like spurs. This gives them an amusing resemblance to rows of baggy trousers—hanging upside down from a line.

D. canadensis, also North American and somewhat similar, grows naturally in moist, black, forest mould or other damp situations. The flowers have a sweet scent and the clusters of yellow bulblets—like grains of wheat—under the soil have earned it the name Squirrel Corn. When these are dried they have medicinal uses as a tonic and alterative.

D. eximia is the North American Fringed Bleeding Heart, a useful border perennial which combines fern-like foliage with racemes of 30–45 cms (1–1½ ft) of rosy-purple flowers. There is also a white form *alba* and a cultivar 'Bountiful' which grows 45 cms (1½ ft) high with thirty to forty fuchsia-red flowers double the size of the type species. Equally at home in sun or shade (providing the soil is moist), the plant can be introduced to odd garden corners where a little colour is needed.

Corydalis are sappy-stemmed perennials with underground tubers, one of the best known being the European (including British) *C. lutea.* This has clusters of bright yellow, spurred flowers on stems 30 cms (1 ft) tall and deeply cut leaves. It flowers for months on end providing its requirements are met in the matter of sun and good drainage.

C. cheilanthifolia is a charming little Chinese species with long and very deeply cut foliage and long spikes of yellow blossoms on stems up to 25 cms (10 ins) tall. It makes a choice rock plant for rich, well-drained stony soil in sun.

C. solida (often confused with *C. bulbosa* in gardens) is a European (including British) species for sun or shade, forming wide tufts of ferny foliage from a mass of solid, bulb-like tubers. Above the leaves rise the flowering stems 30–40 cms (12–16 ins) high, carrying large purple-spurred flowers. The tubers are eaten in Siberia.

C. cashmeriana from Kashmir is a beautiful plant 15 cms (6 ins) tall with rich blue flowers 1 cm. (½ in.) across which have darker tips. It requires a cool, peaty soil to thrive.

Very similar to *Corydalis* is *Fumaria,* a genus with many European representatives, often of a weedy nature. Commonly known as Fumitory, they have compound leaves with oblong-oval segments which are usually glaucous. The small pink, whitish or reddish-purple flowers are borne in dense clusters.

In Kent these are called Wax Dolls from their doll-like appearance. Farmers regard the plant as an indicator of good rich soil. In the 14th century the juice was used for curing skin diseases.

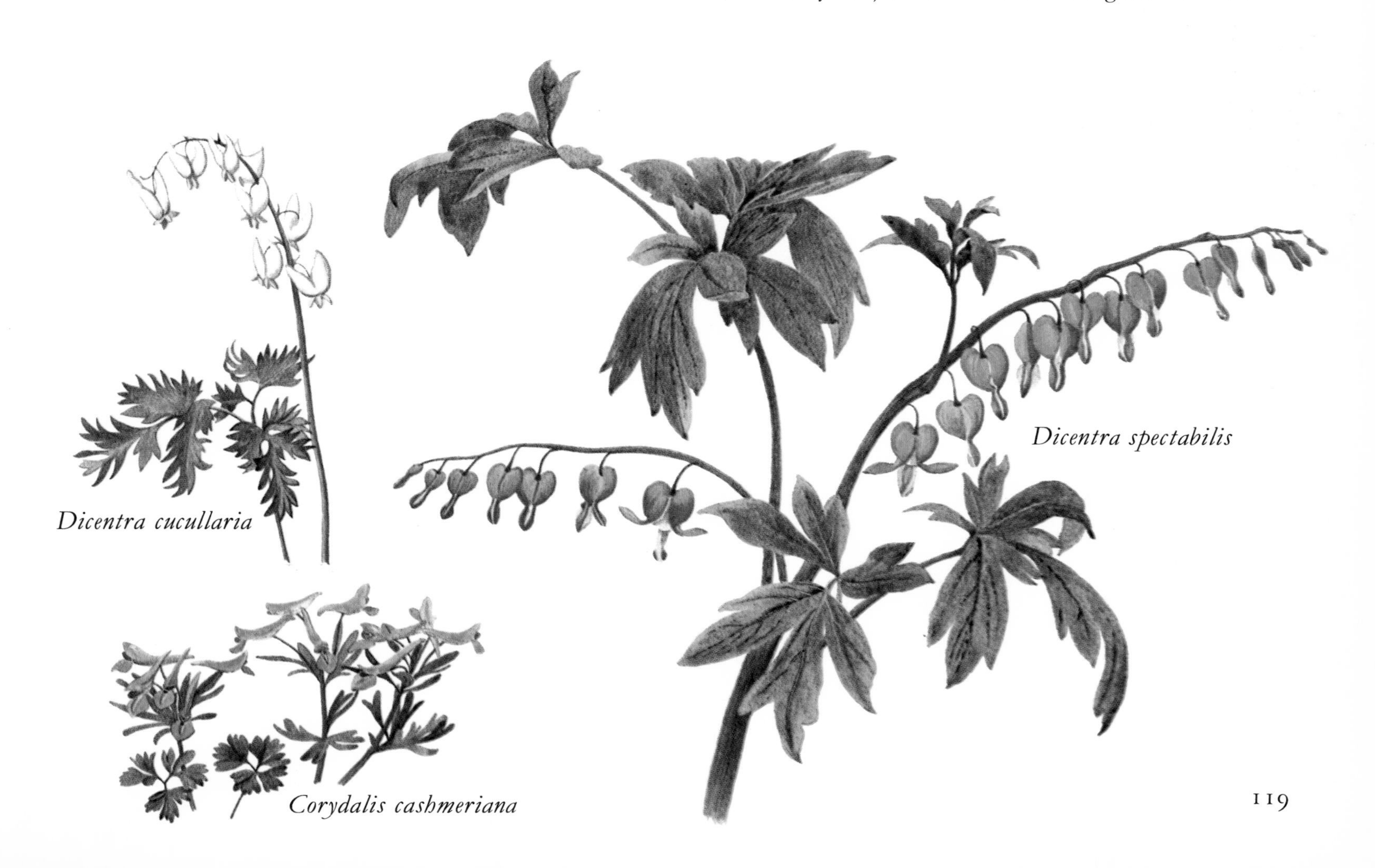

Dicentra spectabilis

Dicentra cucullaria

Corydalis cashmeriana

Garrya elliptica

Garryaceae

1 genus and 18 species

Garryas, native to the warmer western parts of North America, are evergreen shrubs—or sometimes small trees—with opposite, entire, simple leaves and unisexual flowers. Male and female blossoms are borne on separate plants.

The males are carried on catkin-like, long, pendulous inflorescences; each flower has four stamens and usually four perianth segments. The female flowers have shorter tassels, inferior ovaries and as the season advances bear strings of purplish berry-like fruits.

For garden purposes the best species is the Californian *Garrya elliptica,* a splendid winter-flowering shrub which is hardy in Britain if grown against a warm wall. It has dark green oval leaves which are silvery beneath and rather leathery and the male plant bears long suede-green catkins of tightly packed flowers. These are tipped with gold, so that the catkins appear to be segmented like the body of a green and gold caterpillar. Given a good form, and a sheltered situation they may extend to 30 cms (1 ft) in length.

Since seedlings are unreliable in case one obtains the less desirable female plants, propagation is best carried out by means of cuttings. Root these in warmth from half-ripe shoots taken in midsummer. In favoured climates the Silk Tassel Bush (as this *Garrya* is sometimes called) grows to 4.75 m (16 ft) but 1.75–3.5 m (6–12 ft) is more usual. It resents root disturbance, so young struck cuttings should be grown along in pots and turned out and carefully planted when well rooted. The bushes tolerate pruning so they can be kept compact and tidy. The bark and leaves contain a bitter principle which is sometimes employed as a home remedy to relieve intermittent fevers.

The plant was named by David Douglas (who sent it home to the Horticultural Society's Garden in 1834) after Nicholas Garry of the Hudson's Bay Company. Garry helped Douglas considerably in his plant-hunting expeditions in western North America.

Gentianaceae

80 genera and 900 species

This is a family of dicotyledonous plants with a tremendous range of habitats and a wide geographical distribution. From the arctic to the equator, the high mountains and low valleys, in brackish pools and fresh-water marshes species can be found. Some of Europe's most beautiful plants are included amongst their ranks.

Although there are a few shrubs they are mostly herbaceous and often perennial; some genera (like *Voyria*) being saprophytic (obtaining their nutriment from dead organic matter and living without any green colouring matter). Many contain a bitter principle with medicinal properties.

The leaves are opposite, sometimes sessile and usually entire; the flowers solitary or clustered, regular and perfect with four to five sepals and four to five petals in the form of a bell, funnel or sometimes a salver shape. There are as many stamens as there are petals, and the flowers have a superior ovary and a capsule fruit with many small seeds, or occasionally a berry (as *Chironia*). The blossoms are insect-pollinated.

Gentiana with 400 species is the largest and most important genus and takes its name from Gentius, King of Illyria, who is supposed to have been the first to discover the medicinal properties of these bitter plants.

The European species *Gentiana cruciata* was at one time called *Ladislai Regis herba* because of a strange legend. This relates how the land of Ladislas, one time King of Hungary, was swept by plague. Distressed by the miseries of his people, the king prayed to the Almighty and, shooting an arrow into the air, requested that it should be guided towards some herb which might alleviate their misery. The arrow plunged into the heart of a Gentian plant, the root of which was immediately tried and found to possess remarkable curative properties against the plague.

According to Robert Turner, a seventeenth-century astrological botanist, Gentian possesses extraordinary powers such as to 'resist poisons, putrefaction and the pestilence, and help digestion: the powder of the dry roots helps bitings of mad dogs and venomous beasts, opens the liver, and procures an appetite. Wine, wherein the herb hath been steept, being drunk refreshes such as are overwearied by travel, or are lame in their joynts by cold or bad lodgings'.

Many Gentians are alpines of tufted habit whose cultivation provides a challenge even to experienced gardeners. Yet it is a challenge which many accept for the rewards are so great; as Reginald Farrer remarked 'If a Gentian will thrive with you you cannot go wrong with it; if it does not thrive, not all the king's horses nor all the king's men will induce it to do so' (*My Rock Garden*). Probably he was over pessimistic, but with Gentians it is impossible to dogmatize. Yet one fact is certain: the alpine

Gentiana excisa (*G. acaulis*)

kinds like moisture and are impatient of root disturbance, so a thriving colony should be left well alone.

In the European Alps Gentians are found in the high mountains where they have full exposure to the sun, are covered with snow for months on end in winter and can draw on a continuous supply of underground water in summer. These conditions are difficult to emulate in a garden and possibly the reason these alpine sorts succeed best in moist cool climates (as in the west and north of England) rather than in warm dry ones.

Gentians make lovely rock garden plants; the easier sorts such as *G. asclepiadea*, *G. excisa* (*G. acaulis*), *G. septemfida* and its variety *lagodechiana* and *G. pneumonanthe* can be colonized in rock pockets, used as edgings or allowed to form drifts in suitable beds. All the planting areas should be well drained and filled with light but moist soil; during dry weather watering may be necessary.

Perhaps the best known is that well-loved flower of the high alpine meadows of the European Alps, *Gentiana excisa* (*G. acaulis*). It has large, trumpet-shaped flowers of vivid blue in the best forms—although this colour may vary in intensity—and tufts of neat, oval leaves coming directly from the ground. Where *G. excisa* thrives it makes glorious carpets of colour and is indescribably beautiful. It also makes a wonderful garden edging but, if it is not happy, all one gets year after year are leaves. When this happens it is best to lift the plants and try again, choosing a cool, moist but well-drained situation. Other good species are *G. angustifolia, G. kochiana* and *G. clusii*. *G. clusii* has dark blue trumpets with green-spotted tubes; *G. alba* is white with green veining; *G. kochiana* is azure blue, spotted inside with green; and *G. angustifolia* has long, narrow leaves and open mouthed trumpets of deep blue, on stems 10 cms (4 ins) high.

G. asclepiadea, the Willow Gentian, is a much easier plant, its purple-blue—rarely azure—flowers narrower and less arresting than *G. excisa*. Nevertheless it is a reliable good-tempered perennial which blooms year after year, the flowers produced in late summer and autumn on long, arching stems of 60–90 cms (2–3 ft). It is freely reproduced from seed and will grow happily in any moist, shady spot. There is a white form which is equally attractive.

G. farreri, an outstanding Gentian from W China, is again not the easiest of plants. It forms tufts of narrow, linear leaves from which rise the sky-blue flowers—very blue and very beautiful. These have white throats and are striped with violet. *G.* x *macaulayi,* a hybrid of this species crossed with *G. sino-ornata,* is very similar.

The Vernal Gentian, *G. verna,* is one of the most arresting flowers of the High Alps in early spring. It is a loosely tufted plant about 7 cms (3 ins) high, with brilliant blue, white-throated flowers. These are produced singly at the tops of the shoots, but unless the species is grown in quantity its brilliant effect is lost. Grit, sand or broken limestone should be mixed with the soil, which at all times must be moisture retentive.

G. kurroo is a Himalayan plant, making basal rosettes of linear-oblong leaves with very long, narrow, wide-mouthed, sheer blue flowers in early autumn. It is slightly lime-tolerant and the roots have tonic medicinal uses in its native habitat.

G. sino-ornata from W China and Tibet is another autumn bloomer with an almost prostrate habit, although the flowering stems turn up at their tips. The blooms are deep blue, paler at the base and usually stemless. This species, which requires an acid soil to thrive, is not unlike *G. farreri* except for its deeper blue colouring.

There are many other dwarf Gentians, some more accommodating than others; these include *G.* 'Inverleith', a splendid cultivar with Cambridge-blue flowers in autumn; *G. newberryi* from North America, pale blue; *G. bavarica,* a moisture-loving species from Europe, deep blue; and *G. saxosa,* a New Zealander, with white flowers veined with purple. The last forms a clump of rosettes, consisting of thick and fleshy leaves which measure 1–2 cms (½–¾ in.) in length. They are spoon-shaped and dark green tinged with purplish-brown. The flowers, borne on stems of 7–10 cms (3–4 ins), carry one to five flowers, each about 2 cms (¾ in.) in diameter.

G. pneumonanthe has leafy stems 15–30 cms (6–12 ins) high with narrow, linear leaves and terminal, deep to purplish blue flowers which are lighter near the bases. It has a wide distribution through Europe and temperate Asia, mostly in moist, boggy heaths, and the flowers are the source of a blue dye. This is the Calathian Violet, worth cultivating according to Gerard because 'the gallant flowres hereof bee in their bravery about the end of August' and he tells us 'the later physitions hold it to be effectual against pestilent diseases, and the bitings and stingings of venomous beasts'.

Other reasonably easy and taller Gentians for summer and autumn flowering include *G. septemfida*, the Crested Gentian, a Caucasian plant bearing, on stems of 15–30 cms (6–12 ins), clusters of cylindrical flowers which widen and are fringed towards the mouths. These are blue-white inside and deep blue without and there is a variety *lagodechiana,* 22–38 cms (9–15 ins) tall, with oval or heart-shaped leaves and terminal deep blue flowers. *G. trichotoma* (*G. hopei*) from W China has branching stems 30–60 cms (1–2 ft) tall, with clear blue flowers in threes and *G. freyniana* from Asia Minor is a bog perennial with stems of 15–30 cms (6–12 ins) carrying many sessile purplish-blue flowers.

There are also a number of twining Gentians, the best perhaps *G. heleni* (*Crawfurdia trailliana*), a climber of 2–6 m (6–20 ft) with simple leaves and many rich heliotrope funnel-shaped flowers. It was first found by George Forrest in Yunnan in 1905.

Gentiana asclepiadea

Gentiana verna

Gentiana x *macaulayi*

Gentiana lutea

Frank Kingdon Ward, who rediscovered it on several occasions, relates an amusing account of attempts to bring it into cultivation while he was in N Burma with an American expedition. 'The Government botanist had just offered a fabulous reward . . . for ripe seeds of the species, and a young frontier officer was competing for it. He had sent a native collector miles and miles to discover and mark down some plants, and seemed quite surprised when I showed him masses of it on the main mule road within a hundred yards of his own bungalow! But, of course, it is easy to see what others miss if you happen to be a botanist and have your eye 'in' for plants.'

G. lutea from Europe and Asia Minor is a distinct yellow-flowered species for a moist sunny spot. It is reproduced from seed and the young plants must be carefully transplanted without damaging the roots, The star-shaped flowers are more or less in whorls on tall stems up to 90 cms (3 ft) high; the foliage forms a wide rosette of large, crinkled, basal leaves with small, simple leaves on the stems. *G. lutea* is the source of Gentian Bitter, used to improve the appetite and stimulate gastric juices and which is manufactured into liqueurs. This principle comes from the dried roots which are collected in the late summer and autumn. Local names for the species are Felwort, because it was thought to cure fels (galls), Bitterwort and Baldmoney.

In South America and New Zealand there are other yellow kinds and also species with red flowers, but these do not take kindly to cultivation and are rarely if ever seen in gardens. A Gentian which attracted much attention in the 1930s was *G. scarlatina,* collected in Peru by Miss G. Stafford of Enfield, England. This was particularly noteworthy on account of its globular golden-yellow flowers, which were streaked and splashed on the outside of each petal with scarlet. Unfortunately there are several defects which account for its scarcity in cultivation. The blooms never fully open and it has proved difficult to grow and propagate although plants have been raised and flowered in the Royal Horticultural Society's gardens at Wisley.

Two delightful biennials in Gentianaceae are *Exacum affine* from Socotra and *E. macranthum* from Ceylon. Both must be cultivated under glass except in frost-free places, seed being sown in early spring in a sand/peat compost. A temperature of 15–18°C (60–65°F) is necessary for germination although as the season progresses this may fall to 13°C (55°F) at night. *E. affine* makes a branching plant 22 cms (9 ins) high and as much across, with smooth, stalkless leaves and dozens of lilac-blue flowers with golden stamens. These come in the axils of every leaf and have a powerful Lily of the Valley fragrance. *E. macranthum* has larger violet flowers, 5 cms (2 ins) across, but needs more heat.

Chironias are South African plants used in tropical gardens which are well drained, as in sandy soil near the sea. They all need water during the growing season however. Elsewhere they are usually treated as pot plants and grown under glass.

Lisianthus nigrescens

Exacum affine

Sabatia angularis

Gentiana septemfida

Gentiana sino-ornata

Gentiana saxosa

Chironia baccifera is the Christmas Berry or Wild Gentian, an evergreen perennial growing about 45 cms (1½ ft) high with upright stalks and narrow fleshy leaves. These are covered late in the year with bright pink, star-shaped flowers and soon afterwards followed by pea-sized scarlet berries. It is the only berrying species and looks most effective when grouped near a pond or in a rock garden.

C. humilis is an annual, less than 30 cms (1 ft) high, with pinkish-magenta flowers and *C. transvaalensis,* a perennial 50 cms (20 ins) tall, has soft pink blossoms.

Sabatia (*Sabbatia*) are annual or biennial plants from North America with erect branching stems, opposite, ovate to heart-shaped, stalkless or stem-clasping leaves and many white or rosy flowers. As they are generally found near water they make good subjects for the pool side, although *S. chloroides*—with deep rose flowers on stems of 30–60 cms (1–2 ft)—is sometimes used for bedding purposes. The plants contain a bitter principle and when dried are used as a tonic.

S. angularis is the Rose Pink, a pretty little plant of pyramidal habit, with abundant rose-pink or occasionally white flowers on stems of 45–60 cms (1½–2 ft). They are 2–5 cms (1–2 ins) across and sweetly scented. *S. difformis* (*S. lanceolata*) has five-petalled white flowers.

Canscora parishii is a square-stemmed annual from Moulmein, with white, funnel-shaped flowers. It grows about 60 cms (2 ft) tall and likes moist soil with chalk or lime rubble in the compost. All the Canscoras are greenhouse plants except in the tropics.

Centauriums (often listed under *Erythraea*) are mostly N temperate plants, used in rock gardens or moist borders. They have many small pink, yellow or (rarely) white flowers in a terminal inflorescence and stalkless, opposite leaves. *Centaurium venustum* with deep rose to white flowers 2 cms (1 in.) across is the species most frequently grown.

Lisianthus (*Eustoma*) *nigrescens,* the Funeral Flower of Mexico, has pendulous, nearly black, five-lobed flowers 4 cms (1½ ins) long, in bunches and simple, opposite, entire, oval-lanceolate leaves.

Geraniaceae

5 genera and 750 species

This is a family of dicotyledonous plants, many of which make valuable ornamentals for massed bedding in gardens, greenhouses, window boxes, growing indoors and container planting. Pelargoniums are particularly important for all these purposes, with showy, colourful blooms which are produced over long periods.

Most of the species are herbaceous, although some have woody stems which are often hairy. The leaves may be opposite or alternate, the flowers regular or irregular and in general bisexual although some are unisexual. There are five sepals, five petals, five, ten or fifteen stamens, a superior ovary and a long style with five stigmas. The fruits have three to five divisions (each containing one seed), culminating in a long beak or bill. This is most marked in the genus *Geranium* and accounts for its common name of Cranesbill.

The last is not to be confused with the greenhouse 'Geranium' of cool climates, which belongs to the genus *Pelargonium.* The genus *Geranium,* with some 400 species, has a wide distribution throughout the temperate zones. The blooms are regular and somewhat salver-shaped, borne on branching inflorescences and the leaves simple or palmately divided. In the ripe fruit the carpels split violently, the awn twists upwards and the seeds are shot out. This circumstance allows free seeding so that some species become weedy unless the developing fruits are occasionally removed.

It is a fascinating genus and ranges from stalwarts 90 cms (3 ft) or more tall to tiny creeping species of no height at all. Amongst them may be found species for the rock garden, border, woodland glade and even the bog garden.

The majority of Geraniums flower freely in any light, well-drained soil and may be propagated by seed, or in the case of cultivars by division of the roots.

G. endressii from the Pyrenees has small, three- and five-lobed Buttercup-like leaves and light rose, five-petalled flowers lightly marked with red. Growing up to 30 cms (1 ft) in height it flowers most of the summer, making a good carpeting plant in sun or light shade. 'Wargrave Variety' is clear pink without the red tinge of the species. Two useful species for the herbaceous border are *G. psilostemon* (*G. armenum*), an Armenian species 60 cms (2 ft) tall, with brilliant magenta flowers, each petal of which has a basal black spot and *G. pratense,* the Blue Meadow Cranesbill.

The latter comes from N Europe, including Britain—where in the west country great drifts can be seen by the roadsides. It is a very beautiful perennial, 45–75 cms (1½–2 ft) tall with rich blue flowers 2.5 cms (1 in.) wide and long-stalked, seven-lobed leaves. The plant enjoys a number of common names, including Odin's Grace, Bassinet Geranium and Loving Andrews.

There are many garden varieties with white, blue, mauve, pink and purple flowers, also bicolors and doubles in nearly all these shades.

Two really good blue Geraniums are *G. grandiflorum* (now *G. meeboldii*) *alpinum* and *G. wallichianum* 'Buxton's Blue'. The Bloody Cranesbill, as the British *G. sanguineum* is commonly known, may be too intensely magenta for some tastes but it has a white form and also a dwarf salmon-rose variety (which some authorities consider a distinct species), known as *G. sanguineum lancastriense.* This invaluable plant forms flat carpets, festooned with great saucers of rosy flowers and looks particularly well atop a great boulder or in a rock-garden pocket.

G. versicolor (*G. striatum*), a European plant, rejoices in the quaint name of Queen Anne's Needlework, presumably because of the red veins on its rosy-pink flowers. It is similar to and often crosses with *G. endressii*. *G. ibericum* from the Caucasus grows 30 cms (1 ft) high with open panicles of blue to violet flowers. The plant grown under this name in gardens is usually *G.* x *magnificum* a hybrid of this species with *G. platypetalum* which resembles it closely but differs in minor botanical characters.

There are also several silver-leafed alpine species, notably *G. argenteum* from N Italy, 7 cms (3 ins) high, with pink flowers and hairy divided leaves and *G. cinereum,* its Pyrenean version, slightly taller and with a white form *album*.

The genus *Erodium* (commonly known as Storksbill) is distinct from *Geranium* in that the leaves are pinnate, not simple and lobed. The fruits have the same explosive mechanism, the awn contracting in dry air and uncurling in a humid atmosphere. This hygroscopic property is a weather indicator and is made use of in the manufacture of certain types of hygrometers, employed for measuring amounts of moisture in the atmosphere.

Erodiums are plants for sunny, well-drained spots in rock gardens, where both flowers and foliage are seen to advantage. A few species are dioecious which only matters if seed is required, when both male and female plants must be grown. They are lime tolerant, have no special cultural problems and can be propagated by seed or root cuttings.

The following are all European plants suitable for the rock garden: *E. petraeum glandulosum* (*E. macradenum*) with pale violet flowers spotted on its lower petals with dark purple; *E. chrysanthum,* sulphur-yellow with finely cut, silvery leaves; *E. corsicum,* rose with red veining; and *E. absinthoides,* which has loose heads of violet, pink or white flowers.

E. manescavii from the Pyrenees grows taller with leafless stems 30–60 cms (1–2 ft) high, carrying many purplish-red flowers with purple blotches on their upper petals. This species is sometimes planted in herbaceous borders.

The fleshy roots of the North African *E. hirtum* are used by the North African Tuaregs as food.

Sarcocaulon is a small genus of South African xerophytes with fleshy stems which store water, so that the plants can resist drought, if necessary for several years. These have a corky layer secreting an aromatic resin or wax which can be ignited, accounting for the name Bushman's Candle being given to *S. burmannii.* Only after rain do the small fleshy leaves and chalice-like flowers appear, the latter showy with five petals which often have dark blotches at the bases. When the leaves fall the stalks harden to persistent thorns, which protect the plant against marauders. *Sarcocaulon* species can be grown under greenhouse conditions in a peat/sand/loam mixture or in tropical countries as rock plants.

S. burmanii with white to purple flowers grows to about 30 cms (1 ft), as does *S. multifidum* which has pink flowers with dark red spots. *S. patersonii,* red- to pink-flowered, is 60 cms (2 ft) tall.

Monsonia contains about forty species which are mostly natives of Africa but extend to eastern India. They share the family characteristic of long-beaked fruits, which follow the general pattern of seed dispersal and have fifteen stamens in five groups of three.

M. speciosa has large rose flowers with purple eyes on stems of 15 cms (6 ins) and finely cut leaves; *M. lobata*, about 30 cms (1 ft) high, has purple, red and white flowers which are greenish outside, and *M. burkeana* is white with red veining. The plants need greenhouse conditions in temperate or cold climates.

Variegated-leafed Pelargoniums

'Mrs Morrell'

Pelargoniums have a wide tropical distribution. Their 250 species range from the E Mediterranean to S Arabia, from the Canaries to Tristan and from India to New Zealand. By far the greater number however come from South Africa.

Still universal favourites for outdoor bedding and greenhouse display, they were at the peak of their popularity in Britain in Victorian days.

Zonal Pelargoniums, brightest of all bedding plants, are derived from *Pelargonium zonale* and *P. inquinans*. The first cross was made in 1714, doubles appeared in 1863 and early in the 20th century Monsieur Crampel, a French nurseryman, raised the most famous cultivar of them all—'Paul Crampel'. This astute nurseryman, realizing the worth of his find, built up a large stock before offering plants to the public. To outwit thieves he never allowed them to flower until sure of his market, and then in 1903 sold them readily for £1 a time—no mean sum in those days.

Between 1820 and 1830 Robert Sweet's monograph *The Geraniaceae* appeared in five volumes, but few of the hand-coloured cultivars he portrayed remain in cultivation.

Most Pelargoniums grown today fall into one of the following categories.

ZONAL These are bedding or greenhouse Geraniums; stout plants with hairy leaves and stems and umbels of white, pink, salmon, red, crimson, purple or scarlet flowers. Both singles and doubles are found in the group and all have rounded leaves characterized by a broad ring—or horseshoe mark—of darker shading near the centre.

The famous 'Paul Crampel' belongs here, also such well-known bedding sorts as 'Maxim Kovaleski', orange-red; the semi-double,

Ivy-leafed Pelargoniums

'Crocodile' ('Sussex Lace')

Zonal Pelargonium 'Wembley Gem'

'L'Elégante'

showy vermilion-red 'Gustav Emich'; the pink, semi-double 'Queen of Denmark'; 'Wembley Gem', a deep glowing crimson; and the white 'Hermine'. These flower continuously and have countless garden and pot plant uses.

VARIEGATED-LEAFED Zonal cultivars which are grown less for the flowers than for the brilliant foliage. They may be silvery as in the dwarf, non-flowering 'Madame Salleroi' (often used for edging); gold as in 'Golden Crampel'; tri-coloured as in 'Crampel Master', whose olive-green leaves have a white centre streaked with gold; or quadri-coloured as in 'Mrs Henry Cox', the creamy-gold leaves erratically marked with blotches of purple, red and green.

IRENES A race of American origin from California. They have short internodes, which dwarf the height (although they are not miniatures) and so make them useful for pot work. The flowers come in a range of colours.

DWARFS AND MINIATURES These have flowers, leaves and stems in proportion and are very much smaller than normal Pelargoniums; the dwarfs average 20 cms (8 ins) and the miniatures less than 13 cms (5 ins). They can be used for window boxes, mixed bowls or pot work. Various breeders have produced strains including H. F. Parrett and the Reverend S. P. Stringer of England; the late E. Both of Adelaide, Australia; and Mrs Ana R. Heide and Holmes P. Miller of the USA.

The Rosebud group has small, double flowers which retain a bud appearance and bear some resemblance to Apple blossom.

IVY-LEAFED These are derived from *P. peltatum,* a glossy-leafed species from South Africa with a weak stem, so that the shoots trail over window boxes, hanging baskets and containers. They flower very freely and look most effective in raised or tall containers; most of the cultivars are double, in shades of mauve, red, white and pink. If upright plants are required the growths can be tied up to slender canes.

There are also two fancy leaf-patterned varieties from Australia called 'Crocodile' and 'White Mesh'. Both have a cream mesh effect over the green leaves. 'L'Elégante' has variegated white and green leaves which become rosy when the plant is kept fairly dry.

REGALS In America these are known as Martha Washingtons or Lady Washingtons and are noted for their large deeply coloured and ruffled flowers. They are derived from *P.* x *domesticum* and bloom in early summer, remaining in character about two months; they have woody stems and sharply lobed leaves. In Europe and North America Regals or Show Pelargoniums as they are sometimes called, are invariably grown in pots—their free-flowering habit and the two-colour effects of many having great appeal for show purposes. After flowering they need to be rested, then cut back and restarted.

SCENTED-LEAFED Scented-leafed Geraniums were brought to England from the Cape by early navigators during the reign of Charles I. Their popularity was probably greatest in Victorian times, when they became a familiar feature in cottage windows as well as the great glasshouses of the rich. In tropical countries they are frequently bedded out in 'scented gardens' or used between bright plantings to mute strong colours. In colder climates, however, they are usually treated as pot plants; blind people often grow them and they are popular in sick rooms.

The scent is not apparent until the foliage is rubbed or bruised and this varies from a fragrance of Roses (*P. capitatum, P. radens* (*P. radula*) and *P. graveolens*), to Peppermint (*P. tomentosum*), Nutmeg (*P.* x *fragrans*) and Citron (*P. crispum*).

Specialist nurseries also list kinds claiming the perfumes of Lemon, Orange, Lemon Balm, Pine, Camphor, Mint, Eucalyptus, Citronella, Apple, Coconut, Almond, Apricot, Strawberry, Ginger, Musk, Filbert and Lavender.

An oil, extracted from the foliage of several species—notably *P. graveolens, P. odoratissimum, P.* x *asperum* (*P. roseum*), *P. capitatum* and *P.* x *fragrans*—is important as an extender or substitute for Rose oil and more costly essences. It is the source of fragrance for 'Rose' soaps, shampoos, toilet waters, etc., and replaces the expensive 'Attar of Roses' in composite perfumes.

A leaf of *P. capitatum* gives a pleasant flavour to a milk pudding and the leaves are also used for garnishing dishes and in jellies, jams and similar preserves.

All Pelargoniums like plenty of sun and a well-drained soil compost. Outdoors in very wet seasons they tend to produce leaves rather than flowers, an indication of their need for fairly dry conditions. But this is not to say that they do not appreciate feeding during the growing season; only take care not to use too much on the foliar cultivars, otherwise they will lose their bright colours.

The plants are propagated by nodal cuttings, rooted in very well-drained compost. Pelargoniums should not be put outside until all risk of frost is past and must be brought inside again (or perpetuated by rooted cuttings) before the first frosts of winter.

Regal Pelargoniums

'Axminster'

'Sybil Bradshaw'

'Elsie Hickman'

Gesneriaceae

120 genera and 2000 species

This is a large family of dicotyledonous plants, mostly tropical and subtropical, containing many handsome greenhouse plants which are popular in cool climates.

The leaves are usually opposite (rarely alternate or whorled) and simple in outline. These are entire with toothed or crenate margins, or occasionally pinnately cut.

A number of species have tuberous roots and the flowers are nearly always irregular and bisexual. They have five sepals and five corolla lobes (two of which may form a lip), the corolla frequently tubular and flared at the mouth. There may be two, four or rarely five stamens; the ovary is normally superior and the fruit either a capsule or berry.

Among the most important genera is *Saintpaulia,* especially *S. ionantha,* the African Violet, all species of which are native to tropical East Africa, notably Tanzania, where they are found in varying altitudes from sea level to 2400 m (8000 ft).

African Violets are popular house plants especially in America, for their colours are attractive, they are small enough for most rooms and indoor planting schemes and, if culturally suited, will continue blooming for months. The wild forms of *S. ionantha* usually carry loose sprays of deep violet flowers on stems of 7–10 cms (3–4 ins), the simple, long-stalked, hairy leaves forming a surrounding rosette.

Countless cultivated varieties exist with pink, white, red, mauve and purple flowers; also bicolors, doubles and others with variegated foliage. The Diana and Rhapsodie strains are especially free flowering and hold their flowers well under the drying conditions of a living room. The plants require good light (but not sunlight), a humid atmosphere and a temperature which does not fall much below 16°C (60°F). The best compost is light but moisture retentive, with good drainage, such as is obtained by mixing three parts peat with one each of loam and sharp sand. Propagation is by means of leaf cuttings.

The first Streptocarpus was introduced to Europe by James Bowie in the early 1820s. Bowie, once a London seedsman, went to Kew in 1810 and soon joined Alan Cunningham, one of Australia's greatest explorers, in an Orchid-hunting expedition to South America.

Two years later he was sent to South Africa but had to return because of the death of that great benefactor of Kew, the 'Unofficial Director of the budding botanic garden' (to quote Miles Hadfield in *Pioneers in Gardening*), Sir Joseph Banks. This had a stultifying effect on the fortunes of Kew; the government of the day cut the plant collector's allowance to half and after a few years Bowie felt he must leave Kew and returned to South Africa in a private capacity, as gardener to Baron Ludwig of Ludwigsberg. About 1841 he relinquished this post to become a professional collector, but died in some misery and poverty about 1853.

In his years of collecting, however, Bowie sent home many good plants from his South African expeditions such as species of *Aloe, Euphorbia* and *Gladiolus,* also *Clivia nobilis* and the Cape Primrose, *Streptocarpus rexii.* Bowie named this for his friend George Rex (a natural son of George III, who was one of Queen Victoria's 'wicked' uncles) on whose property at Knysna on the Cape Coast it was found.

This species has funnel-shaped flowers of bluish-white or mauve and rosettes of long, corrugated, stemless leaves. In the course of time other species reached England, notably the brick-red *S. dunnii* in 1884.

William Watson of Kew was one of the first to hybridize *Streptocarpus* species, so that new colours began to emerge—blue, mauve, magenta, rose, white, pink and salmon. Modern cultivars have larger and taller flowers and some have well-developed 'pencil' marks in the throat, although this characteristic is absent in others. 'Constant Nymph' is a very fine blue form. Most of these complex hybrids are collectively grouped under the name *S.* x *hybridus.*

Some *Streptocarpus* species have only one leaf, sometimes very large as in *S. wendlandii,* where it may be up to 75 cms ($2\frac{1}{2}$ ft) long and 60 cms (2 ft) across. This plant has scapes of 30 cms (1 ft) carrying up to thirty, bluish-white, Foxglove-like flowers.

Streptocarpus are summer-blooming plants for the cool greenhouse in temperate climates, requiring a temperature of about 10°C (50°F) in winter and 13–16°C (55–60°F) in summer. The seed is among the smallest of any plant and for that reason should be handled with the greatest care. Sow it very thinly and barely cover it with soil (if at all) afterwards.

The plants like a humus type compost containing peat, sand and loam and throughout their life must never be exposed to bright sunshine.

Saintpaulia ionantha 'Blue Star'

Streptocarpus 'Constant Nymph'

The Gloxinias of florists, now called *Sinningia speciosa,* are among the loveliest of greenhouse subjects but enjoy slightly more warmth than *Streptocarpus.* They have rosettes of large, velvety leaves and multicoloured, funnel-shaped flowers, larger and more plushy than the blooms of the latter genus. The plants can be grown from seed and with good treatment will bloom within six months of sowing. Their main desiderata are a warm, moist atmosphere, light, well-drained, peaty soil and shade from bright sunshine. The name *Stroxinia* is a 'bastard' name sometimes incorrectly applied to forms of *S. speciosa* with upright flowers. *S. regina* from Brazil is closely related to *S. speciosa* but has darker green leaves, white-veined above and deep red below, with trumpet-shaped, nodding, violet flowers of 5 cms (2 ins).

Sinningia species are Brazilian plants with tuberous rootstocks and can be propagated not only from seed and division but by leaf cuttings. The flowers have a velvety texture and come in brilliant shades of royal purple, scarlet, crimson, mauve and white; but they are often spotted, marked or outlined with other colours. They make delightful pot plants or can be grown outside in tropical gardens in shade (or, at most, positions receiving only very early morning sun) and under cover from rain. Rain and mud spoil the flowers and after blooming, when the plants die down, the tubers must be rested and kept fairly dry or they will become mouldy and rot.

Two epiphytic genera in Gesneriaceae which make delightful hanging-basket plants for living rooms, porches, glasshouses or positions outdoors in the tropics are *Columnea* and *Aeschynanthus.* The former have neat small leaves borne in opposite pairs down the long trailing stems, in the axils of which come the showy, tubular flowers. There are some 200 species native to tropical America of which *C. gloriosa* has fiery red flowers with yellow throats; *C.* x *banksii* is scarlet; *C. oerstediana,* orange-red; *C. microphylla,* orange-scarlet; and a hybrid called 'Stavanger', orange-red and yellow.

There are also upright or climbing species in cultivation notably *C. erythrophaea* and *C. scandens,* both with scarlet flowers.

Aeschynanthus are usually more straggling with larger, stouter leaves, often 10–13 cms (4–5 ins) long, and carry their flowers in clusters at the ends of the branches. They need more room than *Columnea* species but nevertheless make good basket plants or are sometimes grown wired to moss-covered wooden blocks or in crevices in trees. *A. parasiticus* (*A. grandiflorus*) from the Khasia Mountains of India has deep crimson, tubular flowers with orange bases. *A. parvifolius* (*A. lobbianus*) from Java grows partly upright then droops over the container. It has bright scarlet, slightly fragrant flowers with purple calyces and very dark green leaves. *A. speciosus* from Java has a somewhat straggling habit with leafy stems 60 cms (2 ft) tall carrying clusters of downy, orange flowers, which are individually about 10 cms (4 ins) long.

Aeschynanthus speciosus

Another plant sometimes used in hanging baskets—although its more upright growth makes it a splendid pot plant—is the South American *Hypocyrta glabra* or Clog Plant. This has small oblong-oval, fleshy, dark green leaves, in the axils of which come the small, orange-red pouched flowers. These remain long in character and are very profuse. The plant rarely exceeds 20 cms (8 ins) in height and tolerates hard pruning in spring. *H. nummularia* has a more creeping habit and will grow on rotting logs.

Sinningia speciosa cultivar

Sinningia regina

Rechsteineria (*Gesneria*) *cardinalis* is a South American plant with velvety-green, heart-shaped foliage and spectacular clusters of horizontal scarlet, Foxglove-like flowers. Its variegated-leafed form 'Foliis Aureis' has green and gold leaves and scarlet blooms. *Gesneria blassii* has scarlet tubular flowers.

Smithiantha (named for Matilda Smith, a botanical artist who once worked at Kew) is a genus of Mexican perennials of upright habit with large, plush-like leaves covered with red hairs and spectacular spikes of drooping, orange tubular flowers. Modern hybrids come in a wide range of colours—various shades of scarlet, crimson, yellow, orange, pink, carmine, pure white and cream.

Both *Gesneria* and *Smithiantha* have scaly rhizomes which have to be rested after flowering. They are restarted by standing the rhizomes on boxes containing a sand/peat/loam compost (in equal parts). The tubers are only partially covered, then watered and placed in a temperature of 10–13°C (50–55°F) to sprout. They are then separately potted and fed and grown on in the usual way; stopping the young plants by pinching out their growing points defers the flowering period for five to seven weeks.

Achimenes longiflora is the Hot Water Plant, so called because it has to be started under warm conditions, such as near greenhouse hot water pipes, and not because it has to be watered with hot water as is often erroneously stated. It is a South American, summer-flowering perennial, growing about 30 cms (1 ft) high (if supported on canes, otherwise drooping), with bright, funnel-shaped flowers which flare to a wide mouth. These come in shades of blue, pink, purple and white and make attractive pot or hanging-basket subjects. The plants are started in spring from dried off tubers in boxes of sandy compost in a warm propagating frame. When growth starts plant several together in a pot or basket, pinching the first shoots to induce bushy growth. Dry off after flowering.

Episcia with forty species, all from tropical America and the West Indies, appreciate good drainage with humus type compost. In Ecuador and Colombia they can be seen growing on steep sunless banks, the habit creeping, with rooting runners—after the fashion of the Strawberry (*Fragaria vesca*). The foliage lies close to the ground and is often richly coloured, above which may be glimpsed the bright funnel-shaped flowers.

In moist soil and sheltered situations away from bright sun the plants make good ground cover in tropical gardens. In cooler climates they may be grown on flat beds or over moss and rotting logs in the high temperature greenhouse; some kinds make beautiful basket plants.

E. reptans (*E. fulgida*) grows about 15 cms (6 ins) high with rather large, egg-shaped, woolly leaves. These are wrinkled and purple underneath, but dark green above picked out in silver-green along the veins. The flowers are vermilion-red and about 5 cms (2 ins) long. *E. lilacina* (*E. chontalensis*) is somewhat similar but with green foliage and pale lilac blooms. *E. dianthiflora* has soft green, downy leaves and pure white flowers charmingly fringed into hair-like segments at the edges of the tubes. It has slender creeping stems which root at the joints. *E. cupreata* of somewhat similar habit, has scarlet flowers with fringed margins and coppery leaves. The last two species make pretty hanging-basket plants or can be grown on mounds of moss and humus type soil. In order to bring out the russet colouring of the

Smithiantha hybrid

Rechsteineria cardinalis

Episcia cupreata
'Acajou'

foliage, shade from bright sun is essential and the plants thrive in a compost made up of equal parts fibrous loam, peat and sand with a little charcoal.

Other trailing *Episcia* species for tropical gardens or greenhouses are *E. maculata,* yellow flowers spotted with orange-brown; *E. bicolor,* glossy leaves with hairy stalks, the white flowers spotted and bordered with violet; *E. punctata,* yellowish spotted with purple; and *E. melittifolia,* a more erect species with large, toothed and wrinkled leaves and crimson flowers.

Nautilocalyx is closely related to *Episcia* and the species again are South American. *N. forgetii* has pale yellow, rather downy, tubular flowers and leaves with wavy margins which are green above and red-veined underneath. *N. pallidus* has creamy-white, axillary flowers, hairy outside with purple inner blotches and *N. bullatus,* straw-coloured flowers crowded in the leaf axils.

Kohleria eriantha (*Isoloma erianthum*) from Colombia is an erect plant, 60–120 cms (2–4 ft) tall with oval, opposite leaves in the axils of which come many long-stalked, tubular flowers. These are orange-red with yellow spots on lower corolla lobes. Whole plant-flowers, leaves and stems have a plush-like quality due to a dense covering of short reddish hairs. *K. amabilis,* also Colombian, has deep rose flowers with darker spots which are whitish at the throat. Cultivate as *Gesneria.*

Chirita is a genus of about eighty species of mostly evergreen plants, the majority from Ceylon and India. They have large, oblong to elliptical, very hairy leaves which are often marbled or patterned with cream, and tubular flowers of various colours, for example cream, brown-red, purple, lilac and orange. Shade from strong sunshine is essential, but the species make attractive plants for a moist position in a tropical garden or can be grown on a bench in a warm greenhouse—or even underneath it if the position is not too dark.

C. trailliana, a Chinese species with violet flowers, and *C. marcanii* from Ceylon, with deep orange and yellow flowers, can be cultivated like Gloxinias.

Three European genera are hardy in Britain, where they make attractive rock-garden plants for a shady situation. These are *Ramonda* (*Ramondia*), a genus of three or four species from the Pyrenees and Balkan Peninsula; *Haberlea ferdinandi-coburgii* from the Balkan Peninsula; and another monotypic genus *Jancaea* (*Jankaea*) *heldreichii,* also from the Balkans.

Best known is *Ramonda myconii* (*R. pyrenaica*) from the Pyrenees, a plant like the African Violet (*Saintpaulia*) with rosettes of ovate to oblong, hairy, dull green, heavily corrugated leaves and regular, five-lobed, Potato-like flowers of purplish-blue, three to four on a stem of 10 cms (4 ins). Like most members of Gesneriaceae the plants resent drips and standing moisture and thrive best when tucked into clefts between perpendicular rocks so that moisture drains from their collars. Since they are always planted in full shade they remain fairly cool during the summer. White and rose-coloured varieties exist so that attractive colour groupings are possible.

R. *nathaliae* from Bulgaria and Serbia is distinct, usually having four lobes to the flowers, and is also freer-blooming; the colour usually clear lavender-blue although sometimes white or pale lavender forms come from seedlings. Reginald Farrer described this species (*The English Rock Garden*) as 'the finest of the Ramondias. . . . A well flowered clump of this, staring glossily from some rich-soiled crevice or ledge (in which every *Ramondia* is as happy as linnets on a text), wipes out all the rest of the race'.

Ramondas can be propagated by seeds (which are extremely fine and need little or no soil covering) or by division or leaf cuttings rooted in a warm propagating case.

Haberlea ferdinandi-coburgii is like *Ramonda* and needs similar growing conditions. The dentate leaves are arranged in a rosette, between which rise scapes of five-lobed, tubular flowers—which however are not regular as in *Ramonda* but have three large and two small segments. They are pale lilac with a sprinkling of gold in a throat which may be darker or lighter than the rest of the flower. Seedlings show considerable variation.

Jancaea heldreichii makes rosettes of leaves like *Ramonda* but they are silvery with a dense covering of white down—a circumstance which makes them even more intolerant of drips or standing moisture. Plants should be established in pockets of sandy-peat soil with plenty of grit, tucked into a hollow in overhung rockwork so that the leaves lie perpendicularly against the stone. The Gloxinia-like flowers are clear lavender, several together on stems 5–7 cms (2–3 ins) high. Although a difficult plant to establish, it repays every care and attention.

Kohleria amabilis
hybrid

Gramineae

620 genera and 10,000 species

These are the grasses which, after Compositae, Orchidaceae and Leguminosae (in that order) represent the largest family of flowering plants, with a world-wide distribution. Monocotyledons, their chief characteristics are usually hollow, jointed stems, part sheathed by long, narrow, flat leaves with parallel veining (occasionally these run obliquely from the mid-rib, as in *Pharus*) and spikelets of brown or greenish flowers.

Many grasses have fibrous roots, the majority are herbaceous and they range in height from 2 cms (¾ in.) to 27 m (90 ft) in the tree-like Bamboos. The flowers are usually bisexual and rather inconspicuous, without petals or sepals but protected by two or three minute fleshy scales. Although the majority are self-fertile, others are not, but in any case pollination is usually effected by wind currents. The seeds are light and again may be dispersed by wind, sometimes carried over long distances.

Grasses provide some of our most important economic plants, particularly cereals, like Wheat (*Triticum*), Maize or Corn (*Zea mays*), Barley (*Hordeum*), Millet (*Panicum miliaceum*), Oats (*Avena sativa*), Rice (*Oryza sativa*) and Rye (*Secale cereale*).

Grasses provide animals with fresh green herbage for food, as well as dried fodder (hay) and straw for bedding and litter for hens. They also have many other uses, for example, paper-making, sugar, fibres, perfumery and edible oils and adhesives. Manna Grass (*Ammophila arenaria*) and certain other grasses play an important part in land reclamation, by binding loose sand with their matted roots and so checking erosion.

The place of grass in lawn-making is of consequence to every gardener and many species and varieties are used for this purpose.

Holcus mollis 'Albovariegatus'

Pennisetum orientale

Briza maxima

In temperate countries the finer types—many of them from New Zealand—are most in demand. The average English lawn contains a mixture of grasses, selected for appearance and/or wear. In hot dry countries the choice is more limited and the grass blades tend to be broader.

Ornamental grasses can be employed in the garden as separate features—as in island beds on lawns, for cutting purposes, as windbreaks and hedges, for edging and to provide contrast in mixed border plantings.

Low-growing sorts used for edging purposes, or sometimes as a ground cover through which small bulbs wend their way, include the dainty green and white *Holcus mollis* 'Albovariegatus', a soft-foliaged perennial only 10–13 cms (4–5 ins) high; *Milium effusum aureum,* Bowles' 'Golden Grass', which has golden leaves; *Festuca glauca* (*F. ovina glauca*), the Sheep's Fescue, about 22 cms (9 ins) with rigid, blue-green needle-like leaves in bristly clusters and *Rhynchelytrum repens* (*Tricholaena rosea*), a short-lived perennial, with loose panicles, 13–20 cms (5–8 ins), of fluffy spikelets, which are rosy when young but mature to a purple shade. These are dried for winter bouquets.

Taller is *Phalaris arundinacea* 'Picta', known to gardeners as Ribbon Grass or Gardener's Garters, a delightful cream-striped grass 75 cms (2½ ft) tall, the leaves arching prettily and with panicles of green or purplish spikelets. It is readily increased by division. *P. canariensis* is the Canary Grass, an annual whose seeds are fed to cage birds. It grows 30–90 cms (1–3 ft) tall and is esteemed for its ovate flower spikes 4 cms (1½ ins) in size. These are green and white but often tinged with violet.

Glyceria aquatica 'Variegata' (*G. spectabilis*) is another variegated grass; the foliage is striped and bordered with cream, and the young growths rosy-pink. It will grow in shallow water and also in heavy soil in sun or shade.

Miscanthus sacchariflorus (*Imperata sacchariflora*) is the Japanese Hardy Sugar Cane, a vigorous species of 1.75–3 m (6–10 ft) which will grow on land or under bog conditions. It has reedy stems and terminal panicles of silky brown spikelets; its leaves 4 cms (1½ ins) wide have a white mid-rib and rustle in the slightest breeze. The species makes a light divider or hedge barrier between garden features.

Among the prettiest annual grasses for garden and dried arrangements are *Hordeum jubatum,* the Squirrel's tail, and *Lagurus ovatus,* Hare's Tail. These common names aptly describe the appearance of the flower heads. Another, *Briza maxima,* carries several pearly, locket-shaped inflorescences, dangling from thread-like attachments on stems of 30 cms (1 ft). If stood for a day in water containing a little red or green ink, they take up the colouring and assume delicate pastel tints.

Coix lacryma-jobi is called Job's Tears on account of the hard, white to lead-coloured fruits, which hang down when ripe like

Phalaris arundinacea
'Picta'

Milium effusum aureum

tears. These are credited by the Chinese with miraculous powers and fashioned into necklaces and rosaries. The Burmese grind them down to flour for food and they are used by the Japanese as a source of tea.

The Pampas Grass, *Cortaderia selloana* (*C. argentea*), which under good growing conditions can reach a height of 2.5–3 m (8–10 ft), has heavy plumes of silvery-white inflorescences and saw-edged, arching leaves. Pink- and purple-tinted forms are available. In its native South America the plant is used in the manufacture of paper.

Zea mays

Other noble grasses are *Panicum violaceum,* 90 cms (3 ft) tall, with broad leaves and pendulous heads of green and violet; *Pennisetum orientale*, with slender arching foliage and twisted plumes of white or purplish flowers, like bottle cleaners on stems of 45–60 cms (1½ ft–2 ft); and the decorative forms of Maize (Corn), with the male spikelets in a tassel-like inflorescence at the top of the stem, and the female 'cobs' lower down, wrapped around with leaf sheaths. Varieties are available with white-striped foliage and also variegated in white, yellow, rose and green, and there are coloured-seeded forms.

But perhaps the most useful and impressive of the ornamental grasses are the Bamboos. Grace and elegance are their characteristics and their airy manner of growth causes them to bend and rustle with every gust of wind. In the tropics Bamboos make dense forests, with a hundred columns or stems springing from a single plant and rising to a height of nearly 30 m (100 ft).

The growth rate of Bamboos is fantastic; some species grow as much as 50 cms (20 ins) in a day. Each shoot springs from the ground armed with a sharp point capable of forcing its way through the thickest masses of branches. Like a smooth spear it drives on until it reaches its ultimate height, then the branches begin to spring and the leaves appear.

Bamboos are valuable to many races in many parts of the world, but particularly in South America and Asia. In parts of Asia they furnish almost all the necessities of life, being used to make houses, floors, ladders, furniture, bridges, boats, paper and kitchen utensils. The hollow stems of the larger species, cut just below a joint, make vessels to hold water, honey, fruit and vinegar. The shoots make an excellent pickle and the black filaments (found in the joints of certain species) are used in Java to execute criminals. Mixed with food the filaments stop at the throat and induce inflammation and ultimately death.

The hardier species are used in temperate countries as hedges, windbreaks or specimen plants. Being evergreen they are attractive at all seasons, although they take a year or two to settle again after transplanting. A rich, deep, cool moist soil suits them all and they do especially well at the waterside. The genera commonly cultivated are *Phyllostachys*; *Sinarundinaria*—especially the fine-leafed elegant *S. murielae* and *S. nitida*; *Arundinaria*; *Pleioblastus*; *Pseudosasa japonica* (*Arundinaria japonica, Bambusa metake*) one of the most vigorous with broad leaves on canes 4 m (12 ft) high; *Thamnocalamus*; *Shibataea kumasasa* (*Phyllostachys ruscifolia*), a dwarf with zigzag stems reaching 90 cms (3 ft); and *Sasa,* certain species of which are extremely rampant so must be planted with care.

Greyia sutherlandii

Greyiaceae

1 genus and 3 species

This is a restricted genus of dicotyledonous shrubs or small trees, all of them South African. The leaves resemble those of *Ribes,* being simple, with toothed or crenate margins and are alternately arranged on gnarled and woody branches. They are deciduous and fall in autumn, the flowers—in terminal racemes—studding the leafless branches in early spring. The flowers are brilliant scarlet, bisexual with five small sepals and five showy petals.

Greyia sutherlandii is known as the Wild Bottlebrush or Beaconwood in South Africa and has flower spikes up to 25 cms (10 ins) long and 10 cms (4 ins) in diameter, with prominent red filaments. It is remarkably beautiful and an outstanding shrub among rocks in the mountainous parts of Natal. The tree only thrives in a mild climate and in full sunshine, which ripens the light and somewhat porous wood; it must also have a season of rest before flowering. It was named in honour of Sir George Grey, Governor-General of Cape Colony at the time of its discovery.

The other two species, *G. radlkoferi* and *G. flanaganii,* are very similar. All make good pot plants and flower when quite small.

Grossulariaceae

2 genera and 150 species
temperate areas

This small family comprises dicotyledonous, often spiny shrubs with alternate simple leaves, which are variously lobed and often have resinous glands. They usually have five sepals, five petals, five stamens and an inferior ovary which develops to a juicy berry. The regular flowers may be unisexual or bisexual.

Ribes contains a number of plants of economic importance notably the prickly Gooseberry (*R. grossularia*), Red Currant (*R. rubrum*) and Black Currant (*R. nigrum*).

The most valuable ornamental species is the thornless *R. sanguineum*, the Flowering Currant, introduced to Europe from W North America by David Douglas in 1836. This makes a vigorous shrub 2–2.5 m (6–9 ft) high with drooping racemes of rose-madder coloured flowers which have a rather disagreeable 'tomcat' smell. 'Splendens' is a rich red, 'King Edward VII' is another good form and 'Pulborough Scarlet' is extra brilliant. 'Brocklebankii' has young bright golden foliage. The Flowering Currant will grow in practically any soil and aspect, even in shade, although it only flowers freely in a sunny situation.

R. aureum, from the same habitat, is called the Golden Currant because it has yellow flowers. These have a pleasant spring scent. Other interesting species, all Californian, include *R. roezlii* with purplish-red flowers succeeded by round, spiny and bristly, purple fruits; *R. speciosum*, a bristly 2–3 m (6–10 ft) tall shrub with rows of bright red, Fuchsia-like flowers suspended in clusters beneath its drooping branches and *R. niveum* which has white flowers on a prickly bush, up to 2.5 m (9 ft) high.

Grossularia (usually included in *Ribes*), the other genus in this family, with about fifty species scattered over temperate Europe, Asia, North Africa and North America is rarely represented in gardens, except by certain of the species listed above which by some authors are referred to this genus.

Ribes sanguineum
'King Edward VII'

Guttiferae

40 genera and 1000 species

This is a large family of trees and shrubs, chiefly tropical, containing oil glands which sometimes show up as dots on the simple, entire and opposite leaves.

The flowers vary considerably between species but are usually regular and bisexual, although both bisexual and unisexual flowers are sometimes found on the same plant. They have many stamens which in some genera (as *Hypericum* and *Clusia*) add to their attraction.

Some species are tropical trees of great size yielding good timber. When incisions are made in the bark, yellow or greenish resinous juices are exuded; that from *Garcinia hanburyi* (of Cambodia and Thailand) and *G. morella* yielding gamboge, a bright yellow pigment used in water paints or in spirit varnishes for painting on metal. It is also a powerful purgative as is the juice of species of *Clusia* from tropical America. *C. flava* is called Hog Gum because injured swine rub against the plant and the resin heals their wounds.

Garcinia mangostana, the Mangosteen, is a small Malayan tree bearing purplish fruits with a thick rind and four to six seeds, which have white arils with a delicious Peach-like flavour. Connoisseurs call them the 'Queen of Tropical Fruits'. Another tropical delicacy is the Mammee Apple or St Domingo Apricot, *Mammea americana* of the West Indies, with orange-red fruits tasting of Apricot. It is cultivated in S Florida and S California.

Several species are grown as ornamental shade trees in the tropics or warm temperate zones, notably *Clusia grandiflora*, a Surinam tree 3–6 m (10–20 ft) high with waxy, leathery leaves 30 cms (1 ft) long and large, richly scented, white or pale rose flowers like large Camellias. These are filled with golden stamens and the seed pods which follow are as large as golf balls.

Cratoxylum (*Cratoxylon*) *cochinchinense* is the Tree Avens, a Malayan species up to 9 m (30 ft) with drooping clusters, 5–15 cms (2–6 ins), of crimson flowers which come in pairs. These are fragrant and look very pretty against the purple tints of the young leaves.

Mesua ferrea, the Ceylon Iron Wood, grows to 18 m (60 ft) and yields a valuable timber. It has fine evergreen leaves and many fragrant china-white, four-petalled blossoms like single Roses, 7 cms (3 ins) across. These are persistently fragrant and besides being used in perfumery, are often stuffed into pillows and cushions for bridal beds in the Orient.

Hypericum x *moseranum*
'Tricolor'

Hypericum calycinum

Hypericum
'Rowallane'

Calophyllum inophyllum, the Beauty Leaf (Old World Tropics), is a good seaside plant. It grows to about 9 m (30 ft), with twisted branches having leathery, shining leaves and upright sprays of small, white, stamen-filled flowers. These are richly scented and much loved by honey bees.

In temperate gardens the most important genus is *Hypericum,* largely because the species bloom in summer, an in-between season for shrubs, but also because many are shade-tolerant. One of the most useful is the weed suppressing, border or bank retaining Rose of Sharon, *H. calycinum,* an evergreen 45 cms ($1\frac{1}{2}$ ft) tall from S Europe and W Asia Minor which thrives in sunless spots. The bowl-shaped golden flowers 7 cms (3 ins) across resemble pincushions, with hundreds of long yellow stamens sticking out from their centres.

H. patulum from Japan, China and the Himalayas, and its close relatives are free-flowering sorts, 'Hidcote' being especially good. *H.* x *moseranum* (a hybrid between *H. calycinum* and *H. patulum*) is a desirable shrub, usually not more than 45 cms ($1\frac{1}{2}$ ft) with terminal clusters of golden flowers of 6 cms ($2\frac{1}{2}$ ins). 'Tricolor' has smaller blooms but prettier leaves, these being variegated in white and bordered with red. Both make good ground cover plants.

H. 'Rowallane', a cultivar raised by Mrs Armytage Moore, makes a shrub 1.5–1.75 m (5–6 ft) high with bowl-shaped buttercup-yellow flowers 7 cms (3 ins) across and *H.* 'Elstead', another cultivar, is a shrub 90 cms (3 ft) high which does well near towns, with yellow flowers followed by bright orange-red fruits.

The shrubby Hypericums associate pleasantly with blue Hydrangeas, both liking light shade and a well-drained situation with a moist subsoil. They are lime-tolerant and easily propagated by seed or half-ripe summer cuttings.

Hypericum 'Elstead'

Haemodoraceae

14 genera and 75 species

This family contains monocotyledonous plants from South Africa, Australia and tropical America to eastern US with radical, usually long, grassy leaves often with bulbous rootstocks. The flowers are perfect, regular or slightly zygomorphic, with six perianth segments and three or six stamens. The blooms are often hairy and borne as panicled inflorescences on stiff stems.

Anigozanthos with ten species, all from SW Australia, is the most important genus. Their English name Kangaroo-paw refers to the shape of the flowers, which are felted with short hairs and rough to the touch.

The plants make handsome subjects for borders and rock gardens in warm climates or, under cooler conditions, may be grown in soil beds or pots in a greenhouse. They need sharp drainage, plenty of sun and a rest after flowering. The leaves are flat and ribbon-like and the tubular flowers arranged on stout felted stems.

A. flavida has tubular flowers which flare out to six, unevenly placed segments. These are yellowish-green and have prominent stamens. The general height is about 60 cms (2 ft) although they sometimes reach 3 m (10 ft) in the wild.

A. manglesii, one of the most outstanding, has mostly green flowers, except at the bases where they are brilliant red. The blooms of *A. viridis* vary from deep emerald to bright yellow-green and *A. rufa* is reddish-crimson in both stem and flowers, although the perianth edges are greenish. *A. pulcherrima,* the Golden Kangaroo-paw, brings a touch of sunshine to the coastal plains of Western Australia and is one of the best of the genus.

Macropidia fuliginosa, the Black Kangaroo-paw, has bright yellow-green flowers and stems which are densely coated with short black hairs, creating a most unusual and striking effect.

Anigozanthos flavida
green and red forms

Hamamelis mollis

Parrotia persica in February

Hamamelidaceae

22 genera and 80 species

This is a family of dicotyledonous trees and shrubs, chiefly distributed in temperate and subtropical zones of North America, Asia, South Africa, Madagascar and Australia. Many of the genera are represented by a single species (monotypic). The leaves are usually alternate and simple or occasionally palmate; the inflorescences frequently appear as spikes or heads, sometimes with an involucre of coloured bracts. The flowers may be bisexual or unisexual and have four to five sepals, four to five petals and usually four to five stamens; there is a superior or inferior ovary and tough, woody capsules. Some of the species yield useful timbers and resin or have medicinal applications.

Several members of the Hamamelidaceae are shade-tolerant and winter-flowering and also colour up well in autumn—all circumstances which commend them to the gardener. The North American *Hamamelis virginiana* is the source of Witch Hazel and blooms in autumn with many sweetly scented, small, yellow flowers. The Cowslip-scented *H. mollis*, a Chinese species growing about 4.5 m (15 ft) high, blooms in spring with spreading branches carrying many narrow-petalled flowers, like paper tassels. These are bright yellow, but paler (although larger) in the variant 'Pallida' and orange-yellow (with shorter segments) in 'Brevipetala'. The Japanese *H. japonica* is a variable plant, one of its forms 'Arborea' making a small tree up to 4.5 m (15 ft) high. This has golden flowers with purple calyces but in another form 'Zuccariniana' the flowers are pale lemon in green calyces.

There are also several cultivars of mixed parentage such as *H.* x *intermedia* 'Jelena', a very fine hybrid of *H. mollis* and one of the purplish-red-flowered forms of *H. japonica* raised at the Kalmhout Arboretum, Belgium. The flowers have the size and substance of *H. mollis* but are copper-yellow, suffused orange in colour and very beautiful if seen with the January sunlight shining through the branches.

Most *Hamamelis* achieve brilliant foliage displays in autumn,

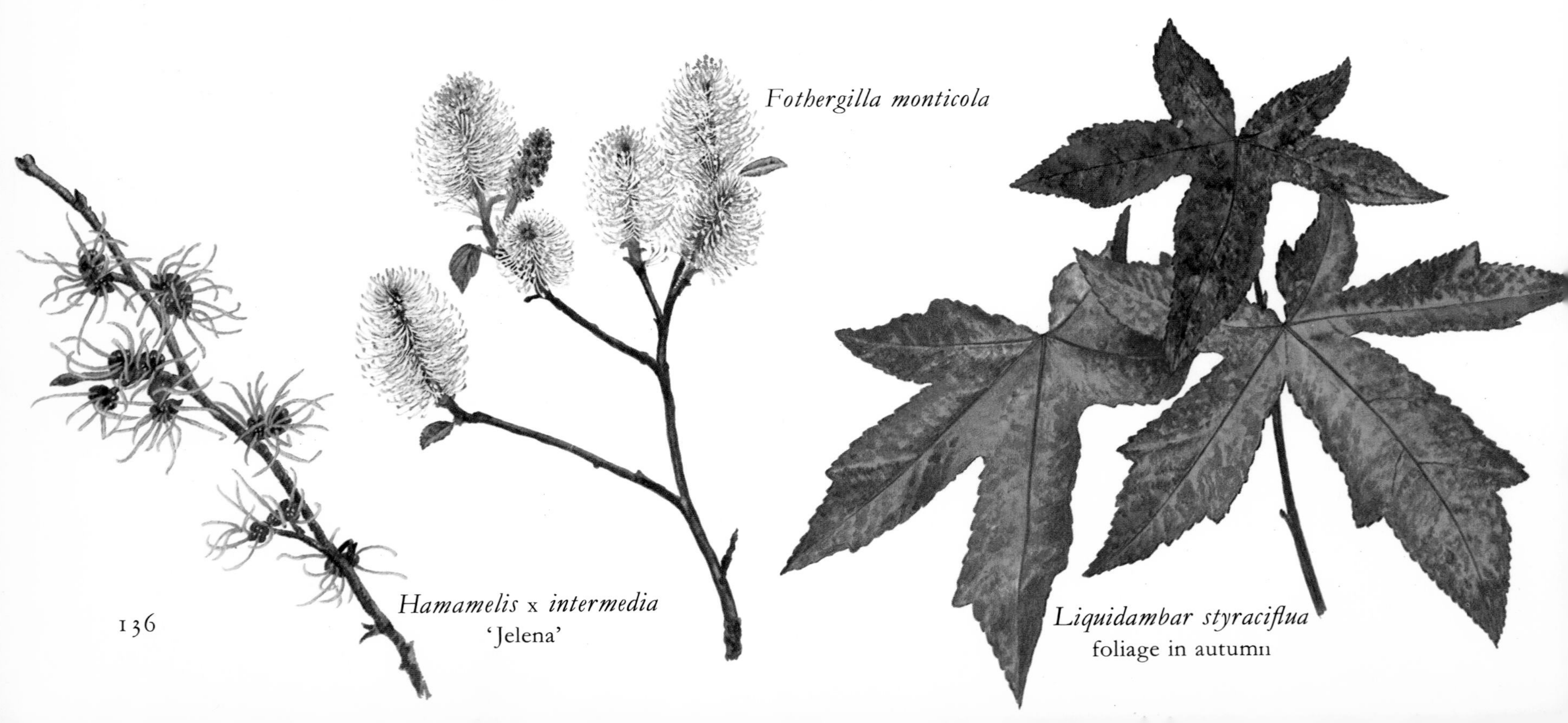
Fothergilla monticola

Hamamelis x *intermedia* 'Jelena'

Liquidambar styraciflua foliage in autumn

Parrotia persica foliage in autumn

when the leaves turn orange, scarlet and yellow before falling. *Parrotia persica* from Persia, a tree 6–9 m (20–30 ft) high, is another remarkable for its autumn colouring. In spring the naked branches are studded with short clusters of flowers having brilliant red stamens set off by brown hairy bracts.

Parrotiopsis jacquemontiana from the Himalayas makes a tree 4.5–6 m (15–20 ft) high with heads 4–5 cms ($1\frac{1}{2}$–2 ins) across of tightly packed yellow stamens subtended by four to six white, petal-like bracts.

Fothergilla monticola, F. gardenii and *F. major,* all North American species, are shrubs to associate with Rhododendrons in a semi-shaded site on humus type soil. The white and fragrant flowers are carried on short rounded heads like Willow (*Salix caprea*) and the simple leaves turn crimson and yellow in the fall.

The Sweet Gum, *Liquidambar styraciflua* (also placed in the family Altingiaceae) is a valuable timber tree, found wild from Connecticut in the eastern US to South America. It has palmate, five- or seven-lobed, Maple-like foliage which assumes varied and brilliant tints in autumn; green, gold, yellow, crimson and scarlet leaves frequently appear together on the same tree. The flowers are greenish-yellow and the branches often have corky wings. The tree grows up to 24 m (80 ft) in its native habitat, but 6–9 m (20–30 ft) is more usual in gardens. It likes moist soil and should be planted in a prominent position where its foliage can be viewed to advantage.

The wood is close grained and heavy with a red tinge and used for such items as veneer, furniture, cabinet work, radio cabinets, also for buildings and pulpwood. A balsam known as Storax is derived from the trunk and used medicinally in various ways, and also for scenting soap and perfumery.

Tea chests are made from wood of the Chinese *L. formosana* and its leaves are a valuable source of food for silkworms.

Heliconiaceae

1 genus and 80 species

This family is sometimes included under Musaceae, the Banana family. Among the more striking plants of the moist forests and damp jungles of tropical America are the Heliconias; large—sometimes gigantic—monocotyledonous perennials with simple stems often sheathed by the large, banana-like leaves.

The inflorescences are compound and alternately arranged, sometimes touching, sometimes zigzagging with lengths of stem between, but always subtended by a large, colourful, spathaceous bract shielding several small flowers. These are individually perfect with six perianth segments, five stamens and an inferior ovary.

The plants are especially popular in West Indian gardens but will succeed anywhere in the tropics given moist rich soil.

Heliconia humilis, the Lobster Claw, is most commanding. It has claw-shaped, bright scarlet bracts edged with gold, tightly packed on opposite sides at the tops of stems 60–120 cms (2–4 ft) high. Individually these are 10–13 cms (4–5 ins) long; the whole inflorescence is 60–120 cms (2–4 ft) in length. The small greenish flowers nestle inside the bracts.

H. acuminata (possibly correctly known as *H. ballia*) has silvery, parallel-veined banana-like leaves and stems 1–2.75 m (3–9 ft) high carrying, with intervals of stem between, alternate inflorescences. The bracts of these are canary-yellow or rarely deep red and orange. *H. subulata* (possibly more correctly *H. hirsuta cannoidea*) is slighter in all its parts, the leaves oblong-elliptic, the inflorescences on slender petioles with mostly red bracts but marked with green and yellow. *H. bihai*, a giant species up to 5.4 m (18 ft), with yellow-tipped, scarlet bracts has edible fruits.

Heliconia bihai

Hippocastanaceae

2 genera and 15 species

The most important genus in this rather restricted family is *Aesculus* with thirteen species, all dicotyledonous shrubs and trees. These have large and usually sticky winter buds, wrapped about with resinous scale leaves, also palmately divided foliage with from seven to nine leaflets and showy inflorescences which look like candles poised on candlesticks of leaves.

The individual blooms are usually bisexual and have five sepals, four or five petals, five to eight stamens and a superior ovary. They are often prettily marked with other colours and not unlike small Orchids in appearance. The fruits are large, often spiky on the outside, and the coat is leathery, with two or three seeds inside.

Aesculus hippocastanum is the Horse Chestnut, a splendid tree with spreading branches, much planted in Europe. A good specimen may be as much as 30 m (100 ft) in height and as much across, with handsome five- or seven-lobed leaves and spikes of variegated white and red flowers, tapering upwards amidst the foliage like so many wax lights. There is also a double-flowered form. The ripe seeds, known to British youngsters as 'conkers', are a source of starch, and sometimes fed to stock—sheep, cattle, horses and deer.

Unfortunately this starch is associated with other substances which make it unpalatable and unsuitable for humans to eat but during the Second World War when other sources of starch were vital for direct consumption, 1500 tons of nuts were collected in Britain and glucose, urgently required for medicinal purposes, derived from them.

The seeds also contain potash and have been used as a substitute for soap. The flowers are occasionally used as tincture for rheumatism and charcoal (for gunpowder) is obtained from the wood. The timber is too light to be of much value, except for packing cases and similar articles.

The species is native to N Greece and Albania and many years back hybridized with *A. pavia*, the Red Buckeye of North America, to produce a fine red-flowered tree now known as *A.* x *carnea*. This has been known since 1818, although its origin is unrecorded. It is hardy to −12°C (10°F) or lower and comes more or less true from seed.

Another large tree, again up to 30 m (100 ft), is *A. indica* from northern India. It flowers later than the other species and has irregular blooms with four clawed petals, the bases often streaked with red.

The American species are commonly called Buckeyes and only one—*A. californica*—has sticky winter buds. This species normally has a slender trunk—15 cms (6 ins) or less in diameter—and dense clusters of white or pale pink flowers on spreading branches. Those without sticky buds include *A. parviflora,* a bushy shrub of 2.5–4.5 m (8–15 ft) with white flowers which have pale pink stamens, and *A. glabra,* 9–12 m (30–40 ft), its greenish-yellow flowers poised on panicles of 10–17 cms (4–7 ins).

A. flava (*A. octandra*), the Sweet Buckeye and another North American, grows up to 9–12 m (30–40 ft) in height and has the largest flowers of the genus. They are yellow with unequal sized petals. The wood is used for artificial limbs, piano keys, wooden utensils and paper pulp.

The other genus in this family, *Billia,* consists of two species of evergreen trees, native to S Mexico and tropical America. These are readily recognized by the fact that the foliage has three leaflets. *B. hippocastanum* is particularly magnificent with its vivid flame-red blossoms in cone-shaped clusters.

Aesculus x *carnea*

Hydrangeaceae

10 genera and 115 species

These are dicotyledonous plants, mostly trees and shrubs, but sometimes climbing or herbaceous. The main family characteristics are opposite or alternate leaves, which may be simple, toothed, or rarely lobed and regular flowers in cymose inflorescences. In some cases, as *Hydrangea* x *macrophylla,* the flowers can be of two types; perfect flowers with stamens and pistils in the centre of the inflorescence, with much enlarged sterile blooms surrounding them. Sometimes all the flowers are sterile as in the Hortensia Hydrangeas.

The true flowers have four to ten sepals, four to ten petals, four or more stamens and an inferior ovary with many seeds. Hydrangeas are invaluable for summer display, particularly in cool—not cold—climates, flowering throughout summer and early autumn with great showy heads of pink, blue or white flowers. All require plenty of moisture so in tropical gardens (except mountain areas) are less successful although some sorts thrive in tubs or in cool shady spots.

From a spectacular point of view the group of garden hybrids and cultivars called Hortensias come first. Michael Haworth-Booth (*The Hydrangeas*) suggests that some of these are derived from a Chinese garden form of *H. maritima,* which was introduced by Sir Joseph Banks in 1739 and others from Thunberg's Japanese plants (themselves of complex origin) for which he retains the names *H.* x *macrophylla* and *H.* x *serrata.* The distinctions are important inasmuch as the parentage of varieties determines their garden value. Forms of *H. maritima,* for instance, flower profusely near the coast so make good seaside plants, but are shy blooming inland. *H.* x *serrata* on the other hand is derived from several woodland species, so understandably its forms do well in shady spots but they are not good subjects for coastal regions.

In a recent botanical study Elizabeth McClintock considers *H. macrophylla* to be the wild prototype from the island of Honshu, Japan, with *serrata* as one of four subspecies.

H. x *macrophylla,* which seems to be a mixture of both the previous types, is good for ordinary gardens inland. Mr Haworth-Booth suggests retaining the name 'Hortensia' to distinguish those Hydrangeas which have round heads of sterile blooms from the sorts with flat heads of fertile blooms, ringed round with sterile flowers. These he terms Lacecaps.

Not unnaturally, therefore, varieties offered by nurserymen vary considerably in hardiness, freedom of flower, colour and a tendency to fade or burn. The nature of the soil affects the colour, for in alkaline soils the flowers will be pink or red, on acid soils blue or purple. Lime in the ground has the effect of locking up aluminium and other elements so that the plants cannot get them, but if aluminium sulphate is watered round the roots the crimson pigments turn blue. Naturally in neutral or slightly alkaline soils this has to be regularly repeated; in very alkaline ground it is better to keep to pink sorts.

There is a widely accepted belief that iron is the element to bring about these changes (tons of rusty nails and tins must lie buried beneath Hydrangea bushes) but this is erroneous for aluminium is the important substance. However on very limy soils the foliage may suffer from shortage of iron and assume chlorotic effects, with green near the veins but the rest of the leaves yellow. This can be corrected by foliar sprays or watering the roots with sulphate of iron dissolved in water (¼ oz to a gallon is the recommended strength).

There are a good many Hortensia varieties for lime soils and also desirable sorts for acid conditions, but as there are constantly new ones and they vary in different parts of the world it is preferable to ask your grower's advice before purchase.

The plants have many uses in the garden as at the fringe of a shrubbery, in key positions on lawns and in borders, or as tub specimens before buildings. The Lacecaps look more at home in a woodland setting, associating particularly well with Rhododendrons, Azaleas, Lilies and Camellias. All favour a cool, rich soil which can be kept fertile by regular mulches of rotted leaves or moist peat.

As regards pruning it is essential to remove dead wood and thin out shoots in spring so that the remainder receive sufficient light and air thoroughly to ripen their wood. Unless this is done the results will be poor and the blooms small. Remove the old flower heads in early winter.

The most brilliantly coloured Lacecap variety is one called 'Bluewave', the sterile flowers of which on acid soils are almost Gentian-blue. It makes a big bush, 1.25–1.75 m (4–6 ft) tall and

Hydrangeas (Hortensia group)

'Westfalen'

'Générale Vicomtesse de Vibraye'

as much across. A good white form is 'Whitewave' and for alkaline soils there is a rose-pink sort called 'Mariesii'. The Lacecaps have fragrant flowers.

H. paniculata, a Chinese and Japanese species, has a distinctive inflorescence—a cone-shaped head of tightly packed flowers. These open white but gradually change with age to pink, then purplish and finally brown. In its native territory it sometimes makes a tree 8 m (25 ft) high, but gardeners usually cultivate a smaller form, introduced from Japan in 1870 as 'Grandiflora'. This has larger heads, sometimes 45 cms (1½ ft) high and 30 cms (1 ft) through at the base, with nearly all sterile flowers. It is one of the most popular shrubs in the US.

The wood of *H. paniculata* is fine grained and very hard; it is used in Japan for such articles as walking sticks, nails and pipes.

H. sargentiana from China (now considered a subspecies of *H. aspera*), requires fairly deep shade. It has large rough leaves, stems thickly coated with rough bristly hairs and plate-like inflorescences made up of rosy-violet perfect flowers surrounded by pinkish-white sterile ones. It was introduced from China in 1908 by E. H. Wilson and named in honour of the then Director of the Arnold Arboretum (USA), Professor C. S. Sargent.

H. quercifolia can be seen wild in sparse woodland from Georgia to Mississippi. It grows 1.5–1.75 m (5–6 ft) tall and is called the Oak-leafed Hydrangea on account of the shape of its leaves. These colour up well in autumn and the flowers are white, deepening to purple with age.

The Chinese *H. villosa* (now considered to be *H. aspera aspera*) has densely hairy stems and shoots, and short bristles on the upper surfaces of the leaves, which makes them rough to touch. It has porcelain-blue flowers with notched sepals and grows about 2.75 m (9 ft) in height.

H. petiolaris (now considered a subspecies of *H. anomala* and also known as *H. scandens*) is the Climbing Hydrangea, a Japanese plant of inestimable value for covering north walls and shady buildings—although it can also be used in more open situations. It will clamber up tree trunks and over stumps or, in the absence of support, run horizontally over the ground and so make ground cover in shady situations. It is self-clinging by means of adventitious roots and deciduous, with small neat leaves and flat heads of white flowers. In moist soils these may be 15–25 cms (6–10 ins) across, with the sterile flowers on the margins.

The young leaves of *H. macrophylla serrata* (*H. thunbergii*) are used for making a sweet tea called Amacha in Japan. They are steamed, rubbed between the hands and dried beforehand and on the Buddha's birthday also used for washing images of the Deity.

Varieties and species are readily propagated from summer cuttings which root easily and soon grow into good plants.

Schizophragma, a small genus of some eight Asiatic species bears a close resemblance to *Hydrangea petiolaris*; so much so that at one time the latter was known as *Schizophragma.* They have aerial roots like Ivy by means of which they clamber over any available support and flat cymes of small white flowers. Although the central blooms of these are small and fertile, each inflorescence develops a large white bract, 5–7 cms (2–3 ins) long (really an enlarged sepal) which makes them strikingly beautiful. It is thus distinguished from *Hydrangea,* the sterile flowers of which have four to five small bracts. The commonest species are *S. hydrangeoides* from Japan, with coarsely toothed leaves and *S. integrifolia,* a Chinese species, both of which under good conditions grow about 9–12 m (30–40 ft) high.

Another self-clinger is the evergreen *Pileostegia viburnoides* from China and Formosa. It is hardy in Britain, extending to 3–6 m (10–20 ft) with opposite, entire, leathery oblong leaves 5–15 cms (2–6 ins) in length and white flowers with prominent stamens,

Hydrangea x *macrophylla*
'Blue Wave'

Hydrangea paniculata
'Grandiflora'

carried in dense terminal panicles. Like most plants in Hydrangeaceae it propagates readily from cuttings and is best suited by a moist loamy soil.

Deinanthe caerulea is a beautiful Chinese plant for a woodland situation. Its bristly leaves rather like Horse Chestnut stand about 30 cms (1 ft) high, above which appear clusters of large nodding flowers. These are of a curious shade of pale violet, with marbled overtones and blue stamens—a delightful combination. *D. bifida* from Japan has flowers of waxy-white with yellow stamens, again in clusters, and great crinkly leaves about 20 cms (8 ins) high. Both need Rhododendron-type soil and partial shade, with shelter from wind and freedom from drought.

Another valuable shade plant, which has the added merit of blooming in late summer when other flowers are scarce is the Yellow Waxbell, *Kirengeshoma palmata*. This is a monotypic genus, the species being native to Japan. It is a beautiful herbaceous perennial, with smooth dark purple stems about 90 cms (3 ft) high, carrying thin, slightly hairy, Acer-like leaves and terminating in clusters of drooping bright yellow flowers. These have oblong petals which are so thick and crisp they look like wax.

The plants should be generously supplied with moist, leafy soil—they will not tolerate drought—and grown in positions sheltered from hot sun. They can be grown in a completely sunless position and since they are frequently difficult to establish a method is to sow seed sparingly in a light wooden box and when they have grown to about 6 cms ($2\frac{1}{2}$ ins), sink the box in the soil where the plants are to grow. In time the containers rot away and the roots find their way downwards.

Dichroa febrifuga is an evergreen shrub, 1–2.15 m (3–7 ft) tall with opposite, toothed, lanceolate leaves and rounded terminal clusters, 7 cms (3 ins) across, of white or violet flowers. It comes from E Asia and likes a moist soil in sun or partial shade.

Kirengeshoma palmata

Hydrophyllaceae

18 genera and 250 species
cosmopolitan except in Australia

These are dicotyledonous plants, mostly herbaceous but with a few undershrubs, closely related to Polemoniaceae and Boraginaceae. The leaves are radical, alternate or opposite, usually hairy and often glandular. The inflorescences resemble those of Boraginaceae, usually with blue or purple regular flowers having five sepals, five petals, a superior ovary and normally five stamens. The blooms secrete nectar and are sometimes cultivated by beekeepers.

Several popular annuals, species of *Nemophila* and *Phacelia,* are grown; those mentioned are all Californian.

N. menziesii, or Baby Blue Eyes, 10–15 cms (4–6 ins) high, has pale green leaves and masses of bright blue, saucer-shaped flowers with white eyes; *N. maculata,* 15–30 cms (6–12 ins) tall, is white-flowered with purple veining and deep purple blotches; *N. phacelioides* is blue and white, and grows 30–45 cms (1–$1\frac{1}{2}$ ft) high. The species do well on any moist soil but prefer a light but rich sandy loam. Heavy rain can damage the fragile blossoms so the plants may need light staking and will grow in light shade. Seed is set freely and if the ground is left undisturbed many self-sown seedlings spring up the following season.

Phacelia tanacetifolia is a hairy plant, up to 90 cms (3 ft) or more in height, with deeply cut leaves and clustered spikes of blue or lavender flowers. *P. minor* (*P. whitlavia*), called the Californian Bluebell, has large blue or purple, bell-shaped flowers and sticky hairy leaves. Both species do well in a fairly heavy soil. *P. campanularia,* a branched annual from S California has deep blue flowers with white basal spots, and is 15–20 cms (6–8 ins) tall.

The Virginian Waterleaf, *Hydrophyllum virginianum,* also known as John's Cabbage, is a perennial for damp, rich woodland soil. It has pinnately compound basal leaves, although those on the flower stalks are fewer lobed. The flowers are white, bell-shaped, with projecting stamens and borne in rounded cymes. The young tender shoots are succulent and were used in salads by early settlers.

Several aquatics occur in the family, notably certain annual or perennial sub-shrubby plants in the genus *Hydrolea. H. spinosa* from South America grows in shallow water reaching a height of 30–60 cms (1–2 ft) with panicles of rich blue flowers, glossy lanceolate leaves and axillary forms. *H. zeylanica* from tropical Asia is a creeping plant of wet places with racemes of pale blue flowers and short-petioled, bright green, lanceolate leaves.

Phacelia campanularia

Nemophila menziesii

Illicium anisatum

Illiciaceae

1 genus and 42 species

This is a small family of dicotyledonous shrubs and small trees from tropical SE Asia, SE United States and the West Indies. The evergreen foliage is leathery, simple and entire, alternately arranged on the stems and often highly aromatic. The flowers are axillary, usually solitary, regular and bisexual, with seven or more perianth segments, four or more stamens and superior ovaries.

Illicium anisatum is a shrub or small tree from Japan and Formosa, hardy in sheltered parts of the British Isles, with many-petalled, greenish-yellow flowers. It is frequently cultivated around Buddhist temples and its bark was formerly burned as incense in Japan. This probably accounts for the alternative name, *I. religiosum*. Its fruits are toxic.

I. verum from S China is the Star Anise, so called since the odour and flavour of the fruits and foliage strongly resemble Anise and also because the fruits have eight parts and a star-like appearance. The leaves are oblong to elliptic, 2.5–5 cms (1–2 ins) wide and the flowers are pink outside and red within.

I. floridanum is the Poison Bay Tree from the southern United States. It makes a much-branched shrub, 1.75–2.5 m (6–8 ft) tall, with leathery leaves which are tapered at both ends. The flowers have twenty to thirty strap-shaped pointed petals of maroon-purple. The whole plant has an aromatic fragrance.

All the Illiciums should be grown in moist vegetable soil and can be propagated from seed or by layering.

Iridaceae

60 genera and 800 species

This is a large family of monocotyledonous plants, mostly herbaceous perennials with underground corms, tubers or rhizomes and long, grassy, parallel-veined leaves. They are native to both tropical and temperate regions but predominate in South Africa and tropical America. The inflorescences are occasionally one-flowered but more usually spike-like and terminal; the flowers are regular or zygomorphic, but bisexual, with six perianth segments united at the base to form a tube, three stamens and an inferior (very rarely superior) ovary.

Some of our best-known garden flowers belong to the Iridaceae and many are highly important in the cut-flower trade.

The genus *Iris* has some 300 species, all native to temperate parts of the Northern Hemisphere. The species are very variable and may have rhizomatous, bulbous or creeping rootstocks. They also have differing soil requirements for while some like lime, others detest it; a few want sun-baked, rather dry soil and a minority prefer bog or even aquatic conditions. Most thrive in good well-drained loam in full sunshine, but there are also woodland sorts like *I. foetidissima*.

Although the blooms are not individually long-lasting, a succession of buds ensures several weeks of continuity and the poise, shape and bright rainbow hues of the flowers (particularly in the Bearded types) are other compensatory virtues. By careful selection, the genus may be had in flower for almost twelve months of the year. The foliage is neat—either thin and grass-like or erect and lance-shaped. Many Irises have fragrant flowers and sometimes rootstocks. The genus was named for the rainbow (Iris), because of the diversity of colours found in the blooms.

Among the bulbous sorts are several gems for the rock garden, some of which flower very early and these should be given a sheltered position in well-drained soil. From Turkey comes *I. danfordiae,* 5–7 cms (2–3 ins) high, with bright yellow flowers and *I. histrioides,* bright blue with large blossoms, 5–7 cms (2–3 ins) across, on stems 10 cms (4 ins) high. These flower so early they often have to push their way through light falls of snow; but in very cold climates they are best grown in pans, under glass.

They are succeeded by the slightly taller, 15 cms (6 ins), *I. reticulata* from N Persia and the Caucasus, a violet-scented beauty with grassy leaves and deep blue or purple flowers. Good cultivars and free-flowering hybrids from *I. histrioides, I. reticulata I. bakerana* include 'J. S. Dijt', red-purple; 'Joyce', rich blue; and 'Pauline', deep violet.

The tall bulbous Irises variously known as English, Dutch and Spanish Iris make excellent cut flowers and also force satisfactorily. As border plants they appreciate a good fertile soil, warmer for the Spanish and Dutch than the English.

I. xiphioides, the English Iris, is in spite of its name native to the Pyrenees and should be planted about 15 cms (6 ins) deep and the same distance apart. The flowers come in various shades

Iris danfordiae

Iris histrioides

Iris reticulata forms

of blue, except for one white form. For hundreds of years it has flourished in the vicinity of Bristol (originally brought from Spain by merchant seamen), so that early botanists—and the Dutch bulb merchants who obtained their first supplies from Bristol—thought the plant was native to Britain. Hence its name.

I. xiphium, the Spanish Iris from Spain and Portugal, has smaller flowers in a wider colour range (white, various yellows, orange, bronze and blue). Often the standards and falls are of different colours and some of the blooms are fragrant. These need planting 10 cms (4 ins) deep. The Dutch Iris are mongrels, being derived from the previous two sorts as well as other species and garden varieties. They have a wide colour range and should be planted 15 cms (6 ins) deep in well-drained soil.

Perhaps the most widely grown Irises are the German or Bearded Iris—often known as Flags and variously referred to *I. barbata* or *I. germanica.* They have rhizomatous rootstocks which should be left half exposed at planting time. All like full sun, well-drained soil and lime (although this is not essential) and, for plentitude of flower, the roots need dividing every second or third season. As border plants they have a short season but the flowers are superb, being large and shapely in various colours or combinations of shades (white, yellow, red, pink, bronze, mauve, purple and blue) and often delightfully scented. The heights vary from 75–180 cms ($2\frac{1}{2}$–6 ft) and there are hundreds of named cultivars.

Iris germanica hybrids

Iris xiphioides

I. florentina, usually considered an albino form of *I. germanica,* has for centuries been the source of the Violet-scented 'Orris-root'. The dried rhizomes are the fragrant part but the roots have to be at least two years old before their full flavour is developed. Orris was included among the rare spices of the Egyptians and the Greeks and Romans esteemed it for perfumery purposes as well as medicine. In medieval times Orris was used to cure ulcers, induce sleep and also regarded as a sovereign remedy for a 'pimpled or saucie face'. It is still employed in various toilet preparations, as 'essence of Violets'.

The broad-leafed dwarf Irises from S Europe and Asia Minor include *I. chamaeiris,* 13–25 cms (5–10 ins) and *I. pumila,* 7–13 cms (3–5 ins), which flower earlier than the Flags and being so small are excellent for rock pockets or the fronts of borders. Blue, purple, yellow and white forms are available.

Moisture-loving Irises include the elegant, grassy-leafed *I. sibirica* from central Europe and Russia, with slender stems of 75–120 cms (2½–4 ft) carrying several bluish-purple flowers. They are good for cutting and very garden worthy, especially such cultivars as 'Caesar's Brother', deep violet-blue; 'Eric the Red', red-purple; 'Perry's Blue', china-blue; 'Marcus Perry', deep Oxford blue; and 'Snow Queen', white.

I. laevigata and *I. kaempferi,* both Japanese, are also moisture-loving and both dislike lime. *I. laevigata* and its variants, being aquatic, need standing water 5–7 cms (2–3 ins) deep, constantly over their roots but *I. kaempferi* only requires moisture during the growing season with drier conditions in winter. The two look somewhat alike with flat, parasol-like flowers, but can be told apart by the leaves; those of *I. kaempferi* have a prominent mid-rib which is absent in *I. laevigata.*

Some of the *I. kaempferi* hybrids are most colourful, and as well as selfs in blue, purple, white, pink or red, the large flat flowers (often doubled) come up in combinations of shades and are often blotched, stippled or striated in fantastic patternings. The cultivars are sold under Japanese names—difficult for Westerners to pronounce—but a mixed batch of seedlings affords a cheap and often highly satisfactory method of starting a colony. They are ideal for waterside planting and most of the plants grow about 60 cms (2 ft).

A well-known plant of pond and streams is the yellow-flowered aquatic *I. pseudacorus,* from Europe (including Britain), Asia Minor and N Africa. This is the Fleur-de-lis of France, a name it is supposed to have acquired from Louis VII, who chose it as his emblem when he joined the Crusaders fighting to drive out infidels from the Holy Land. It is possible that Fleur-de-Louis became gradually corrupted to Fleur-de-luce and then Fleur-de-lis —or sometimes lys (Lys being a river on whose banks the flowers grew in profusion). The variegated-leafed form is especially fine and there is a pale yellow variety called *bastardii.*

The Stinking Gladwin (Gladdon) or Roast Beef Plant is *I. foetidissima* from W Europe (including Britain). It is not a particularly exciting flower, being a washy lilac-blue or occasionally yellow, but the foliage is evergreen and the plant will grow in dry shade. In autumn the large pods split to reveal treble rows of brilliant orange-scarlet seeds. These look like jewels in a casket and are not only ornamental but persistent, and so popular for indoor decoration. The specific name refers to the peculiar smell of the bruised leaves, which has been variously likened to roast beef, boiled milk and wet starch.

To California we owe *I. douglasiana,* a lime hater which forms thick mats of grassy foliage topped by flower stems 30 cms (1 ft) tall. The blooms are variable, being mauve, lavender, apricot, cream or yellow and large in proportion to the height of the plant. Named sorts are available in America and the plants make charming edgings.

According to Edward K. Balls (*Early Uses of Californian Plants*) Indian squaws wrapped their babies in the soft green leaves of these plants, to retard perspiration and save them from thirst while their mothers were harvesting berries. The leaves were also used as a source of fibre for cord, rope, fishing nets and snares,

Iris foetidissima
seed pods

Iris sibirica

the women extracting the fibre, the men twisting it together, but apparently it took weeks to make even a short piece of cord.

I. forrestii from China has clear yellow flowers, as has *I. wilsonii* from the same area and *I. fulva* (*I. cuprea*) from Texas is reddish-orange. All are moisture-loving.

I. orientalis (*I. ochroleuca*), the Yellowband Iris from Asia Minor, comes into bloom in midsummer, with strong stems of 1.25–1.5 m (4–5 ft) carrying several large, yellow and white flowers. It has tough, strap-shaped leaves and succeeds under ordinary garden conditions in sun or light shade.

I. unguicularis, the Algerian Iris, which is often known as *I. stylosa,* is indispensable for its winter flowers. Tucked lightly against a warm wall, where the soil is well drained and rather dry, or protected by a framelight in cold climates, it throws a succession of blooms all winter. The long stems of 30 cms (1 ft) which are really perianth tubes with very long styles inside (the ovaries lie just under the soil) should be pulled—not picked—directly the buds show colour. They soon open up indoors into large sky-blue flowers with gold markings. The evergreen grassy leaves need trimming from time to time. In warm climates this Iris grows and flowers well in the open.

The flowers of Irises are highly popular in oriental and Persian art, being frequently portrayed in needle embroidery, paintings, cloisonné, lacquer and carved crystal.

Some strange uses include eating the tubers of *I. juncea* in the Mediterranean region; using the seeds of *I. setosa* in Alaska and *I. pseudacorus* in Europe as a Coffee substitute; hanging the rhizomes in barrels of wine or beer to keep the contents fresh; as a plaster for bruises and as the source of 'Iris-green' or 'Verdelis', a pigment once favoured by artists.

Closely related to *Iris* is *Hermodactylus tuberosus* (*Iris tuberosa*) from S Europe, sometimes known as the Widow Iris, presumably on account of its strange colouring. The flowers are green with nearly black falls on stems of 30 cms (1 ft). The plant needs a warm, sunny position.

Moraea is the Peacock Flower of South Africa with large flowers, 4 cms ($1\frac{1}{2}$ ins) across, having three wide spreading outer petals and three small inner ones. The outer petals are broad with striking peacock blotches in other shades. Thus *M. pavonia,* often referred to *M. villosa* or *M. neopavonia*, is red on yellow, and has a royal-blue eye; *M. villosa* may be light mauve, cream, yellow or orange with an iridescent blue blotch; and *M. glaucopsis*

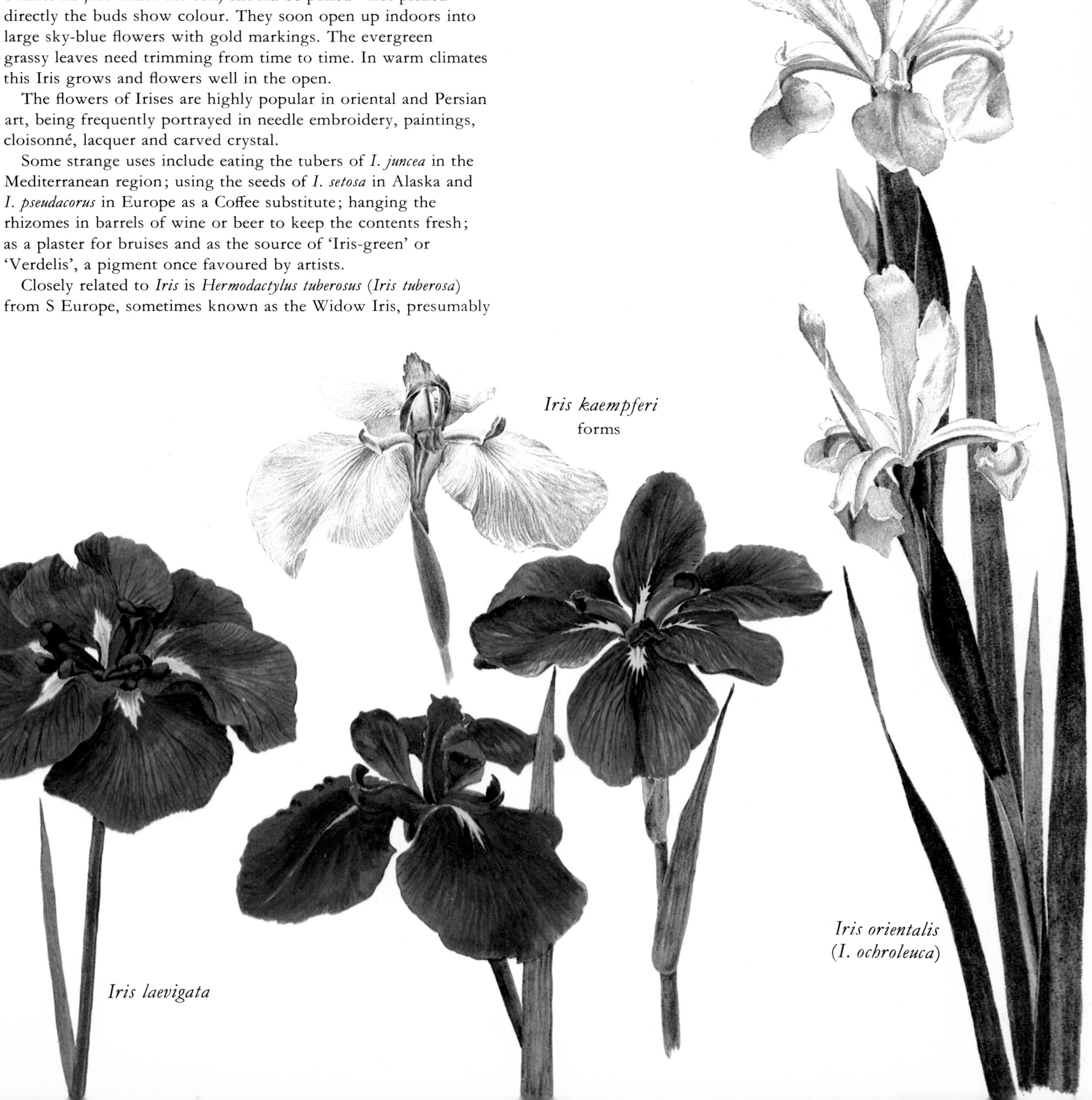

Iris kaempferi forms

Iris laevigata

Iris orientalis (*I. ochroleuca*)

is white or pale blue with peacock-blue markings. In cool countries Moraeas should be grown in pots or protected in cold weather.

Tigridias come from Central and South America. The flowers bear some resemblance to *Moraea,* but have the habit and cultural requirements of *Gladiolus. Tigridia pavonia,* the Tiger Flower, is so called because its flowers, with three large outer segments and much smaller inner segments, come in vivid scarlet, orange, deep yellow and violet-mauve and are spotted inside near the base with contrasting or deeper colours. There are also unspotted white and orange forms.

Gladioli are sun-lovers. They come mostly from South Africa (frequently in mountainous areas) or from moist warm spots in central Africa. In the tropical rain forests which skirt the Victoria Falls, it is possible to find *Gladiolus primulinus* flowering in perpetual spray and drip.

They are accordingly adaptable plants but not hardy in frost areas, so in cold countries the plants must be lifted in autumn, dried and stored during winter and planted afresh each spring. Large corms should be set 10–13 cms (4–5 ins) deep, the smaller species and hybrids 7 cms (3 ins) deep, in well-drained, fertile soil in full sun. The flowers are in great demand for cutting and excepting that there is no true blue, come in a wide range of shades or mixtures.

Staking of the taller hybrids may be necessary in windy situations and a quick way of doing this is to set stakes at intervals of 90 cms (3 ft) around the bed, and criss-cross these from side to side with thick twine, so that each plant grows in a square of twine approximately 60 cms (2 ft) up from the ground.

Most of the plants in cultivation are hybrids, of which there are literally hundreds, with new cultivars coming along all the time. Some kinds are suitable for the rock garden, others for border work but many are cultivated primarily for cutting. The main groups are:

Large-flowered (*Gandavensis*) hybrids Commonly 120 cms (4 ft) high with flowers 10–15 cms (4–6 ins) across.

Primulinus hybrids Smaller, with hooded flowers.

Miniatures 60–120 cms (2–4 ft) high with small flowers; ideal for cutting. They include the Butterfly hybrids, characterized by waved and fluffy petals with rich markings in contrasting shades.

Peacock hybrids Dwarf, multicoloured sorts with reflexed petals ideal for cutting.

Star-flowered A race raised by Unwins of Cambridge, England, with flat, star-like flowers.

There are also a number of species, including the crimson-magenta, hardier-than-most *G. byzantinus*. This plant is fairly common in E Mediterranean regions and is thought to be the flower alluded to by Ovid, which sprang from the blood of Hyacinthus when he was accidentally slain by Apollo. As Hyacinthus lay dying he cried out in agony 'AI AI' and the white marks on the lower petals bear some resemblance to these hieroglyphics.

G. primulinus from the Victoria Falls were first exhibited at the Royal Horticultural Society in London by Sir F. Fox in 1904. He was the engineer responsible for the bridge spanning the

Iris unguicularis

Hermodactylus tuberosus

Gladiolus 'Christabel'

Tigridia pavonia

Zambezi at Victoria Falls. The yellow-flowered species is still plentiful in this area and through the years has been crossed with Gandavensis hybrids to produce large blooms and different colours.

Other species sometimes grown are *G. tristis,* sulphur-yellow and fragrant; *G. cardinalis,* crimson; *G. blandus,* mauve, white or bluish-pink; and *G. alatus,* the Cornflag, a small plant with red and yellow flowers.

The corms of some *Gladiolus* species are a source of food and beverages used by various African tribes.

Also African are the delightful Acidantheras, plants somewhat similar to Gladioli except for the long, straight, perianth tubes behind the flowers. Most garden worthy is *Acidanthera bicolor* var. *murielae,* discovered in Abyssinia in 1936. It has linear leaves and stems of 60–90 cms (2–3 ft) carrying loose spikes of pure white flowers, with purple spots on the lower petals and a sweet penetrating fragrance. Cultivation is as for *Gladiolus.*

Antholyza ringens is an interesting and striking plant with vivid scarlet flowers arranged in zigzag, double-sided clusters on stems 15–30 cms (6–12 ins) high. It grows wild near the Cape, always in sandy ground, but it is no longer common and (except in the tropics) should only be grown in pots. The roots need to be covered with 15 cms (6 ins) of soil. Its common name (Rat's Tail) refers to the terminal portion of the stem above the flowers.

With one exception (a species in Socotra), all *Babiana* species are South African, and were so named because early settlers discovered baboons eating the corms. These are small plants with strongly ribbed, slightly hairy and pleated leaves and open cup-shaped flowers which are predominantly blue but also come in mauve, cream, red, white and yellow. They need full sunshine and the secret of cultivation is planting them deeply, 15–22 cms (6–9 ins) down, in sandy soil. If grown in pots (necessary in cool climates) they should be covered with not less than 24 cms (10 ins) of soil, or they will be stunted.

The most desirable species is *B. stricta,* which grows up to 30 cms (1 ft) with four or five royal-blue flowers 4 cms (1½ ins) across. *B. rubrocyanea* only grows 5–15 cms (2–6 ins) in height, but the flowers are striking, being deep caerulean-blue with bright crimson centres. They are known in South Africa as Wine-cups.

B. plicata (*B. disticha*), the Hyacinth-scented Babiana, varies from violet to pale blue and is sweetly scented. It grows about 30 cms (1 ft) in height and like the other kinds mentioned flowers in spring.

An important genus of small bulbs, especially for the spring garden, is *Crocus* of which there are some seventy-five species and innumerable varieties. They are plants of the Mediterranean region where the summers are hot and dry and the corms get a good baking but also receive plenty of moisture in winter and spring (during the growing season). Crocuses thrive where these conditions can be matched, which is why they are so popular in North America and Europe. They can also be gently forced for early flowers and in the garden may be naturalized in grass—where they soon make wide carpets of colour—or grown in rock pockets and at the fringe of woodland.

In order to extend their season advantage should be taken of the different flowering periods of the various species. The autumn bloomers are at their best when nearly everything else in the garden is dying down and look particularly well between the fading stems of herbaceous plants or at the edge of woodland. *C. speciosus,* the commonest, varies in colour from pale lavender to deep blue and grows about 10 cms (4 ins) high. *C. banaticus* (*C. byzantinus*) has large, rich blue-purple flowers and is a woodland species, so should not be dried out too much; *C. kotschyanus* (*C. zonatus*) and *C. medius* are mauve and *C. niveus* a good white.

C. sativus is the Saffron Crocus, source of a much-prized drug and dye in the Middle Ages. This is obtained from the dried stigmas, 4300 flowers being required to produce a single ounce of Saffron. Before the spices (Nutmeg, Clove, Cinnamon, etc.) were known the industry was highly important, for Saffron was used in many products including medicine, disinfectant, cosmetics and for colouring and flavouring food—Saffron cakes are still eaten in England.

Italian ladies of the past, envying the blond locks of northern lands, would dye their hair with Saffron, a practice which called upon them the fulminations of the early Fathers of the Church.

Christopher Colton in 1591 believed that Saffron benefited the health and mitigated the effects of Britain's humid climate. He observed that 'The saffron hath power to quicken the spirits . . . provoking laughter and merriment', and Gerard commended it for yellow jaundice but said too much 'cutteth off sleep . . . but the moderate use thereof is good for the head and maketh the sences more quicke and lively . . . and maketh a man merry'. Saffron was also used to promote eruptions in measles.

Dyeing linen sheets with Saffron was forbidden during the reign of Henry VIII, when a law was passed prohibiting the practice on the grounds that the dyed sheets were not washed frequently enough!

Hakluyt in *Remembrances for Master S.* 1582 relates a legend describing how the Saffron Crocus reached England from Asia Minor during the reign of Edward II, secreted in the hollow stave of a pilgrim. This was done 'with venture of his life, for if he had been taken, by the law of the country from whence he came he had died for the fact'. From these few corms began an industry in the pilgrim's native Walden, in Essex. This lasted for several hundreds of years until synthetics provided a cheaper substitute. The town is now called Saffron Walden, and *C. sativus* may still be found on occasions in nearby fields. Curiously enough this species now seldom flowers well in Britain.

The winter- and early spring-flowering species include the golden *C. flavus* (*C. aureus*), from which the well-known 'Dutch Yellow' has been derived and *C. chrysanthus,* a fertile parent which has produced many cultivars including 'Snow Bunting', white and cream with lilac patterning; 'E. A. Bowles', old gold flushed with bronze; 'Yellow Hammer', deep buttercup-yellow; 'Warley White', white streaked with grey-purple; 'Blue Peter', blue edged with silver and a golden throat; and 'Zwanenburg Bronze', deep orange suffused mahogany.

C. imperati blooms in the depths of winter, the large globular flowers yellowish-buff veined with purple outside and bright lilac-mauve inside. *C. tomasinianus*—affectionately known as 'Tommies' by gardeners—shoots up all over the place, even in hard pebbles, the flowers wine-purple to pale mauve and white.

Among the spring sorts the garden forms of *C. vernus* are most in demand. The flowers are much larger than those of other Crocus and come in various shades of blue and mauve—either plain or striped, also white. They are favourite sorts for naturalizing and forcing. *C. sieberi,* another good spring species, has deep mauve flowers with orange throats.

Schizostylis coccinea, the River or Kaffir Lily, comes from South Africa but is hardier than many Cape plants and grows outdoors in parts of Britain and Ireland, N Europe and North America. It has rhizomatous roots and likes a moist situation; indeed the late Viscountess Byng of Vimy used to grow plants in shallow water. The species is a late bloomer; the spikes, 45–60 cms ($1\frac{1}{2}$–2 ft) tall, of tubular, six-petalled, rosy-red flowers appear in late autumn. These have a flat, star-like appearance when fully expanded and are useful for cutting. 'Mrs Hegarty' (named for an Irish doctor's wife in whose garden the first seedling appeared) is rose-pink and 'Viscountess Byng' (raised by the author's father-in-law, the late Amos Perry) is shell-pink. The plants may be propagated by seed or division.

Ixia species from South Africa are characterized by flat brightly coloured little flowers on slender stems of 25–30 cms (10–12 ins). The combinations are often striking, for example lemon-yellow with a deep red centre; orange-red and deep yellow; white and red; and violet and green. Most of the garden sorts are hybrids with unknown backgrounds derived from several species but one, *I. viridiflora,* is noteworthy for its unusual and exciting electric-green/blue flowers which are rendered still more conspicuous by having violet-black eyes. These are carried on stems of 30 cms (1 ft). Ixias are normally grown in pots of sandy soil and dried off after flowering. Alternatively they may be planted in late spring in light soil in sheltered, sunny situations and lifted in autumn.

Sparaxis is somewhat similar with stems of 45 cms ($1\frac{1}{2}$ ft) bearing several flowers which are often brilliantly patterned. *S. grandiflora* is variable with red, pink, purple or white blooms having yellow throats and darker central markings. *S. tricolor* shows similar variation in colour. In their native South Africa *Sparaxis* are called Harlequin Flowers or sometimes Velvet Flowers. They should be grown in full sun as the flowers close in cloudy weather.

Lapeirousia (*Lapeyrousia*) from tropical and South Africa is chiefly represented in gardens by *L. laxa* (*Anomatheca cruenta*) and *L. grandiflora*; both are somewhat similar with spikes of scarlet-carmine flowers and narrow-bladed leaves. They look

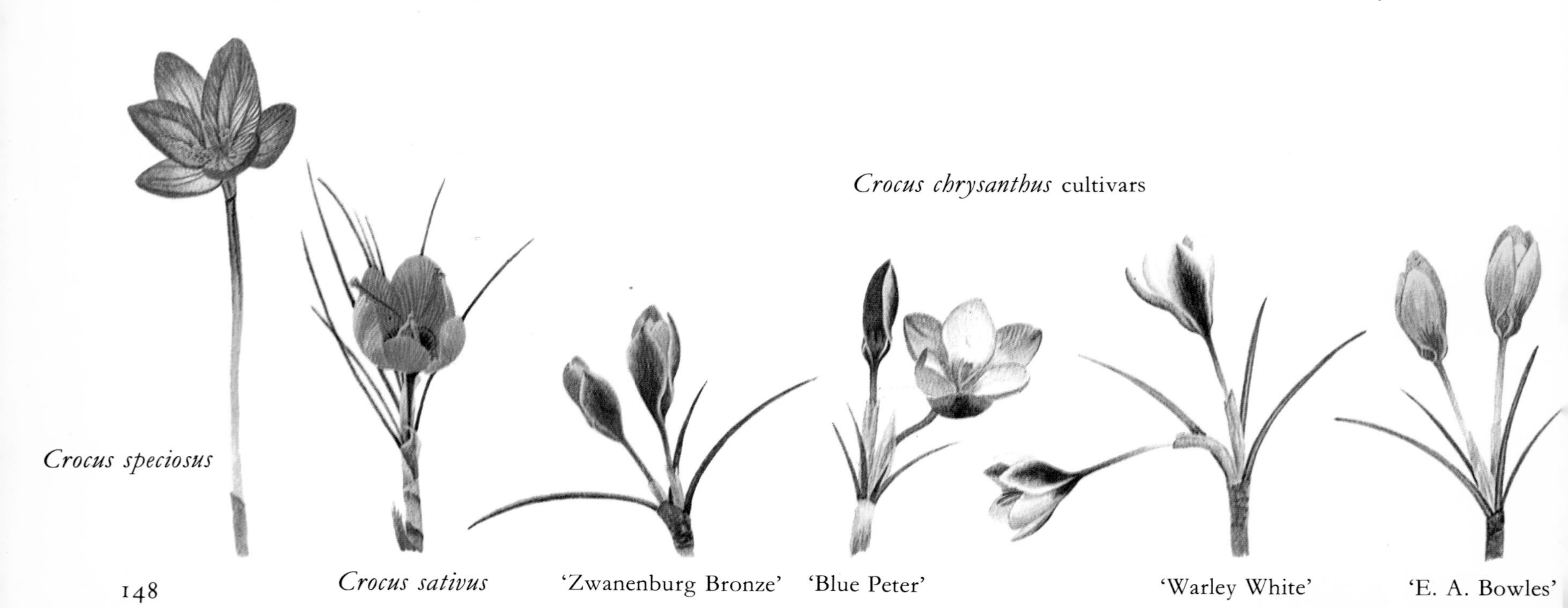

Crocus speciosus

Crocus sativus

Crocus chrysanthus cultivars

'Zwanenburg Bronze' 'Blue Peter' 'Warley White' 'E. A. Bowles' something like diminutive Gladioli, only 22–33 cms (9–12 ins) high. Suitable for growing in pots of sandy soil or well-drained pockets in sheltered or warm situations, the plants often naturalize and flower all through the summer.

A highly popular genus especially for the cut-flower trade is *Freesia*. The plants are South African with small round corms and slender stems 60 cms (2 ft) high bearing blossoms in a one-sided, arching spike. These branch out into one or two smaller flower spikes, carrying trumpet-shaped flowers 5 cms (2 ins) across which flare at the mouth to six perianth segments. These open successively along the branch and are highly fragrant. According to John Gilmour of the Cambridge Botanic Gardens, England, Freesia scent, so penetratingly sweet for most people, cannot be detected by five per cent of the population—even though the same people can smell most other flowers. The reason for this is unknown. Unfortunately modern cultivars seem to be sacrificing scent to size for, although the large blooms and many doubles now becoming available are really spectacular, the scent is more marked in the older cultivars.

Most of the present-day hybrids have been derived from *F. armstrongii* and *F. refracta* and its varieties. Colours available include white, cream, yellow, orange, pink, crimson, blue and mauve.

The corms are easily forced in a cool greenhouse for early spring flowers. The soil should be light but rich with decayed compost or peat; light staking is usually necessary and the flowers need protection from hot sun, frosts and cold winds. They should be dried off after flowering and are readily propagated from cormlets or offsets.

A striking South African perennial for a sheltered herbaceous border is *Homoglossum watsonium* (*Antholyza revoluta*). It has stiff, narrow sword-like leaves and one-sided spikes of brilliant red, tubular flowers on stems of 45 cms ($1\frac{1}{2}$ ft).

Aristea is a genus of about sixty tropical South African and Madagascan perennials with numerous, flat grassy leaves, short woody rootstocks and many-flowered, terminal racemes of (mostly) blue, tubular flowers. *A. ecklonii,* also called Blue Stars, is one of the best, the flowers a rich deep blue on stems of 45 cms ($1\frac{1}{2}$ ft); *A. africana* is shorter, 15 cms (6 ins); and *A. capitata* taller, 90 cms (3 ft). In temperate countries the plants need greenhouse conditions.

Dierama pulcherrima (*Sparaxis pulcherrima*) is the South African Wand Flower, a delightful garden plant with thin, rigid, grassy leaves and slender, willowy flower stems which arch at the tips like fishing rods because of the weight of a series of pendulous

Schizostylis coccinea

Freesia hybrids

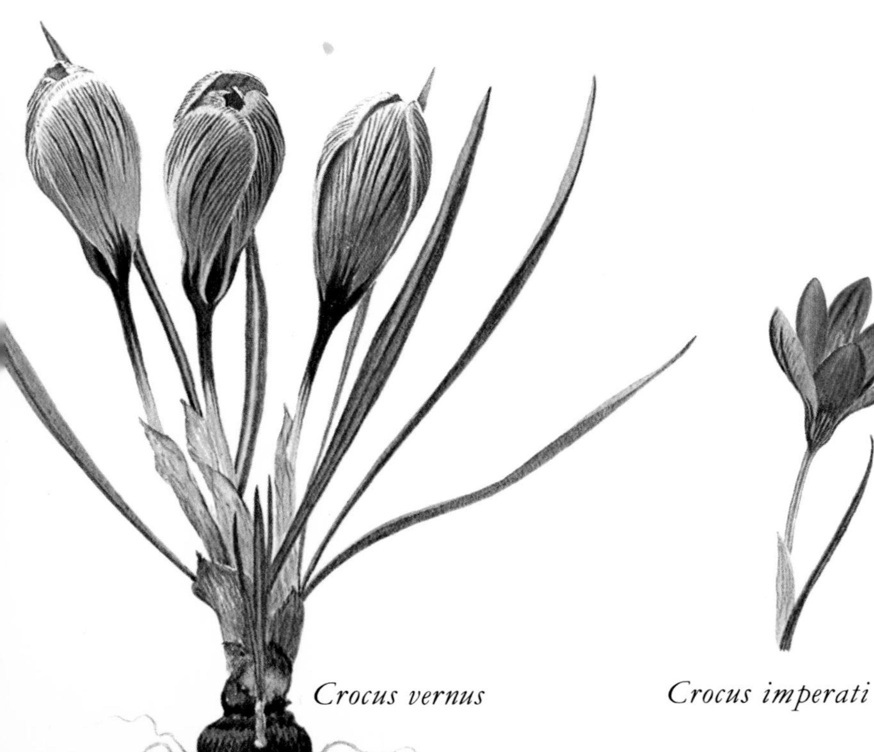

Crocus vernus

Crocus imperati

Crocus sieberi

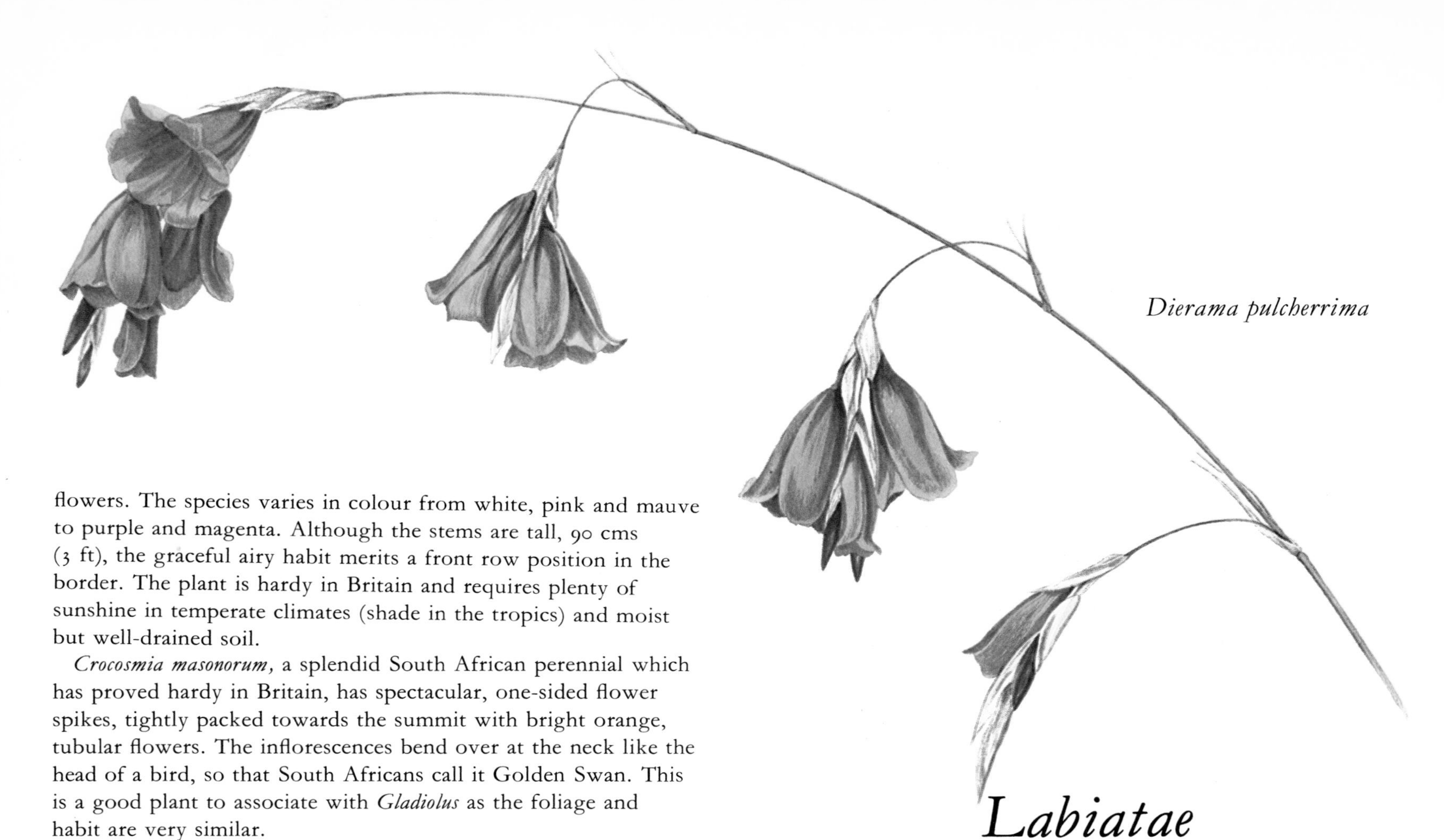

Dierama pulcherrima

flowers. The species varies in colour from white, pink and mauve to purple and magenta. Although the stems are tall, 90 cms (3 ft), the graceful airy habit merits a front row position in the border. The plant is hardy in Britain and requires plenty of sunshine in temperate climates (shade in the tropics) and moist but well-drained soil.

Crocosmia masonorum, a splendid South African perennial which has proved hardy in Britain, has spectacular, one-sided flower spikes, tightly packed towards the summit with bright orange, tubular flowers. The inflorescences bend over at the neck like the head of a bird, so that South Africans call it Golden Swan. This is a good plant to associate with *Gladiolus* as the foliage and habit are very similar.

C. aurea or Falling Stars has brilliant orange flowers with protruding stamens. These look like small Tiger Lilies and the flowers are the source of a yellow dye which has been used as a substitute for Saffron.

At one time *Crocosmia masonorum* was referred to *Tritonia,* a genus which produces its flowers on similar tip-bending one-sided spikes. Both genera provide excellent flowers for cutting and grow readily in well-drained but moist soils. The garden Montbretia (which is a hybrid between *Tritonia pottsii* and *Crocosmia aurea*) belongs here and is the hardiest member of the group.

The *Watsonia* species are showy South African plants rather like *Gladiolus* but less adaptable in cool zones where the corms sometimes rot in winter. In the British Isles they are generally cultivated in greenhouse borders.

There are two types, deciduous and evergreen, the former the hardier as the corms can be rested and lifted if necessary. The plants appreciate plenty of water and sunshine and a rich but well-drained soil.

Watsonias have long sword-like leaves and (according to species) spikes 45–150 cms (1½–5 ft) high of tubular flowers. These may be white, pink, lilac, crimson or purple and flare out into six petal-like segments. They are popular cut flowers in some parts of Africa.

Sisyrinchium, a genus of some 100 species, is mostly found in North and South America. They are frequently low growing with creeping rootstocks, narrow grassy leaves and clusters of funnel-shaped, yellow, blue or red flowers. *S. bermudiana*, the Blue-eyed Grass from Bermuda, grows 13–15 cms (5–6 ins) tall with violet-blue flowers and *S. californicum*—a Californian species, about the same height—is yellow. *S. striatum* resembles a Flag Iris about the foliage but the flower spikes of 30–60 cms (1–2 ft) carry clusters of small, yellowish-white, purple or brown striped flowers. It comes from Chile and self-set seedlings are common when the plant is sited in a sunny, well-drained location.

One species of *Sisyrinchium* (*S. acre*) is the source of a dye used in Hawaii for a form of tattooing which is said to be painless.

Labiatae

180 genera and 3500 species

This large family (also called Lamiaceae) is of cosmopolitan distribution, with greatest representation in the Mediterranean region. The plants are of various types including shrubs, trees, herbaceous plants, bog plants and a few climbers. Many are xerophytic, the leaves protected by tucked-under margins, hairiness, waxy cuticles or reduced foliage; they are often aromatic.

General characteristics include square stems (usually), simple, often hairy leaves and flowers in racemes, cymes or whorls around the stems. Individually they are bisexual and usually zygomorphic and bilabiate, with a five-toothed cleft calyx; a two-lipped, five-lobed (rarely four-lobed) corolla; two or four stamens; one style with two stigmas and a superior ovary. The fruits consist of four one-seeded nutlets or sometimes a drupe.

The two-lipped corolla facilitates nectar gathering by visiting insects, for if—as they must—they use it as a landing platform the essential flower organs are so placed that they brush the insect's back.

Because Labiates secrete volatile oil and have characteristic scents and bitter principles many have considerable economic importance, notably Sages, Mints, Thymes, Lavender and Patchouli (*Pogostemon patchouly*) used for keeping insects from furs, woollens, and other fabrics.

Among the ornamental kinds Lavender holds a special place in the affections of the English. It is a favourite cottage garden plant, the sweetly scented mauve or purple flowers being harvested just before they open for making into lavender sachets and lavender water, or for strewing between linen. At one time many acres were devoted to the cultivation of Lavender for perfumery purposes, and even in the author's native Enfield, just north of London, there is a Lavender Hill where the plant was grown in quantity less than a century ago. *Lavandula vera* (*L. spica*) is especially rich in oil, but *L. stoechas* is also used.

In the garden Lavender requires a warm sunny position in well-drained soil. Plants may be used as solitary specimens (the silvery leaves are evergreen and ornamental all the year), also for edging purposes and low hedges. The name Lavender is derived from the Latin verb *lavare,* to wash, from its use by the Romans to scent their baths but, being a Mediterranean plant, it is not reliably hardy in very cold climates or situations. 'Twickel Purple' and 'Munstead Dwarf' are two good cultivars and there are variants with white and rose-pink flowers.

Another fragrant evergreen is *Rosmarinus officinalis,* a Mediterranean shrub which likes full sun and well-drained soil. It does especially well near the sea. Sprigs of Rosemary in olden times were worn at funerals (hence Rosemary for remembrance) also at weddings (Anne of Cleves wore a crown of Rosemary at her wedding to Henry VIII), and at Christmas joints of roast beef were formerly decked with Rosemary and Bay. Sprigs of Rosemary give a pungent flavour to certain meat dishes and among a number of odd uses ascribed to the plant is that of wrapping its twigs round the limbs to relieve gout (*Bancke's Herball* of 1525); rubbing the ash of burnt twigs on the teeth to cure toothache; burning branches in churches as a substitute for incense; making combs of the wood to prevent giddiness; and using the steeped leaves as a hair tonic to discourage baldness. There are varieties with deeper blue flowers than the species, and also a columnar form called 'Miss Jessop's Upright' which originated in Enfield and makes an attractive wall shrub.

Cedronella triphylla is the Balm of Gilead, a fragrant-foliaged shrub 90–120 cms (3–4 ft) tall, with spikes of white or pale purple flowers carried in loose whorls, and foliage with three leaflets, sometimes used for making tea. It is native to the Canaries and Madeira so only suitable outdoors in warm temperate or tropical climates.

Physostegia is a North American genus, of which *P. virginiana* makes a useful late summer garden plant. It is a herbaceous perennial of sturdy habit, with leafy stems and spikes of fuchsia-rose, tubular flowers which are twisted and puckered at the open ends like those of a nettle. The long tapering leaves are toothed. A curious characteristic of the plant is the way in which the blossoms may be moved from side to side on the flower spikes and then remain as placed. This feature has earned it the cognomen of 'Obedient Plant'.

Physostegias will grow in sun or light shade but should not suffer from drought during the growing season. 'Vivid' is a particularly bright and glowing cultivar but at a height of 60–75 cms (2–2½ ft), nearly 30 cms (1 ft) shorter than the type. There is also a white-flowered form. The plants are propagated by spring division.

Physostegia virginiana
'Vivid'

Phlomis comprises a genus of shrubs and deciduous perennials esteemed no less for their silvery foliage than their unusual flowers. The latter resemble giant Dead-nettles, with long tubular blossoms borne in conspicuous axillary whorls on the stems. Usually they are yellow, but there are also white and purple forms. The plants like a fairly dry position and full sunshine. *P. fruticosa* from the Mediterranean region, the Jerusalem Sage, is one of the hardiest of the shrubby species. It is a branching shrub 60–120 cms (2–4 ft) tall with grey-green, wedge-shaped leaves and yellow flowers; *P. samia* from North Africa, 60–90 cms (2–3 ft), is creamy-white; and *P. cashmeriana* from Kashmir, 60 cms (2 ft), lilac-purple. The leaves of *Phlomis* according to Phillip Miller (1759) are 'greatly recommended by some Persons to be used as Tea for Sore Throats'.

Several of the Deadnettles, *Lamium* species, are used in gardens as ground cover plants, notably a variegated form of the British Archangel *L. galeobdolon variegatum* (now *Lamiopsis galeobdolon* or *Galeobdolon luteum*). This is invasive but useful for covering the ground in deep moist shade and has yellow flowers and silver-marked foliage. *L. maculatum,* of Europe including Britain, has light and deep purple leaf-patternings and a form called 'Aureum' is golden-leafed. Both have pink or purplish flowers. *L. orvala* from S Europe grows about 60 cms (2 ft) high, with showy heads of bright rose or white flowers.

Nepeta mussinii and *N. cataria* are both known as Catmint but in gardens the hybrid *N.* x *faassenii* is more common. A useful edging plant with aromatic spikes of small mauve flowers and neat silver-grey leaves, it grows from 30–45 cms (1–1½ ft) tall and has a curious fascination for cats, which roll on the plant, chew it and are eventually brought by it to a state of acute inebriation. Cats apart, it is an attractive subject for a warm, well-drained situation and easily propagated by cuttings.

Glechoma (*Nepeta*) *hederacea variegata* is a variegated form of the common British Ground Ivy. It has a creeping stem with

Lamium maculatum
pink form

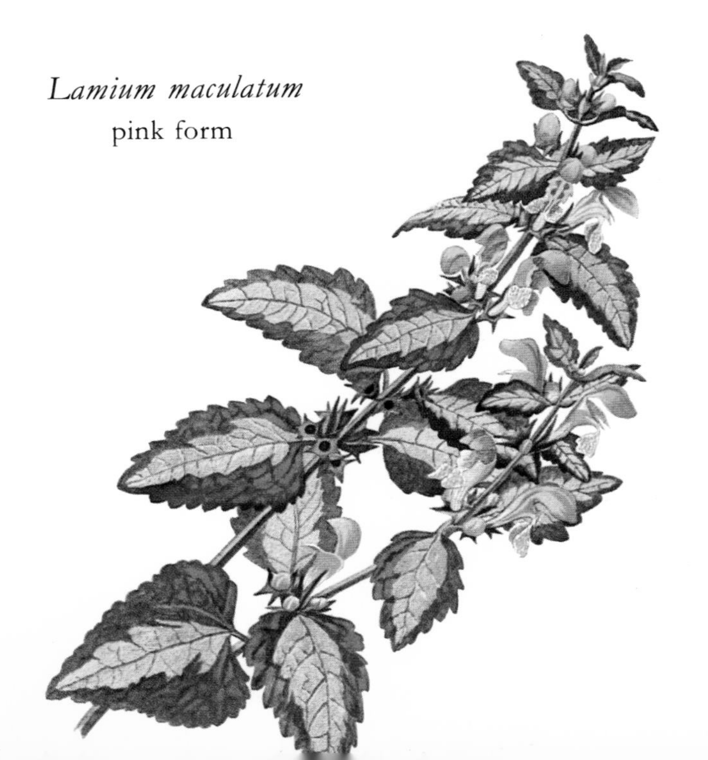

small, kidney-shaped leaves with crenate edges and cream centres and occasional whorls of blue flowers. Variously known as Gill-over-the-Ground, Haymaids, Cat's-foot, Robin-run-away and Ale-hoof it makes a useful hanging-basket plant, with the stems trailing down. The green-leafed type plant was formerly valued as an antiscorbutic and until the reign of Henry VIII was commonly used for making beer.

Dracocephalums or Dragon's Head are herbaceous plants, the perennial species easily cultivated in well-drained, loamy soil. A partially shaded position is best as the flowers quickly fade in a sunny situation. The blooms resemble Catmint, but are larger and on long spikes; the foliage is almost entire, but sometimes toothed or deeply cut.

D. hemsleyanum from Tibet has purplish-blue, Sage-like flowers in loose spikes on stems of 45 cms (1½ ft) and oblong, stalkless leaves. *D. isabellae,* also Chinese, is violet-blue; *D. ruyschianum,* a Siberian species, has oblong smooth leaves and broken spikes of purplish-blue or purple flowers; and *D. sibiricum* (*D. stewartianum*; *Nepeta macrantha*), also Siberian, is like a giant Catmint, 90–120 cms (3–4 ft) high. It does not overwinter well in damp low-lying localities unless planted on raised beds, so that surplus moisture can drain away.

Calamintha (or *Acinos*) *alpina* is the Alpine Savory, a low-growing, S European plant with hairy, aromatic leaves and stems of 15 cms (6 ins) bearing whorls of small purple flowers.

Monarda is the Bee Balm or Bergamot, a moisture-loving genus from North America to Mexico. The plants have square stems, extremely aromatic Nettle-like leaves and brightly coloured flowers in close heads or whorls on stems 60–90 cms (2–3 ft) high. The fragrance is as compelling as, and the blossoms more arresting than, Lavender. Monardas associate splendidly with *Iris sibirica, Hemerocallis* and *Astilbe.*

Provided the soil is moist, Bergamot will grow in full sun or partial shade; under dry conditions water frequently in summer or mulch the ground with peat or compost. *M. didyma* from North America is the best garden species. It is a robust plant with bright scarlet flowers and has produced several good garden forms such as 'Blue Stocking' (violet-purple), 'Cambridge Scarlet' (a brilliant scarlet), 'Croftway Pink' (clear rose-pink), 'Pillar Box' (red), 'Snow Maiden' (white) and 'Perfield Crimson' (crimson). Another species, *M. fistulosa,* was brought to England from North America by John Tradescant the younger. It has deep violet-purple flowers.

Bergamots have pleasantly scented leaves which are sometimes made into tea (Oswego Tea) and also dried for pot-pourri.

A closely related plant is the Mexican Bergamot. This is *Agastache mexicana* which was introduced by Dr E. K. Balls and Dr W. Balfour Gourlay from Mexico in 1938 as *Brittonastrum mexicanum.* It resembles a slender *Monarda* but is less hardy than that genus and has aromatic leaves and rosy-crimson, Sage-like flowers on spikes of 60 cms (2 ft).

Monarda didyma
'Cambridge Scarlet'

Ajuga reptans
'Rainbow'

Coleus thyrsoideus

Ajugas are ground cover plants from damp temperate zones, spreading by means of dense carpeting runners. They are useful on heavy soils in light woodland or between the coarser types of perennials.

Ajuga reptans, the N European and British Bugle, has green leaves and spikes, 15–30 cms (6–12 ins), of deep blue flowers. In olden times the species was regarded as a remedy for gout, jaundice and other ailments. Its foliage variants are the most attractive, particularly 'Variegata', with leaves edged and splashed with creamy yellow; 'Atropurpurea', purplish; and 'Rainbow', which has green leaves marbled with dark red, purple and yellow. The Geneva Bugle, *A. genevensis,* likes more sun, when it flowers freely with spikes, 15–30 cms (6–12 ins), of blue, rose or white.

The genus *Mentha* is rich in fragrant plants. From a culinary standpoint the most useful species are the smooth-leafed Spearmint, *M. spicata* (*M. viridis*), and the Round-leafed or Hairy Mint, *M. rotundifolia.* The young leaves of both are in constant demand for Mint Sauce, flavouring new Potatoes and Peas, Mint Vinegar and Mint Jelly.

Peppermint is obtained from *M.* x *piperita,* a British native which is cultivated commercially for its oil which is used in confectionery, liqueurs (as crème-de-menthe), chewing gum, tooth paste and medicines.

Other Mints with distinctive flavours or characteristics are Apple Mint, white shoots and leaf patternings; Ginger Mint, with gold variegations; Lemon Mint, Pineapple Mint and the strongly scented Eau-de-Cologne Mint. Some of these are used in pot-pourri.

One minute species, *M. requienii* from Corsica, looks like moss spangled with minute mauve flowers. Planted between crazy paving stones it releases a strong smell of Peppermint when walked on.

M. pulegium is the European Pennyroyal which, on account of its carpeting habit, is also known as Run-by-the-Ground, Lurk-in-the-Ditch and even Pudding Grass—from its former use in flavouring Haggis Pudding. It is still used medicinally for stomach ailments and in the manufacture of soap.

M. aquatica, commonly called Water Mint or Fish Mint, is an aquatic plant used in fish pools for its spikes, 30–90 cms (1–3 ft), of lilac flowers. The leaves are hairy and aromatic and the plant has a wide distribution in Europe, North and South Africa and parts of Asia. It has become naturalized in America.

Several members of Labiatae which are not hardy enough to grow outside in temperate zones (North America, N Europe, etc.) have become popular subjects for greenhouse and room decoration. The most important is *Coleus blumei*, a Nettle-like plant from Java with soft but brightly coloured leaves in a bewildering array of shades and patternings. Given good rich soil and auxillary feeding they grow to a height of about 60 cms (2 ft) in summer, the dwarf forms making good summer bedding edging plants.

Coleus blumei

Some *Coleus* leaf forms

Coleus are easily raised from cuttings taken while the shoots are still young, or they may be raised from seed. As the seedlings appear discard the green ones, retaining only those showing promise of colour. The flowers are not important in this group as they detract from the foliage, so they should be picked out as and when seen.

Nevertheless there are two notable sorts grown for their winter bloom. *C. thyrsoideus* from central Africa has bright blue flowers borne on a strong spike of 90 cms (3 ft) with anything from ten to twenty spikes on a single plant. It remains in bloom for weeks and has strongly scented leaves. *C. frederici* from Angola is still more decorative but has a shorter flowering period.

Another tender plant is the South African Lion's Ear, *Leonotis leonurus,* so named because of a fancied resemblance between a lion's ear and the flowers. It is a handsome shrub 1–2.15 (3–7 ft) high, which flowers in autumn with whorls of showy, bright orange flowers separated by sections of silver-grey, downy stem. The oblong-lanceolate leaves (which are also downy) are dried and smoked by natives in South Africa in the same way as *Cannabis sativa*. Fortunately they lack the narcotic properties of *Cannabis*.

There are many other species native to Africa, particularly Kenya, some of which are cultivated.

Plectranthus oertendahlii and *P. fruticosus* from tropical and South Africa are soft-wooded, evergreen sub-shrubs or perennials with oval leaves and numerous spikes of two-lipped, tubular, pink or lilac flowers. They are used in greenhouses or hanging containers (when the growth trails down), or in the tropics, planted in tubs or shady spots outdoors. Under such conditions they form broad carpets of blossom and are in character for weeks on end. The South Africans call them Spur-flowers and some species have edible tubers, known as Kaffir Potatoes.

Melissa officinalis is the European or Lemon Balm, a perennial 30–60 cms (1–2 ft) high, with simple toothed leaves which have a delightful lemon fragrance and whorls of white or pale yellow flowers. The golden-leafed form 'Aurea' is the most attractive.

Melittis melissophyllum, also European, is called the Bastard Balm. It has wine-purple and white, Nettle-like, axillary flowers with a musky fragrance on stems of 30 cms (1 ft), and heart-shaped, puckered leaves.

Thymus vulgaris is the well-known herb Thyme, a common sub-shrub of the Mediterranean region. It is the source of Oil of Thyme and cultivated as a condiment for flavouring meats, soups and sausages. Mount Hymettus honey owes its aromatic flavour to the wild Thyme which abounds there. *T.* x *citriodorus,* the Lemon Thyme, is also used for culinary purposes.

T. serpyllum, found in all of Europe except Iceland, is mat-forming with tiny leaves and purplish (sometimes pink or white) flowers. It is occasionally used for lawns or planted on banks and between crazy paving stones where it emits a pleasant fragrance when walked on. The very similar *T. drucei* often does duty for it.

Stachys is a genus which thrives in sun or light shade, often succeeding in very poor soils, particularly the Caucasian *S. olympica* long known as *S. lanata*. This is commonly called Lamb's Tongue or Donkey's Ears because of the plush-like texture of its silver felted leaves. It makes pleasing ground cover (especially the non-flowering clone 'Silver Carpet') under crimson Roses. The flower spikes of 45 cms ($1\frac{1}{2}$ ft) on normal plants are a curious shade of purplish-rose, half hidden by silver bracts and not very attractive.

S. macrantha (*Betonica macrantha*) or Big Betony, also from the Caucasus, forms compact plants with long, wrinkled, crenate-

Leonotis leonurus

Salvia officinalis 'Tricolor'

Salvia patens

edged leaves and stems of 60 cms (2 ft) carrying many-flowered whorls of violet blossoms. Garden forms exist with deeper and richer violet blooms, also pink.

The Marsh Woundwort (*S. palustris*), a British native, is sometimes used in bog gardens. It has tapering, crenate leaves and spikes of pungent-smelling, purplish-rose flowers on stems 30–90 cms (1–3 ft) tall. The English cognomen refers to its reputed healing properties, which Gerard alludes to in his *Herbal.*

The Chinese Artichoke (*S. affinis* or *S. sieboldii*) with edible tubers is included in this genus.

Salvias are popular garden plants, particularly the Scarlet Sage, *Salvia splendens* from Brazil, 60–90 cms (2–3 ft) tall. Although perennial in the tropics it is usually treated as annual in temperate countries, where it makes a good bedding subject and is also useful for pot work. Modern cultivars tend to be more compact, approximately 30 cms (1 ft), especially variants like 'Tom Thumb' and 'Blaze of Fire'. The vivid scarlet flowers are set off by equally vivid bracts and borne in close spikes. Forms exist with white, mauve and brick-red flowers.

S. sclarea is the herb Clary, used in high-grade perfume and for European wines and liqueurs such as muscatel and vermouth. It is a European plant, at its best the second year from seed, with branching spikes 90 cms (3 ft) or more of pale blue flowers surrounded by rose and white bracts. The leaves are covered with white woolly hairs and have toothed margins.

S. horminum is grown for its brightly coloured bracts. It comes from S Europe and has stems 30–45 cms (1–1½ ft) tall bearing soft, hairy, oval leaves and spikes of small purple or lilac flowers set off by large, velvety, rich purple bracts. Cultivars of different colours are available such as 'Alba' (white), 'Bluebeard' (rich violet-blue) and 'Rose Bouquet' (pink). It is grown as an annual.

S. patens, the Gentian Sage, a beautiful Mexican perennial, has stems of 45–75 cms (1½–2½ ft) with soft arrow-shaped leaves and slender spikes carrying a number of large, brilliant blue flowers which are startling in their intensity. It blooms the first year from seed and makes a delightful bedding or conservatory plant. 'Cambridge Blue' is a pale blue. *S. argentea* from the Mediterranean area is called Silver Clary because of its beautiful foliage. This forms clustered rosettes of large, wavy-edged leaves densely covered with long silver hairs. The white, cream and purple flowers are carried on spikes of 60–90 cms (2–3 ft). It may be propagated by seed.

Salvia splendens
'Blaze of Fire'

S. nemorosa 'Superba' (*S. virgata nemorosa*), a plant of questioned origin, is a splendid border perennial, hardy in Britain and similar climates and in flower for weeks. The sage-green leaves are aromatic and the branching flower spikes of 60–70 cms (2–2½ ft) support many small violet flowers with reddish-violet calyces. 'Lubeca' is similar but more compact, and only 45 cms (1½ ft) tall.

Other good Salvias for border work include *S. nutans*, the Nodding Sage from E Europe, which is violet-blue; *S. uliginosa,* the Bog Sage from eastern North America, a graceful plant with spikes, 1.25–1.5 m (4–5 ft), of rich sky blue flowers late in the season; and *S. farinacea,* the Mealy Sage from Texas, with flowers rather like Lavender, of deep violet with white woolly bracts and stems. It is excellent for cutting and grows about 90 cms (3 ft) with oval, aromatic leaves. *S. haematodes* is European, an impressive plant best treated as a biennial, with large, heart-shaped corrugated leaves and panicles, 90 cms (3 ft), of bluish-violet, funnel-shaped flowers.

The garden Sage, *S. officinalis,* belongs here and has several attractive foliage forms for the ornamental garden, such as 'Icterina', green and gold variegations, and 'Tricolor', green splashed with red and white.

Perovskia (*Perowskia*) *atriplicifolia* is a semi-shrubby plant from Tibet and Afghanistan, useful in mixed herbaceous or shrub borders on account of its late summer flowers. In a sunny, well-drained situation it grows 90–150 cms (3–5 ft), with coarsely toothed leaves of silvery-grey and soft blue, Nepeta-like flowers. The whole plant is silvery and smells of Sage when bruised.

Molucella laevis is the Molucca Balm or Shell Flower, an interesting plant from W Asia grown for the large shell-like, pale green calyces with delicate, feathered white veining. Inside these are the small white or pale pink, fragrant flowers. The spikes, which grow to about 90 cms (3 ft), can be dried for winter decoration and are called Bells of Ireland by flower arrangers.

Sideritis candicans is a plushy plant which children like stroking. The stems and wedge-shaped leaves are covered with silver hairs, as are the whorled inflorescences except for the small yellow flowers. It comes from Tenerife, grows 60–90 cms (2–3 ft) tall and needs greenhouse treatment in cool climates.

Ballota pseudodictamnus from Crete is superficially similar, except that the leaves are more rounded and the flowers white and purple. It will grow outside in warm temperate situations, provided the site is reasonably dry in winter.

Origanum (*Amaracus*) *dictamnus* is the Dittany of Crete, a plant of proverbial virtues including the power of drawing out splinters. The Ancients believed that if an arrow hit a goat and the beast fed on Dittany the arrow would fall out again—hence the human application. It is a small shrubby plant with drooping pink flowers and woolly leaves. *O.* (*Marjorana*) *onites* is the Pot Marjoram, at one time very popular in England for culinary purposes and credited with the ability to drive away ants. *O. marjorana* (*Marjorana hortensis*) is the Sweet Marjoram, a popular herb with purplish flower heads. Its leaves have an aromatic though slightly bitter taste and are used in salads and for various flavouring purposes. It was also much valued years ago as a strewing herb for the pungent fragrance of its bruised leaves.

Lardizabalaceae

7 genera and 35 species

Members of this dicotyledonous family are Asiatic and Chilean, often twining shrubs with palmate or pinnate leaves. The flowers come in racemes and have six perianth segments, six stamens and a superior ovary with many seeds.

Two *Akebia* species make unusual climbing plants for covering pergolas, railings, etc., or will clamber over other shrubs and trees. They produce male and female flowers separately, but on the same raceme. These are deep violet and not very showy, the main attraction being the smooth, palmate leaves with their long-stalked leaflets and the large, sausage-shaped, greyish-violet fruits. When the latter split in autumn they reveal rows of black seeds embedded in white, cotton-wool like pulp.

A. trifoliata (*A. lobata*) and *A. quinata* are native to China and Japan, the blossoms of *A. quinata* being spicily aromatic. They are vigorous plants which require sheltered conditions in temperate climates, for the flowers come early and if frosted will not produce their edible fruits.

Decaisnea fargesii from W China is a small free-standing shrub of 2–3 m (6–10 ft) with large pinnate leaves 60–90 cms (2–3 ft) in length. The inconspicuous six-parted, yellowish-green flowers appear on the young growths and are followed by striking cylindrical fruits about 10 cms (4 ins) long, of a curious metallic-blue shade. This shrub is suitable for a shady position in rich moist soil. The fruits are edible.

Laurus nobilis

Lauraceae

32 genera and 2000–2500 species

This family consists of dicotyledonous trees and shrubs with evergreen, occasionally deciduous, alternate, rarely opposite leaves; bisexual or unisexual, regular flowers with (usually) six perianth segments in two whorls, stamens in three or four whorls and a superior ovary with dry, berry or drupe fruits. The chief centres of distribution are South-east Asia and Brazil.

Cassytha species are parasitic on the young green shoots of other plants.

Lauraceae contains a number of plants of economic importance, notably the Avocado Pear (*Persea gratissima*); Cinnamon (*Cinnamomum zeylanicum*) obtained from the twig bark; Camphor (*C. camphora*) secreted in the roots, shoots and indeed every part of the tree; Cassia Bark (*C. cassia*) used to adulterate Cinnamon; Sassafras (*Sassafras albidum*) used in medicine; Stinkwood (*Ocotea bullata*); Greenheart or Sweetwood (*Nectandra rodiaei*) with its useful timbers; and Sweet Bay or Bay Laurel (*Laurus nobilis*).

The Bay Laurel is a S European shrub or small tree with aromatic leaves and axillary umbels of yellowish flowers. During the reign of Queen Elizabeth I the floors of the better sorts of houses were strewn with Bay leaves instead of rushes and in the palmy days of Rome, Bay was considered the emblem of victory. Dispatches announcing victories were wrapped between Bay leaves, victorious generals were crowned with it, and in triumphal processions every private soldier carried a sprig in his hand.

In the Middle Ages honoured poets—and later men of letters—were crowned with a wreath of berried Bay Laurel—hence the term 'Poet Laureate'. Students studying for degrees at universities were called Bachelors, from the Latin *baccalaureus* (laurel berry) and were forbidden matrimony lest this enticed them from their studies. And so it was in course of time that all unmarried men were styled 'Bachelors'.

Small clipped trees are often grown in pots on patios and terraces, especially in those parts of the world where the neat evergreen leaves suffer winter damage. Free-standing trees should be grown in full sun in well-drained soil. Bay leaves have many culinary uses, giving a pleasing flavour to fish dishes, stocks and milk puddings for example, and can be picked fresh from the tree or dried.

Leguminosae

600 genera and 12,000 species

Leguminosae, the third largest family of flowering plants, has almost cosmopolitan distribution and embraces a wide range of types. These include small annuals and herbaceous perennials, climbers, shrubs, trees, water plants and even xerophytes. They are dicotyledonous, usually with alternate and compound leaves which in some genera may be reduced as in *Ulex*, or exhibit sleep movements, that is close at night or when kept from light as in Clover (*Trifolium*). Several are touch-sensitive like *Mimosa* and *Neptunia*, the leaflets folding and the petioles collapsing when handled.

The Telegraph Plant (*Desmodium gyrans*) at temperatures above 22°C (72°F) is never still during the day, the leaves gyrating round and round in an elliptical orbit. These do, however, adopt sleep movements at night. This violet-flowered annual, 30–90 cms (1–3 ft) tall, comes from the East Indies.

The stems of Leguminosae may be erect or climb by various methods such as leaf tendrils as in *Vicia*; stem tendrils as in *Bauhinia*; hooks as in *Caesalpinia*; or by twining as in *Wisteria*. Many like *Acacia* have thorns, while *Carmichaelia* and several related genera have flat leaf-like stems with tiny scale leaves.

The inflorescences are racemose or in heads, in simple spikes or panicles and the flowers regular or irregular. A characteristic of many Leguminosae is a symbiotic relationship between the roots and certain nitrogen-fixing bacteria by means of which the host obtains nitrogen from the bacteria and the latter carbohydrates from the host. When a Pea plant is dug up, these organisms can be discerned as small nodules on the roots.

The fruits are very variable, for example seed-filled pods or legumes of Peas and Beans; hooked pods in *Medicago*; inflated pods are found in *Colutea*; and sometimes the fruits split explosively and scatter the seeds, as in *Cytisus*.

Leguminosae contains a number of plants of economic importance, especially the pulses (Peas, Beans, etc.) used for human consumption and cattle food. There are also plants with edible pods; timber is obtained from a number of species and others yield gums, resin, oil, dyes and fibres.

The family is broadly divided into three subfamilies, the first Mimosoideae, having regular flowers, four to ten or many stamens and frequently bipinnate leaves.

An important ornamental genus for tropical gardens in this subfamily is *Acacia*, a large group of evergreen trees and shrubs, mostly native to Australia. There they are known as Wattles, because early settlers used them for makeshift 'wattle and daub' dwellings. The walls were constructed of interwoven *Acacia* branches and then plastered with mud. In East Africa Acacias are always called Thorns but in Europe and North America are frequently known as Mimosa, a name which rightly belongs to a different but closely related plant. The flowers are regular and usually yellow in crowded globose heads or cylindrical spikes, a tree in full bloom looking really magnificent with its silvery, finely cut foliage particularly against a blue sky background. The foliage is usually bipinnate, with a feathery appearance. *A. spadicigera,* a species from Central America is called the Bull's Horn Acacia because of the swollen spines on its trunk and branches, which are at first filled with a sweetish pulp which attracts ants. The latter consume this then live inside the hollowed-out thorns, protecting the tree to some extent against other visitants. Shrubby species in Kenya can be similarly tenanted.

Apart from their decorative uses and employment as shade trees, the leaves of some Acacias are fed to livestock, the pods eaten in South America, the seeds in Australia and the wood is made into such things as spears, boomerangs, boats, railroad ties, handles, wheels, bullock yokes, gates, furniture and tobacco pipes. The pods of certain species yield a detergent for washing silk and woollens, and also for shampooing the hair and cleaning silver. The flowers are gathered and spread for their fragrance between linens and used in perfumery; gum arabic is obtained from certain species, and fibres from the young stems; ink, honey and tannin are other byproducts.

Acacia podalyriifolia

Acacia juniperina

There are countless species of which the following are representative: *A. podalyriifolia,* a shrub 3 m (10 ft) tall, hardy in Cornwall; *A. juniperina,* 1.5–2.15 m (5–7 ft); *A. armata,* the

Delonix (Poinciana) regia

Kangaroo Thorn, with rounded heads of flowers; *A. baileyana,* a small but leafy tree; *A. longifolia,* often called the Sydney Golden Wattle, a tree 6–9 m (20–30 ft) tall; and *A. decurrens* var. *dealbata,* the Silver Wattle, with silvery, finely cut leaflets and fragrant yellow flowers which are in great demand as winter cut flowers. This one will grow to a height of 30 m (100 ft) and all the species mentioned are Australian.

Albizia (often misspelt *Albizzia*), from the warmer parts of the Old World, are deciduous trees and shrubs—some of which yield good timber. One of the most garden worthy is *A. lophantha* from W Australia, a tree growing to 14 m (45 ft) with beautiful 'bottlebrush' inflorescences made up of countless yellow or white flowers with numerous long stamens.

Mimosa pudica is the Sensitive Plant, a small South American species grown for its interesting touch-sensitive foliage. The leaves collapse beneath the feet when walked on, but after a while they expand to their normal position. The plant grows about 30 cms (1 ft) tall, and has Acacia-like foliage, prickly stems and rounded heads of mauve-pink flowers. Similar sensitivity is shown by *Neptunia oleracea,* a floating aquatic with yellow flowers, which is also South American.

Calliandra is a genus of about 100 tropical evergreen trees and shrubs with bipinnate leaves and exquisite rounded heads of flowers which have numerous long stamens. They look like swansdown powder-puffs and although of short duration are

Bauhinia variegata candida

Bauhinia purpurea

repeatedly produced, so that the shrubs remain attractive over long periods. *C. fulgens* and *C. haematocephala* are especially beautiful with brilliant red blossoms; *C. gracilis* is creamy-yellow and *C. portoricensis* white.

In Caesalpinioideae, the second subfamily, the flowers are usually irregular with ten or less stamens—or, very occasionally, numerous stamens.

Many large trees of the tropical forests come in this group, and such splendid ornamental shrubs as *Caesalpinia* (*Poinciana*) *pulcherrima,* the Barbados Pride. This makes a prickly plant 3–3.5 m (10–12 ft) tall with bipinnate leaves, and erect terminal racemes carrying orange-yellow flowers from which dangle long tassels of scarlet stamens. *C. gilliesii* from Argentina and *C. japonica* from Japan are hardier and have yellow flowers. *C. sappan* furnishes the sappan or peachwood of commerce and several others yield tannin.

Also outstanding is the Madagascar Flamboyant, *Delonix* (*Poinciana*) *regia,* a deciduous tree growing to 12–15 m (40–50 ft) with large feathery leaves on spreading branches. The flowers appear before the foliage (or in some areas at the same time) as brilliant scarlet racemes at the ends of the branches. Each bloom has four scarlet petals and a fifth variegated in red and yellow or red and white. It is widely planted in the tropics. *D. elata* from Abyssinia is similar but with white flowers which deepen to yellow with age.

Amherstia nobilis, Pride of Burma, caused tremendous excitement when it was first discovered by Dr Nathaniel Wallich (1786–1856) in the early 19th century. Its pendulous racemes of blossom may be 60–90 cms (2–3 ft) long, each vermilion and yellow flower averaging 20 cms (8 ins) in length and 10 cms (4 ins) across. It is difficult to establish.

Bauhinias are small tropical trees with simple but deeply notched leaves the shape of camels' hoofs. The five-petalled flowers are borne in racemes, one petal in each bloom being veined or striated with other shades.

Bauhinia variegata, the Orchid Tree, used for street planting in Trinidad is popular in Florida and cultivated in tropical

Caesalpinia pulcherrima

Calliandra haematocephala

gardens. It is native to India and grows about 6 m (20 ft) high, the flowers light purple with the fifth a deeper purple streaked with cream and carmine. Its variety *candida* has four pure white petals and a fifth patterned with green. Both kinds are richly fragrant. *B. purpurea,* another Asiatic species, is a shrub of 1.5–1.75m (5–6 ft) with fragrant flowers from purple to white. Some Bauhinias are climbing lianas with flattened, corrugated or twisted stems and occasionally tendrils.

Cassias are native to both hemispheres, nearly always in tropical or warm temperate regions. They are shrubs or trees (occasionally herbs) with alternate, pinnate leaves and five-petalled, normally yellow flowers.

One of the most popular for tropical gardens is *Cassia fistula,* the Shower of Gold or Pudding-pipe Tree. Indigenous to India it grows freely from seed and has become widely distributed all over the tropics. The canary-yellow fragrant flowers are borne on drooping racemes of 30–60 cms (1–2 ft). Each has ten yellow filaments, three of which turn up at the ends after the fashion of skis. The pods, which are 30–60 cms (1–2 ft) long, are green at first and then black and hang down from the tree like fat cigars.

Many and varied are the medicinal uses claimed for the seeds, pods, flowers and bark. Various parts of the plant are reputed to cure ringworm, destroy insects and are used for cough medicines and febrifuges. The fruits are the Cassia pods of commerce, a purgative which also, according to the English naturalist Hans Sloane (1660–1753), 'helps Mad People to sleep'.

There are many other species among which the Candlestick Senna, *C. alata,* a Brazilian shrub 1.75 m (6 ft) or so in height, has erect racemes of golden flowers which are black in the bud. *C. auricularis* from Ceylon makes a bush about 1.5 m (5 ft) high, flowering freely and continuously; the individual blooms are yellow and about 5 cms (2 ins) across.

Cercis are small trees, attractive in spring when the Pea-shaped flowers stud the naked branches and at times burst out from the bark of the limbs and trunk. The smooth simple leaves have rounded bases and the pods are thin and black. All the species should be established when very young as mature plants sometimes fail after being transplanted. They thrive in moist soils and appreciate plenty of sunshine.

C. canadensis is the Redbud of North America, a striking shrub or small tree with stalkless clusters of rosy-pink flowers.

Lathyrus odoratus hybrids

When viewed in the wild in spring, along with the white-flowered Dogwoods (*Cornus florida*) they present an unforgettable picture. There is a white form, perhaps still more beautiful, and a double variety. The flowers were eaten by early settlers in salads.

C. siliquastrum from the Mediterranean region and Asia is called the Judas Tree because legend has it that Judas hanged himself from a tree of this species. Its purplish-rose flowers are larger than those of the Redbud and it grows to a height of about 12 m (40 ft). Both plants are suitable for temperate gardens.

Tropical again is *Brownea grandiceps,* the Rose of Venezuela, an evergreen, small to medium tree which will reach 9–12 m (30–40 ft) in height. It has alternate, pinnate leaves about 30 cms (1 ft) in length and spectacular globe-like inflorescences which have been likened to Rhododendrons. These are up to 20 cms (8 ins) across and made up of many tightly packed red or deep pink flowers. *B.* x *crawfordii* is a garden form with clusters of deep rosy-red flowers.

From the Mediterranean region comes the Locust Tree, *Ceratonia siliqua,* an evergreen 12–15 m (40–50 ft) tall with pinnate leaves, catkins of petal-less flowers and leathery pod fruits. The latter contain a sweet nutritious pulp much appreciated by animals and small boys, and small flattened bean-like seeds said to have been the original 'carat' weights used by ancient apothecaries

Cercis siliquastrum

and goldsmiths. The species is also known as Carob Tree and St John's Bread, this last because of a widely held belief in the East that the fruits were the 'locusts' (and wild honey) eaten by John the Baptist in the Wilderness.

In temperate countries, the most important subfamily is Papilionoideae. The species may be climbing, woody or herbaceous but are easily recognized by the papilionaceous (butterfly-like) flowers. These are characterized by one large, erect and spreading petal at the back called the standard, and two lateral petals, the wings, which in turn stand each side of the keel formed by two more petals joined at their lower margins.

One of the most popular is the Sweet Pea, *Lathyrus odoratus,* which reached England from Sicily around 1699. The species, with small but fragrant flowers, has maroon wings and blue standards. It has now practically disappeared from its native habitat and is scarce in cultivation, but just recently has been reintroduced from stock found in Quito, Ecuador, taken there presumably centuries ago by the Spaniards.

Modern Sweet Peas, which are derived from this species, have six or seven blooms on strong stems and there are distinct strains such as the Early (that is, winter) Flowering; Giant Frilled or Spencer; the Cuthbertsons; Multifloras; Dwarf Cupids; Knee-Hi and Bijou, the latter useful for edging purposes. Colours range from white and cream through pink, salmon, orange, scarlet, mauve, blue and purple.

The first use of Sweet Peas is undoubtedly as cut flowers; they were popular in Edwardian times and a great favourite with Queen Alexandra. Successful cultivation demands deep rich soil and a sunny situation. The plants are annuals and climb by means of tendrils, so need the support of strings, wire or netting. For exhibition work the plants are usually restricted to a single stem and all side shoots are removed. Feeding, watering in dry weather and the regular removal of dead flowers is essential to keep the plants flowering.

Two perennial sorts commonly grown are the Everlasting Pea, *L. latifolius* from S Europe, and *L. grandiflorus.* The former has clusters of small but very bright rose flowers on winged stems and a large fleshy root. It grows readily over buildings or arbours and has a white form *albus.*

L. grandiflorus has two deep rose flowers the size of an ordinary Sweet Pea on each stem and paired leaflets which go on to produce forked tendrils. It is one of the hardiest but is not known to set seed in cultivation so may be a clone. Nevertheless it spreads readily by means of underground stems and makes a striking hedge 1.5–1.75 m (5–6 ft) high if supplied with the wherewithal (for example peasticks) to climb.

Cassia fistula

Wisterias are beautiful deciduous climbers, the flowers superficially like Laburnum except that they are never golden. Usually they are mauve or purple although white and rose forms and doubles exist. Wisterias make good wall shrubs for warm, sheltered situations (provided they are kept away from tiles or guttering where they can cause damage) and can also be grown over pergolas and trees or trained as standards. Deep rich soil is essential otherwise the plants fail to flower satisfactorily.

The most important are the purplish-blue *Wisteria floribunda,* a vigorous climber from Japan and its variety *macrobotrys* which has the longest racemes of any Wisteria. These are a rather light purple and 60–90 cms (2–3 ft) in length although reputed to reach up to 1.75m (6 ft) in Japan when specially trained. There is also a double variety 'Violacea Plena' and a pink form 'Rosea'.

W. venusta (*W. brachybotrys alba*) from Japan is white-flowered (violet in the wild form) and fragrant, with downy leaves and short flower trusses of 10–15 cms (4–6 ins). *W. sinensis* from China is the best for colour, and is a vigorous violet-blue species

Brownea x *crawfordii*

which has been known to climb 30 m (100 ft). However, by restricting all extension shoots and annual hard pruning, the stem may be induced to thicken and make a standard of 1.75–2.5 m (6–8 ft).

In temperate climates Laburnums are popular for their striking clusters of golden, Pea-shaped flowers in spring. They make small trees up to 6 m (20 ft), and are often called Golden Chain or Golden Rain. Suitable for practically any soil and situation they are easily raised from seed, although the better kinds such as *Laburnum* x *watereri* 'Vossii' with larger and longer racemes, are usually budded or grafted on to stock of the S European *L. alpinum* or *L. anagyroides*.

Laburnocytisus adamii is a graft hybrid between *Cytisus purpureus* and *Laburnum anagyroides*. Years ago it was fashionable to obtain standard Brooms by grafting *Cytisus* on Laburnum Stock. At the beginning of the century one plant, belonging to a French nurseryman called Adam, became damaged; all but a small part of the scion broke off, but this continued to grow, as did the Laburnum stock. The resultant plant—a chimera—shows three kinds of flowers: ordinary yellow Laburnum, purple Broom (usually bursting out through the bark) and pink Laburnum (which is derived from a fusion of the two plants).

Among a number of herbaceous perennials used for border work in temperate countries Lupins are the most important. Before the Second World War they were very ordinary plants with blue, purple or white flowers. The chief species grown at that time was *Lupinus polyphyllus,* a North American perennial of sturdy habit, able to hold its own in any garden soil or situation.

Today this plant is rarely seen for it was superseded—almost overnight—with the introduction of Russell hybrids. George Russell, a simple countryman who had grown Lupins for years on a small allotment garden, staggered the experts when they first saw his plants. These had shapely spikes and larger and better blooms in an extensive range of shades, pinks, reds, and bicolors included. Nevertheless Russell Lupins have not the stamina of the old time species, possibly because of mixed parentage. In his quest for brighter colours Russell worked with various annual forms such as *L. laxiflorus, L. lepidus* and *L. mutabilis* which may account for some degree of reduced permanency.

In order to prolong their life, never allow *all* the spikes to set seed. Remove most of the old flower heads as the blooms pass and never save seed from blue varieties (the dominant colour). They flower in twelve months from seed and in cold situations are best planted out in spring. The plants like sun but not too hot a situation, and also a moist but well-drained soil. Numerous varieties exist with white, yellow, pink, red and purple flowers; also bicolors showing attractive combinations of shades. They usually grow around 75 cms (2½ ft) tall.

Baptisia australis, the False or Blue Indigo from North America, has the habit of a Lupin, with Vetch-like leaves and deep, rich blue, Pea-shaped flowers on stems of 1.25–1.5 m (4–5 ft). *Thermopsis montana* from NW United States has rather similar, but more sparsely clothed, spikes of bright yellow flowers.

Other perennials of similar habit are the European *Lathyrus luteus* var. *aureus* (*Orobus aurantiacus*), with rich orange flowers on upright stems of 60 cms (2 ft), and *Galega officinalis,* the Goat's Rue, also European. This has soft green pinnate leaves and spikes of white or mauve keeled flowers. Garden forms include 'Her Majesty' (lilac-blue) and 'Lady Wilson' (mauve and white).

Among the hardier shrubs Brooms (*Cytisus* species) hold pride of place. They have the merit of growing rapidly in very poor soils and of looking interesting even when out of flower. Since they only transplant satisfactorily when the roots are undisturbed it is usual to raise them in small pots and plant out from these. They have wiry green stems, tiny leaves and myriads of small Pea-shaped flowers studding the branches. Unfortunately many Brooms become straggly with age, so to keep them compact, prune them immediately after flowering taking all side shoots back practically to the second year wood. Never cut into the older branches or these will die.

Cytisus require full sun and well-drained soil and are propagated by seed, or the named varieties by half-ripe summer cuttings.

Ulex europaeus, Gorse or Whin, is somewhat similar except that it is extremely prickly. In Britain and W Europe it is widespread and often found on poor sandy heathbeds but it is adaptable and will grow in most soils. It makes a useful bank shrub. The plant is evergreen, 60–120 cms (2–4 ft) high, with attractive, golden, richly scented flowers. It flowers spasmodically throughout the year but most heavily in early summer. There is a double form called 'Plenus'.

Spartium junceum is not unlike a tall green-stemmed Broom, with large, fragrant, golden flowers in late summer; it grows to 3 m (10 ft). It comes from the Canary Isles and Mediterranean regions but is now widespread in tropical parts of South America, Africa and S Europe.

Genista again is a closely related genus. *G. aetnensis,* the Mt Etna Broom, makes a small tree, 3.5–4.5 m (12–15 ft) high, with drooping branches carrying a few tiny leaves and many small yellow flowers. *G. hispanica* is the Spanish Gorse, a spiny shrub 30–90 cms (1–3 ft) high, which is pleasingly compact when young but tends to straggle with age. It needs sun and well-drained soil and in some situations makes a useful edging plant to deter trespassers. The yellow florist's Genista, which is popular as a pot plant grown in early spring is *Cytisus* x *spachianus* (*C. racemosus*).

Robinia pseudacacia, the False Acacia, is a noble tree from North America now naturalized in many parts of Europe, especially the southern Alps where it has become an important timber tree. It has deeply furrowed bark, soft green leaflets and short pendulous racemes of fragrant white flowers. Growing 20–24 m (70–80 ft) it has a number of forms including one with a columnar habit called 'Fastigiata' and another with rose-pink flowers 'Decaisniana'. False Acacia is known as Black Locust in the US.

Colutea arborescens is the Bladder Senna, a shrub with Pea-shaped yellow flowers succeeded by inflated seed pods which pop when squashed. It is native to S Europe (including the slopes of Vesuvius) and grows 2.5–3 m (8–10 ft) high. It has similar properties to Senna (*Cassia* species).

Cladrastis lutea (*C. tinctoria*) is the Yellow Wood, a North American tree 12 m (40 ft) high with long pendulous panicles of fragrant white flowers and leaves with seven to nine leaflets. The heartwood is the source of a yellow dye and the close-grained yellow wood is used for gunstocks.

Lupin hybrids

Clianthus formosus

Strongylodon macrobotrys

Indigofera tinctoria is one of several species once yielding the blue dye called indigo. The species has pink and red flowers. It grows 1.25–2.5 m (4–6 ft) tall and comes from the East Indies. More ornamental for the garden is *I. pendula,* a shrub 2.5–3 m (8–10 ft) tall, with pinnately divided leaves 20–25 cms (8–10 ins) long and drooping Wisteria-like bunches of rosy-purple flowers 30–45 cms (1–1½ ft) long. *I. decora* is white and crimson and 30–45 cms (1–1½ ft) tall and *I. kirilowii* is a plant of 90 cms (3 ft) with almond-pink flowers which have darker spots at the petal bases.

Plants of economic importance with papilionaceous flowers include *Arachis hypogaea,* variously known as Monkey Nut, Peanut and Groundnut. A native of Brazil, it is widely grown in many parts of the tropics particularly in Africa, India and the southern United States on account of its edible seeds. After fertilization the white Pea-like flowers grow downwards, drawing the two-seeded pods beneath the soil to ripen.

The Lentil (*Lens esculenta*) is a food plant of great antiquity; French Beans (*Phaseolus vulgaris*) in spite of their name are really South American (seeds have been unearthed from Peruvian graves); *Derris elliptica* is an important insecticide; Broad Bean (*Vicia faba*), Pea (*Pisum sativum*) and Soya Bean (*Glycine max*) are other Leguminous plants. Soya Beans have been cultivated in China for thousands of years. They are rich in protein and have a high fat content.

Liquorice is obtained from the roots of *Glycyrrhiza glabra,* a Mediterranean plant; Clover (*Trifolium* species) one of the world's most important forage plants and Lucerne (*Medicago sativa*), called Alfalfa in the US, are also members of the family.

Several plants from warmer regions are prized exotics for growing under glass in Europe and North America.

One of the most interesting is the Parrot's Bill (*Clianthus puniceus*), which was planted by Maoris near their homes before the coming of the white man and is now found only in cultivation. It is an evergreen climber reaching 3.5 m (12 ft) or more with glossy pinnate leaves having eight to fourteen pairs of leaflets. The flowers are borne in axillary clusters of six to fifteen on a pendulous stalk and are brilliant scarlet, shaped like a parrot's bill, which accounts for the common name.

Australia possesses a distinct but more tender species, having long trailing stems and pinnate leaves which are almost entirely covered with long silky hairs. This is *C. formosus* (*C. dampieri*), Sturt's Desert or Glory Pea, with clusters of large scarlet flowers with conspicuous purple-black blotches. The species makes a good subject for a hanging basket or window box but it is not always easy to cultivate on its own roots and for this reason is sometimes grafted on seedling *Colutea arborescens.*

Chorizema cordatum is a W Australian shrub of loose habit, with slender branches bearing rough heart-shaped leaves and small, vivid red and orange flowers in loose racemes. It must be grown under glass in cool climates, likes a well-drained soil and flowers in spring.

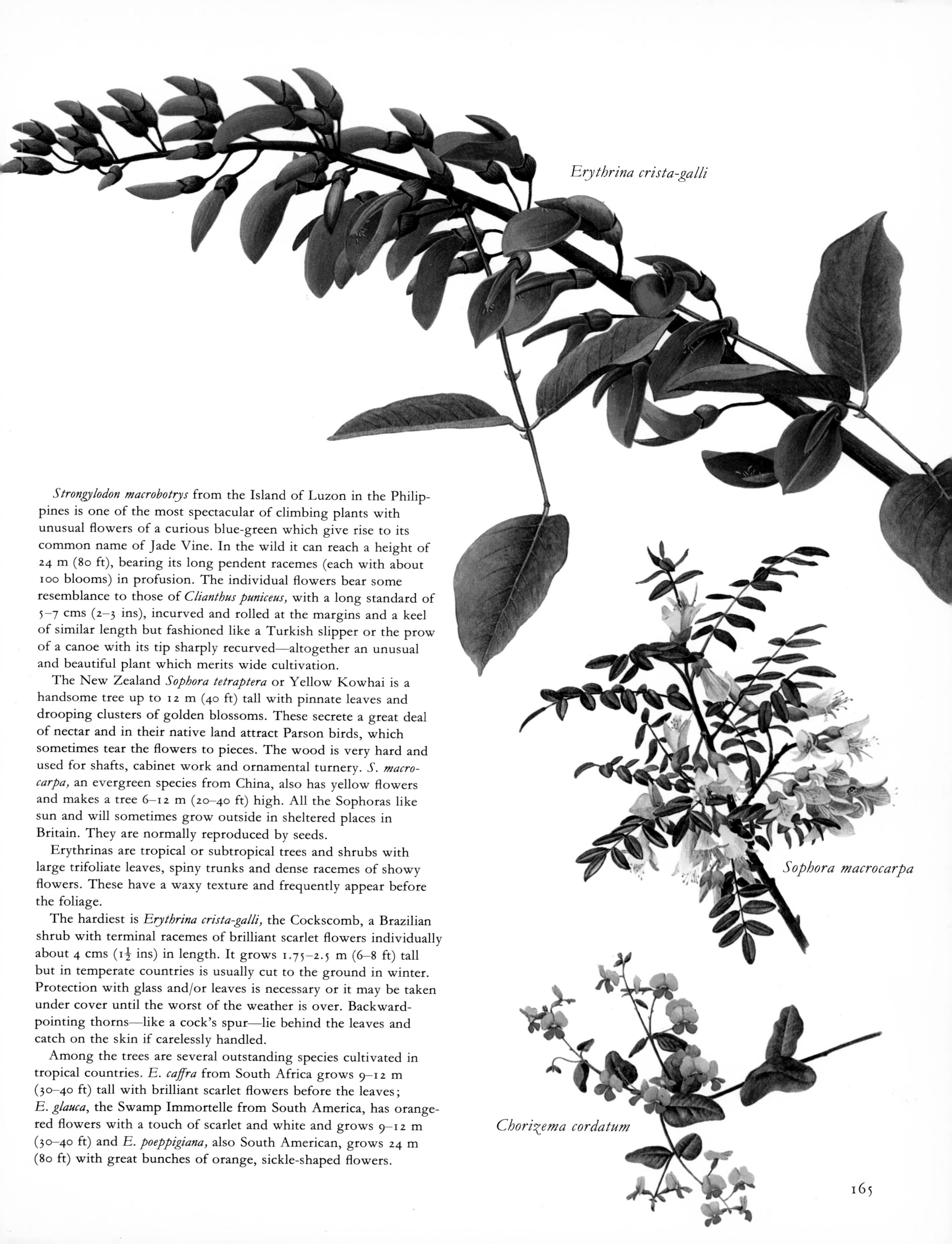

Erythrina crista-galli

Sophora macrocarpa

Chorizema cordatum

Strongylodon macrobotrys from the Island of Luzon in the Philippines is one of the most spectacular of climbing plants with unusual flowers of a curious blue-green which give rise to its common name of Jade Vine. In the wild it can reach a height of 24 m (80 ft), bearing its long pendent racemes (each with about 100 blooms) in profusion. The individual flowers bear some resemblance to those of *Clianthus puniceus,* with a long standard of 5–7 cms (2–3 ins), incurved and rolled at the margins and a keel of similar length but fashioned like a Turkish slipper or the prow of a canoe with its tip sharply recurved—altogether an unusual and beautiful plant which merits wide cultivation.

The New Zealand *Sophora tetraptera* or Yellow Kowhai is a handsome tree up to 12 m (40 ft) tall with pinnate leaves and drooping clusters of golden blossoms. These secrete a great deal of nectar and in their native land attract Parson birds, which sometimes tear the flowers to pieces. The wood is very hard and used for shafts, cabinet work and ornamental turnery. *S. macrocarpa,* an evergreen species from China, also has yellow flowers and makes a tree 6–12 m (20–40 ft) high. All the Sophoras like sun and will sometimes grow outside in sheltered places in Britain. They are normally reproduced by seeds.

Erythrinas are tropical or subtropical trees and shrubs with large trifoliate leaves, spiny trunks and dense racemes of showy flowers. These have a waxy texture and frequently appear before the foliage.

The hardiest is *Erythrina crista-galli,* the Cockscomb, a Brazilian shrub with terminal racemes of brilliant scarlet flowers individually about 4 cms (1½ ins) in length. It grows 1.75–2.5 m (6–8 ft) tall but in temperate countries is usually cut to the ground in winter. Protection with glass and/or leaves is necessary or it may be taken under cover until the worst of the weather is over. Backward-pointing thorns—like a cock's spur—lie behind the leaves and catch on the skin if carelessly handled.

Among the trees are several outstanding species cultivated in tropical countries. *E. caffra* from South Africa grows 9–12 m (30–40 ft) tall with brilliant scarlet flowers before the leaves; *E. glauca,* the Swamp Immortelle from South America, has orange-red flowers with a touch of scarlet and white and grows 9–12 m (30–40 ft) and *E. poeppigiana,* also South American, grows 24 m (80 ft) with great bunches of orange, sickle-shaped flowers.

The flowers of Erythrinas are used as seasoning in Colombian cooking and in salads and soups in Guatemala. The seeds, leaves and bark, however, are toxic in some species and used to stupefy fish and poison rats. *E. variegata* (*E. indica*) is the Indian Coral Tree, a plant 18 m (60 ft) high with scarlet blooms which yield a red dye after boiling. The young leaves are eaten in curry and in Indo-China used for wrapping minced meat.

The New Zealand *Carmichaelia* species have distinct foliage differences from most members of Leguminosae, and for these shrubs reduction of leaf surface has been carried to extreme lengths. Although the seedlings usually show small leaves these disappear with maturity, and it is possible that this protects the plant from moisture losses. Photosynthesis is carried out by the stems which are green and more or less flattened—especially when young. The flowers are frequently highly fragrant and carried in short, many-flowered clusters on the stems. Occasionally the blooms are solitary. *C. australis* grows 1–2.75 m (3–9 ft) tall and has lilac flowers with lighter or darker stripings. *C. nana* only grows approximately 10 cms (4 ins) high and has red flowers but no leaves and *C. petriei,* with fragrant violet-purple flowers, makes a bush of 60–180 cms (2–6 ft).

Mucuna species are handsome tropical climbers with drooping clusters of Pea-shaped blooms with upturned keels. They look particularly fine growing over buildings or draping a fence. *M. bennettii,* a perennial from New Guinea, with brilliant orange-scarlet flowers in pendulous trusses is one of the best. Some species have stinging hairs on the seed pods.

Parochetus communis, the Shamrock Pea from the mountains of tropical Africa and Asia, is a delightful prostrate-growing, Clover-like plant with solitary, clear blue, Pea-shaped flowers. In frost-free areas it makes good ground cover in moist soil, or alternatively can be grown under a greenhouse bench.

Another blue-flowered legume is *Clitoria ternatea,* a slender climber from India with showy, clear blue, Sweet-pea like flowers which only last a day. There is a double form and also a white. In Malaya the blossoms are sometimes used to give a blue colouring to cooked rice.

Pinguicula macrophylla (*P. caudata*)

Pinguicula grandiflora

Lentibulariaceae

4 genera and 170 species
cosmopolitan distribution

These are aquatic or moisture-loving, dicotyledonous insectivorous plants which trap their prey by means of various foliar devices. The flowers are in spikes, racemes or sometimes solitary. Individually they are bisexual but irregular in shape, with two-lipped corollas, two stamens and superior ovaries.

Pinguicula vulgaris is the Bog-Violet or Butterwort, a European species usually found in acid bogs. It forms rosettes of simple, oblong, yellowish-green leaves with slightly upturned margins. These are covered with glands which exude a clammy substance that traps insects, and later secretes a ferment which dissolves and then digests their bodies after the fashion of *Drosera.* The flowers are variable in colour from white and mauve to deep purple, the latter reminiscent of Violets (*Viola odorata*). *P. grandiflora*, also European, is similar but with larger flowers of a deeper blue.

North of the Arctic Circle nomadic Lapps can be seen milking reindeer cows, pouring the milk—while still warm—through sieves lined with *Pinguicula* leaves. Some rennet-like property in the leaves causes the milk to solidify after a time and this the Lapps use instead of butter, a circumstance which probably accounts for the name Butterwort being applied to this species.

P. macrophylla (also known as *P. caudata* and *P. bakerana*) is a handsome species from Mexico with carmine to purple flowers; in *P. lutea*, native to SE North America, the blooms are orange-yellow and nearly 4 cms (1½ ins) across; and *P. alpina* from arctic Europe has white flowers with yellow throats.

Pinguiculas can be grown in well-crocked pans of sphagnum and fibrous peat with a little loam and coarse sand. This compost must be fairly loose. Moisture is supplied by standing the pans in saucers of water and giving them an occasional spray. Soft water must be used—lime kills the plants.

Utricularias come from both tropical and temperate zones, the latter all aquatic. *Utricularia vulgaris*, a Eurasian and American species, is the Bladderwort, sometimes cultivated in ornamental ponds. It is a rootless plant sending out leafy stems 15–45 cms (6–18 ins) long, the leaves divided into hair-like segments. These are more or less thickly furnished with small, opaque, ovoid bladders which trap water fleas (*Daphnia*) and similar small creatures. Once inside the victims cannot escape and presumably their decaying bodies nourish the host. Most of the time the plant lies well submerged but in summer the bladders fill with air and bring it near the surface to flower. The blooms are rich yellow on spikes of 7–22 cms (3–9 ins), but once they are over the bladders fill with water and the plant descends to ripen its seeds at the bottom of the pool.

Some species are land plants with remarkable runners, which develop bladders amongst the moss or compost in which they grow. They can be cultivated in large pans of peat and sphagnum partially plunged in water, so that the top soil is always moist. Examples so grown are the South American *U. humboldtii,* which has racemes of bluish-purple flowers with brownish sepals and *U. longifolia,* an exotic looking species with spikes, 60 cms (2 ft) long, of mauve flowers which have golden blotches.

Other species are epiphytic and have tuberous branches which act as water storage organs. These are often grown in baskets of peat and sphagnum, kept constantly moist and hung up in a warm greenhouse. Examples are *U. alpina* from South America with large white flowers with a touch of yellow and green, and the Costa Rican *U. endressii,* pale lilac with yellow markings.

The other genera, *Polypompholyx* from Australia and *Genlisea* from Central America and South Africa, are not cultivated.

Liliaceae

250 genera and 3700 species

This is one of the largest families of flowering plants, monocotyledonous and of cosmopolitan distribution. While the majority are herbaceous from bulbs or rhizomes, many are xerophytic, some are of a fleshy nature (for example *Gasteria* and *Aloe*) and there are also climbing types as with *Gloriosa*.

The inflorescences are generally racemose, although solitary flowers are found in Tulips and certain other genera. The blooms (usually insect-pollinated) are normally regular and perfect, their floral parts in units of three (rarely two, four or five). Thus there are six (three plus three) perianth segments, the same ratio of anthers, a superior ovary having three divisions and many seeds.

The most important genus for the garden is probably *Lilium,* the true Lilies, all of which are N temperate bulbous plants with leafy stems and large flowers (usually) in terminal racemes. They are among the oldest of cultivated plants with several distinct types, some only producing basal roots, others with basal and stem roots—the latter a fine mass of feeding roots immediately above the bulbs.

Most Lilies make a fresh bulb each season, generally within the older one and assuming its concentric shape. They may have in addition smaller bulblets around the parent bulb or, in the case of stem rooters, on the stem. Many American species (like *L. superbum*) produce bulbs at the end of an underground stolon (stoloniferous); other species (like *L. pardalinum*) have the scales clustered—not in a single bulb but along a horizontally branching rhizome (rhizomatous).

Most Lilies require plenty of moisture during the growing season together with free drainage at all times, for excessive wet rots the bulbs. In general they thrive in a neutral soil, although most prefer slightly acid conditions and a few tolerate lime.

Leafmould is generally beneficial but it must be well rotted; peat which retains moisture, and bonemeal (3 to 4 ounces per square yard) which provides a slowly released source of phosphate are also useful. Fresh manure is an thema to Lilies; in fact excessive feeding of any kind (while it may increase the size of the bulbs) makes them 'soft' and encourages fungoid diseases. Although the flowers appreciate sunshine the roots should always be kept cool, which is the reason Lilies do well when planted amongst herbaceous perennials or in light woodland.

Those which prove difficult or tender in outdoor situations should be grown in pots or tubs, the bulbs of stem-rooting kinds buried deeper than the others.

Non stem-rooting Lilies should be planted so that they have twice their own height of soil above them; that is, a bulb 7 cms (3 ins) high should be set 15 cms (6 ins) deep. The one exception *L. candidum,* the Madonna Lily, should be planted during a short period in late summer when it is leafless and dormant and then only the tip of the bulb should be covered with soil.

The Madonna Lily with its stems, 1.5–1.75 m (5–6 ft) high, of immaculate white flowers, golden-yellow anthers and a delightful fragrance is probably the oldest cultivated Lily. We know it was grown by the Cretans and Egyptians centuries before Christ and representations of the flowers on vases and other objects date back to 1750 B.C. The monks of the Middle Ages cherished this Lily and many beautiful paintings of Renaissance times link the species with the Virgin Mary.

Although its distribution is uncertain, it has been recorded from the Balkans where in 1916–17 bulbs of a form known as *salonikae* were discovered by Mr J. Norman Ambler. This seems to be hardier than the original stock, which sometimes tests the patience of gardeners. More recently C. D. Brickell of Wisley has found it on barren hillsides in Turkey and K. Aslet in N Greece. It is also known from Mt Hermon in Lebanon and N Israel. Where *L. candidum* thrives it is a splendid plant but one to leave alone as it resents disturbance. It is one of the few Lilies that prefer a chalky soil.

Other European Lilies (which for the most part have few or no stem roots) are *L. chalcedonicum,* the Scarlet Martagon, a native of Greece introduced to Britain during the reign of Elizabeth I. It likes partial shade, tolerates lime and grows 90–120 cms (3–4 ft) tall with up to ten pendulous, reddish-orange flowers which turn upwards at the edges like a Turk's cap.

Lilium candidum

L. martagon has the same shaped flowers and is widespread in Europe. It needs a woodland setting, grows 90–150 cms (3–5 ft) tall and has a rather unpleasant scent. The species is a dirty purplish-pink but the white-flowered *album* with twenty to thirty flowers on a stem is very fine and another called *cattaniae* has dark burgundy-red blooms.

L. monadelphum, a handsome and easily grown Caucasian Lily, has pendulous, clear yellow flowers with black spots near the throat. It is lime-tolerant but takes a year or two to establish. *L. szovitsianum* (*L. monadelphum* var. *szovitsianum*) is a Caucasian Lily with twenty to thirty pendulous, campanulate flowers on a stem. These are yellow spotted with black. *L. pyrenaicum,* a vigorous species from the Pyrenees, has heavily scented, yellow Turkscap flowers spotted with black. It is best suited to the woodland garden where, if happy, it soon establishes well.

Among the American species *L. canadense* grows 90–120 cms (3–4 ft) tall with pendent bell-shaped flowers of yellow or orange. It should be grown in light woodland and like *L. candidum* requires summer planting. The Swamp Lily, *L. superbum,* has brilliant orange flowers on stems of 1.20–2.5 m (4–8 ft). It will not tolerate lime and likes moist (but not wet) soil.

Among the stem-rooters are such splendid plants as *L. auratum,* the Golden-ray Lily of Japan, with immense flowers up to 30 cms (1 ft) across, white, streaked with gold and spotted in crimson. They are very fragrant with twenty to thirty flowers on each stem 1.5–1.75 m (5–6 ft) high. Cultivars are available with a miscellany of markings. *L. regale* is a Chinese species discovered by E. H. Wilson in 1904. It is very fragrant with many funnel-shaped blooms of creamy-white and often seeds itself about the garden. This Lily should be grown in full sun in moist but well-drained soil, but becomes soft if overfed.

L. speciosum from Japan has large, white, reflexed flowers spotted with crimson which are intensely fragrant. This species again has many cultivars which make good pot plants. *L. tigrinum* (now *Lilium lancifolium*) is the Tiger Lily, the flowers rich fiery gold with curved and pointed petals and dotted with deep purple. It has been grown for centuries in China, Korea and Japan for its edible bulbs. The Mid Century group of Lilies are derived in part from this species and make excellent plants for naturalizing in sun or partial shade.

Hybrid Lilies are now of great importance as garden plants as they are reliable and free flowering. The main breeders are Mr de Graaf of Oregon, USA, and Miss Isabella Preston of Ottawa, Canada. The Preston Hybrids are sturdy stem-rooting plants with flowers in various shades of orange-red and yellow. The Stenographer Group and Fighter Group come in this range.

Bellingham Hybrids have up to twenty terra-cotta-red to yellow flowers on each stem, 1.25–2.5 m (4–6 ft) high. They are stem-rooting and particularly suitable for woodland planting.

A closely allied genus to *Lilium* is *Fritillaria* and the most widely distributed is *F. meleagris,* with a habitat which extends right across Europe from Britain to the Balkans. It naturalizes readily in damp lime-free spots and has slender stems 30 cms (1 ft) high bearing scattered linear leaves and solitary, nodding, bell-like flowers. These are rich red-purple chequered with a darker shade but white and paler forms are frequent. Common names for it are Snake's Head Lily and Leopard Lily, the latter said to be a corruption of Leper Lily from the similar shape of the flowers to that of the warning bell of the leper.

Lilium regale

Lilium szovitsianum

Quite distinct is *F. imperialis,* the Crown Imperial, one of the oldest known cultivated plants, and a native of northern India, Persia, Afghanistan and the Himalayas. It blooms in spring on stout stems of 90–120 cms (3–4 ft), the large pendent flowers arranged in rings near the top of the stems. These are surmounted by tufts of leaves something like Pineapple tops. Many of the old Dutch masters depict Crown Imperials in their flower paintings.

A charming legend relates how the blossoms of this species were once white and turned upwards. It grew in the Garden of Gethsemane along with many other flowers which, as Christ passed by, hung their heads in humility. All except the Crown Imperial which was too full of pride on account of its unique crown. Observing this, our Lord turned and rebuked it whereupon it hung its head, blushed a rosy-red and tears came into its eyes. The usual colour is reddish-orange (although yellow forms exist) and the white, unshed tears can plainly be seen within the bells. These drops of nectar (not tears) are very difficult to dislodge even with a vigorous shaking.

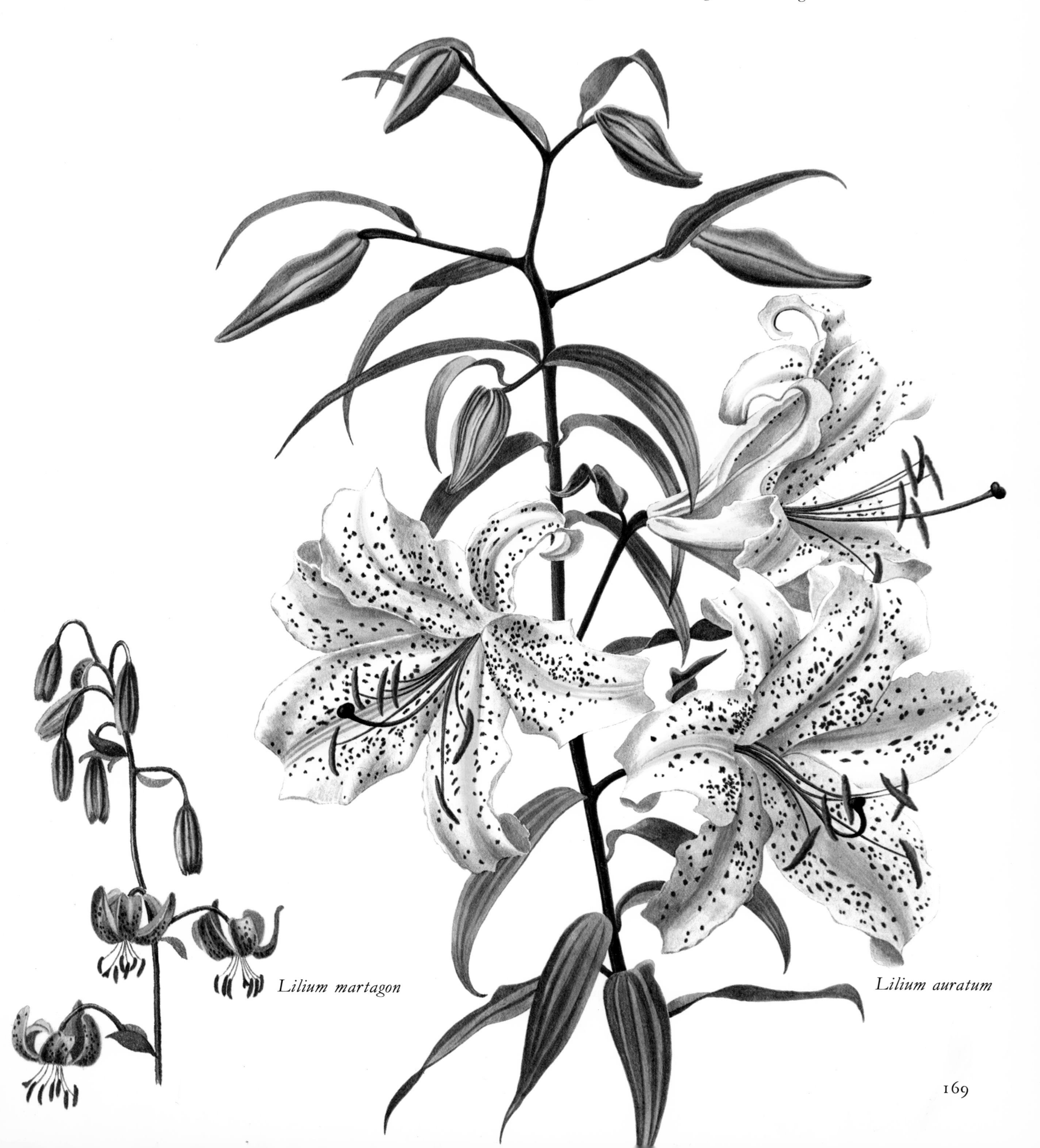

Lilium martagon

Lilium auratum

The bulbs need deep planting, summer being the best time.

Nomocharis are indigenous to certain parts of Asia and have nodding, saucer-shaped flowers of great beauty. They are suitable for cool climates as in southern England, needing partial shade and a moist but well-drained situation in lime-free soil. Their cultivation is somewhat similar to that of *Lilium*. *N. mairei* is soft pink spotted with crimson and *N. pardanthina candida,* white with maroon markings. They grow 60–90 cms (2–3 ft) tall.

Hemerocallis are the Daylilies, good garden plants from Europe and Asia (especially Japan), with graceful arching foliage and stout flower spikes carrying a number of large, fragrant, trumpet-shaped flowers. Hardy in temperate regions, dependable and long lasting they grow in sun or shade and practically any soil or situation, as at the edge of a shrubbery, in the herbaceous border, by the lakeside or sometimes in shallow water. They thrive in the tropics and yet can even be seen flowering in arctic Lapland. The short duration of the individual flowers may be a slight drawback but modern varieties often last several days and produce a succession of bloom so that the plants remain weeks in character. The dried flowers are used by the Chinese in soups, various meat dishes and with noodles. Packages sold in the Orient (and even America) go under such names as gum-jum (golden needles) or gum-tsoy (golden vegetables). They are somewhat gelatinous and when chewed are supposed to cure toothache.

The closely related Hostas (Funkias) are striking Asiatic plants with beautiful foliage and handsome spikes of mauve, white or violet flowers. They are ideal for key positions where fine foliage effects are required—as in stone vases, by a woodland seat, in shady borders or flanking a water garden. Here they thrive and can be seen to advantage, the smooth, heart-shaped or lanceolate leaves frequently crimped and in some instances margined or variegated with attractive shades of green or cream or white. The clumps improve with age and should be left undisturbed. They need rich humus soil which is unlikely to dry out; full sun overhead is detrimental as it burns the foliage. Propagation is effected by division in spring just about the time new growth is starting.

Most of the species occur wild in Japan but the first to reach Europe was the Chinese *Hosta plantaginea* (*H. subcordata*) in 1789. This has large bright yellowish-green leaves and white, scented blooms but others worth noting are *H. ventricosa,* dark glossy green; *H. sieboldiana,* glaucous green; *H. decorata,* white-margined; and *H. fortunei* 'Albopicta' with creamy-yellow variegations in the centre of the leaf.

Eremurus are stately herbaceous perennials from the steppes of western and central Asia. They have long, narrow, radical leaves and naked flower stems, sometimes 2.5–3 m (8–10 ft) tall, densely packed towards the tops with white, yellow, rose or golden flowers. The tubers are octopus-shaped, with radiating thongs, and need planting 15–20 cms (6–8 ins) deep, resting on sand.

These Foxtail Lilies or Desert Candles as they are commonly called associate splendidly with Delphiniums and Irises. The leaves of some species are eaten in Afghanistan as a vegetable. For garden purposes *E. stenophyllus* (*E. bungei*), one of the smaller species 60–90 cms (2–3 ft) tall with bright golden-yellow flowers is very useful, as are its hybrids such as 'Highdown Gold' (deep gold) or 'Dawn' (deep pink with black stems). The Shelford hybrids are a group of garden origin with orange, buff, pink, white or yellow flowers on stems of 1.35–1.8 m ($4\frac{1}{2}$–6 ft). *E. robustus* from Turkestan grows 1.8–3 m (6–10 ft) with soft pink

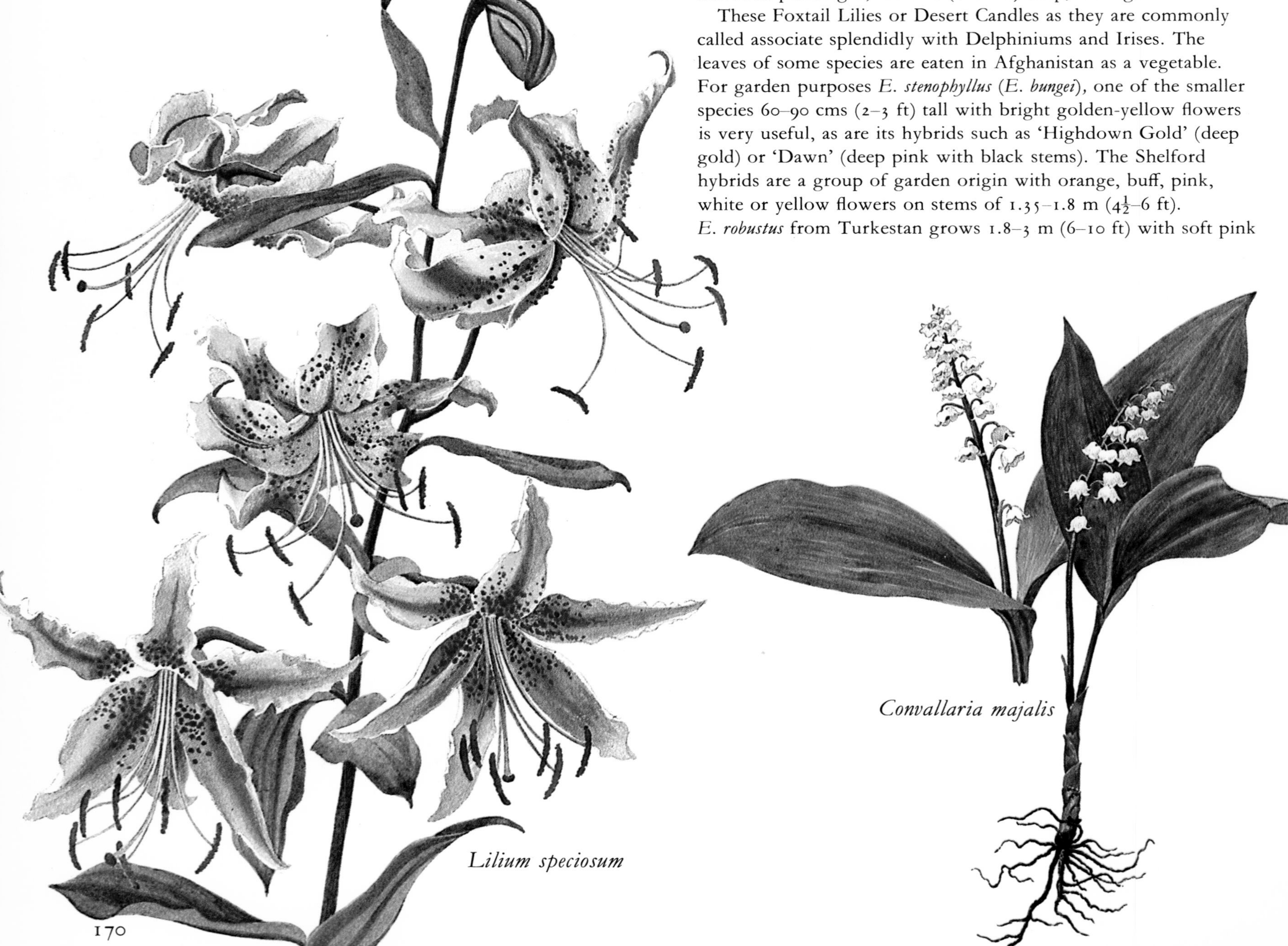

Lilium speciosum

Convallaria majalis

blooms and *E. olgae*, 1.25 m (4 ft) in height, is pink or white.

A favourite plant for temperate gardens is the European Lily of the Valley, *Convallaria majalis*. This is the National Flower of Finland and requires a cool situation in leafy, humus type soil with light shade. It has a running rootstock, oval-oblong leaves arranged in pairs and sprays, 15 cms (6 ins) long, of pendulous, white, very fragrant flowers. The cultivar 'Fortin's Giant' is larger and taller and there are forms with gold-striped leaves, 'Variegata'; pink flowers, 'Rosea'; and two doubles 'Prolificans' and 'Plena'.

Smilacina racemosa, the North American False Spikenard, has smooth slender stems 60–90 cms (2–3 ft) tall with alternate leaves and stout, densely flowered panicles of white blooms. It needs shade and rich moist soil.

Polygonatum multiflorum is the Solomon's Seal, a European (including British) plant with arching stems 60 cms (2 ft) tall carrying oval-oblong leaves and clusters of small green and white hanging bells. The plant usually grown in gardens is *P.* x *hybridum*, and the parents *P. odoratum* and *P. multiflorum* are less commonly seen. There are a number of forms—tall kinds and dwarf ones, also one with double flowers 'Flore Pleno' and another having variegated leaves, 'Striatum'. Solomon's Seals are ideal for shady spots and associate pleasingly with other woodland subjects. *P. canaliculatum* is the best of the eastern US species.

Tulips have been prized as garden flowers for centuries. They were first brought to Europe in 1554 by O. de Busbecq (the Austrian Ambassador to the Sultan of Turkey), who saw them growing in a garden near Constantinople. De Busbecq acquired a few bulbs 'at great price' and took them to Vienna and about the same time the botanist Conrad Gesner also obtained bulbs from a garden at Augsberg.

Suddenly everyone wanted Tulips. Their cultivation spread across Europe to the Netherlands, where the Dutch really took them to their hearts. New colours were produced, their popularity spread and by 1634 'Tulipomania' swept the country. Fortunes were made and lost (sometimes for bulbs which never existed) and a single bulb of the rare 'Semper Augustus', once changed hands for 13,000 guilders (approximately £1500).

Eremurus
Shelford hybrid

Fritillaria imperialis

Fritillaria meleagris
forms

The production of Tulip bulbs is still a major industry in Holland, the main groups used for forcing being the Duc van Thol Tulips—about 15 cms (6 ins) high—and the early single and double types. A little later come the Mendels, Triumphs and May-flowering (which include the Darwins), Lily-flowered forms, Cottage Tulips with long pointed petals, Parrots (with curled petals splashed with green, rose and other colours) and Rembrandts (broken Darwins). Colours range from white and cream to pink, red, scarlet, crimson and deep blue-purple.

There are also many fine species including *Tulipa kaufmanniana,* the Water-lily Tulip from Turkestan, and *T. fosterana* from Samarkand which is brilliant scarlet and very early, and countless small kinds for the rockery. Tulips generally need rich but well-drained soil and plenty of sunshine.

Hyacinths are equally popular for spring bedding and also in great demand for forcing. The large-flowered kinds are mainly derived from the southern European *Hyacinthus orientalis* and its variety *albulus,* the Roman Hyacinth, a charming little plant with slender spikes of small, sweetly scented, white flowers. The rather stiff habit of the large-flowered sorts makes them unsuitable for naturalizing; they are best used for formal plantings in beds, window boxes or containers. After the leaves have shrivelled they should be lifted and dried off and replanted in late summer. White, yellow, salmon, pink, dark and light blue varieties are available, and several doubles.

H. amethystinus (*Brimeura amethystina*) from the Pyrenees with spikes 20 cms (8 ins) long of clear blue flowers and the smaller *H. azureus* (*Hyacinthella azurea*) from Asia Minor, which is sky-blue, are early kinds for the rock garden.

Galtonia (*Hyacinthus*) *candicans* from South Africa is a distinct plant; the species grows up to 120 cms (4 ft) high with pendulous, white, bell-like flowers which are waxy and fragrant in mid-summer. The bulbs can be left undisturbed for years and associate pleasingly with scarlet Red Hot Pokers (*Kniphofia* species) in well-drained, sunny borders.

Muscari are the Grape Hyacinths, pretty spring-flowering bulbs with stems of 15–30 cms (6–12 ins) carrying dense heads of small, blue, grape-shaped flowers. They are also known as Starch Lilies because the mucilaginous bulbs were once used for stiffening linen. The most commonly cultivated are the deep blue *M. botryoides* from central and SE Europe which grows about 15 cms (6 ins) tall, and *M. armeniacum* 'Heavenly Blue', an exceptionally fine variety with rich blue fragrant flowers each spring and several whites. Grape Hyacinths naturalize readily in grass or light woodland and can also be used for edging flower beds. They make good companions for yellow Violas and white Narcissi. One curious species, *M. comosum monstrosum,* is called the Feather Hyacinth on account of the bluish-mauve sterile pedicels which festoon the flower stems 30 cms (1 ft) tall and take the place of bells in other species.

Small bulbs for naturalizing or planting in rock-garden pockets include Chionodoxas which, since they flower early in the spring, are often called Glory of the Snow. *Chionodoxa luciliae* (of gardens) has clusters of small funnel-shaped flowers on stems of 15–20 cms (6–8 ins). They are bright blue with a white base but deep blue throughout in *C. sardensis,* also from Asia Minor.

Puschkinia scilloides has powder-blue flowers with a deep blue stripe up each petal. It grows 10–15 cms (4–6 ins) tall and comes from the Orient.

Scilla sibirica is the Siberian Squill, a dainty beauty, 10 cms (4 ins) high, with blue flowers which are particularly vivid in the variety 'Spring Beauty'. *Endymion* (*Scilla*) *non-scriptus,* the English Bluebell is a matchless plant for naturalizing in thin woodland. It has pendent, deep blue bells on stems of 30–45 cms (1–1½ ft) and there are also pink and white forms. *E. hispanicus* from Spain and Portugal, a stouter plant with larger flowers and sturdier stems, thrives under similar conditions and also has pink and white forms.

Ipheion uniflorum comes from South America and is variously known as *Milla, Triteleia* and *Brodiaea uniflora*. It produces in spring many solitary, white, pale mauve or lilac (violet in the form *violaceum*) sweetly scented, star-shaped flowers on stems of 15 cms (6 ins) and has grassy leaves with a faint odour of garlic. In sunny, well-drained soil it increases rapidly.

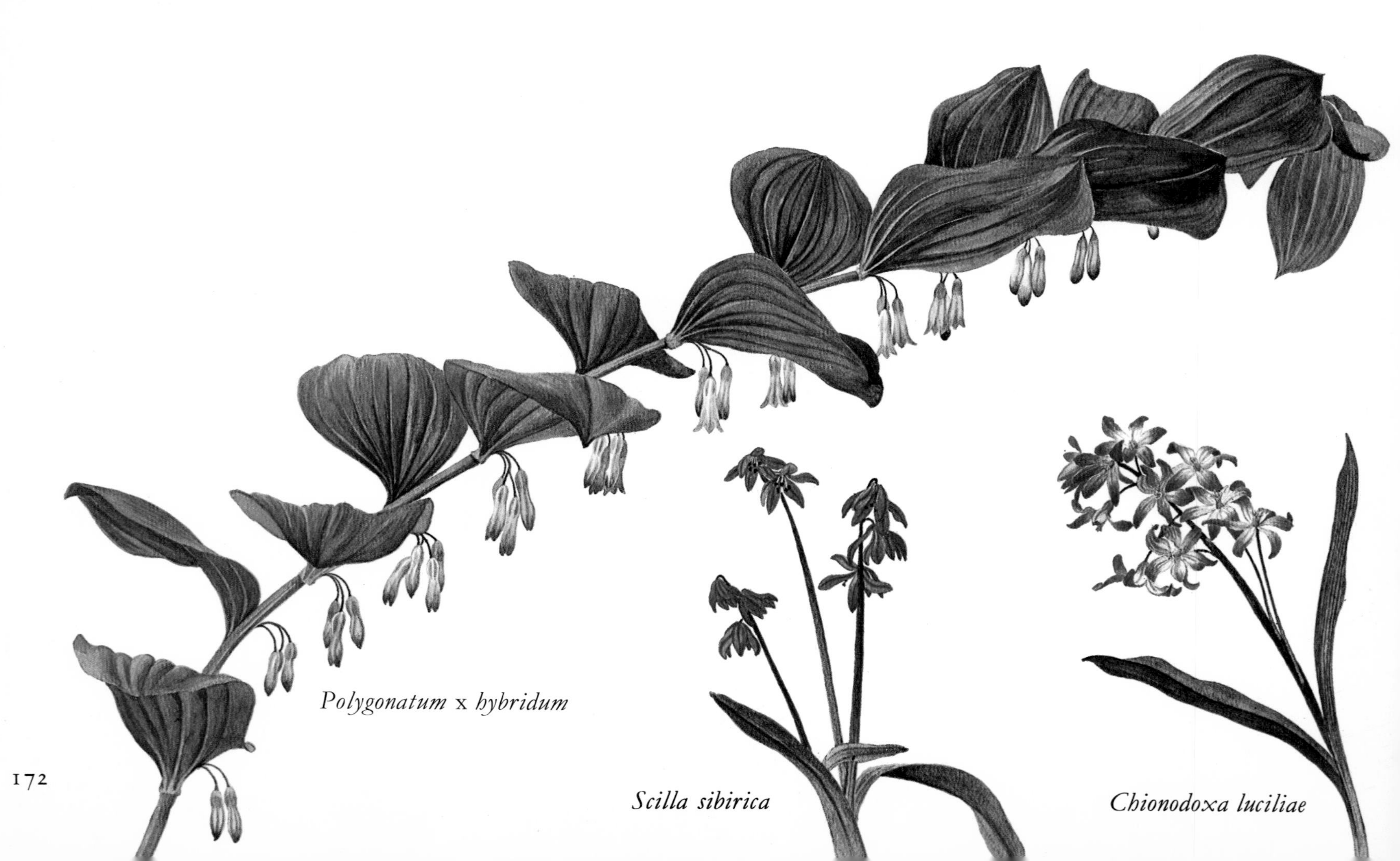

Polygonatum x *hybridum*

Scilla sibirica

Chionodoxa luciliae

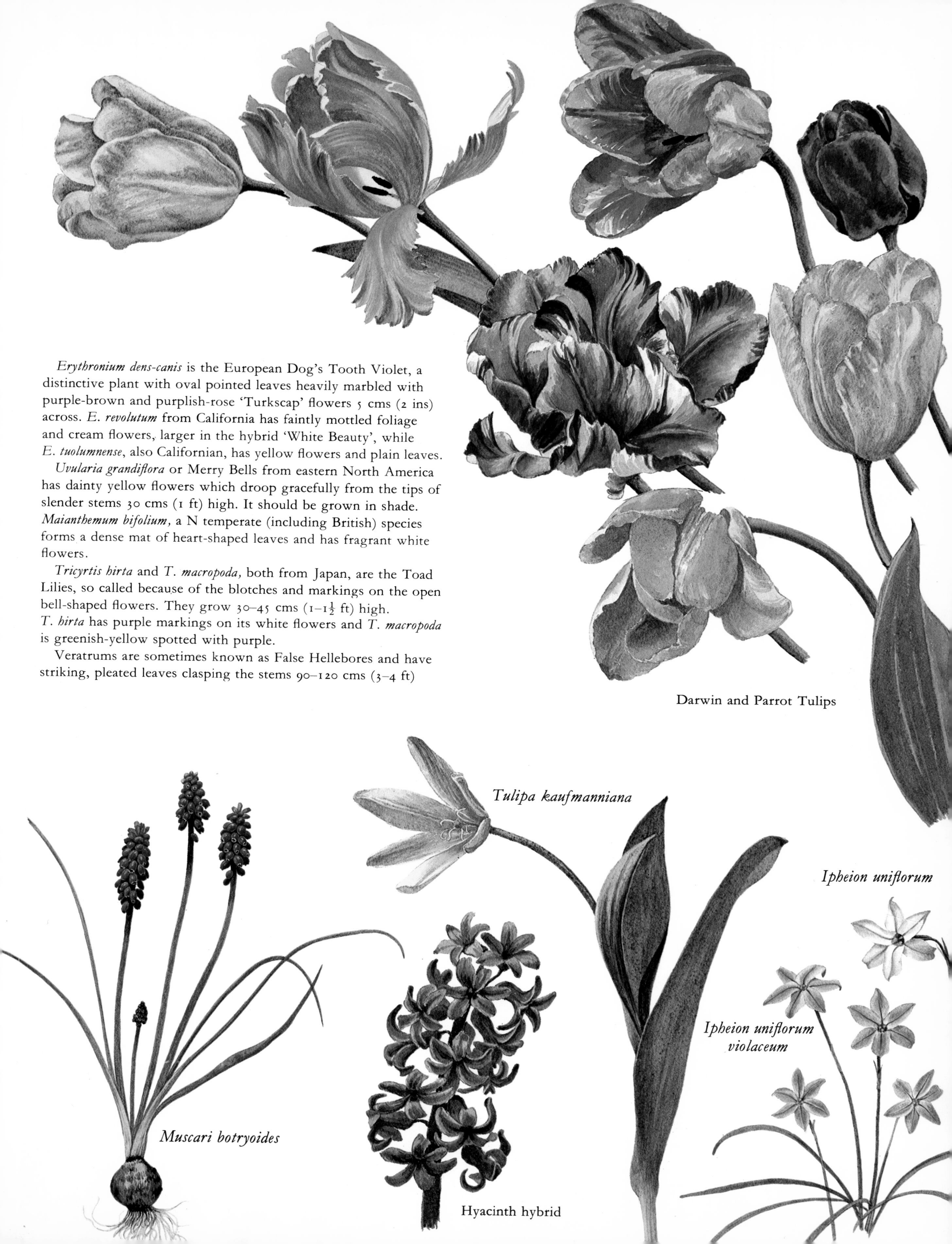

Erythronium dens-canis is the European Dog's Tooth Violet, a distinctive plant with oval pointed leaves heavily marbled with purple-brown and purplish-rose 'Turkscap' flowers 5 cms (2 ins) across. *E. revolutum* from California has faintly mottled foliage and cream flowers, larger in the hybrid 'White Beauty', while *E. tuolumnense*, also Californian, has yellow flowers and plain leaves.

Uvularia grandiflora or Merry Bells from eastern North America has dainty yellow flowers which droop gracefully from the tips of slender stems 30 cms (1 ft) high. It should be grown in shade. *Maianthemum bifolium,* a N temperate (including British) species forms a dense mat of heart-shaped leaves and has fragrant white flowers.

Tricyrtis hirta and *T. macropoda,* both from Japan, are the Toad Lilies, so called because of the blotches and markings on the open bell-shaped flowers. They grow 30–45 cms (1–1½ ft) high. *T. hirta* has purple markings on its white flowers and *T. macropoda* is greenish-yellow spotted with purple.

Veratrums are sometimes known as False Hellebores and have striking, pleated leaves clasping the stems 90–120 cms (3–4 ft)

Darwin and Parrot Tulips

Hyacinth hybrid

tall. The flowers are borne in large panicles, green in the North American *Veratrum viride,* white in *V. album* and reddish-purple in *V. nigrum*; the last two are European. The dried roots of *V. viride* are used as an insecticide and medicinally as a nerve sedative and for lowering blood pressure; *V. album* roots are the source of Hellebore Powder, used medicinally and for insecticidal sprays and preparations.

Camassias are North American bulbous plants with many star-shaped flowers on spikes of 60–90 cms (2–3 ft). The leaves are broadly strap-shaped and the plants are suitable for naturalizing in the herbaceous border but do best in moist soil. The best blue species are *C. cusickii* and *C. quamash* (*C. esculenta*) with edible bulbs. *C. leichtlinii* is more variable with white, cream, blue or purple flowers and has a form with double yellow, star-shaped flowers.

Calochortus venustus from California is the Mariposa Lily with Eschscholzia-like flowers on stems of 45 cms (1½ ft). These are white or pale mauve with conspicuous central blotches of deep red-purple.

Colchicums, often mistakenly called Autumn Crocus, are distinguished by large coarse leaves on tall leafy stems in spring. The blooms of most species appear in autumn when the plants are leafless and look very like large white or rosy-purple Crocuses. The names Son before the Father and Naked Boys refer to their leaflessness while flowering. *Colchicum autumnale* is a European species with narrow, rosy-lilac segments. There are also white, purple and double forms. *C. speciosum* is another useful species from which such good cultivars as 'Water-lily' and 'The Giant' have been derived.

Kniphofias are the Red Hot Pokers or Torch Lilies, noble grassy-leafed plants from East and South Africa with conspicuous spikes of brilliant flowers. Sun, good drainage and an occasional mulch keeps them going for years without any need for transplanting. The foliage is evergreen. *Kniphofia caulescens* has a Yucca-like trunk and foliage and spikes 1.25–1.5 m (4–5 ft) long of salmon flowers. *K. galpinii* is flame coloured and 75 cms (2½ ft) tall and *K. uvaria* is the parent of many cultivars such as 'Royal Standard' (bright red and yellow); 'Maid of Orleans' (ivory-white); 'Lord Roberts' (bright red); 'Bee's Lemon' (citron-yellow); and 'Yellow Hammer' (yellow).

Tender in N temperate countries, where plants have to be protected in winter or grown in a greenhouse is *Veltheimia viridifolia* (often grown as *V. capensis,* a related species), an

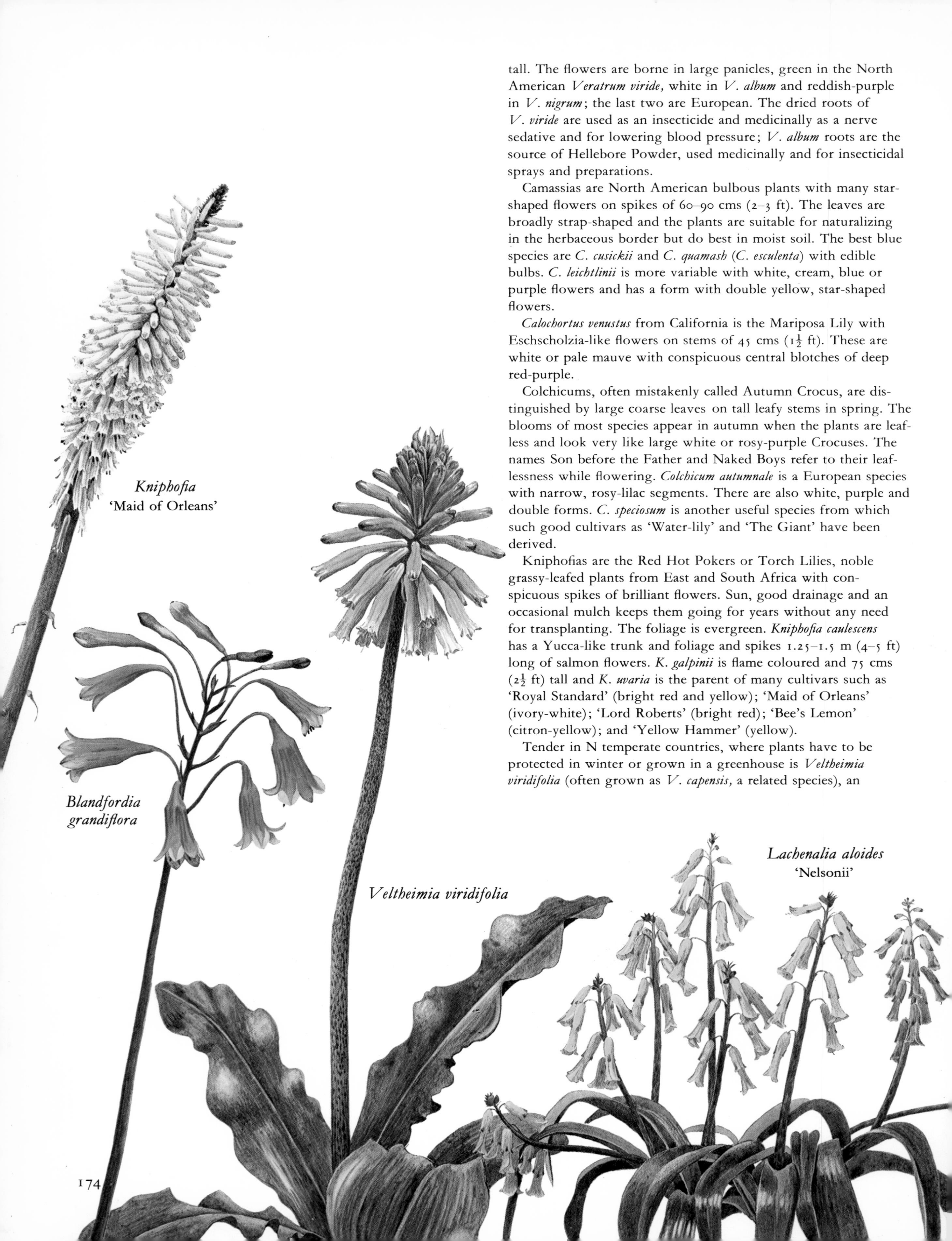

Kniphofia 'Maid of Orleans'

Blandfordia grandiflora

Veltheimia viridifolia

Lachenalia aloides 'Nelsonii'

attractive bulb from the Cape with dense spikes 30 cms (1 ft) long of tubular reddish flowers and strap-like leaves. Another is the Japanese *Aspidistra elatior* (incorrectly grown as *A. lurida* from China), a plant much loved by the Victorians who called it Cannon-ball Plant on account of its tolerance of shade, fluctuating temperatures, dust, smoke and general neglect. The flowers are insignificant and it is the broad strap-shaped leaves, either green or variegated, which give it character.

Gloriosa rothschildiana from tropical Africa is the Climbing Lily with large, claw-like blossoms which open bright yellow and red and then change through orange to claret. The climbing is achieved by 'finger-tip' tendrils at the tops of the leaves. *G. superba* is deep orange and red and climbs to 1.75 m (6 ft).

Aloe comprises some 275 species of shrubby or arborescent xerophytes mostly from tropical and South Africa, Madagascar and Arabia. Many are tall, 6–9 m (20–30 ft), with striking scarlet, Kniphofia-like flowers but others, such as the Partridge-breasted Aloe (*A. variegata*) can be used for house decoration. This species 30 cms (1 ft) high is stemless with green and white triangular leaves and loose spikes of red flowers. *Gasteria,* another succulent, produces thick tongue-like pointed leaves in ascending rows, often covered with white spots (tubercles) and *Haworthia* forms rosettes of leaves which are often very ornamental, although the blooms are insignificant. Both genera are South African.

Lachenalia aloides (*L. tricolor*) is the Cape Cowslip, a plant for the cool greenhouse in temperate climates. It has soft green, ribbon-like leaves mottled with purple and flower stems 30 cms (1 ft) tall bearing about twenty tubular, drooping blooms of pale green edged with red and yellow. 'Nelsonii' is bright yellow tinged with green.

Ornithogalum umbellatum, the Star of Bethlehem from Europe (including Britain) and North Africa, spreads rapidly in grass and shrubberies, with clusters of white flowers striped with green. *O. thyrsoides,* the South African Chincherinchee, has thick shiny stalks 15–60 cms (6–24 ins) long, each terminating in a tapering spike of white starry flowers. These last a long time in water and travel well in bud.

Dipidax triquetra is the South African Star of the Marsh, a pretty water plant growing about 30 cms (1 ft) high with spikes of white starry flowers which have purple dots at the petal bases. The bulbs go down very deep in South Africa and will stand about 30 cms (1 ft) of water above them, but in temperate countries are best grown as ordinary pot plants in a greenhouse.

Camassia leichtlinii 'Plena'

Camassia leichtlinii 'Lady Eve Price'

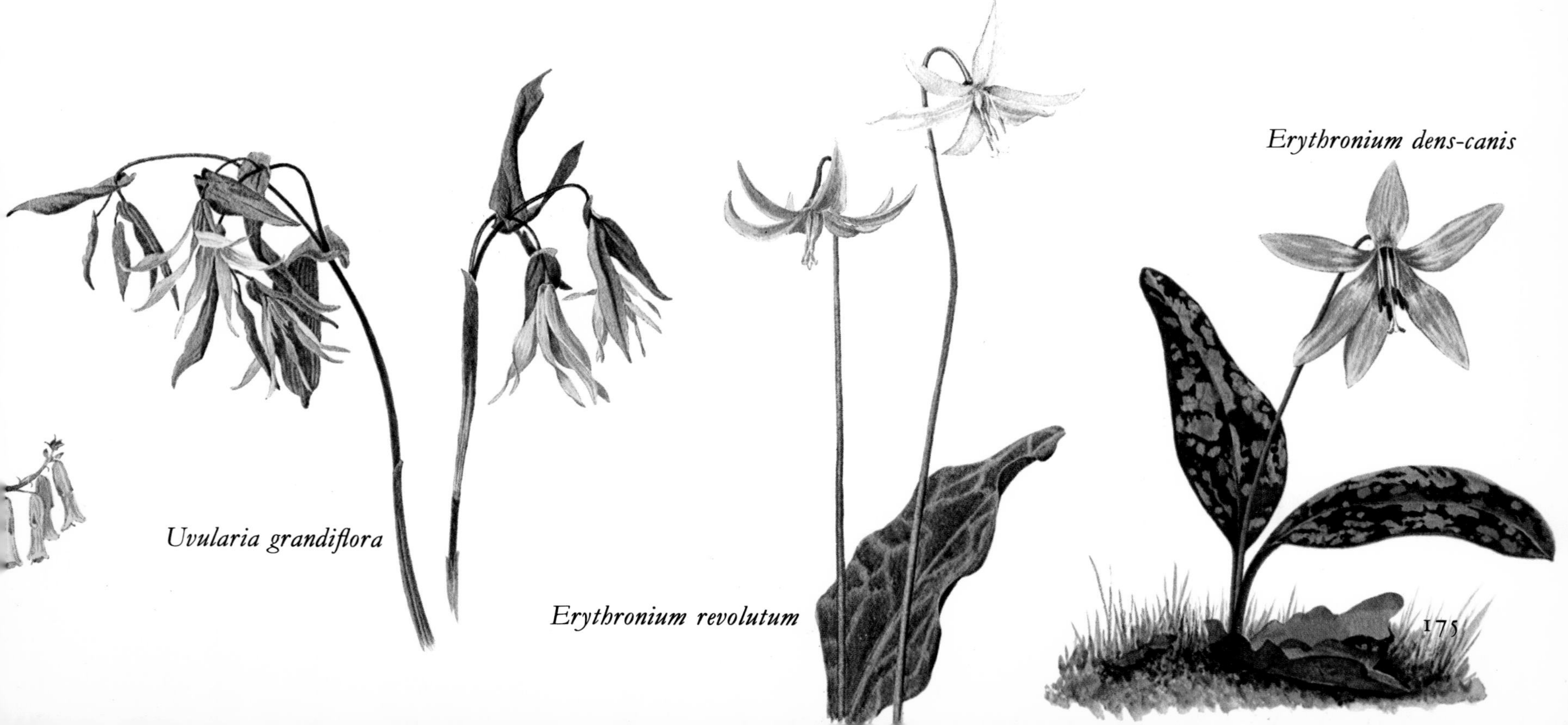

Erythronium dens-canis

Uvularia grandiflora

Erythronium revolutum

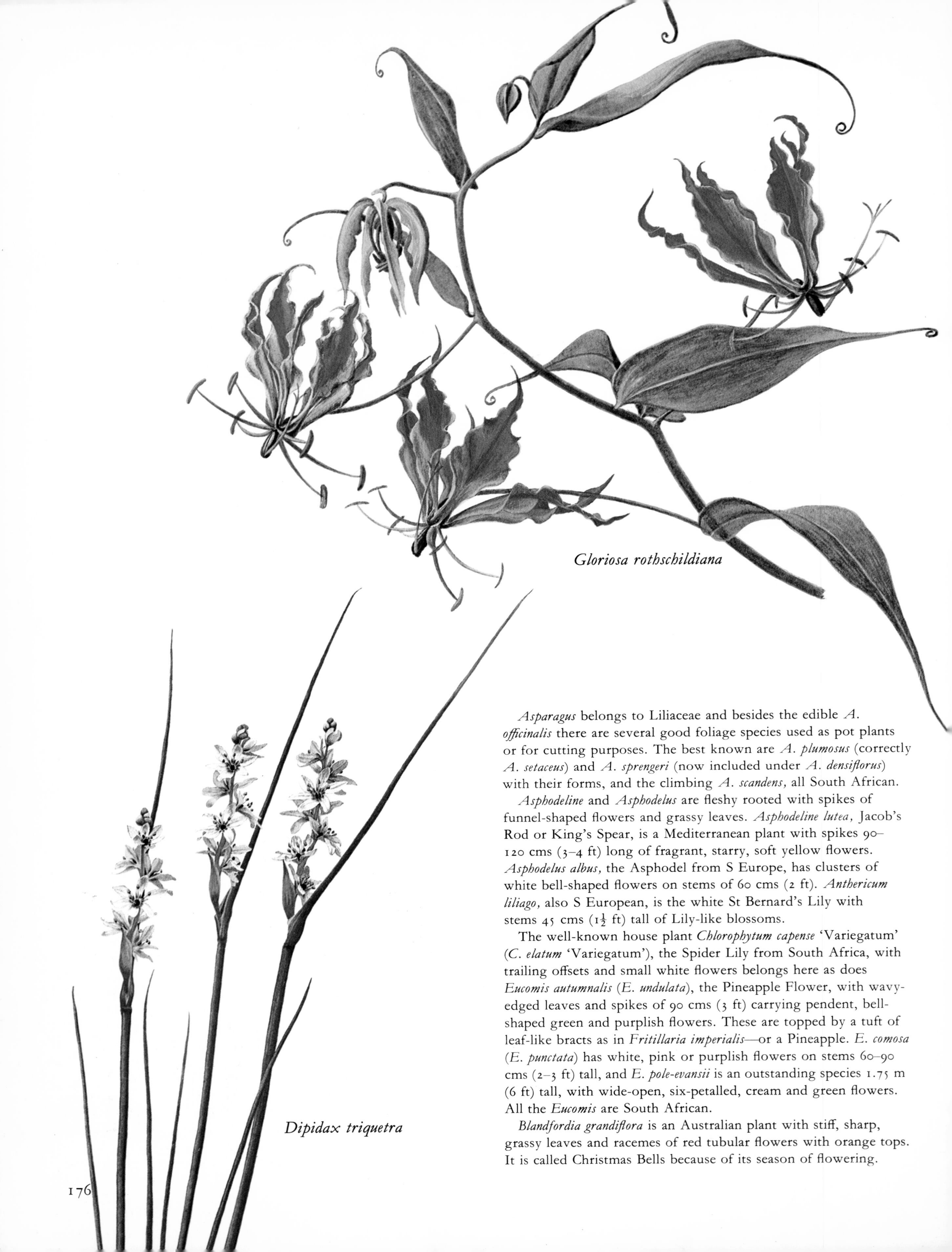

Gloriosa rothschildiana

Dipidax triquetra

Asparagus belongs to Liliaceae and besides the edible *A. officinalis* there are several good foliage species used as pot plants or for cutting purposes. The best known are *A. plumosus* (correctly *A. setaceus*) and *A. sprengeri* (now included under *A. densiflorus*) with their forms, and the climbing *A. scandens,* all South African.

Asphodeline and *Asphodelus* are fleshy rooted with spikes of funnel-shaped flowers and grassy leaves. *Asphodeline lutea,* Jacob's Rod or King's Spear, is a Mediterranean plant with spikes 90–120 cms (3–4 ft) long of fragrant, starry, soft yellow flowers. *Asphodelus albus,* the Asphodel from S Europe, has clusters of white bell-shaped flowers on stems of 60 cms (2 ft). *Anthericum liliago,* also S European, is the white St Bernard's Lily with stems 45 cms (1½ ft) tall of Lily-like blossoms.

The well-known house plant *Chlorophytum capense* 'Variegatum' (*C. elatum* 'Variegatum'), the Spider Lily from South Africa, with trailing offsets and small white flowers belongs here as does *Eucomis autumnalis* (*E. undulata*), the Pineapple Flower, with wavy-edged leaves and spikes of 90 cms (3 ft) carrying pendent, bell-shaped green and purplish flowers. These are topped by a tuft of leaf-like bracts as in *Fritillaria imperialis*—or a Pineapple. *E. comosa* (*E. punctata*) has white, pink or purplish flowers on stems 60–90 cms (2–3 ft) tall, and *E. pole-evansii* is an outstanding species 1.75 m (6 ft) tall, with wide-open, six-petalled, cream and green flowers. All the *Eucomis* are South African.

Blandfordia grandiflora is an Australian plant with stiff, sharp, grassy leaves and racemes of red tubular flowers with orange tops. It is called Christmas Bells because of its season of flowering.

Linaceae

12 genera and 290 species

This family consists of dicotyledonous shrubs or herbs with alternate leaves and branched inflorescences of regular and bisexual flowers, usually with five sepals, five petals, five, ten or more stamens and staminodes and superior ovaries.

The most important genus is *Linum,* the flowers of which are often beautiful in a rich shade of blue. Most garden worthy amongst the herbaceous perennials is *L. narbonense* from S Europe, especially its cultivars 'Six-Hills' (sky-blue with violet veining) and 'Heavenly Blue' (gentian-blue). The usual height is around 45 cms (1½ ft) and the plants associate pleasingly with yellow Lupins or golden Irises.

L. perenne is a European (including British) species with sky-blue flowers, white in var. *alba*.

L. grandiflorum rubrum, a showy annual from Algeria, has rich red flowers on stems 30–45 cms (1–1½ ft) tall. By successive sowings it can be had in bloom right through the summer. *L. campanulatum* is a perennial from S Europe, the same height, with yellow flowers. Similar is the shrubby *L. arboreum* from Crete, and also *L. flavum,* another golden-flowered species, again both 30–45 cms (1–1½ ft) tall. The latter is native to Austria.

L. monogynum is a New Zealand species, less hardy than the others, with large white flowers in summer. Like all Linums, it does best in full sun, often remaining closed in shade or dull weather, and likes good light soil.

L. usitatissimum is Flax, a pale blue European species growing about 60 cms (2 ft) tall. Linseed oil, which is used in paints, varnish, linoleum, soap, etc., is obtained from its seeds; the residue is made into cattle cake.

The stem fibre (source of linen) was one of the first textiles to be extensively used by Man. Egyptian mummies dating back over 4000 years have been found wrapped in linen, which retained its exclusive position as the leading fibre in the Western World until the 19th century when cotton took its place.

Reinwardtia indica (R. *trigyna*) is an Indian shrub, 90–120 cms (3–6 ft) high, with many branchlets. The oval-oblong leaves are minutely toothed and the five-petalled yellow flowers come in axillary or terminal inflorescences. In tropical countries and even under glass they are often extremely free so that the shrub is weighted with flowers. It needs good soil and full sunshine. *R. cicanoba* (R. *tetragyna*) from the East Indies grows about 90 cms (3 ft) high and produces pale golden flowers in great abundance.

Reinwardtia cicanoba (R. *tetragyna*)

Viscum album

Loranthaceae

36 genera and 1300 species
tropical and temperate

This is an interesting family of dicotyledonous semi-parasites; mostly small shrubs attached to their hosts by means of modified adventitious roots. The W Australian Christmas Tree, *Nuytsia floribunda*, is a root parasite forming a small tree.

Best known is *Viscum album*, the Eurasian Mistletoe, a plant which often branches inside the tissues of its host, and then breaks through the bark at different places. It has woody stems and long, opposite, pale green leaves. The insignificant flowers are unisexual, in groups of about three, both sexes being necessary to produce the translucent white (red in some species like *V. cruciatum*) 'berries'. These contain the seeds, surrounded by sticky viscous pulp.

V. album grows on a variety of hosts, but is commonest on Apple, Pear, Poplar, Lime and more rarely Oak. If the ripe fruits are crushed against the lower side of a young branch the viscid pulp fixes the seeds, where it later germinates and thus propagates the plant.

Mistletoe has long association with British folklore for from time immemorial it was the object of superstitious veneration, especially among the Druids. Magical powers ascribed to Mistletoe included prophetic dreams, a panacea for children's ailments, a universal remedy against poisons and diseases and a charm against witches.

The arrow with which the Norse god Baldur was slain by Loki was supposed to have been made of Mistletoe, and Japanese gardeners used to chop the leaves with Millet and other seeds (after prayers had been said over them) to ensure good harvests. In Austria it was believed that a twig laid on the doorstep safeguarded the inmates from nightmares. In comparatively recent times Worcestershire farmers would feed Mistletoe to the first cow calving in the New Year, to ensure good luck in the dairy.

The custom of kissing under a Mistletoe bough hung at Christmas is however peculiarly English. It may have arisen from a Scandinavian belief that the plant was sacred, so that if enemies met beneath it in the forest they laid down their arms and maintained a truce all day. From this the practice of hanging Mistletoe over a doorway was adopted to denote peaceful intent and the visitor on entering was greeted with a kiss of friendship.

Lythraceae

25 genera and 550 species

This is a widespread family of dicotyledonous herbs, shrubs and trees. The leaves are entire, simple and usually opposite, with the flowers in racemes, cymes or panicles. Individually these are perfect, although they may be regular or zygomorphic in shape, with (usually) four to eight divisions (sepals, stamens, petals).

The Purple Loosestrife (*Lythrum salicaria*) is a plant 60–120 cms (2–4 ft) high of wide distribution in N temperate regions (including Britain) and is also found in Australia. It inhabits moist grassland and ditches, thriving in damp heavy soils, but will also flower freely in ground that is not especially wet. The reddish-purple flowers are borne in whorls around the stems.

Named cultivated forms include 'Brightness' (rosy-pink); 'Lady Sackville' (bright rose); and 'The Beacon' (intense rosy-red). *L. virgatum* from SE Europe has purple flowers in groups of three in the leaf axils; 'Rose Queen', with light pink flower spikes is a dainty graceful form.

Lawsonia inermis, which is the source of henna dye, also belongs to this family. The dye comes from the leaves and young shoots. It is a tropical shrub widespread in East Africa and the Middle East, with panicles of small, fragrant, red and yellow flowers.

Two members of the Lythraceae are grown as water plants in Britain. *Decodon verticillatus* (*Nesaea verticillata*) from castern North America is the Water Willow, a shrubby perennial with willowy stems of 30 cms (1 ft), axillary purple flowers and lanceolate leaves. In autumn the latter assume brilliant crimson tints.

Rotala indica, sometimes used as a submerged aquarium plant, is really a bog perennial from wet places in tropical Asia. It has small, round, opposite leaves and terminal spikes of bright pink flowers. The general height is 7–10 cms (3–4 ins).

Cuphea is a large genus of herbs or sub-shrubs from the warmer parts of America. All the species mentioned here are Mexican. *C. procumbens* blooms profusely on horizontally branched stems 45 cms ($1\frac{1}{2}$ ft) tall. The flowers are violet-purple, and the whole plant covered with soft purple, glutinous hairs. It is an annual, easily raised from seed and makes a good bedding plant in the tropics. *C. micropetala* is a shrub of 60 cms (2 ft) or more with narrow lanceolate, grey-green, shiny leaves and tubular, white or yellow flowers, 2.5 cms (1 in. across), with red calyces. *C. ignea* (*C. platycentra*) is the Cigar Plant, an attractive little evergreen about 30 cms (1 ft) tall, with reddish stems bearing lance-shaped, dark green leaves and tubular flowers. These are bright scarlet with ash-grey margins and a black ring behind—like a cigar band.

Lagerstroemia indica, the Crape Myrtle, despite its scientific name is not indigenous to India. It is a Chinese shrub or small tree with oblong leaves, which may be opposite, alternate or in whorls of three. It bears terminal clusters of beautiful pink or mauve flowers with crinkly petals and woody capsular fruits. Another, *L. speciosa* (*L. flos-reginae*) from the damp jungles of Assam, Burma and Ceylon, is a splendid tree—up to 24 m (80 ft)—with quantities of deep mauve flowers which change with age to pink or white. The tree remains in flower for several weeks.

Lagerstroemia indica

Lythrum salicaria
'Lady Sackville'

Magnolia campbellii

Magnoliaceae

12 genera and 230 species

This family consists of dicotyledonous trees and shrubs from temperate and tropical parts of America and E Asia. The simple leaves, often quite large, are arranged alternately. The flowers too are frequently big and spectacular, usually solitary and either terminal or axillary. In some instances they are bisexual, in others unisexual, with deciduous enclosing scales and three petal-like sepals (petaloids), six to fifteen free petals, many stamens and superior ovaries. Large stipules enclose and protect the young growths.

Magnolia is a genus of about eighty trees and shrubs with fine flowers and a distribution ranging from the Himalayas to Japan, Borneo and Java, and E North America to the West Indies and Venezuela. The majority do well in warm or sheltered positions in rich, free-draining soil containing plenty of humus. Some like peat and others tolerate lime. Generally the Chinese species tolerate lime and the Japanese do not, but there are exceptions like the Japanese *M. kobus* which succeeds well on lime.

There are both deciduous and evergreen species, with kinds suitable for sunny or shady situations. Some of the most spectacular flower early in the year, with the attendant risk of frost in temperate countries. This browns the petals so that frost pockets should be avoided when selecting sites. Magnolias are sometimes difficult to establish after transplanting; the roots are thick and fleshy and often decay after disturbance, a circumstance which causes them to sulk (while they make new roots). They then look very miserable and are practically leafless—sometimes for several seasons. The best time to move them is in early spring, just as the roots start to grow. They should be lifted with a good ball of soil and frequent spraying over the foliage with soft water the first season helps them establish. Mulching the ground with leaves keeps the roots cool and also provides a reservoir of food. All Magnolias like abundant moisture and in poor soil deep planting pockets, 45 cms ($1\frac{1}{2}$ ft) deep and 2–4 m (6–12 ft) across, should be prepared for each tree beforehand. Fill this with good loam, peat and rotted leaves.

Propagation is by means of seed (a slow and not always successful method), layers or grafts (one of the best stocks being the American *M. acuminata*).

Botanists consider that Magnolias are among the oldest plants in the world, for five million-year-old fossils are common over a wider territory than today's native habitat would suggest.

The American Magnolias include the well-known *M. grandiflora,* the Laurel Magnolia, and the only evergreen species hardy in Britain. It is the State Flower of Louisiana and Mississippi. A well established tree will reach 20 m (70 ft) or more in height, with large, oval-oblong, glossy green leaves which are rusty-brown beneath and enormous, 20–25 cms (8–10 ins), globular, creamy-white, thick-petalled flowers with a strong lemon aroma. In Britain it is usually grown as a wall specimen.

The leaves skeletonize readily and are popular at Christmas for winter bouquets while the bark has been employed as a stimulant and tonic.

Another American species is *M. acuminata,* the Cucumber Tree, so called on account of the shape and colour of the young fruits. Later these turn dark red, succeeding dull greenish-yellow, fragrant flowers. The tree grows 18–27 m (60–90 ft) tall. *M. virginiana*, the Sweet Bay, makes only a small tree in Britain, but reaches 15 m (50 ft) in America. It has delicately scented, creamy-white, globular flowers in character over a period of several months. These develop to bright red fruits.

The Asiatic Magnolias start the year with *M. stellata,* a Japanese species which flowers when quite small and remains a compact albeit spreading shrub for many years. It blooms on the naked wood, the white starry flowers having twelve to eighteen strap-shaped petals. With age these develop purplish tinges; they are sweetly scented (something like lemons) and the young bark is aromatic.

M. kobus, another deciduous Japanese species, reaches a height of 9–12 m (30–40 ft) in time. Its white flowers are smaller than most of the garden Magnolias. The Ainus of Japan use the bark to cure colds and the soft wood is employed in the Orient for cabinet work, engraving and in the manufacture of matches.

Another spring sort is the Yulan, *M. denudata,* from China which is generally reckoned one of the most beautiful. Growing about 9 m (30 ft) tall it sometimes flowers when as small as 90 cms (3 ft), the large white flowers very solid and deliciously fragrant.

A commonly cultivated hybrid derived from this species with *M. liliiflora* (also Japanese) is *M.* x *soulangiana,* often erroneously called the Tulip Tree (a name which in fact rightly belongs to

Magnolia grandiflora

Magnolia wilsonii

Liriodendron tulipifera) on account of its large, Tulip-shaped flowers. These are white, sometimes stained with purple throughout and make a magnificent spectacle in early spring with hundreds of blooms on a single tree. Although most flowers appear in spring on the leafless branches, spasmodic sexless blossoming continues throughout the summer. This is one of the easiest and most adaptable of Magnolias. Other hybrids of the same parentage include 'Lennei', with rich purple-pink flowers, and 'Brozzonii', white, shaded purple.

M. liliiflora is a deciduous shrub with straggling branches and large white flowers which are purple outside. Presumably it is from this species that *M.* x *soulangiana* and other hybrids get their colour.

M. campbellii from the Himalayas makes a tree which in its native terrain may reach 45 m (150 ft). It is deciduous with huge blossoms. These are soft rose-pink to crimson with very thick petals. Unfortunately the tree does not bear flowers until it is about twenty years of age.

Later come several pendent-flowered Magnolias like the Chinese *M. wilsonii,* named after the famous collector E. H. Wilson. This has fragrant, white, cup-shaped blossoms 7–10 cms (3–4 ins) across with numerous rich red stamens and purplish-pink pistils which later develop to cucumber-shaped fruits. To enjoy this species one must stand underneath it and look up at the flowers. It is deciduous and grows up to 8 m (25 ft). Another pendent kind is *M. sinensis,* a shrubby plant, which was also found by Wilson in China. The white, fragrant, saucer-shaped flowers are 10–13 cms (4–5 ins) across and have conspicuous red stamens.

M. sieboldii (*M. parviflora*) is Japanese and makes a small tree with semi-pendent, fragrant white flowers packed with rosy-crimson stamens. *M.* x *watsonii* is considered a hybrid between *M. sieboldii* and *M. obovata,* another Japanese species. It crops well in early summer and then intermittently until the end of the season. The flowers are white, tinged with pink, occasionally with a central boss of crimson stamens and always fragrant.

A closely related plant is *Michelia champaca,* an evergreen tree from the Himalayas, largely planted in the tropics on account of its rapid growth and the rich scent of its Narcissus-like flowers. These are deep yellow with the petals in three rows. In Thailand the fragrant blossoms are used in cosmetics, hair oils and perfumes.

M. figo is a Chinese shrub with yellow-green flowers stained with purple. These are extremely fragrant, smelling of bananas and are used by the Chinese for scenting hair oil. The species is sometimes called the Banana Shrub.

Magnolia x soulangiana

M. doltsopa comes from the E Himalayas and normally makes a tall evergreen tree up to 27 m (90 ft) high although sometimes the branches spread outwards so that it adopts a shrubby habit. The elliptic-oblong leaves are nearly 30 cms (1 ft) in length and the highly fragrant flowers are white or cream with greenish bases.

Talaumas have flowers like Magnolias, very large and fragrant, with thick velvety petals. In the United States *Talauma hodgsonii* is sometimes cultivated on account of its fist-sized flowers which are white inside and purplish without. The long evergreen leaves are reminiscent of Magnolia, being dark green and shiny, 30–45 cms (1–1½ ft) in length.

Liriodendron tulipifera is the North American Tulip Tree, a magnificent deciduous species with saddle-shaped simple leaves which have wedged tops—as if someone had bitten off the ends. In the autumn these turn brilliant yellow. The flowers appear in midsummer and are upright and tulip-shaped with oblong petals. They are green and yellow with orange spots at the petal bases and have numerous golden stamens surrounding green pointed pistils.

Several interesting horticultural forms exist with wavy leaf margins, 'Contortum'; yellow foliage variegations, 'Aureo-maculatum'; and a conical shape, 'Pyramidale'.

The wood is soft and easily worked and frequently used for such things as boxes, shingles, radio cabinets, etc. It is fine-grained and does not split easily and is often erroneously called Yellow Poplar, or sometimes White wood. The bark has a pleasant pungent scent, the inner bark being acrid and the source of tulipiferine, an alkaloid which has been used as a stimulant after heart attacks. The flowers provide a rich source of honey for bees. *Liriodendron* makes a good shade tree and is much planted for this purpose.

L. chinense, the Chinese Tulip Tree, is less vigorous than *L. tulipifera,* which in its native North America often attains 45–57 m (150–190 ft). The Chinese species grows 15–18 m (50–60 ft) high and has similar shaped leaves but smaller flowers.

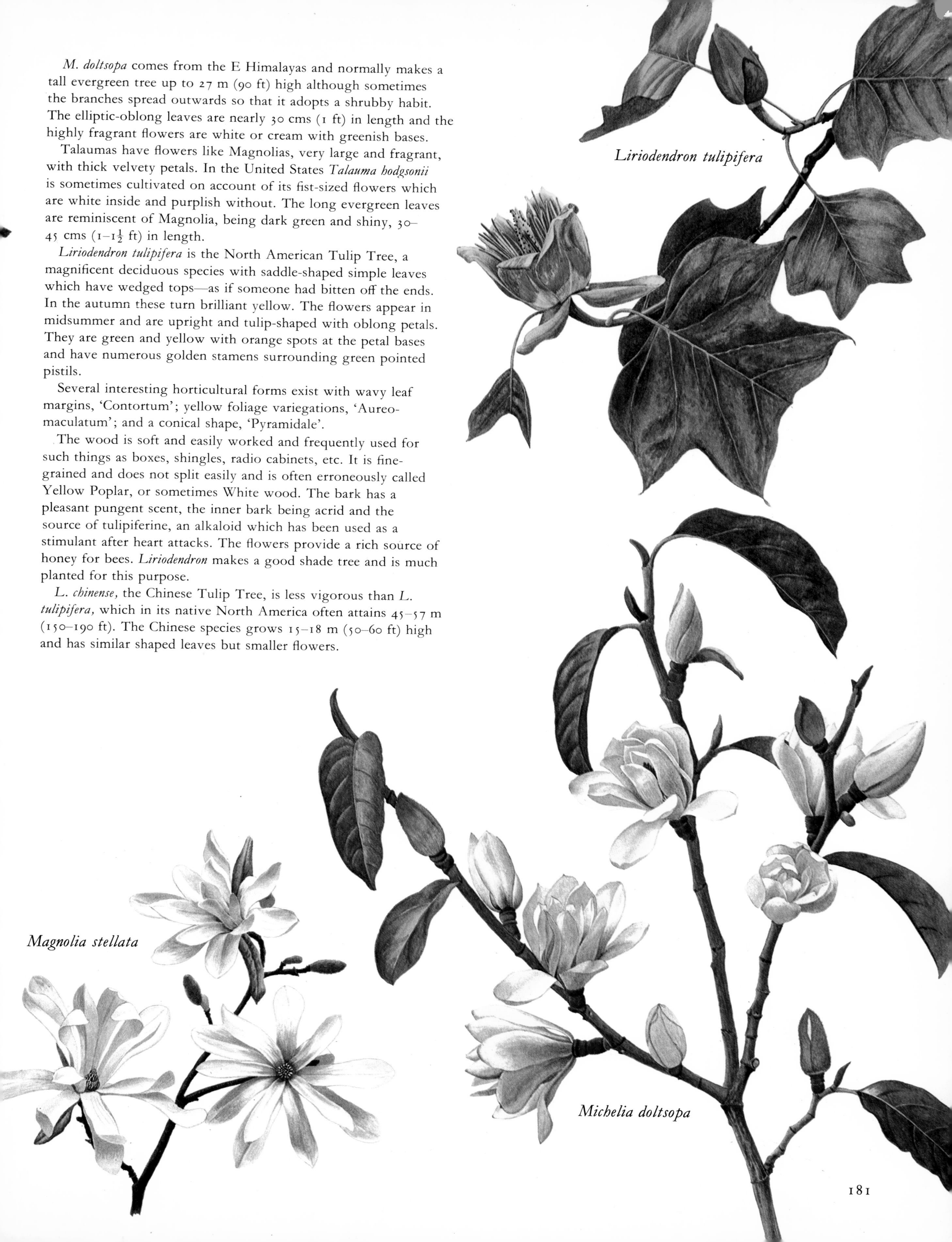

Liriodendron tulipifera

Magnolia stellata

Michelia doltsopa

Malvaceae

75 genera and about 1000 species

This family consists of tropical and temperate dicotyledonous trees, shrubs and herbs whose general characteristics include alternate, palmately-lobed leaves and solitary or compound, cymose inflorescences made up of perfect flowers, often tubular at the base, with five sepals, five petals, many stamens and a superior ovary. A number are important ornamentals, especially for tropical gardens.

Malva, for which the family is named, is a N temperate genus with spikes of rose-coloured flowers and handsome foliage. *M. moschata,* the Musk Mallow, of Europe (including Britain) has Buttercup-shaped leaves cut to narrow segments; these emit a musky fragrance when bruised. The species, which was formerly much planted on graves in Europe, has purplish-pink flowers on stems 30–60 cms (1–2 ft) tall and bristles on the foliage.

Alcea (Althaea) rosea

Althaea officinalis is the Marsh Mallow, a native of N Asia to E Europe (including Britain) and naturalized in east US. The plant has a fleshy rhizome, lobed leaves and five-petalled, flesh-coloured flowers. The roots contain a mucilaginous lubricant once valued as a cough medicine and also applied to burns and animal bites. These medicinal virtues are of ancient origin and mentioned in the writings of Pliny, Virgil and Dioscorides. The root extract blended with sugar, white of egg and gum arabic is the source of Pasta Althaea (Marsh Mallows); the leaves when boiled were eaten by the Romans and the stems yield a fibre which has occasionally been used for paper making.

From a garden point of view one of the most outstanding species is *Alcea (Althaea) rosea,* the Hollyhock, a very old garden plant with a history of cultivation extending back more than five hundred years. It is thought to have originated somewhere in the eastern Mediterranean region and not China as is frequently stated.

Hollyhocks grow in most reasonably fertile soils and are useful at the back of a border or in key situations; in England they are common cottage garden plants. The flowers, borne on stems 1.8 m (6 ft) tall, have a diameter of 10 cms (4 ins) and come in every conceivable colour (except blue) from white, cream and yellow, through pinks, reds and purples to near black. Both singles and doubles are available, the latter coming fairly true from seed although the singles are more variable. The leaves are normally rounded, but fig-shaped and with five to seven angles in the Antwerp Hollyhock (*A. ficifolia*). In spite of being perennial, Hollyhocks are best treated as biennial in order to discourage the spread of Hollyhock Rust, a disfiguring disease to which they are prone.

Most members of *Hibiscus* are native to warm or tropical parts of the world so they have to be grown under glass in temperate countries. In the tropics they are grouped or used as specimen plants in borders or among shrubs and some make fine hedges. Most have a ring of small bracts immediately below the calyx and a very long stamen tube (bearing the stamens) which ends in five spreading appendages. The large showy flowers are borne in the leaf axils and the leaves are simple and variously shaped.

H. rosa-sinensis, the Shoeflower from China, is a well-known shrub in the tropics where it is often planted singly and flowers for most of the year. Unpruned it will eventually grow 4–5 m (12–15 ft) high but it also makes a good hedge or, in cold countries, tolerates the root restriction of pot culture. There are many varieties with showy pink, red or yellow flowers, usually with prominent yellow stamens and red stigmas. Some sorts are double and there is one with cream leaf variegations which makes a charming low hedge. All over the tropics women use the flowers for adornment. In Hawaii particularly they are fashioned into the necklaces of blooms (leis) used to greet visitors.

H. tiliaceus grows near the coast in both Old and New World tropics—including the Galapagos where it thrives on almost pure lava and coral. It has leaves shaped like the Lime Tree (*Tilia*) and pure yellow flowers with overlapping petals which turn red-maroon with age. The plant is the source of a strong fibre used for fishing nets, sails and tow.

H. esculentus, now known as *Abelmoschus esculentus,* is an annual, also widespread in the tropics with heart-shaped leaves and yellow flowers with crimson centres. The long pod-like fruits called Okra, are eaten as a vegetable. *H. schizopetalus,* a charming East African shrub has pendent, carmine flowers with deeply cut petals.

The Blue Hibiscus, *H. syriacus*, hardy in Britain and US, is a deciduous shrub 1.8 m (6 ft) high covered in late summer with bell-shaped flowers of 7 cms (3 ins). These are very variable with white, pink, red, blue and purple flowers—often having two or

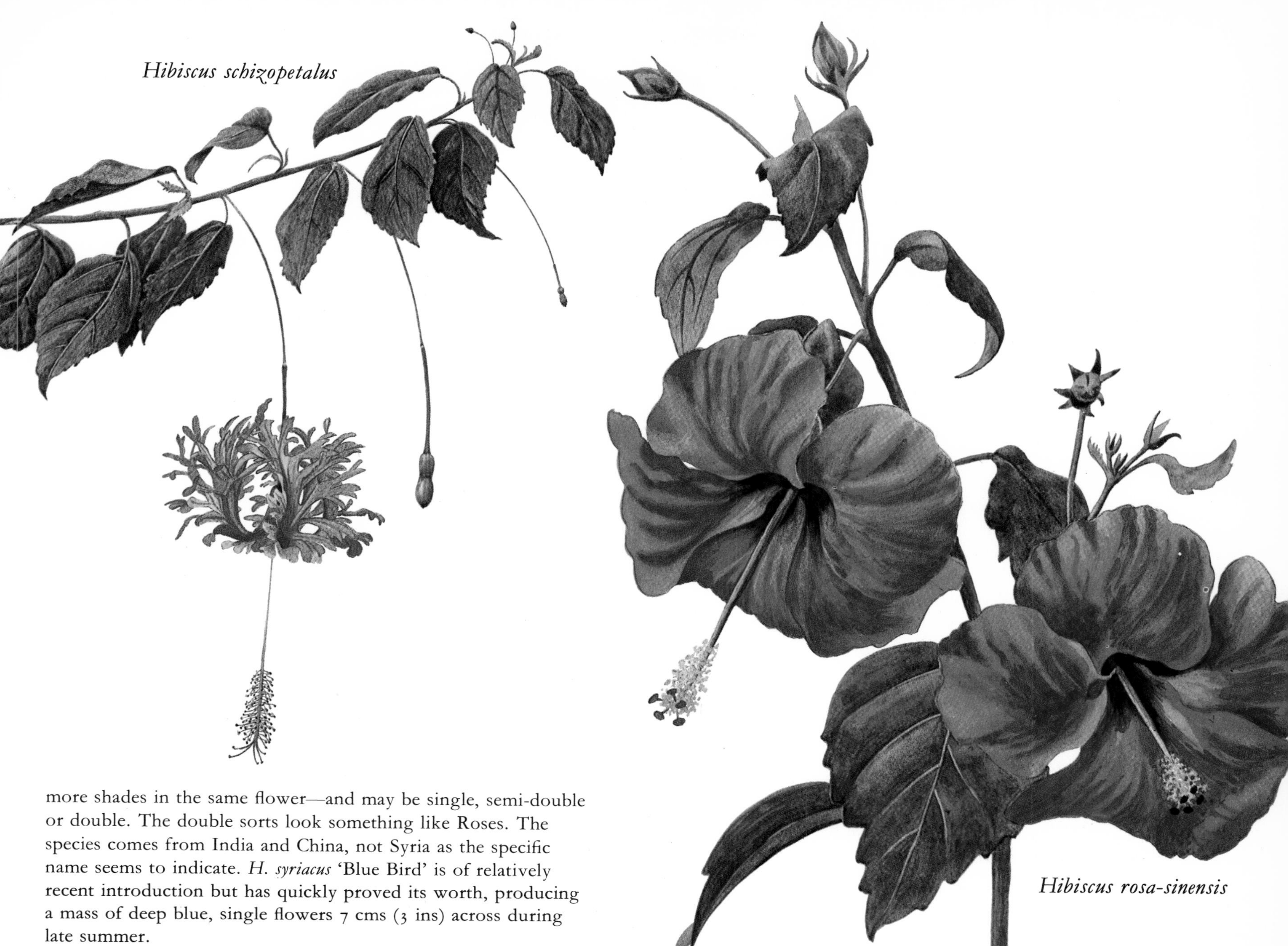

Hibiscus schizopetalus

Hibiscus rosa-sinensis

more shades in the same flower—and may be single, semi-double or double. The double sorts look something like Roses. The species comes from India and China, not Syria as the specific name seems to indicate. *H. syriacus* 'Blue Bird' is of relatively recent introduction but has quickly proved its worth, producing a mass of deep blue, single flowers 7 cms (3 ins) across during late summer.

Other species include the Jamaican Sorrel or Roselle, *H. sabdariffa,* from the Old World tropics. This yellow-flowered species has a fleshy calyx and edible fruits used for making Cranberry-flavoured jelly. A fibre obtained from the stalks resembles jute. *H. cannabinus,* also from the Old tropics, is the source of Deccan Hemp, another fibre. It has sulphur-yellow flowers with deep purple or maroon eyes. Since the dawn of history it has been cultivated in India for the sake of its fibres and more recently grown for the same purpose in S Russia, Java, the West Indies and South Africa.

Abutilons comprise a large genus of shrubs and herbs with heart-shaped or slightly lobed leaves and axillary, solitary, pendent flowers. Among the most popular is *Abutilon megapotamicum* from Brazil, a slender willowy shrub with small leaves, which can be grown against an outside or greenhouse wall or is sometimes trained over wire (balloon or conical) shapes. It has many small, hanging, scarlet and gold flowers with prominent protruding stamens.

A. striatum 'Thompsonii' is one of the best golden variegated plants for subtropical summer bedding. To that end the plants are regularly raised from cuttings and planted outside when all risk of frost is past. They reach 1–1.5 m (3–5 ft) at the end of a season and associate particularly well with blue-flowered plants such as Heliotrope and *Salvia patens.* The flowers are orange with deep red veining.

Hibiscus syriacus
'Blue Bird'

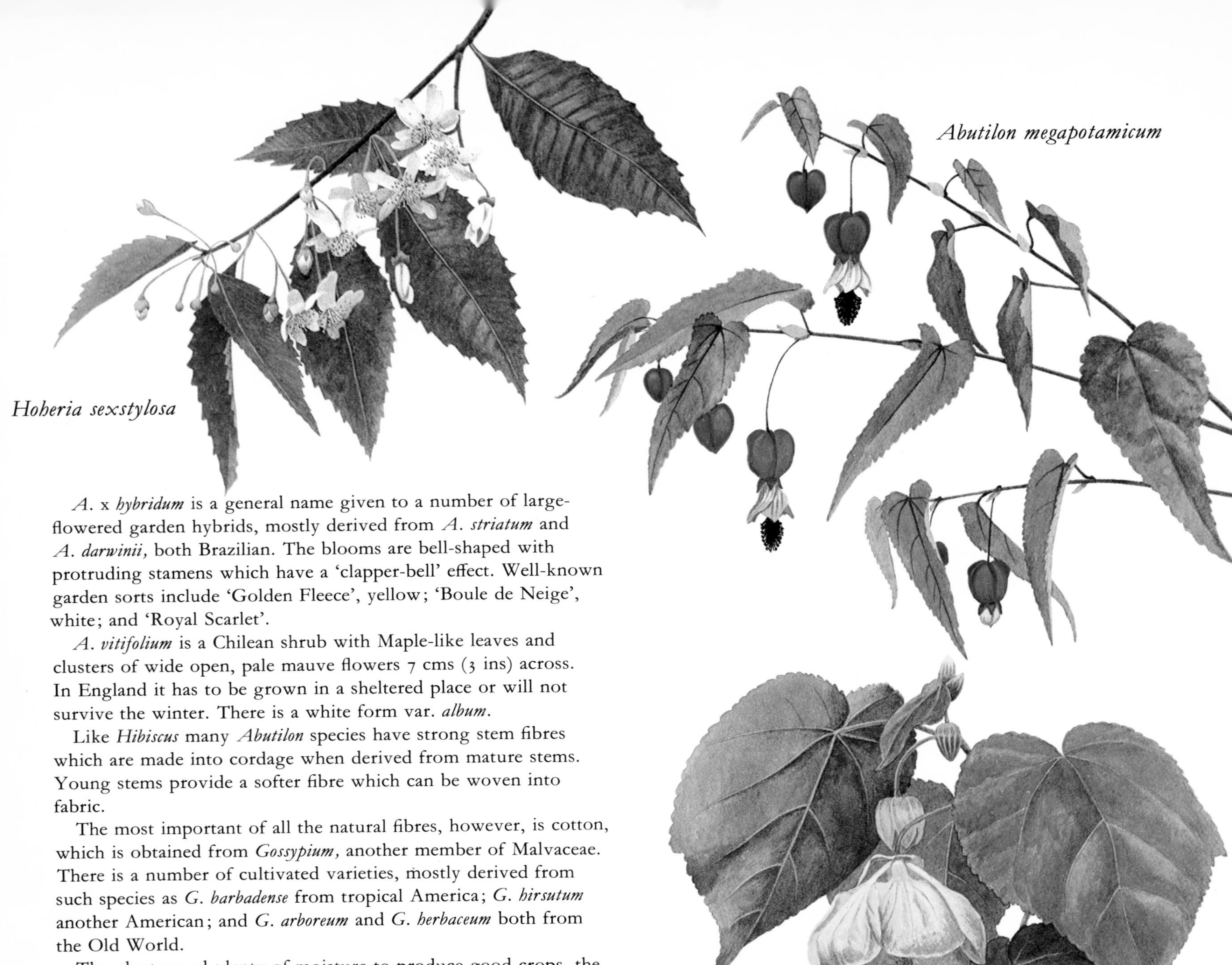

Abutilon x *hybridum* 'Golden Fleece'

A. x *hybridum* is a general name given to a number of large-flowered garden hybrids, mostly derived from *A. striatum* and *A. darwinii,* both Brazilian. The blooms are bell-shaped with protruding stamens which have a 'clapper-bell' effect. Well-known garden sorts include 'Golden Fleece', yellow; 'Boule de Neige', white; and 'Royal Scarlet'.

A. vitifolium is a Chilean shrub with Maple-like leaves and clusters of wide open, pale mauve flowers 7 cms (3 ins) across. In England it has to be grown in a sheltered place or will not survive the winter. There is a white form var. *album.*

Like *Hibiscus* many *Abutilon* species have strong stem fibres which are made into cordage when derived from mature stems. Young stems provide a softer fibre which can be woven into fabric.

The most important of all the natural fibres, however, is cotton, which is obtained from *Gossypium,* another member of Malvaceae. There is a number of cultivated varieties, mostly derived from such species as *G. barbadense* from tropical America; *G. hirsutum* another American; and *G. arboreum* and *G. herbaceum* both from the Old World.

The plants need plenty of moisture to produce good crops, the cotton being procured from the long white hairs which surround and envelop the seeds. Cotton plants are shrubs with lobed leaves and yellow flowers having purplish bases.

Cotton has been manufactured into yarn for cloth since time immemorial and has had a profound influence on the lives and fortunes of many races including the unfortunate African slaves who worked the American cotton plantations. Because of it many families in the deep south of America amassed great fortunes and the cotton mills of England (which dominated nineteenth-century economy) made it possible for England to extend her market and influence to India and Africa. The crop has been vital to Egyptian economy, and a source of revenue to Brazil, India (the leading producers for 3000 years), Russia and China.

Egyptian cotton is derived from *G. barbadense,* which is also the source of Sea Island cotton. This is of excellent quality and possesses longer fibres than any other species.

Apart from its importance as a fibre-producing plant there are several byproducts, particularly from the seeds. These produce an oil used for frying and making into soap and cosmetics and after this has been extracted the residue is turned into fertilizers, paper, cellulose and fuel.

Hoherias come from New Zealand and are shrubs or small trees with simple leaves having toothed margins. These are arranged alternately on the stems with the white flowers carried in showy, axillary clusters.

All Hoherias require well-drained soil and a sheltered situation, especially in temperate zones. Although hardy in S Europe, S United States and the milder parts of Britain, greenhouse protection may be necessary in areas having cold winters. They sometimes grow to a height of 12 m (40 ft), although 5–9 m (15–30 ft) is more usual under cultivation. Propagation is effected by seed, cuttings or layers.

Such names as Lacebark, Ribbonwood and Thousand-jacket were frequently applied to them by early settlers on account of the perforated lacy appearance of the inner bark fibre. This is very tough and suitable for cordage, also for such ornamental purposes as basketry.

Two of the most beautiful species are *Hoheria* (*Gaya*) *lyallii* and *H. populnea.* The latter comes from North Island, New Zealand, and has evergreen, ovate, deeply toothed leaves, and bunches of pure white flowers with numerous stamens. These are united into tubes (as is common in Malvaceae) and then separate into several distinct bundles of five or six stamens.

H. lyallii is a deciduous species with yellow leaf tints in autumn. The flowers are larger, 4 cms (1½ ins) across, and pure white or yellowish with many stamens. *H. sexstylosa*, has pink-tinged stamens to the white flowers and evergreen lanceolate leaves.

The closely related *Plagianthus betulinus,* also from New Zealand, makes a small to medium-sized tree, 9–18 m (30–60 ft) high, with terminal panicles of white flowers. The leaves are simple and shaped like those of the Birch (*Betula*). The tree possesses the ribbon-like bark of *Hoheria* and was used by the Maoris for making into rope, twine and fishing nets.

Lavatera arborea is the Tree Mallow, found wild in Europe in coastal crevices. It grows 1.8–3 m (6–10 ft) tall with clusters of pale to deep purple flowers 5 cms (2 ins) across. *L. olbia,* the Tree Lavatera from Portugal to Sicily, is a fine but short-lived shrub, since large specimens are often killed in a cold winter. It grows about 1.8 m (6 ft) high with hundreds of reddish-purple flowers which remain in character throughout the summer and autumn.

L. trimestris is an annual from the Mediterranean region with quantities of exquisite, satin-rose flowers. There are many garden varieties including white forms and one called 'Loveliness' with deep rose-pink flowers 10 cms (4 ins) across. The general height is 90–120 cms (3–4 ft).

Pavonia multiflora is a quaint shrub from Brazil with narrowly oblong, long-pointed, toothed leaves and terminal corymbs of purple-red flowers. These have red hairy bracts, the corolla being narrowly segmented which gives the impression of a flower made up of many thread-like petals. The long stamen tube and stamens protrude beyond the bloom.

Another species *P. cancellata,* is a prostrate, very hairy plant with showy yellow flowers having purple basal blotches. It is native to sandy savannahs in South America.

Other species yield bark fibre used for cloth making and cordage. The mucilaginous root of the Rhodesian *P. hirsuta* is put into milk to hasten (after shaking) butter production.

Malvaviscus arboreus is a tropical American shrub of branching habit, 1–2 m (3–6 ft) high, with dense foliage. The leaves are heart-shaped and serrated, the flowers red with protruding stamens. In the tropics it is usually group planted but in temperate places has to be grown in a cool greenhouse. *M. mollis* from Mexico has three-lobed velvety leaves and long-stalked red flowers. It attains a height of around 90 cms (3 ft) and is often seen in gardens in Bermuda and the West Indies.

The Portia Tree *Thespesia populnea* is a common street tree in southern Florida. It is evergreen with leathery, ovate, long-stalked leaves and yellow Hibiscus-like flowers which have purple bases. It is native to the Old World tropics and is probably naturalized in the SE United States.

Sidalceas are graceful and long-flowering herbaceous perennials suitable for growing in any good garden soil, in sun or light shade. They have palmate, often deeply cut leaves and small Hollyhock-like sprays of pink, rose or purplish flowers in late summer.

Sidalcea malviflora from California is the Checker Bloom or Prairie Mallow, a variable plant with bright pink to rosy-purple flowers. Many good named forms have been derived from the species including 'Sussex Beauty', soft satiny-pink; 'Listeri', pink with fringed petals; 'Rev'd Page Roberts', light rosy-pink; and 'Brilliant', rich carmine; all are about 60 cms (2 ft) tall.

S. candida from Colorado, USA is the White Prairie Mallow, with pretty white flowers about 2 cms (1 in.) across loosely arranged on slender stems 1 m (3 ft) tall.

Callirhoe papaver from the southern USA is the Poppy Mallow, a scrambling or sometimes erect herbaceous perennial with reddish-purple Poppy-like flowers on stems of 60 cms (2 ft) and delicate Mallow-like leaves. It needs light soil and full sunshine.

Kitaibela (*Kitaibelia*) *vitifolia* is native to the lower Danube area in SE Europe. It is a tall herbaceous perennial with stems 1.8–2.5 m (6–8 ft) high clothed with five-lobed leaves and showy white or rose-pink Mallow-like flowers.

Malvastrums are evergreen, rather tender shrubs chiefly valued for their extended period of flowering, which under warm conditions (as in South Africa) is almost continuous throughout the year. The pink, red or yellow flowers resemble small Hibiscus with five overlapping petals forming a shallow cup.

The Cape species such as *Malvastrum scabrosum* (*M. capense*), an aromatic, hairy and rather sticky species 1.8 m (6 ft) tall with deep rose flowers and *M. grossulariifolium,* with pink, fragrant blooms on bushes 3 m (10 ft) high are two of the best.

M. campanulatum from Chile grows 30–45 cms (1–1½ ft) tall and has purplish-rose flowers.

Pavonia multiflora

Lavatera trimestris 'Loveliness'

Maranta leuconeura erythroneura

Calathea makoyana

Calathea ornata 'Albo-lineata'

Marantaceae

30 genera and 400 species

This is a family of monocotyledonous herbaceous perennials chiefly from tropical America, grown for their fine foliage. The leaves are two-ranked and sheathing with one side larger than the other. The flowers are asymmetric and bisexual with three sepals, three petals, one fertile stamen and several petaloid structures and an inferior ovary.

Calatheas are attractive plants with tuberous roots and finely marked foliage often grown for home decoration. They should not be placed in too strong a light or the leaves curl and the temperature should not drop below 10°C (50°F). Constant humidity is required, particularly in summer. *Calathea makoyana,* which is known as the Peacock Plant in the United States, has green and silver leaves with darker green veining and purple undersides; *C. zebrina* has leaves striped in two shades of green; and *C. ornata* 'Albo-lineata' has green, pink and cream leaves. All are South American.

Marantas are sometimes called Prayer Plants because of the manner in which the young leaves fold together. They are stocky little plants with bright green, oval leaves variously spotted or patterned. *Maranta leuconeura* has a tuberous root and green, white and dark green leaves with purple undersides. There are several good named forms such as the red-veined *erythroneura.* This is a warm room plant needing shade from bright light and disliking lime in the soil; it needs very little water in winter.

Thalia dealbata from the southern United States is a handsome aquatic growing 90–150 cms (3–5 ft), with long-petioled Canna-like leaves and erect panicles bearing many drooping purplish flowers. The plant is slightly glaucous due to a dusting of white mealy farina.

Ctenanthe oppenheimiana is the Never Never Plant with long-petioled, oblong leaves about 13 cms (5 ins) long, banded with cream, dark and light green and with purple undersides. The plants thrive best in acid peaty soil in shallow containers.

Melastomataceae

240 genera and 3000 species

This is a large family of dicotyledonous tropical and subtropical herbs, shrubs and trees, frequently recognized by their longitudinal leaf veining, which diverges from the base and converges again at the apex. There is no true mid-rib and the leaves are simple and usually entire. The inflorescences are cymose, the flowers normally perfect and regular with four to five sepals, four to five petals, twice as many stamens (that is, eight to ten) and ovaries which may be superior or inferior.

The family provides some fine ornamentals (chiefly trees and shrubs) for tropical gardens and a number which in cooler climates are grown in warm greenhouses.

Tibouchina is a large genus of some 200 species all native to tropical America. They have oval, velvety leaves with the characteristic veining and showy terminal bunches of violet or purple flowers. These are five-petalled and wide open, with long curling anthers which resemble a spider on its back. The formation and prominence of the latter probably accounts for the name Spider Flower being applied to the Brazilian *T. urvilleana* commonly cultivated incorrectly as *T. semidecandra.*

This species is a branching shrub of 3–6 m (10–20 ft), very showy in flower, with many royal purple blooms of 7–13 cms (3–5 ins) which have purple anthers borne on magenta-red filaments. The plush-like quality of the leaves is especially marked in the young foliage; later they may become suffused with red or turn entirely crimson, particularly when they are dry at the roots.

T. laxa from Peru is smaller and *T. gayana,* also Peruvian, has white flowers. *T. granulosa,* the Purple Glory Tree, of Brazil is popular for street and garden planting in South America, South Africa and Florida, USA. It frequently has two flushes of blossom—in spring and again in early autumn. It makes a tree up to 14 m (45 ft) with deep violet flowers although there

Tibouchina urvilleana

are pink ('Rosea') and rosy-purple ('Cyanea') forms. In the Cypress Gardens at Winterhaven, Florida, this species grows with its roots practically in water.

All the Tibouchinas need rich growing conditions with plenty of water in summer but they should be given rather less during the winter months.

Another popular ornamental is *Medinilla magnifica,* an evergreen shrub about 180 cms (6 ft) high from the Philippines, with showy inflorescences of 30 cms (1 ft), consisting of large bracts and many rosy-red flowers which have yellow stamens with purple anthers. The rich green, oval leaves are 20–25 cms (8–10 ins) long, prominently veined and stalkless. The species makes a handsome pot plant, especially when stood above water so that the long trailing flower trusses are seen to advantage and mirrored in its surface. There are many other species, some climbers and others epiphytic, found in the Malay Archipelago, W Africa and the Pacific Islands. The young leaves of *M. hasseltii* are eaten with rice by natives of Sumatra and the crushed fruits in fish dishes.

Medinillas should be grown in a compost of equal parts fibrous loam, peat and decayed leafmould with a half-part silver sand. They are at their best in a temperature which does not fall much below 16°C (60°F) and are propagated from spring cuttings, rooted singly in small pots.

Miconia is a large genus of some 700 species, mostly with handsome, strongly veined leaves and panicles or corymbs of small flowers. They are native to tropical America but also found in the West Indies and there is one W African species.

M. robinsoniana girdles Santa Cruz, one of the islands in the Galapagos Archipelago, separating the grassy summit 1500 m (5000 ft) high from the fern belt, which in turn lies above the intermediate zone and the arid coastal strip. It makes a big bush with glossy leaves and small purple flowers—somewhat reminiscent of a tropical Rhododendron.

M. argentea, a tree 9 m (30 ft) high, is cultivated in the tropics on account of its pyramidal panicles of white flowers and handsome leaves. When the wind blows the white undersides of these are exposed so that the tree appears to be spangled with white blossoms. The common name Two Faces alludes to this circumstance.

M. magnifica from Mexico has large 30–45 cms (1–1½ ft) long, oval leaves, of velvety green, with reddish-purple undersides and white veining.

Bertolonias are Brazilian plants grown for their handsome leaves. They are mostly dwarf herbaceous or creeping, with white or purple flowers in terminal cymes and stalked, oval-cordate leaves. These are frequently rich purple underneath and may have chocolate or white markings above, especially along the veins. *Bertolonia marmorata* is particularly well patterned and grows about 15 cms (6 ins) tall.

Sonerila from tropical Asia is also grown for its foliage. *Sonerila margaritacea argentea,* a varietal form of a Javanese species, is particularly pretty, with silvery leaves patterned near the veins in dark green. The rose-pink flowers are in loose sprays and have three petals and usually three stamens. The two genera (*Sonerila* and *Bertolonia*) are not unlike and in fact have been crossed to

Medinilla magnifica

Sonerila margaritacea argentea

Centradenia floribunda

produce bi-generic hybrids known as x *Bertonerila houtteana*. The specific name of the latter refers to Van Houtte of Ghent, Belgium, who made the first crosses in the 19th century. Some varieties are velvety-green with white or darker markings round the five main veins; others have a metallic sheen and silver and pink spots.

All these plants require a humid atmosphere, shade from sunlight and plenty of root moisture during the growing season.

Centradenia floribunda is a charming little shrub under 45 cms ($1\frac{1}{2}$ ft) in height, with opposite but irregular leaves—one usually considerably larger than the other. The dainty sprays of lilac-rose flowers appear in winter and last well when cut.

C. grandifolia, growing about 60 cms (2 ft) high, has light rose flowers and red undersides to the leaves. Both species are Mexican.

Dissotis are herbs or small shrubs which like moist conditions so are grown in the vicinity of the bog garden in tropical countries. In cooler climates they should be kept rather drier and cultivated in pots of peaty compost.

D. incana, the Dwarf Glory Bush from South Africa, is a small sub-shrub about 60 cms (2 ft) tall with many broad-petalled, rose-purple flowers 4 cms ($1\frac{1}{2}$ ins) across, in crowded terminal panicles. The small lanceolate to oval leaves are strongly ribbed and are velvet-textured. The plant is resistant to light frosts but not prolonged bad weather such as occurs in Europe and North America. It can be raised from seed and also sends out long runners which may be separated for propagation purposes. *D. princeps* is similar but has larger flowers.

Osbeckia is a genus of about 100 herbs and shrubs with a wide Old World tropical distribution, and large flowers like *Tibouchina* with four or, more frequently, five petals and curved anthers.

O. stellata makes a shrub 1–2 m (3–6 ft) high with ovate-lanceolate, rather hairy leaves and clusters of pink or lilac four-petalled flowers. It is native to India and also China.

O. chinensis with purple flowers and *O. yunnanensis,* bright magenta, both come from China; the deep purple *O. rubicunda* is native to Ceylon and the red or purple *O. glauca* to India.

The hardiest member of Melastomataceae and the only one which can be grown outside in Europe—and even then it needs winter protection—is *Rhexia virginica*. It comes from eastern North America where it is found in wet sandy soil. It is of compact and bushy habit and grows 30–45 cms (1–$1\frac{1}{2}$ ft) high, the stems having four prominent ridges along their length. The leaves are oval-cordate and opposite and the flowers deep rose with four lopsided petals and long arching bright yellow stamens. They are succeeded by red fruits. The species, which is variously known as Deer-grass and Meadow Beauty, is a pretty subject for the bog garden. It can be propagated by seed or by division of the tubers, but the plants should not be disturbed too often.

Melastoma belongs to the Old World tropics, all the species being evergreen shrubs and often hairy. The entire, oblong or lanceolate leaves have a leathery texture and the flowers are purple or rose, or very occasionally white. They are succeeded by fleshy berries which stain the flesh when picked or eaten.

Species which are sometimes cultivated include *M. candidum* from S China, with pink or white cymes of flowers at the ends of the branches; *M. malabathricum,* purple, from the East Indies and *M. denticulatum,* white, which is native to New Caledonia.

Melastomataceae has few representatives of economic importance (apart from their garden uses) although a few members of this family do have slight medicinal properties.

Menyanthaceae

5 genera and 33 species

This family consists of dicotyledonous water plants with alternate, simple or tripartite leaves and bisexual flowers with five sepals, five petals, five stamens and a superior ovary.

Menyanthes trifoliata is the Bog Bean or Buckbean, a Eurasian and American aquatic with snake-like rhizomes which run in and out of the water, trifoliate leaves and clusters of delicate fringe-petalled flowers which open pink and then fade to white. They look like stars cut out of Turkish towelling. The scraped rhizome is eaten as bread by the Lapps for its tonic properties and was at one time used in the making of beer.

Nymphoides peltata (*Limnanthemum nymphaeoides*) which is also European, is the Water Fringe with small floating leaves (like those of a Water-lily) having wavy margins and yellow fringed flowers standing just above the water. This plant also has medicinal uses.

Villarsias are white or yellow-flowered; *Villarsia reniformis* from Australia, is an erect marshy plant, 30–60 cms (1–2 ft) high, with rounded leaves and panicles of five-petalled flowers. *V. ovata* from South Africa grows 15–30 cms (6–12 ins) tall. This species has clusters of oval leaves and slender scapes carrying panicles of attractive fringed citron-yellow blooms.

Menyanthes trifoliata

Nymphoides peltata

Morinaceae

1 genus and 17 species

This dicotyledonous family has but a single genus—*Morina*—of which the most widely grown species is *M. longifolia,* the Himalayan Whorlflower of Nepal. There it grows at altitudes between 2133–3660 m (7000–12,000 ft). At the higher altitudes the natives believe it restricts breathing.

The plant has long, spiny, thistle-like leaves and flowers arranged in whorls on stems of 45 cms ($1\frac{1}{2}$ ft). Individually they are pink and red and white, with long tubular blooms which flare out at the mouths into five petals. The ovary is inferior.

Normally in gardens it is sensitive to prolonged wet and cold so should be given a protected position, with sun and a light but well-drained soil. Propagation is by seed which should be sown soon after harvesting.

Musaceae

2 genera and 42 species

This family of gigantic, monocotyledonous, herbaceous plants is native to the tropics of Asia, Africa and Australia. It has branching rhizomes from which rise large, sheathed leaves, which are rolled in the bud and have stout mid-ribs and many parallel veins. The leaf blades between the veins are easily torn by wind and rain.

The finger-shaped flowers are arranged in racemes and have brightly coloured bracts, six petals (five of which are joined), six stamens and an inferior ovary. They contain much nectar and are visited and pollinated by bees and birds.

Musa x *paradisiaca,* the Plantain, and its subspecies *sapientum,* the Banana, are important tropical food plants with juicy stems sometimes 4.5–6 m (15–20 ft) high, with large leaves and inflorescences having purple bracts. The fruits are eaten raw, boiled or roasted and also ground into meal and made into alcohol. The fibres of the leaf stalks are the material from which the gossamer muslins of India are manufactured and the coarse fibre of other species (as *M. textilis*) makes a kind of hemp—Manilla Hemp—used for binder twine, papier-mâché, wrapping paper and a lustrous cloth.

The cropping of Bananas is successive; after a ripe cluster has been gathered the stem dies, but new shoots rapidly spring from the roots and take its place.

A scarlet-bracted, red-fruited Banana (possibly *M. coccinea*) growing about 120 cms (4 ft) tall, can sometimes be seen in tropical gardens or may be grown in pots. It is striking and attractive and deserves wider cultivation.

Musa textilis

Morina longifolia

Myrsinaceae

35 genera and 1000 species

This family of dicotyledonous tropical and subtropical trees or shrubs, comes mainly from tropical Asia and America but is also represented in South Africa and New Zealand. The leaves, which possess resin glands, are entire and leathery, usually alternately arranged on the stems. The flowers come in umbelled or panicled inflorescences and are regular, sometimes unisexual, with four or five sepals and petals, four or five stamens and usually superior ovaries. The fruit is a drupe or berry.

A popular member for subtropical bedding or as a greenhouse subject in cooler climates is *Ardisia crispa*, East Indies to Japan which makes a small evergreen shrub 90 cms (3 ft) high. It is slow growing but flowers and fruits at a very early stage; the flowers are white and fragrant, borne in whorls round the stems, and followed by scarlet, Holly-like berries. These are very persistent so that established plants bear flowers and berries at the same time. The long, narrow leaves have crimped margins.

The slightly hardier *A. japonica* from Japan is similar except for smaller leaves, white or pink flowers and bright red berries.

Taller species of Ardisia may make trees 6–12 m (20–40 ft) high and are frequently planted in the tropics. They include *A. lanceolata* from Malaysia with large terminal panicles of rich mauve-pink flowers 2 cms (1 in.) wide and *A. paniculata,* an Indian species with pyramidal panicles of rosy-lilac at the ends of the branches. Blossoms and red fruits appear together and are both very persistent.

Myrsine africana from Africa, China and India belongs to another genus of trees and shrubs, widely scattered in Asia, in parts of Africa and tropical America. It is evergreen, 60–120 cms (2–4 ft) high with angled and downy shoots, with small, shiny, narrow leaves and pale brown axillary flowers which are succeeded by blue-purple berries. Since the flowers are unisexual, both sexes must be represented to ensure fruiting.

Some species of *Myrsine* have tough hard timber which can be used for furniture and turnery; the berries are eaten by natives and the bark of certain New Zealand species has been employed in that country for tanning purposes in the preparation of skins.

Myrtaceae

100 genera and 3000 species

This is a family of dicotyledonous plants, for the most part native to the warmer parts of Australia and tropical America, which includes some of the tallest trees in the world (*Eucalyptus*) as well as many small shrubs and creepers.

The foliage is evergreen and contains, together with the young shoots, flowers and fruit, essential oils stored in special glands. The various valuable herbs and medicinal plants in Myrtaceae owe their importance to this circumstance. Usually the leaves are entire and opposite, although occasionally they are alternate. The perfect flowers are borne in cymes and have four to five sepals and four to five petals, although their most attractive feature lies in the innumerable stamens with their long coloured filaments. The ovary is inferior and develops to a capsule or juicy berry or drupe.

Amongst the most important genera is *Eucalyptus* with 500 species. They are fast growing, drought-resistant trees much cultivated in tropical and subtropical areas, as Africa and South America, for shade and timber. With few exceptions they all come from Australia where some grow to great size. Specimens of *E. regnans,* for example, have been officially recorded as reaching 97 m (323 ft) in height and 8 m (25 ft) in girth, although Ferdinand Müller in 1880 quoted (perhaps erroneously) fallen specimens of *E. amygdalina* as 138 m (460 ft) and 153.6 m (512 ft).

Eucalyptus leaves have twisted petioles so that the leaves hang vertically instead of horizontally and both surfaces are identical (having no distinct upper and undersides).

The inflorescence is usually an umbel, the flowers spectacular with many conspicuous red or white filaments; the calyx is a round nut-like swelling which at first glance looks like a fruit. The bark is variable, having a smooth finish (which often peels off in patches) in the Gum Trees, is scaly all over in the Bloodwoods, set in long fibres in the Stringy-barks and with black furrowed bark in the Iron-barks. These common names describe such characteristics.

Eucalypts grow very quickly; seedlings attain about 1.5 m (5 ft) in one year and 10.5 m (35 ft) in ten. They thus provide a quick source of timber, which is valuable to the economy of such places as East Africa and the Andean regions. The wood is durable and resistant to damp but splits easily; it is used for such things as railway sleepers, parquet floors, shingles, wheel spokes, posts and shipbuilding. The power of Eucalypts to assimilate

Eucalyptus leucoxylon

water has proved important in swampy areas such as the Pontine marshes near Rome and parts of North Africa. Here Eucalyptus trees not only are used to reclaim land but are planted to eradicate the breeding places of malaria-carrying mosquitoes.

All parts of the plants are rich in essential oils and particularly the fresh leaves of *E. globulus*. This is the source of Oil of Eucalyptus, an antiseptic and expectorant which is used in pharmaceutical preparations and medicines. The Tasmanian Blue Gum, as it is commonly known, is much used for street planting in warm climates or, under temperate conditions, young plants are often employed in subtropical bedding schemes.

Some Eucalypts produce several stems from one root and adopt a shrubby habit. These are known as Mallees and are typical of the more arid parts of Australia; they thrive in similar hot dry situations elsewhere in the world and help to prevent soil erosion. There are countless forms, most possessing brightly coloured flowers.

Among the larger Eucalypts *E. ficifolia,* the Red Gum, is a favourite for park and home planting in Australia. Its flowers vary from white through pink to deep crimson and it rarely exceeds 9 m (30 ft) in height. *E. leucoxylon* is the Yellow Gum or White Ironbark, a tree 4.5–6 m (15–20 ft) high with long pendulous leaves and pink, white or crimson flowers.

E. citriodora is called the Lemon-scented Gum because of its fragrant leaves. It has slender white stems and white flowers. Hardiest in Britain are *E. gunnii,* the Tasmanian Cider Gum with white flowers, *E. macarthuri,* the Camden Woollybutt, a spreading tree from New South Wales with narrow leaves and creamy flowers, and *E. pauciflora.*

Eugenias are not hardy in Britain or most of North America where they generally have to be grown under glass. In warmer parts of the world they make fine ornamental trees or are used for tall hedges and windbreaks. They have opposite, evergreen leaves, small flowers with prominent yellow stamens, attractive young foliage and heavy crops of showy berries. *Eugenia luehmannii,* the Brush Cherry, is one of the most attractive with delicate pale pink young foliage and rose-pink berries; *E. smithii* is the Lilly-pilly Tree with handsome fruits varying from white to blue; and *E. pendula* has purple berries. These are all Australian but *E. michelii,* the Brazil Cherry, from tropical America has red fruits. All the Eugenias are rich in nectar. *E. jambos,* the Rose Apple, from the East Indies makes a tree 6–12 m (20–40 ft) high with white flowers and creamy fragrant fruits used in confectionery. The genus contains about 1000 species, many with edible fruits.

Another important plant in Myrtaceae is *Syzygium aromaticum* (*Eugenia caryophyllus*), the dried flower buds of which provide the Cloves of commerce. It comes from the Moluccas, near New Guinea, but is much cultivated in various parts of the tropics. *Psidium guajava,* a shrub or small tree from Central and South America has large white flowers succeeded by small, green, pear- or apple-shaped fruits—Guavas—full of vitamins.

Metrosideros diffusa

Myrtus communis

Leptospermum scoparium 'Crimson Damask'

The genus *Leptospermum* consists of about fifty species of shrubs and small trees, mostly natives of Australia, New Zealand and Tasmania. Their common name, Tea Tree, refers to the fact that early colonists made tea from the pungent leaves. The whole plant is highly aromatic and has been claimed as the colonial counterpart of the English Broom (*Cytisus*). The small leaves are hard and leathery; the scentless, five-petalled, white, pink or red blooms 6 mm–2.5 cms ($\frac{1}{4}$–1 in.) across, crowded on long shoots. There are many garden forms with larger, more richly coloured or even double flowers. Leptospermums make good pot plants or can be grown outside in mild climates; all are rich in nectar which produces a dark, strongly flavoured honey. *L. scoparium,* the Manuka from New Zealand, and its varieties are among the best known in Europe.

Metrosideros diffusa is a straggling New Zealand climber with ragged bark and terminal cymes of red flowers. *M. fulgens* (*M. scandens*), the Rata Vine, has yellow, pink or red blooms. Bushmen quench their thirst by slitting the wood; the clear pink liquid tastes like cider. *M. excelsa* is the Christmas Tree of New Zealand.

Callistemons are the Bottlebrushes from Australia, evergreen shrubs with stiff leaves and compact colourful, upright or occasionally pendulous flower heads mainly consisting of fluffy stamens. The plants make fine loose hedges in suitable climates or can be grown in greenhouses. *Callistemon citrinus* (*C. lanceolatus*) is called the Lemon Bottlebrush because its leaves have a faint lemon odour, but its flowers are crimson and packed tightly together at the ends of a stiff stem like a bottlebrush cleaner. *C. salignus* has yellow flowers.

Pimenta dioica (*P. officinalis*), another member of Myrtaceae, is the source of Allspice which is obtained from the unripe fruits. It is a small West Indian tree, as is *P. racemosa* which yields Bay Oil—an important constituent of Bay Rum.

The only European representative is *Myrtus communis,* the classical Myrtle, a small shrub with small white flowers full of stamens, blue-black berries and small fragrant leaves. It is native to the Mediterranean area and in ancient times this plant was dedicated to Venus, the goddess of love, an association which has lingered in the use of Myrtle sprays in bridal bouquets.

Callistemon citrinus

Nelumbonaceae

1 genus and 2 species

This small family of large aquatics is distributed in various warm parts of Asia, Australia and America. The long Banana-like rhizomes of *Nelumbo*, the only genus, have a peculiar leaf arrangement, sending up at intervals, first a scale leaf on the lower side of the internode, then one on the upper side immediately followed by a foliage leaf. After another section of internode the process is repeated. From the axil of the second (upper) scale leaf springs the flower and from the true foliage leaf axil a branch. These circumstances give the plant great potential for spreading and flowering.

The leaves are entire, round like tea-trays and usually borne well above the water. In the young state, or in very deep water they may float, the upper leaf surface covered with a protective wax coating which keeps them dry even in the most inclement weather. Raindrops roll from the surface like globules of mercury. The foliage is glaucous-green and has the petiole centrally placed as in the common Nasturtium (*Tropaeolum majus*).

The flowers are solitary at the ends of long stems and have four sepals and numerous petals and stamens. They are usually large and showy, not unlike a full blown Rose or Paeony with numerous golden stamens surrounding a large seed pod. The latter, shaped something like a large Poppy head or the rose of a watering can, contains edible seeds.

From very early times Man has looked for symbols of a higher destiny, finding these in the Sun and the world around. Those he selected became objects of veneration, a distinction which was even applied to plants if they had the right qualities. The Lotus was one so favoured and in its time became greatly esteemed not only in Ancient Egypt but in India, Tibet, Japan and China.

At the same time there seems to be some doubt as to whether *Nelumbo nucifera,* sometimes called the Sacred Lotus of the Nile, is the Lotus portrayed in Egyptian art and sculpture; or whether *Nymphaea lotus* and *N. caerulea* (the former white, the latter blue) are the plants represented (see page 196). Both of the last are indigenous to Egypt but *Nelumbo nucifera* was introduced into Egypt from India centuries ago, probably somewhere about 525 B.C. (the time of the Persian invasion).

Herodotus (born 484 B.C.), who was the earliest Greek historian, described it as the Rose Lily and following extensive travels in that country, gave this account of its use by the Egyptians. 'When the river is full, and has made the plains like a sea, great numbers of lilies, which the Egyptians call Lotus, spring up in the water; these they gather and dry in the sun; then having pounded the middle of the Lotus, which resembles a poppy, they make bread of it and bake it. The roots also of this Lotus is fit for food, and is tolerably sweet, and is round and of the size of an apple. There are also other lilies, like roses, that grow in the river, the fruit of which is contained in a separate pod that springs up from the root, in form very like a wasp's nest; in this there are many berries fit to be eaten, of the size of an olive-stone and they are eaten both fresh and dried.' While the first description probably refers to Nymphaeas, the latter Lilies can only be *Nelumbo nucifera.*

The Egyptians too made bread from the seeds, pounding them into flour, which they moistened with milk or water and baked.

Nelumbos reproduce by means of the long creeping rhizomes, which under ideal conditions may run 6–9 m (20–30 ft) in a season. For this reason in small gardens they may need to be confined in some sort of receptacle, and will require good planting compost. The smaller kinds may be conveniently grown in tubs to which they adapt quite happily, but incipient seed pods should be removed to encourage new flower growth. Nelumbos are occasionally grown outdoors in Britain but rarely successfully, as the summers are not sufficiently warm to ripen the tubers and they invariably rot away during the winter. They succeed reasonably well in southern Europe (especially Italy) and have been found hardy in the eastern United States as far north as New Jersey.

N. nucifera (*Nelumbium speciosum*), the Hindu Lotus, has a wide distribution, especially in Asia and NE Australia. It has large, peltate, slightly hairy leaves with a metallic sheen and vivid rose-coloured, globular flowers which become paler with age. White, carmine, purplish, pink and double forms exist, and also dwarf types and kinds with striped or colour-tipped petals.

Being rich in farinaceous matter Nelumbos are cultivated in the Orient for food. The seeds—known as Lotus nuts—can be eaten raw or are used in soups and various Japanese and Chinese dishes. The spiral fibres extracted from the leaf stalks make lamp wicks to burn before idols. The rhizomes are also edible, the latter sometimes preserved with salt and vinegar or sugar for winter use.

The boiled young leaves are eaten as a vegetable and in Indo-China the stamens are used for flavouring tea. Years ago in China, dishes prepared from Lotus seeds and slices of the root, mixed with the kernels of apricots, walnuts and alternate layers of ice were laid before British ambassadors at some of the breakfasts given by the principal mandarins.

In the Orient the plant is held sacred to Buddha who is thought to have been born in the heart of a Lotus blossom. The order of the Star of India comprises a Lotus flower, the Rose of England and two crossed Palm branches. Incidentally Hindus compare their country to the Lotus, the petals suggesting Central India and the leaves the surrounding provinces.

N. pentapetala (*Nelumbium luteum*) is the Duck Acorn or Water Chinquapin, found in lakes and waterways from the E United States to Colombia. It is a somewhat hardier species, with pale yellow flowers 13–25 cms (5–10 ins) across and round leaves, raised above the water. The seeds were used as food by the Indians and the entire plant—especially the large starchy rhizome—is edible. If the seeds are eaten they should be roasted while fresh, otherwise they dry very hard and are difficult to break down, even in water. They taste rather like boiled chestnuts.

Nelumbo nucifera

Nepenthaceae

2 genera and 68 species

This family of dicotyledonous carnivorous plants is peculiar to Asia, the Seychelles, tropical Australia and New Guinea. It contains mo tly herbs of boggy places or else epiphytes, which manage to climb by means of tendrils formed from a continuation of the mid-ribs in the long narrow leaves. Most of these go on still further to develop a pouch-like pitcher, with a lid projecting over the opening. Although the latter only fits tightly over a young pitcher, it nevertheless still protects a mature specimen from excessive wet and safeguards the numerous honey glands near its entrance.

The latter bait the trap for insects already drawn by the bright colours of the pitchers, which curve inwards at the necks to reduce the openings. The prey alights near the mouth, sups the honey sweetness and gradually wends its way downwards until it reaches the slippery base of the openings. Inevitably it slips and after ineffectual attempts to escape ultimately drowns in water at the bottom of the pitcher. The decomposed remains later serve to nourish the host plant, after the fashion of the related Droseraceae (page 102) and Sarraceniaceae (page 270).

The pitchers are often very showy in shades of red, claret, cream and green with violet, crimson or purple spots. At one time there was great interest in *Nepenthes* and many hybrids were raised in Europe for cultivation in warm houses. Most of the species and hybrids are not difficult to grow provided they are kept in a moist warm temperature of around 21–27°C (70–80°F) in summer and 18°C (65°F) in winter. They succeed in a compost of sphagnum and peat fibre mixed with a few pieces of charcoal.

Sometimes the pitchers are quite large, for example 17 x 7 cms (7 x 3 ins) in *N. khasiana* from the Khasia Hills in India or 20 cms

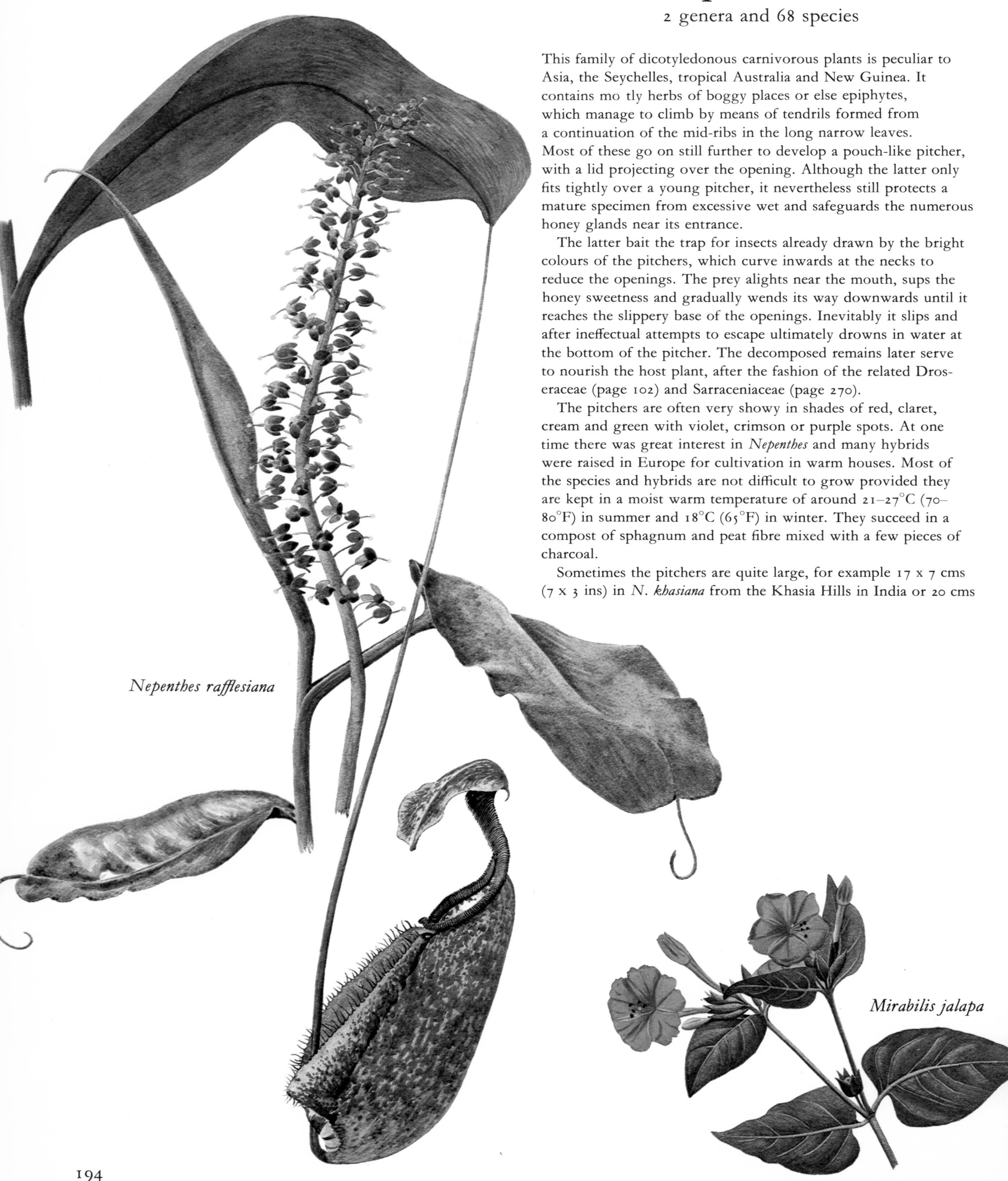

Nepenthes rafflesiana

Mirabilis jalapa

(8 ins) in *N. maxima* from Borneo. Dr Hooker in Borneo noted a species with 'pitchers which, including the lid, measure a foot and a half and the capacious bowl is large enough to drown a small animal or bird'.

They also provide refreshment for travellers and Knight and Step (*The Living Plant*) describe the plight of two Englishmen travelling with their tame monkey in southern Ceylon who 'overcome by thirst, sank down, weak and despairing at the foot of a palm tree and waited for death'. Providentially at that moment, their pet scented and led them to a group of *Nepenthes* with pitchers containing an abundant supply of water and thus they were saved.

The highly nectariferous flowers of *Nepenthes* are borne in racemes or panicles. They are male or female, regular, with the perianth segments green, yellow, claret or red and in two whorls of two, with four to sixteen stamens and a superior ovary. The seeds occur in a leathery capsule and can be germinated in the same compost as recommended for adult plants. Propagation is also carried out by means of shoot cuttings rooted in a warm propagating frame.

All *Nepenthes* are known as Pitcher Plants. *N. rafflesiana* from India has greenish pitchers marked with brown and *N. rajah* from Borneo possesses the largest pitchers in the genus—up to 90 cms (3 ft) with the leaf in cultivated plants, but reputedly twice that length in the wild. Two garden hybrids are *N.* x *mastersiana,* claret-red with purple and cream spots and *N.* x *mixta,* which is cream flushed with green and crimson.

Nyctaginaceae

30 genera and 290 species

Most members of the Nyctaginaceae are dicotyledonous trees, shrubs or herbs native to the tropics, particularly tropical America.

They have alternate or opposite leaves and cymose inflorescences; the individual blooms are set off by several bracts which in some cases are large and coloured and constitute the showy parts of the flowers. These flowers may be bisexual or unisexual, with five-lobed calyx, between three and ten stamens to each (but usually five) and a superior ovary. There are no petals.

Bougainvillea with eighteen species is a South American genus named for Louis de Bougainville, the first Frenchman to cross the Pacific. They are robust and showy scandent shrubs which owe their colour to the bright hued bracts, the real flowers being small and insignificant. Some varieties flower intermittently but others continuously and in tropical countries such as Kenya, where, supported by hooked thorns they climb to almost any height, they are used to drape walls, trees, pillars and buildings, or cut hard back to make hedges or large shrubs.

The chief species are *B. glabra,* with bright purple bracts, *B. spectabilis,* lilac-rose, and its variety *refulgens,* deep rich purple. All are Brazilian, but the many hybrids come in other colours, as for example 'Asia', cyclamen-purple; 'Audrey Delap', rust; 'Mrs Helen McClean', bright orange; 'Ardenii', bright red; and 'Snow White', white. The plants are easily propagated by cuttings of half-ripe wood rooted in sandy soil.

Mirabilis jalapa is the Marvel of Peru or Four o'Clock Plant, the last name referring to its habit of opening about 4 p.m. and remaining open all night. The plant is a herbaceous perennial about 60 cms (2 ft) high and of bushy habit with clusters of red, pink, white, yellow or streaked flowers. These are tubular flaring out to a wide mouth and faintly scented. In the tropics *Mirabilis* is often used for bedding purposes or in mixed borders, and flowers continuously if the roots are kept moist.

The tuberous roots were formerly used as jalap (a purgative medicine); true jalap, however, comes from *Ipomoea purga.*

Another genus, *Pisonia*, with some fifty species, includes a small tree from Malaysia, *P. alba,* the young tender leaves of which are boiled and eaten by natives. Growing 6–12 m (20–40 ft) high it has a crooked trunk and large oblong-ovate leaves which are creamy-white when young and gradually turn to yellowish-green. The flowers are insignificant but the foliage very beautiful, especially when contrasted with the darker greens of other trees.

Another small tree *Heimerliodendron* (*Pisonia*) *brunoniana* is the New Zealand Parapara. This has thin oblong leaves 10–38 cms (4–15 ins) in length and many-flowered cymes of small greenish flowers, which are very viscid so that small birds—and even reptiles—coming into contact with them are first ensnared, then glued down and eventually killed. The leaves and roots, however, have medicinal uses.

Bougainvillea glabra

Nymphaeaceae

2 genera and 75 species

These are dicotyledonous plants which, except for New Zealand, have a cosmopolitan distribution.

The main genus *Nymphaea* comprises the true Water-lilies which vary greatly in form and habit, some possessing short stout rhizomes of erect growth, like a Canna, others rhizomatous after the fashion of *Iris germanica*. There are viviparous species with young plantlets developing at the point where the leaf blades join the petioles, bulbous rootstocks in the tropical species which can be dried for months and yet grow when restarted, and one species with runners (the pale yellow *N. mexicana*), which increases after the fashion of a Strawberry plant.

The flowers may be floating or poised on stout stems above the water surface; in the various varieties they are occasionally as large as a soup plate or small enough to slip through a wedding ring. The colours vary from white, cream and yellow to copper-red, pink, scarlet, crimson, blue, mauve and purple, and whereas the majority open their flowers during the day, closing them in late afternoon, many of the tropicals wait until evening before expanding their blooms and remain open until the following morning.

The *odorata* section of the hardy Water-lilies is richly fragrant, as are many of the tropical sorts, particularly the night bloomers which use scent to attract night-flying moths which pollinate the flowers while foraging for nectar.

Botanically the Water-lily is characterized by large, nearly circular floating leaves which are entire and leathery with a centrally placed leaf stalk. The flowers are bisexual and regular with four sepals and many petals, usually in spirals of four and many stamens. In some cases the latter are petaloid. The large berry fruits contain many seeds, each with a spongy aril which entangles air bubbles and allows it to float for a time before sinking. This provision carries the seeds away from the parent plants and starts new colonies.

Water-lilies hold pride of place among aquatic plants and have an interesting historical background. They were well known to primitive Man who, groping for signs and portents of the Deity, saw in their blooms living symbols of regeneration and purification. Their manner of rising with the return of the rains, pure and undefiled from the mud and slime of dried up watercourses suggested immortality, purity and resurrection.

Although the plant has interested many primitive peoples, it is its association with the Ancient Egyptians which is particularly fascinating. As the Lotus of the Nile the Water-lily played an important part in their lives from the Fourth Dynasty (approximately 4000 B.C.), especially in religious observances.

Petals of the white *N. lotus* and blue *N. caerulea* were found by Schweinfurth in the funeral wreaths of Rameses II (1580 B.C.) and Amenhotep I. The flowers appear on many mural decorations of the time, and on furniture, pottery and other objects associated with Ancient Egyptians. It is possible that the reference in Kings I:26 to the pillars of King Solomon's Temple, 'and it was an hand's breadth thick, and the brim thereof was wrought like the brim of a cup with flowers of lilies' refers to Water-lilies.

The blooms were extensively cultivated for temple ceremonies and for use in the home. When an Egyptian nobleman of rank entertained, slaves presented arriving guests with a Lotus flower. These the visitors were expected to hold in their hands or wear twined in the hair; it is possible that this custom symbolized the peaceful intentions of the guests and that the Lotus was chosen because of the esteem in which it was held.

The use of flowers at funerals was very prevalent; the custom with priests and Pharaohs particularly being to lay wreaths in concentric semicircles from the chin downwards until the sarcophagus was piled with floral tributes. Some of these have been so well preserved that botanists can identify not only the genus but the species. Because of their religious significance Water-lilies were popular for the purpose, and it is interesting to note that botanists have not been able to detect any material differences between Water-lily flowers found in sarcophagi and those of plants still growing in Egypt today. This means that their mutation or development is very slow or non-existent, and that climatic conditions in Egypt can have undergone few changes in the last 4000 years.

The name Lotus has been applied to two plants, only one of which is a *Nymphaea*. There are two Water-lilies native to Egypt, one a white, broad-petalled, nocturnal bloomer called *N. lotus* and the other, *N. caerulea,* a narrow-petalled, sky-blue species which opens during the day. The latter is frequently depicted on murals and was much cultivated, but it is conceivable that the white species was also grown on account of its sweet scent and night blooming qualities.

The other plant known as Lotus is also an aquatic but belongs to a different genus and family. It is an Asiatic plant called *Nelumbo nucifera* (see page 193) which has interesting associations with Buddhism.

Years ago Water-lilies had several economic uses. The seeds contain starch, oil and protein and were consumed by Europeans,

Nymphaea 'Escarboucle'

Nymphaea tetragona 'Helvola'

Nymphaea tuberosa 'Richardsonii'

Asians and Africans in times of emergency. The dormant tubers (of the tender species) were also ground down and boiled or roasted like potatoes. They contain starch, mucilage and sugar which renders them both palatable and nutritious.

N. alba was at one time employed by the French in the preparation of beer and the rootstocks, which contain a dye, were used in Ireland and the Scottish Highlands for dyeing wool blue-black.

The porous leaf stalk of the Australian *N. gigantea* is peeled and eaten raw or roasted by aborigines and the powdered roots of *N. lotus* are used for dyspepsia, dysentery and various other intestinal disorders.

Culpepper says of the white *N. alba* that 'the leaves both inwards and outwards are good for agues, the syrup of the flowers produces rest and settles the brain of frantic persons. The distilled water of the flowers is effective for taking away freckles, spot, sunburn and morphew from the face and other parts of the body. The oil of the flowers cools hot tumours, eases pains and helps sores'. Again, the leaves being cooling and softening have been used in the past as a dressing for blisters.

Professor Goodyear (*Grammar of the Lotus*) ascribed to Water-lilies a high place in the arts of thirty centuries before Christ. From its twisted sepals he traces the Ionic capital and from that the Greek fret or meander. This doubled becomes the swastika, earliest of all symbols and according to the way it faces the representative of good or evil, male or female, light or darkness. The well-filled seed pods suggest the cornucopia, emblem of fertility and another old symbol. Many mentions of Nymphaeas are found in early writings, from Sanscrit literature to Theophrastus and Pliny.

All Water-lilies need a fairly stiff planting compost but organic substances should be avoided as they tend to turn the water murky or green. Firm planting is necessary and the roots need dividing from time to time when overcrowding forces the foliage above the water and hides the flowers. Covering the soil around newly planted Nymphaeas with clean shingle stops fish from rooting in the soft mud and fouling the water.

Although the tropicals seed freely few hardy Water-lilies set fruit, so that their propagation is normally by root division in spring. Most of the beautiful hybrids available today have been raised by the French firm of Marliac, particularly the founder M. Latour-Marliac.

Born in 1820, the son of a prosperous landowner, Bory Latour-Marliac took an early interest in Water-lilies and during a long lifetime (he died in 1911) crossed and recrossed any species which came his way until the parentage of many of his hybrids became obscured. He kept his methods secret and to all intents and purposes they have passed away with him. In the tropical Water-lily field George Pring of the Missouri Botanic Garden, USA, matches Marliac's record. Both men have brought the genus from semi-obscurity to today's important position as 'Queen' of the Water Garden.

Among the hardy Lilies are several to which Marliac gave his patronymic such as 'Marliacea Rosea'. 'Escarboucle' is a deep purplish-red; *N. tuberosa* 'Richardsonii', white; 'Sunrise', a many-petalled yellow of American origin; and *N. tetragona* (*pygmaea*) 'Helvola', a miniature yellow. The blue *N. capensis* (sometimes confused with *N. caerulea*) comes from South Africa, and *N. rubra* is a night bloomer from India. There are many others in both sections.

The other genus *Nuphar* has the same thick leathery floating leaves, and beautiful submerged foliage—crisped, translucent or membranous. The flowers are smaller than those of Water-lilies, mostly yellow or orange and have a strong vinous odour. They will grow in deep, shaded or running water—all situations which do not suit Nymphaeas—so have a place in large lakes and rivers.

The tuberous rhizomes are very thick and strong and contain tannic acid. They have been used for tanning purposes and fed to stock and the seeds can be roasted like popcorn. The common British species is *N. luteum* but the North American *N. advenum* is the better plant with handsome leaves and rich yellow flowers.

Nymphaea rubra

Nymphaea capensis

Nymphaea 'Sunrise'

Ochna serrulata

Ochnaceae

40 genera and 600 species

These are tropical dicotyledonous plants, usually trees and shrubs with simple alternate leaves (pinnate in the genus *Godoya*) and racemes, cymes or panicles of flowers. The flowers are regular and bisexual with five sepals, five petals, five, ten or many stamens and a superior ovary. The fruits are frequently fleshy.

Most of the eighty-five species of *Ochna* are found in tropical and South Africa with a few in tropical Asia. They are shrubs or small trees cultivated in gardens—or as greenhouse shrubs in cooler climates—for their ornamental fruits and flowers.

One of the most popular, especially in Europe is *Ochna serrulata* (*O. multiflora*), a shrub 1.25–1.5 m (4–5 ft) high native to Natal. It has narrowly elliptic bright green leaves and five-petalled yellow flowers. The calyces behind these are at first green but turn bright red as the fruits develop from green Pea-like berries to black. The contrast between sepals and fruits is then very striking.

O. atropurpurea, the Carnival Bush from South Africa, is a slow-growing shrub which commences flowering when about 45 cms (1½ ft) high and ultimately reaches 1.25–1.5 m (4–5 ft). The flowers are bright yellow with a purple calyx and about 2 cms (¾ in.) across with a central boss of golden stamens. Later these drop; the green sepals turn scarlet and the green seeds become glistening black. They then hang on the bush for several months, so that flowers and fruit are frequently in character together. The bark is covered with rough, fawn coloured spots.

Ochnas need pruning to keep them shapely and also thinning to expose the berries. They are propagated from half-ripe cuttings in summer. The seeds smell unpleasantly and normally only germinate satisfactorily when they have passed through the alimentary canals of birds, which eat them greedily. The process can be hastened artificially by soaking them in acid before sowing.

Tylerias, all native of Venezuela, are trees with showy, usually pink flowers which are either solitary or in large terminal panicles.

Ouratea oliviformis comes from Brazil. It makes a small tree 3–4.5 m (10–15 ft) tall with simple, toothed, leathery leaves and large, much-branched, terminal panicles of bright golden flowers. It is a showy plant but has a tendency to straggle so must be kept well trimmed, especially when young. The wood of some species of *Ouratea,* being tough and durable, is used in Asia for building purposes and in the making of boats, agricultural implements and furniture.

Lophira alata (*L. lanceolata*), the African Oak, is a W African timber-tree with magnificent, fragrant white (or occasionally yellow) flowers, the clusters 15–25 cms (6–10 ins) across. These smell like Musk and have two irregularly elongated sepals. The heavy red wood is very hard and tough, so called Ironwood by timber-men. The simple, strap-like leaves are arranged alternately and resemble those of the Hartstongue Fern (*Phyllitis scolopendrium*).

Oleaceae

29 genera and 600 species

This is an important family of dicotyledonous shrubs and trees with many ornamentals, particularly for temperate gardens. The leaves are usually opposite and simple but occasionally pinnate; the inflorescence is panicled, fasicled or cymose. Individual flowers are perfect (or rarely unisexual) and regular, with four sepals, four petals, two stamens and a superior ovary. The ovary, often with one to four seeds, may take the form of a berry, drupe or capsule.

Among the most delightful is *Forsythia,* a genus of slim-stemmed shrubs which in late winter have their leafless state hidden by thousands of golden, bell-shaped, four-petalled flowers which are followed by small, oval-oblong, toothed leaves.

The plants make good wall shrubs or can be used for hedging or as specimen bushes in the shrubbery. The chief kinds are *F. suspensa,* the Golden Bell from China, which was originally described by Thunberg as a kind of Lilac. It is a drooping species which needs hard pruning following flowering. *F. viridissima* was found by Robert Fortune in a mandarin's garden in Chusan and has twisted greenish-yellow flowers. The best garden plants are of hybrid origin, especially *F.* x *intermedia* 'Spectabilis' and a bud sport from this—'Lynwood'. These have larger, brighter flowers.

Forsythia x *intermedia* 'Spectabilis'

Abeliophyllum distichum

Jasminum officinale

Jasminum nudiflorum

Jasminum rex

The genus is named for William Forsyth (one-time gardener to George III), a brilliant but strange character who was famous in his lifetime and infamous afterwards. As a man's misdeeds live after him so he is remembered for the way he bluffed Parliament into parting with £1500 (with promise of more), for a worthless soap and ash mixture claimed to cure tree decay.

Forsyth, Director of the Royal Gardens at Kensington Palace, had produced a treatise describing the success of his recently invented 'Forsyth's Plaister'. In those days ships were made of oak and the authorities were becoming concerned at the vast amount of timber being rejected because of its unsuitability for naval purposes. Accordingly a Parliamentary Commission was set up to examine Forsyth's claims. It was satisfied that the 'Plaister' was efficacious and recommended a Parliamentary grant, but later, especially after Forsyth's death in 1804, a bitter controversy took place. This occurrence marred his very real abilities; he was one of the seven founder members of the Horticultural Society of London (now the Royal Horticultural Society), and for thirteen years Curator of Chelsea Physic Garden.

Abeliophyllum distichum, a monotypic genus from Korea, is somewhat similar to *Forsythia* but with white and fragrant flowers instead of yellow. It only grows 60–90 cms (2–3 ft) high and is hardy in Britain, flowering best following a hot summer.

Jasminum is a genus with some 300 species mostly of tropical and subtropical distribution although none are native to North America: some are climbers, some shrubs with white, yellow or occasionally rosy flowers. Those with white blooms are usually very fragrant, while the yellows are practically scentless.

One of the hardiest for temperate gardens is the sweet-scented Jessamine or Jasmine, *J. officinale*, found from Iran to China and introduced along the ancient trade routes to Britain in 1548. It is a rambling plant which does best against some form of support such as a wall. It will grow 3–6 m (10–20 ft) but can be kept lower and has small white flowers and deeply cut leaves. *J. nudiflorum* is the Winter Jasmine from China, a popular plant in Europe which blooms when little else is in character, the bright yellow flowers studding the green, leafless branches. This too is best supported, but can also be used to trail over rocks or trained as a loose hedge.

Less hardy but one of the most fragrant is *J. polyanthum,* a strong climber, deciduous or evergreen (according to climate) which will reach a height of 6 m (20 ft). It makes a good conservatory plant and is Chinese like *J. mesnyi* (*J. primulinum*), another slightly tender species, with primrose-yellow often double or semi-double flowers. *J.* x *stephanense* is soft pale pink and the scentless *J. rex* from Thailand has large white blossoms.

Several Jasmines are cultivated primarily for the perfumed oil, which is extracted by a process known as enfleurage, that is, the

Olea europaea

Syringa vulgaris
'Edmond Boissier'

flowers are pressed into shallow trays of wax and replaced daily until the wax is impregnated with oil. These are then dissolved in alcohol and the essence distilled. *J. grandiflorum*, the Spanish or Italian Jasmine, is one of the most important for this purpose. The flowers of *J. lanceolarium* (*J. paniculatum*) are used in China for scenting tea.

Syringa is the botanical name for Lilac, although also used as a popular title for *Philadelphus.* It comprises a genus of deciduous bushes with smooth, heart-shaped leaves and heavy panicles of four-petalled, tubular flowers. These come in early spring and are sweetly fragrant, particularly after rain. *Syringa vulgaris,* the common Lilac from SE Europe, grows up to 6 m (20 ft) with lilac-purple flowers, but is largely passed over in gardens in favour of its many cultivars. These may be single or double, with white, soft yellow, mauve, purple or reddish-purple flowers. 'Edmond Boissier' is a cultivar with deep purple single blooms. They are frequently grafted on common Lilac—a bad practice as the stock is prone to suckering, but they are also budded on Privet (which does not sucker) or grown on their own roots from layers or cuttings.

Several species are grown in mixed shrubberies, especially the Chinese *S. reflexa,* a shrub 3.5 m (12 ft) high with panicles of fragrant, long-tubed, rich pink flowers, and also *S.* x *persica,* the Persian Lilac, which has broad panicles of Lilac on stems of 2–2.15 m (6–7 ft). *S.* x *prestoniae* 'Isabella' is a good Canadian cultivar growing about 3.5 m (12 ft) high with mallow-purple flowers.

Osmanthus delavayi is a beautiful Chinese evergreen with small, glossy, pointed leaves and clusters of white, tubular and very fragrant flowers. It grows about 2 m (6 ft) tall and blooms in spring. Autumn-flowering kinds include the Chinese *O. armatus,* creamy-white, and the Japanese *O. heterophyllus* (*O. ilicifolius*) with Holly-like leaves and clusters of white blooms. When purchasing *Osmanthus* make sure they are on their own roots and not grafted on Privet.

Other members of Oleaceae include Privet (*Ligustrum*), a well-known hedging plant with strong-smelling, small white flowers in clusters, black berries and neat oval leaves, and *Olea europaea,* the Olive. This is a small tree, cultivated in Mediterranean regions since ancient times. Olive oil, expressed from the fruits, is used for food, cooking, canning sardines, soap and lubricants. The cured fruits are used as pickles and the wood for turnery, brushes and other items. The European Ash Tree, *Fraxinus excelsior,* and *Chionanthus virginicus,* the North American Fringe Tree, also belong to Oleaceae. The latter grows 6–9 m (20–30 ft) tall with masses of slightly fragrant, white, fringed flowers and oblong-ovate leaves. The root bark has medicinal properties.

Onagraceae

21 genera and 640 species

This is a widely distributed family of dicotyledonous, chiefly herbaceous plants, plus a few shrubs and trees. Although not entirely absent from the tropics the majority prefer the temperate or subtropical zones, where some like *Epilobium* have a wide representation—from the arctic to Australia and New Zealand, as well as South America and South Africa.

Family characteristics include alternate, opposite or whorled leaves (usually simple); bisexual flowers having four sepals, four petals (or occasionally none at all); eight stamens in two whorls (usually) and an inferior ovary. These may be solitary in the leaf axils, or grouped in spikes, racemes or panicles. Generally the flowers are regular but at times irregular.

The most important genus from an ornamental viewpoint is *Fuchsia*. This has some 100 species, mostly native to Central and South America, although they are also represented in New Zealand and Tahiti. The plants are of a woody nature, sometimes making vigorous shrubs in their native jungles, the branches draped with attractive red and purple, pendent flowers. These are tubular with four long coloured sepals framing short petals, the latter frequently of a different shade.

There are countless cultivars, some having their natural characters exaggerated, so that the sepals are long and flying or the petals frilled and doubled like bunched petticoats. The small entire leaves are usually smooth and many kinds have juicy berries which are edible. Jam from these is excellent. The leaves and bark of *F. magellanica* have medicinal uses and the blue pollen of the New Zealand, *F. excorticata,* is used by Maori women for face adornment.

The genus varies from ground-creeping species like *F. procumbens* to small trees. Some grow upright, a few like *F. tunariensis* are epiphytic, and in the hands of skilled growers cultivars can be fashioned to a miscellany of shapes, for example upright or weeping standards, pyramids and cordons.

The place of Fuchsias in the garden is largely dictated by climate. In the temperate regions of North America and N Europe

Fuchsia magellanica

Fuchsia 'Rose of Castile'

the South American species are rarely hardy; in colder climates they will not survive at all. Over much of the world the majority have either to be wintered indoors under protected conditions or else grown all the time under glass. It is usual to keep them fairly dry in winter and rest them until spring.

The New Zealand species and *F. magellanica* from the mountains of Magellan are among the hardiest and often grown outside in Britain against walls or similar supports. They are almost invariably cut to the ground in winter, but if the roots are undamaged new growths arise each spring.

Drooping Fuchsias make exquisite hanging-basket and window-box subjects and the more upright kinds can be grown in urns, tubs or large containers stood about in patios and courtyards. The smaller cultivars of *F. magellanica* like 'Pumila' and 'Riccartonii' and the species *F. microphylla* and *F. procumbens* are occasionally planted in rock gardens.

Among the species note should be made of several of the hardiest, as these are frequently grown outdoors in sheltered situations. *F. procumbens* from New Zealand is a prostrate creeper with small, petal-less tubular flowers of greenish-yellow having purple tips. It berries freely.

F. magellanica grows to a height of 2.5 m (8 ft) in its natural environment and has small red and purple flowers with protruding stamens, *F. coccinea* from Brazil grows to 1 m ($3\frac{1}{4}$ ft) with reddish stems; the leaves are in groups of three and the flowers red and violet. *F. excorticata,* the Tree Fuchsia of New Zealand, makes a spreading tree about 9 m (30 ft) high. The leaves are shiny and white underneath, the flowers green, developing to purple.

F. fulgens from Mexico must be wintered indoors. It grows 1.25–2 m (4–6 ft) high, the flowers with their long tubes of 5–7 cms (2–3 ins) a rich scarlet.

In addition there are of course many cultivars with single or double flowers or with pendulous tendencies (for basket work) and varieties with ornamental foliage.

The first known hybrids were raised in 1828 from crosses made between *F. coccinea, F. magellanica* and *F. arborescens.* Many British nurserymen around that time became particularly interested in the genus, as were several Continental breeders. Between them they raised many varieties but today nearly all the new cultivars come from California, where fresh kinds are raised every year.

There are several legends concerning the introduction of Fuchsias to England. One rather charming tale relates how a widow living in humble circumstances at Wapping had a sailor son who, knowing his mother's fondness for flowers, frequently brought home bits and pieces from his wanderings. On one occasion, while in South America he chanced to see Fuchsias growing and, thinking his mother would like one, dug up a small plant. In those days of sailing ships voyages took many weeks but eventually the widow had her plant and flowered it in her cottage window.

Then in 1793, the great nurseryman of the day, James Lee of Hammersmith (whose nursery stood on ground now occupied by Olympia), heard of this new plant and visited the widow. He offered to buy it but she refused, even when the tremendous sum of eighty golden guineas was mentioned. But Lee persisted, he promised the money and three new plants and so the widow relented—she got her money, Lee got his plants and the gardening public had Fuchsias.

Zauschneria is a small genus from Mexico and the western United States, commonly known as Californian Fuchsia or Hummingbird's Trumpet. There are four species, all dwarf perennials, mostly sub-shrubby but valuable on account of their late flowering habit and the brilliance of the long, scarlet, tubular flowers. As autumn arrives, these start to bloom, so that they are also grown in pots in greenhouses for fall decoration.

Normally Zauschnerias are subjects for light well-drained soil and make good drape plants over rocks. They spread underground like Epilobiums and are remarkably resistant to drought. Propagation is by means of cuttings or division.

Z. californica is the commonest species with branching stems of 30–60 cms (1–2 ft) carrying many long narrow flowers and small, lanceolate to oblong, greyish, hairy leaves. *Z. cana* is similar but with narrow, thread-like leaves and smaller blooms.

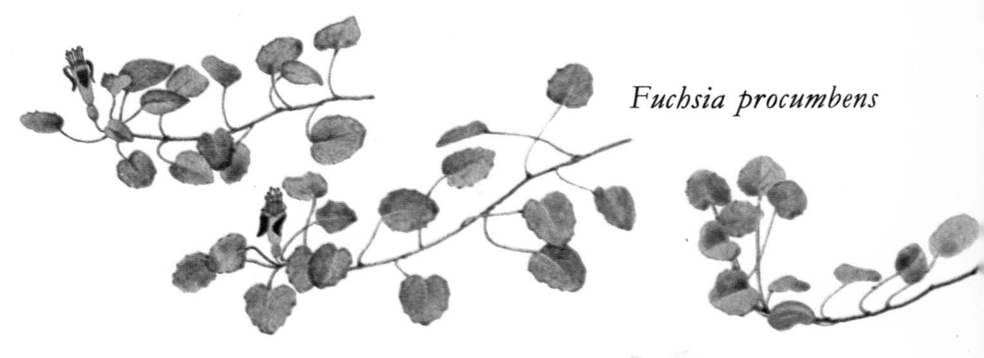

Fuchsia procumbens

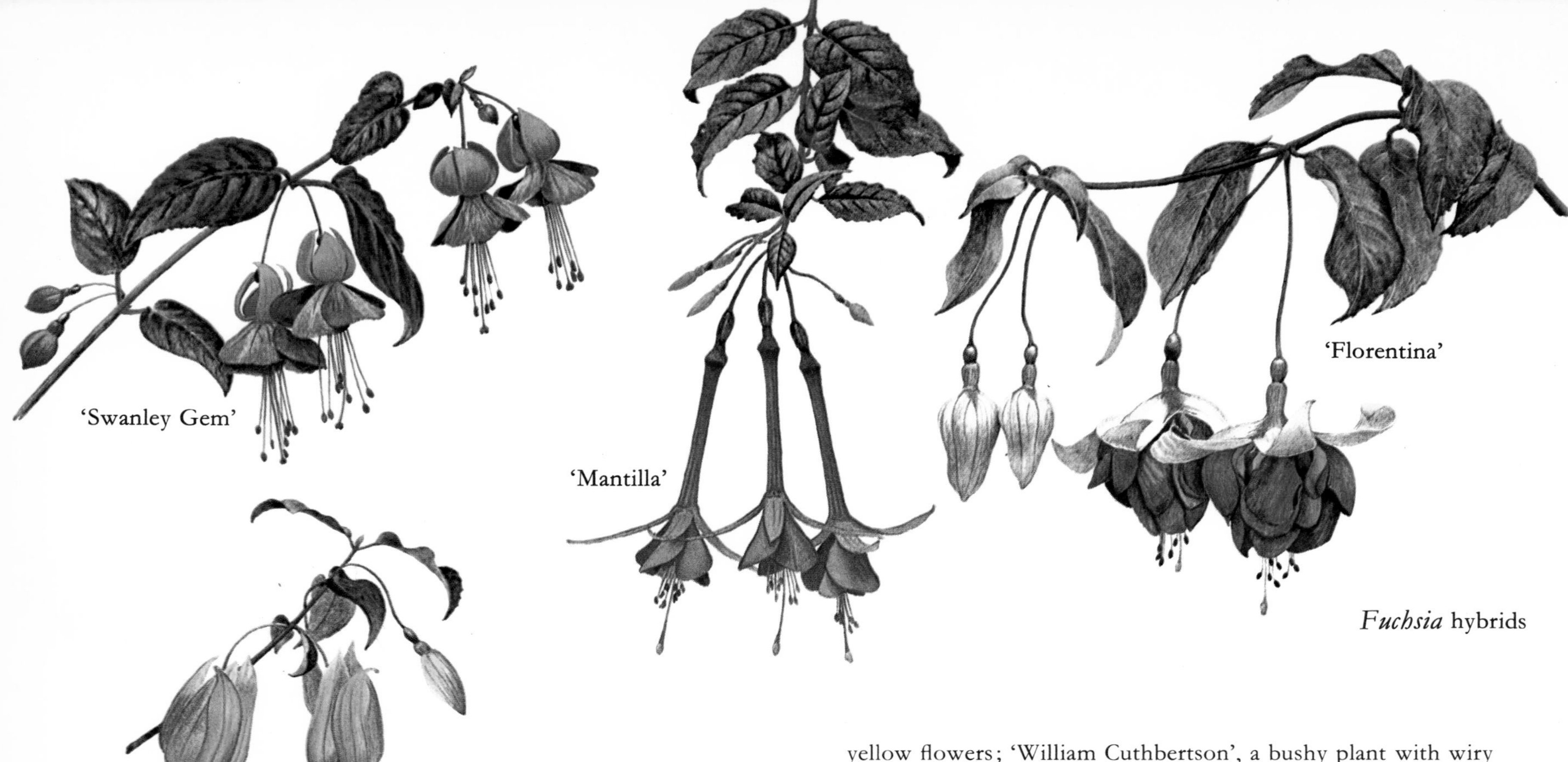

Fuchsia hybrids

Oenotheras may be biennial or monocarpic plants (which perish after flowering), or annuals or perennials. The biennials are frequently nocturnal and very fragrant. These are called Evening Primroses, the day bloomers being known as Sundrops.

The taller species make fine border plants, especially in the wild garden, with the smaller kinds suitable for edging purposes or rock-garden pockets. They all like a well-drained soil in an open position and can be propagated by seed, division or spring cuttings.

Oenothera biennis, the common Evening Primrose, now naturalized in Britain, is really a North American species which reached England from its native Virginia around 1619. It is a biennial of branching habit, growing about 1 m (3 ft) high with large, fragrant, primrose-yellow flowers. These open towards sunset and close an hour or two after sunrise. Other names for the plant are Tree Primrose and Evening Star.

Another night bloomer is *O. odorata* from South America, a plant 45 cms (1½ ft) tall with wavy-edged, lanceolate leaves and fragrant yellow flowers which die off red. *O. speciosa* has flat to basin-shaped, fragrant white flowers with pale greenish-cream centres. As the blooms pass they fade to soft rose. Growing 38–45 cms (1¼–1½ ft) high this is one of the best for the front of the border or rock garden. *O. missouriensis*, Ozark Sundrops, is a charming small species with reddish stems and bright silver-yellow, funnel-shaped flowers. Like *O. speciosa* it is native to the S central USA.

O. fruticosa, Nova Scotia to Florida, is the true Sundrops, and this species and its varieties are among the prettiest border plants in their particular shade of lemon-yellow. Forms include 'Yellow River' with larger yellow flowers; var. *angustifolia* (*O. linearis*), an attractive short-lived perennial from 30–45 cms (1–1½ ft) high with spoon-shaped leaves and clusters of buttercup-yellow flowers; 'William Cuthbertson', a bushy plant with wiry stems and pale yellow blossoms; and 'Youngii', which carries flowers in terminal clusters. All these are day bloomers.

O. caespitosa has large white or pink, scented flowers each about 7 cms (3 ins) across. It comes from North America and forms clusters of long narrow leaves covered with soft downy hairs. It opens in the evening and grows 30–60 cms (1–2 ft) tall.

Ludwigias grow in wet places particularly in tropical America, some actually in water, the stems and leaves floating on its surface. Under these conditions they develop two kinds of roots, one set for anchorage and the others of erect spongy habit with aerating tissue. The latter enable them to float and breathe under aquatic conditions. Several make popular underwater aquarium subjects, living for months completely submerged.

Some species have the leaves in pairs, others alternately; they may be smooth, shining and leathery or matt with a downy surface. The flowers may be petal-less and insignificant or showy with four or five (usually yellow) petals and twice as many stamens.

Those with the boldest flowers are known as Primrose Willows on account of the shape of the leaves and the flower colour. Most of them were previously grouped under *Jussiaea*, but are now often transferred to *Ludwigia*. They include *L. grandiflora*, a

Zauschneria californica

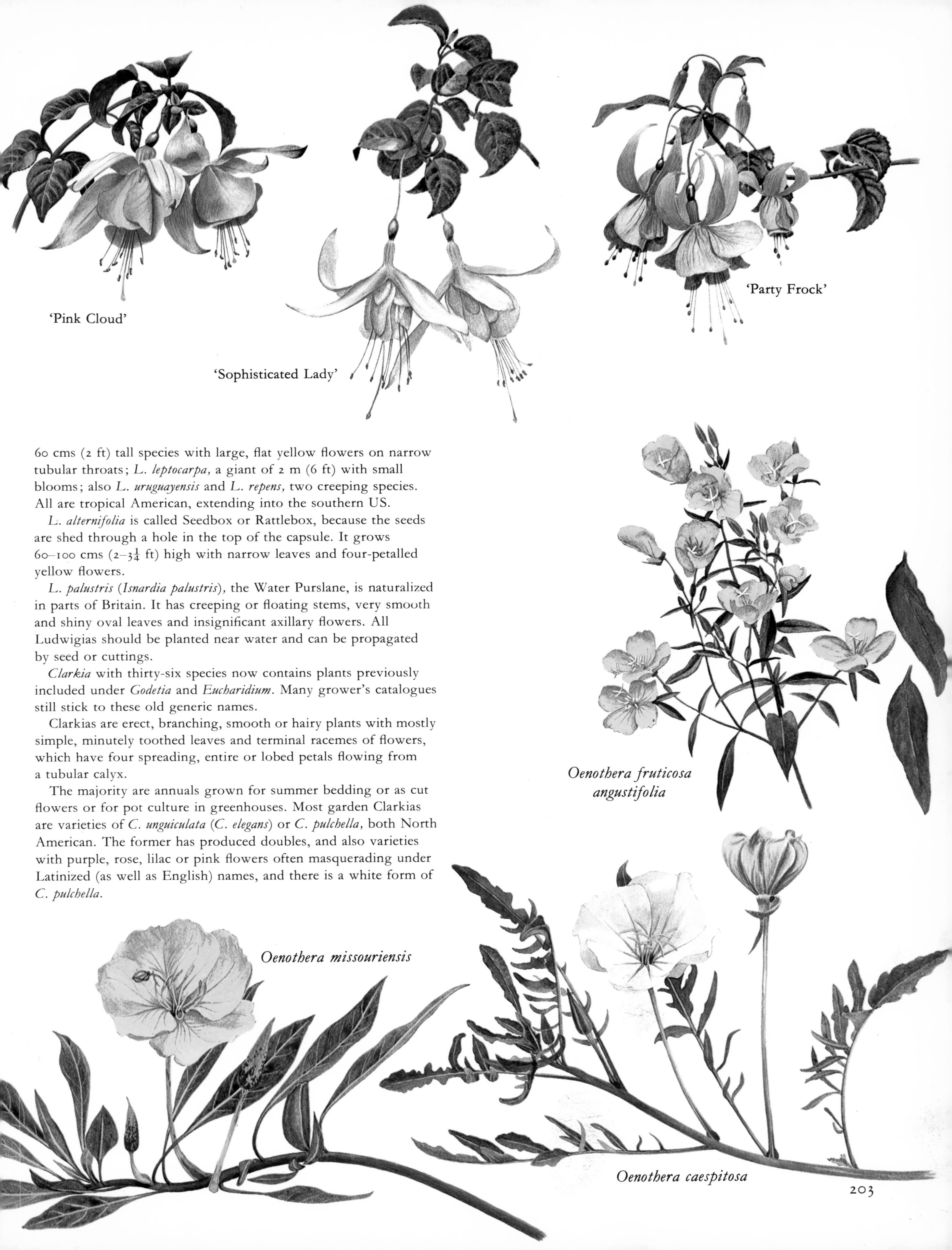

60 cms (2 ft) tall species with large, flat yellow flowers on narrow tubular throats; *L. leptocarpa,* a giant of 2 m (6 ft) with small blooms; also *L. uruguayensis* and *L. repens,* two creeping species. All are tropical American, extending into the southern US.

L. alternifolia is called Seedbox or Rattlebox, because the seeds are shed through a hole in the top of the capsule. It grows 60–100 cms (2–3¼ ft) high with narrow leaves and four-petalled yellow flowers.

L. palustris (*Isnardia palustris*), the Water Purslane, is naturalized in parts of Britain. It has creeping or floating stems, very smooth and shiny oval leaves and insignificant axillary flowers. All Ludwigias should be planted near water and can be propagated by seed or cuttings.

Clarkia with thirty-six species now contains plants previously included under *Godetia* and *Eucharidium*. Many grower's catalogues still stick to these old generic names.

Clarkias are erect, branching, smooth or hairy plants with mostly simple, minutely toothed leaves and terminal racemes of flowers, which have four spreading, entire or lobed petals flowing from a tubular calyx.

The majority are annuals grown for summer bedding or as cut flowers or for pot culture in greenhouses. Most garden Clarkias are varieties of *C. unguiculata* (*C. elegans*) or *C. pulchella,* both North American. The former has produced doubles, and also varieties with purple, rose, lilac or pink flowers often masquerading under Latinized (as well as English) names, and there is a white form of *C. pulchella*.

Clarkia (Godetia) 'Vivid'

Clarkia unguiculata (*C. elegans*)

C. (*Godetia*) *amoena* is called Farewell to Spring and also Summer's Darling, which gives some indication of its flowering time. It is a graceful plant with stems of 60–90 cms (2–3 ft) bearing long, narrow, pointed leaves and loose spikes of large reddish-pink flowers of 5 cms (2 ins) or more. The plant grown in gardens as *Godetia grandiflora* is a form of this species which has produced some outstanding cultivars. The type plant is compact and bushy, about 30 cms (1 ft) tall with long tapering leaves and dense clusters of large flowers. In the cultivars like 'Vivid' depicted here, these measure up to 13 cms (5 ins) across in shades of satin-pink, scarlet, carmine, white, lilac, crimson and pink. Doubles are common and each cluster is set off by several leaves. The various hybrids are generally classified in seed catalogues under Tall Double Varieties, Semi-double or Whitneyi Varieties and Dwarf Varieties.

C. (*Eucharidium*) *concinna* is a Californian species about 30 cms (1 ft) high, popularly known as Pink Ribbons because of its rose-pink, deeply cut, ribbon-like petals.

Gauras are annuals, biennials and perennials from the warmer parts of North America grown for their small but pretty flowers. *Gaura lindheimeri* from Texas is sometimes planted in England in herbaceous borders, although it is not reliably hardy in bad winters. However, since it flowers freely the first year from seed, replacements are easy. It is a bushy plant with lance-shaped leaves and long graceful spikes of rosy-white flowers. It grows to a height of 90–120 cms (3–4 ft) and likes well-drained sandy soil and a sheltered site.

Chamaenerion (*Epilobium*) *angustifolium* is the Willow Herb or Fireweed, with magenta-pink flowers in loose spikes 90–120 cms (3–4 ft) long. These are followed by red-brown seed pods, which open and release the seeds which are clothed in silky fluff. The slightest breeze wafts them far away and the plant then increases rapidly, particularly on land cleaned by burning—hence the name Fireweed. During and after the Second World War, the bomb sites of London became pink with its rosy spikes. Although a pretty plant, this free seeding debars it from all but the wild garden, although a white-flowered form (*album*) is often seedless.

Orchidaceae

740 genera and 18,000 species

This is a tremendous family—the second largest in the Vegetable Kingdom—which, as new areas of the world become systematically botanized, may one day prove to be very large indeed.

The genera and species show great variation—in size, colour, flower shape, mode of growth and habitat. But they have certain common denominators; they are all monocotyledonous and all perennial herbs.

The geographical range is wide. There are Orchids in the semi-desert and in the arctic; at sea level and at 4200 m (14,000 ft) in the high mountains of the Andes. The ground species are found in marshland, in boggy areas, in sandy soil or among lush grass, frequently under the trees in the darkest parts of the jungle and the epiphytes may be perched precariously on rocks or on trees up to 30 m (100 ft) above the ground. They may be terrestrial or epiphytic, saprophytic or autotrophic, but are never parasitic. A complete plant may be no larger than a man's thumbnail but some Orchids have flowers with petals 60 cms (2 ft) long or sprays of fist-sized blooms. The colours of these are as varied as the rainbow; sometimes self-hued but more frequently spotted, outlined or striated with other shades. Although blue, mauve, purple, white, green, brown, cream, yellow, pink and red flowers are common, there are no truly black Orchids.

In spite of legend and widespread fable Orchids *do not* trap insects or people, nor do they have aphrodisiac or medicinal qualities. Indeed, notwithstanding its size, the family has scant economic importance beyond its great ornamental value for cutting and cultivation. Vanilla is derived from the seed pods of *Vanilla planifolia* (*V. fragrans*) and the tubers of various European Orchids such as *Dactylorhiza majalis* are made into a nutritious drink called Salep, but these are exceptions.

Many tropical species are epiphytic and because they rely on irregular rainfall possess special water-storage organs. These are known as pseudo-bulbs and do indeed resemble clusters of bulbs. Others have thick leathery leaves which reduce transpiration losses and there are a few leafless saprophytes. The latter take their food from the decaying organic matter found on forest floors and have rhizomes—sometimes with, sometimes without roots buried in the humus.

Most of the temperate species are of a terrestrial nature and grow out of the ground like other forms of land plants.

Orchid flowers may be solitary as in some species of *Paphiopedilum* or they may be panicled or racemose, frequently in spikes. They are always zygomorphic and have three sepals, two petals, a lip and a column containing the essential organs. However the overall shape varies considerably in different species and also the arrangement of these floral organs so that there is no standard appearance.

Pollination is mostly effected by insects or occasionally hummingbirds, the flowers possessing various means to ensure success. Thus, in *Paphiopedilum* the creatures become daubed with pollen when foraging for nectar and because the shape and make up of the flowers encourages a cake-walk entry from one direction and exit via another they cannot avoid fertilizing the strategically placed stigmas.

With the great majority of Orchids such as *Vanda,* the individual pollen grains are fused into waxy groups called *pollinia.* These are sticky and adhere to the visitors in various ways—

frequently on the foreheads of bees searching for honey. They are thus carried to other flowers. Sometimes, as in various *Orchis* species the dryness of the outside atmosphere causes the *pollinia* to droop, so that by the time the bee reaches another bloom they are brought down to the exact height where they will hit—and fertilize—its stigma.

Another interesting feature of Orchids is the quantity of seed they produce. A scientific count made on a single pod of the Venezuelan *Cycnoches chlorochilon* revealed 3,700,000 seeds. Naturally these are infinitely small and extremely light so that breezes often waft them miles away from the parent plant. Such extreme prodigality is necessary because so many seeds perish before they germinate. Unlike most flowering plants Orchid seeds do not have their own food reserves. As a result there is no measure of independence in the early stages.

Instead every species acquires a symbiotic relationship (partnership) with certain fungi and if the seeds do not immediately make contact with the right fungi to start them into growth they perish.

Frequently such Orchids are epiphytes with pseudo-bulbs and a few tough white roots. The fungus grows on the bark of trees in a fine, almost invisible cobweb mesh and penetrates the thick porous roots of the Orchid like a parasite. However it does not rob its host entirely but works with it, giving the plant organic salts derived from humus collected in the tree bark. The Orchid also has the power to digest the fungus if necessary, thus keeping it in check. This complicated process explains why big plants with scanty roots are able to survive in windswept or unpromising situations and also why epiphytic types are naturally more plentiful on rough bark than smooth.

Various saprophytic Orchids also live in symbiosis with fungi. *Corallorhiza trifida,* the Coral Root, and *Neottia nidus-avis,* the European Bird's Nest Orchid, are both woodland plants which form a mycorrhizal association with certain root fungi found in damp humus. They cannot germinate without these. The seeds produce a tuber-like organ which the fungus penetrates, after which roots and stems develop. The fungus helps feed the Orchid seedling and lives on in the outer cells, although constantly being digested by the inner.

These curious circumstances account for one-time difficulties in germinating certain kinds of Orchid seed. Some would only develop in the parent pod, others proved intractable to all known methods. Nowadays all seed is placed in sterile containers with a preparation of Agar-Agar*, sugar, yeast and various salts. If this is carefully prepared and the seeds properly sown they will germinate as readily as Mustard and Cress. Completely sterile conditions are necessary otherwise harmful moulds and fungi take over. After the containers have been innoculated with the seed on its Agar-Agar jelly mixture base they are kept in the dark in a temperature around 22°C (72°F). Following germination they are given more light (but never brought into full sunlight) and after six or twelve months in the container are taken out and pricked off into pans of prepared compost. Individual potting is carried out when the plants are large enough—but avoid over-potting (giving them too big a container) when growing Orchids.

* A non-nitrogenous substance of a gelatinous nature made from the seaweed *Plocaria lichenoides,* used for the cultivation of bacteria and in the Far East for jellies.

Cattleya
Nellie Roberts

In semi-desert areas epiphytic Orchids sometimes grow on Cacti, and some conception of their remarkable colonizing powers can be gauged from the fact that after the volcanic eruption of 1882—which destroyed all forms of life on the Pacific Island of Krakatoa—Orchids (from wind-blown seeds) were among the first plants to reappear.

In general Orchid compost should be light with an organic base and have free drainage. Chopped osmunda and certain other fern roots, sphagnum moss and charcoal form the basis of many mixtures but perlite and many forms of shredded, filamentous and granulated plastics are often used nowadays and peat, leaf-mould and soot can also be added. It is impossible to generalize for different Orchids have different needs and it is advisable to check on each kind beforehand. Usually they require slightly acid conditions and soft water. Overwatering is the cause of more failures than underwatering—the plants are much more adaptable than is sometimes realized. They have to be to survive in some of their strange habitats.

In Europe the commonest Orchids are various species of *Orchis,* especially *O. mascula*, *O. morio* and *Dactylorhiza maculata* and *D. majalis*. They are all terrestrial with tuberous roots, some of which are used in the making of Salep or for sweetmeats in the Orient. Several have representation in Asia and North Africa.

Often European Orchids possess curiously shaped and marked flowers which bear fanciful resemblance to monkeys, birds, men, wasps, bees, flies, butterflies and other insects and creatures. Such English names as Fly Orchid, Lizard Orchid and Man Orchid reflect these characteristics.

The European (including British) *O. mascula,* the Early Purple Orchis, often has its leaves marked with purple spots which, legend says, were first acquired by plants growing near the foot of Calvary. In Cheshire (England) it is sometimes called Gethsemane. The plant likes to grow in moist pastures, attaining a height of 30 cms (1 ft) with oblong, stem-clasping leaves and erect racemes of purple flowers mottled with light and dark shades. These sometimes have a rather offensive odour, especially towards evening.

O. morio, the Green-winged Meadow Orchis, is similar but with some green in the flowers; *Dactylorhiza incarnata,* the Early Marsh Orchid, is purple or red and *D. maculata,* the Heath Spotted Orchid, usually has spotted leaves and light purple flowers variously marked with dark lines and spots.

Neottia nidus-avis, the Bird's Nest Orchid already referred to, is a leafless plant about 30 cms (1 ft) high with brown flowers and stems covered with numerous, sheathing brown scales. The tuber is surrounded by a confused mass of short thick roots resembling the tangled material of a bird's nest. It is saprophytic and native to Europe (including Britain and Russia). Pollinating insects are attracted by its honey-like odour.

An attractive plant for moist shady spots in a sheltered woodland garden is *Bletilla striata,* often known as *Bletia hyacinthina.* Where the climate presents doubts as to hardiness the plants can be grown in small pots of peaty soil and then plunged into the ground to flower. In autumn they can be lifted and kept in a cool place until the following spring. The species is Chinese and grows about 30 cms (1 ft) tall with grassy, slightly pleated basal leaves and erect stems carrying six to ten brilliant purple flowers. The latter are very variable and occasionally show white, pink or striped blooms.

Pleiones are beautiful little Orchids from E Asia—particularly India, Burma, Formosa and China. They have small, squat or flask-shaped pseudo-bulbs which only persist for one season, deciduous plicate leaves and showy, Cattleya-like flowers, 5–7 cms (2–3 ins) across on short stems. They usually flower before the foliage appears in early spring.

Pleione flowers have gradually widening, slipper-like lips with crested margins and are set off by narrow and separated petals and sepals. The colours vary through light and deep shades of orchid-purple and pink; there are several white cultivars and *P. forrestii* has flowers of a rich yellow.

Pleiones should be grown in shallow pans of fern fibre, sphagnum moss, leafmould and sand—which makes a moist but free-draining compost. They need to be kept cool but not cold and in nature are frequently found growing among moss and humus or in the crevices of moss-clad trees or rocks.

The main species are *P. pricei* and *P. formosana* (now considered forms of the very variable *P. bulbocodioides*), *P. maculata*, *P. hookerana, P. praecox* and *P. humilis,* all of which can be propagated by separation of the pseudo-bulbs.

Several Cypripediums are hardy in temperate gardens. Because of the flower shapes these are commonly known as Moccasin Flowers or Slipper Orchids, names also applied to *Paphiopedilum* which at one time was included in *Cypripedium.*

Many can be grown in the vicinity of the bog garden for they like moist conditions and if left alone form imposing clumps. They bear one or several terminal flowers on sturdy stems, each having free lateral sepals and petals and an erect dorsal sepal with a brightly coloured pouch or labellum. This is usually in a different shade from the sepals and petals. The foliage is plicate, usually deciduous and often hairy.

Cypripedium calceolus, an extremely rare British native, but widely distributed in parts of Europe and N Asia and North America has a thick, short-jointed creeping rhizome with numerous roots. Each plant has three to five stem-clasping leaves (resembling those of Lily of the Valley) and large flowers with bright yellow pouches and brownish-purple sepals and petals. They have a sweet smell rather like oranges.

Cypripedium reginae

Neottia nidus-avis

Pleione bulbocodioides

Cypripedium calceolus

Bletilla striata

C. candidum has waxy-white pouches veined inside with purple and greenish-brown sepals and petals. *C. acaule* has a deeply cleft rosy-purple pouch with brownish-green segments and grows 15–30 cms (6–12 ins) tall, much shorter than *C. reginae* which may reach 30–90 cms (1–3 ft) according to cultural conditions. The latter is a beautiful species with large white-petalled flowers and striking rose-mouthed pouches atop leafy, hairy stems. These three species are North American.

C. calceolus var. *pubescens* from the same area has yellowish-green flowers patterned with purple and large pale yellow pouches. The dried roots are used as a nerve stimulant in the United States of America although the fresh plant should be handled cautiously. Some people are allergic to its sap which can cause intense irritation.

All the fifty odd species of *Paphiopedilum* come from tropical Asia so have to be grown under glass in temperate areas. Some require more warmth than others (according to their place of origin) and for those needing high temperatures 21–24°C (70–75°F) day heat in summer—6°C (10°F) lower at night—and 15–18°C (60–65°F) by day in winter—with 3°C (5°F) less at night—is recommended.

Paphiopedilums are great favourites with Orchid growers because they are good for cutting and do not take up much room and show a wide range of colours and patternings. The leaves may be plain or variegated.

The first species known to European growers was *Paphiopedilum venustum* which was brought to England from NE India by Dr Wallich in 1819. It has bluish-green leaves mottled with grey-green above and dull purple underneath and solitary flowers. These have greenish-white or pink petals and sepals, striped with bright green, the petals fringed and with purple markings and the yellowish-green pouches suffused with purple.

P. Hunters Moon is an example of an artificial hybrid made fairly recently. It is a hybrid between *P.* Hancar and *P.* Banchory and was entered with the International Registration Authority for Orchid hybrids by Mr Faines in 1952. In its ancestry are *P. insigne, P. villosum, P. spiceranum, P. callosum, P. lawrenceanum, P. bellatulum* and *P. druryi*.

Several Paphiopedilums have interesting backgrounds. Veitch of Chelsea, for example, acquired a species in 1878 which took the Orchid world by storm. This had attractive violet, green and ivory flowers and was traced back to a Mrs Spicer whose son—a tea planter in Assam—had forwarded some to England in a consigment of Orchids. Finding it again in the wild presented great problems for it grew in a tiger-infested jungle, but a German collector called Forstermann eventually ran it to earth. Apparently this collector destroyed all the wild plants he could not transport in order to preserve the rarity of the species. It is still grown—under the name *P. spiceranum*.

Another legendary Orchid is *P. fairieanum,* a beautiful plant with white and purple silky-hairy petals and purple-green pouches with brown lines. It first turned up in some London sale rooms around 1855 and was later shown at a Royal Horticultural Society Show in 1857 by a Mrs Fairie of Liverpool. Dr Lindley, the great botanist, named it after Mrs Fairie but for many years collectors tried in vain to find it again in the wild. It became known as the 'Lost Orchid' and although a reward of £1000 was offered for its reintroduction, it was not until 1904 that a party of surveyors rediscovered it again in the mountains of Bhutan.

A century ago fortunes were paid for rare species of Orchids so that collecting became big business. Professional searchers scanned likely places with great secrecy and often at great personal danger in the quest for new species. This is one of the reasons Orchids have become the aristocrats of the plant world.

Angraecum (*Mystacidium*) *se qu pedale* is an epiphytic species native to Madagascar. It has thick fleshy roots and dark green, stem-sheathing, strap-shaped leaves 25–32 cms (10–15 ins) long and 2.5–5 cms (1–2 ins) wide. The very fragrant flowers are spectacular being waxy white, 13–20 cms (5–8 ins) across with similar, pointed sepals and petals and a long tapering spur like a goose quill. The latter may grow to a length of 30–45 cms (1–1½ ft) and greatly intrigued the great naturalist Charles Darwin, who felt that such a flower must have a moth with an unusually long proboscis to pollinate it. This proved to be correct and later a moth of this description was found in Madagascar.

The species is the finest in the genus and lasts months in flower It is tolerant of cultivation in a temperature around 13°C (55°F)

There are also a number of striking Phragmipediums (once included in *Cypripedium*) from South America, all characterized by long ribbon-like petals. These must be grown in tropical gardens or warm greenhouses.

Phragmipedium longifolium var. *hincksianum* is a South American species with many flowered spikes. The blooms have whitish-green sepals with darker green veins, light greenish, brown-spotted petals and a green and brown spotted lip.

Paphiopedilum venustum

Paphiopedilum Hunters Moon

Zygopetalum mackaii is typical of a genus peculiar to South America and Trinidad. It has been in cultivation since 1826, succeeding in a temperature around 13°C (55°F). This robust plant has several large and showy flowers to a stem, each with similar sepals and petals and a three-lobed lip. The leaves are stiff, sometimes with raised veins, occasionally part pleated. *Z. mackaii* is native to Brazil and has five to seven fragrant, greenish-yellow sepals and petals with purple or brown spots and broad, fan-shaped white and violet lips.

For many people Orchids are Cattleyas. The reason is understandable. Their big—almost vulgar—flamboyant flowers cannot be overlooked, particularly in their richest shades of cyclamen-purple. They are widely grown for the florist trade so the flowers are often seen and cannot be likened to any other plant.

Cattleya takes its name from one William Cattley, a keen North London amateur of the early 19th century who took particular interest in the genus. When he died in 1832 his collection went to Joseph Knight and Thomas Perry, famous nurserymen of their day, who were later succeeded by James Veitch & Son. The latter became great orchid specialists and they sent collectors to many parts in search of new species. Today their successors are Messrs Black & Flory of Langley, Buckinghamshire.

There are approximately sixty species of *Cattleya,* all epiphytes from South and Central America and the West Indies. They come from a wide range of habitats and heights in the jungle, being mostly found on forest trees. Often such places are extremely humid and this led to errors in early cultivation attempts. Many plants died in the heavy atmosphere of overheated, overwatered greenhouses but nowadays they are grown under much drier conditions with greater success.

For cultivation under glass Peter Black recommends summer day temperatures of 21–27°C (70–80°F) (which usually means keeping the ventilators closed) and night temperatures 6°C

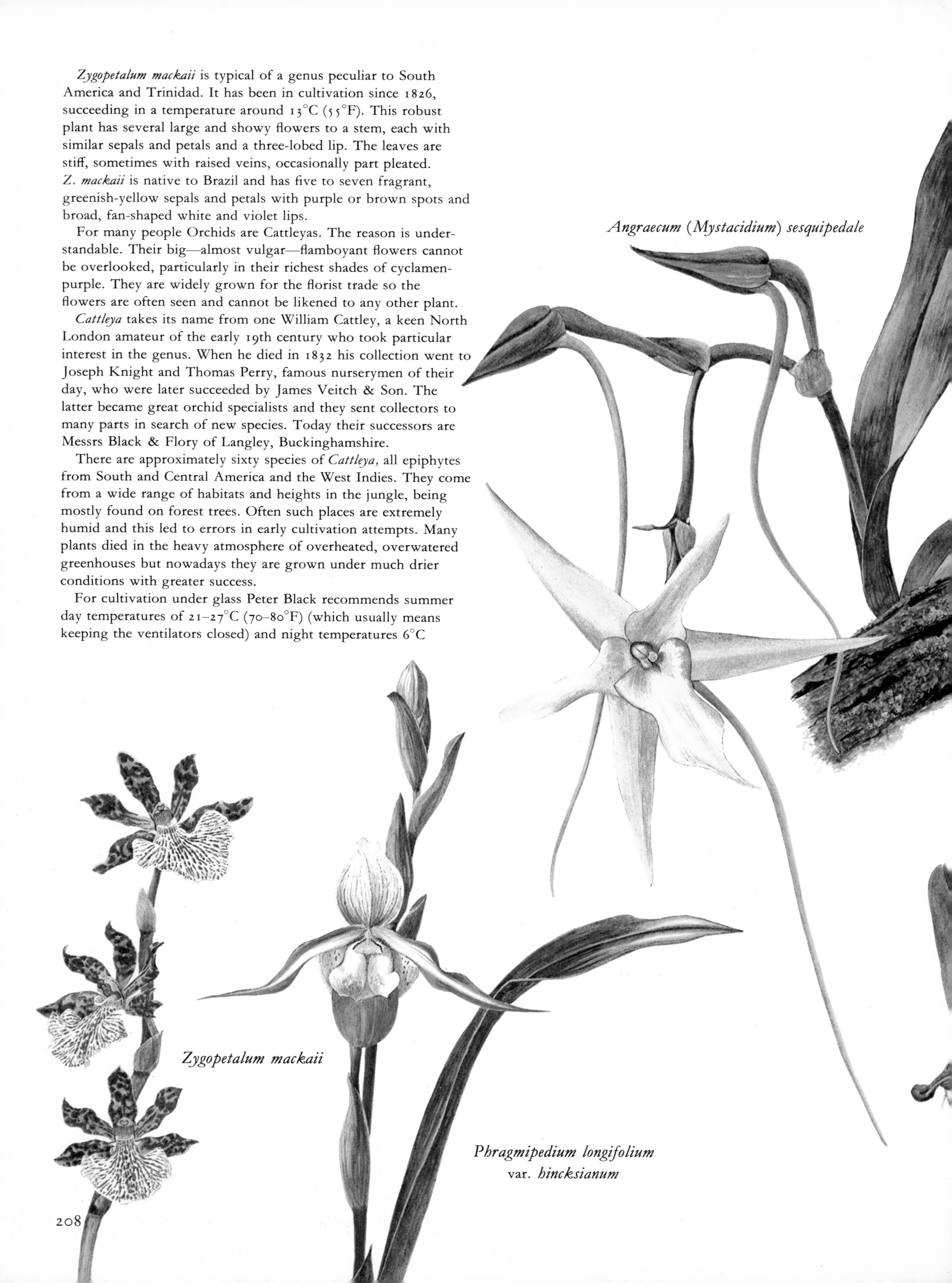

Angraecum (*Mystacidium*) *sesquipedale*

Zygopetalum mackaii

Phragmipedium longifolium var. *hincksianum*

(10°F) lower. In autumn they need more air to mature the lush growth and by winter will tolerate 13°C (55°F) at night and up to 18°C (65°F) during the day. Light is important, especially in winter and one must learn to sense or smell out the correct atmospheric conditions. In a well-run greenhouse one can detect a sense of well-being by a distinctive odour generated by growing plants. It is a vegetation smell composed of a mixture of moisture, earth and plants.

Cattleyas are mostly grown in composts based on osmunda fibre and sphagnum moss. There are hybrids in a wide range of shades—Nellie Roberts depicted at the beginning of this section being one such hybrid (*C.* Bow Bells x *C. dowiana*). The species portrayed on page 209 is *Cattleya bowringiana,* a sturdy plant from Central America with five to twenty light and deep rosy-purple flowers individually about 6 cms (2½ ins) across.

Cirrhopetalum mastersianum is an epiphyte from Java with pseudo-bulbs and umbels of deep yellow and brownish-purple flowers. The leaves are strap-shaped, 10–13 cms (4–5 ins) long.

Masdevallia fulvescens is one of a large group of epiphytic Orchids with brilliantly coloured and spotted flowers. These usually terminate in tails of various lengths—most frequently yellow. This Colombian species has yellow and light rosy-purple appendages of 5 cms (2 ins), the main part of the flower being white, ochre-yellow and brown.

Phalaenopsis schillerana from Manila in the Philippines is a beautiful plant with flowers of 5–7 cms (2–3 ins) arranged in double rows on a long spike. In a good specimen there may be up to 100 blooms on a single stem 90 cms (3 ft) tall. Individually these have soft rose sepals of varying shades and a three-lobed lip of white or light rose with yellow protuberances at the base. White and deeper coloured forms are in cultivation. *Phalaenopsis* are often grown in suspended baskets or pans with holes round the sides, filled with osmunda fibre and sphagnum moss.

Phalaenopsis schillerana

Cattleya bowringiana

irrhopetalum mastersianum

Masdevallia fulvescens

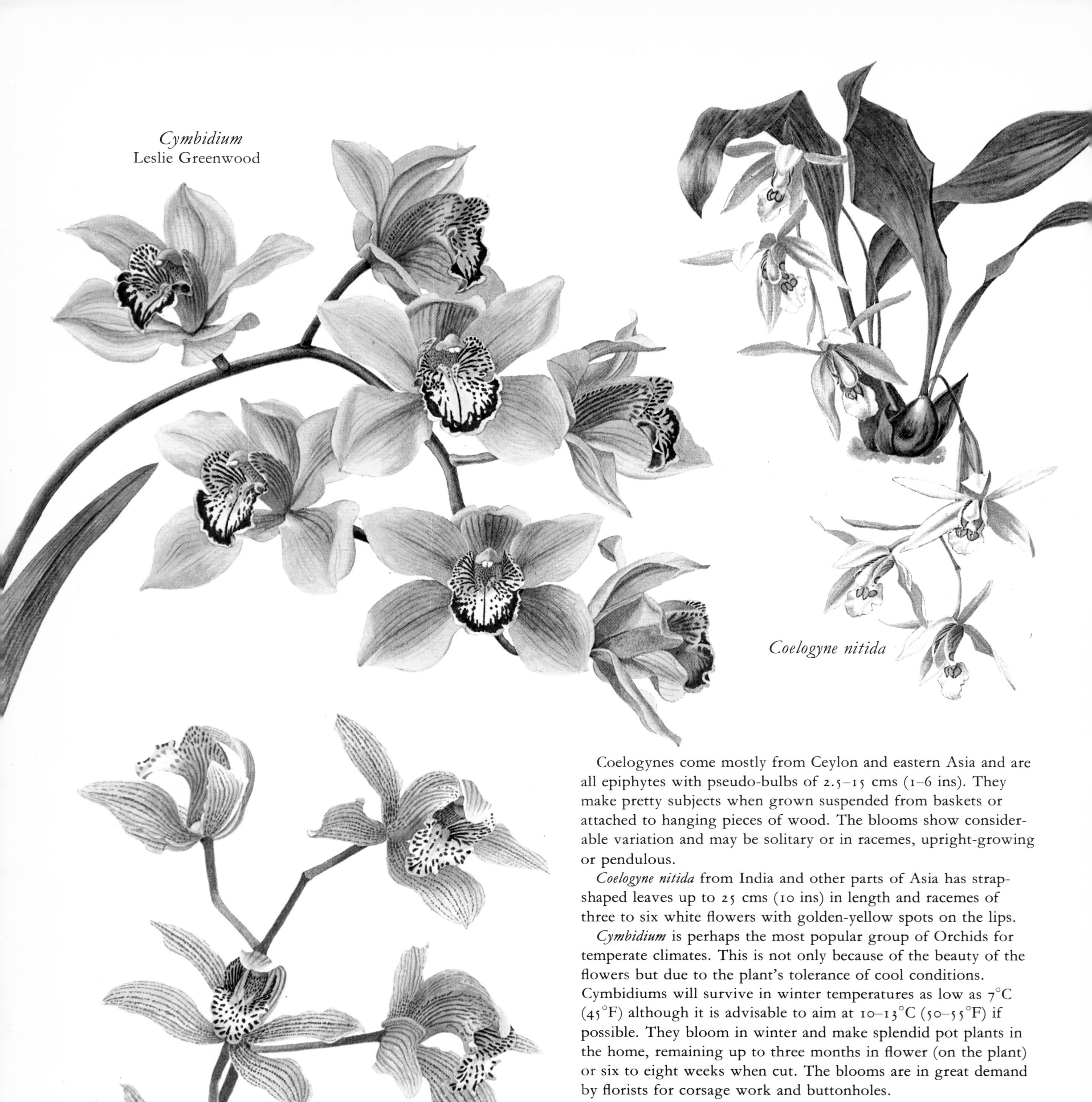

Cymbidium
Leslie Greenwood

Coelogyne nitida

Cymbidium
Queen of Gatton

Coelogynes come mostly from Ceylon and eastern Asia and are all epiphytes with pseudo-bulbs of 2.5–15 cms (1–6 ins). They make pretty subjects when grown suspended from baskets or attached to hanging pieces of wood. The blooms show considerable variation and may be solitary or in racemes, upright-growing or pendulous.

Coelogyne nitida from India and other parts of Asia has strap-shaped leaves up to 25 cms (10 ins) in length and racemes of three to six white flowers with golden-yellow spots on the lips.

Cymbidium is perhaps the most popular group of Orchids for temperate climates. This is not only because of the beauty of the flowers but due to the plant's tolerance of cool conditions. Cymbidiums will survive in winter temperatures as low as 7°C (45°F) although it is advisable to aim at 10–13°C (50–55°F) if possible. They bloom in winter and make splendid pot plants in the home, remaining up to three months in flower (on the plant) or six to eight weeks when cut. The blooms are in great demand by florists for corsage work and buttonholes.

The arching sprays—of which there may be twenty on a large well-grown plant each averaging fifteen to twenty-five blooms—are beautifully poised and exquisitely marked and coloured. They are complemented by narrow, green, strap-shaped leaves and have large pseudo-bulbs.

The forty or more species are native to Asia and Australia, frequently being found at high altitudes. This indicates their liking for airy conditions. Most gardeners grow cultivars (of which there are hundreds) rather than species because the flowers are larger, have a wider colour range and remain longer in character. Many are fragrant and there are miniature species and cultivars which are becoming increasingly popular as house

Odontoglossum
Amabile

plants. The colours range through white, cream, yellow, green, greenish-yellow, brown, pink and red usually with bold lip markings of other shades. *Cymbidium* Leslie Greenwood (named for the artist) and Queen of Gatton here illustrated demonstrate these attractive colourings.

Cymbidiums can be grown in a compost of sphagnum moss and osmunda fibre or perlite and peat are sometimes used. They should be given soft water to keep the pH slightly acid (hard water builds up the lime content). The plants can be left for several years without repotting and they like plenty of light but not strong sunshine as this turns the leaves yellow—especially under glass. In many places it is possible to stand the containers outside during the summer months. Fertilizers may be used in warm climates such as Australia, S California, South and North Africa but are not recommended in cooler situations.

Odontoglossums come from Central and tropical South America and comprise a large genus of some 200 species (mostly epiphytic) and countless natural hybrids. They have beautiful flowers with a wide range of colours and patterns and exhibit considerable variation in size and shape. These are long lasting—particularly the hybrids—and are often crested. Cultivation procedure may differ slightly between some of the species but general principles include maintaining a buoyant atmosphere and giving the plants plenty of light in winter, while protecting them against strong sunshine during the spring, summer and autumn months.

Watering should be geared to keeping the compost just moist. Although temperatures may drop to 10°C (50° F) in winter, 16–18° C (60–65°F) in summer is high enough, particularly in a greenhouse. The plant illustrated, *Odontoglossum* Amabile, is an artificial hybrid between a white, red and yellow species from Colombia and the hybrid *O.* x *crispoharryanum* (*O. spectabile*) itself a hybrid between O. *crispum* and O. *harryanum*.

Glossodia major is a delightful little Australian Ground Orchid with an underground tuber which is replaced each year. It has only one leaf, which is basal and linear in shape with a hairy surface and about 5 cms (2 ins) in length. The flowers are bluish-purple and smell like vanilla when crushed. The species can be grown in well-drained compost as a pot plant.

Pterostylis with about 100 species comes from Australia, New Zealand, New Guinea and New Caledonia. These small terrestrial Orchids have underground tubers from which spring soft and thin leaves (usually in a radical rosette) and slender-stemmed green and white to reddish-brown and white hooded flowers. In some instances the sepals extend to long tails. Such characteristics account for a variety of native names like Greenhoods, Shell Orchids, Parrot-beak or Bird Orchids. *P. curta* (an Australian species) is one of the Greenhoods and has green flowers marked with white and brown.

Except for one Javanese representative all thirty-seven species of *Diuris* are endemic to Australia where, on account of the twin, narrow, deflexed, lateral sepals possessed by many, they are known as Donkey Orchids or Double Tails. They are ground plants with several grassy leaves, an underground tuber (sometimes two) and yellow, orange, white or purple flowers frequently marked with red, brown or purple. Although not commonly cultivated they make good pot plants in a sandy, loamy compost.

D. pedunculata is probably the commonest species in the genus and unlike most of the other species is in no danger of extinction by increasing urbanization and changing agricultural methods. Its soil preferences are catholic, plants having been reported growing in all types of soil from moist clay to sandy loams. It can be cultivated quite easily and a pan of the species is a very pleasing sight. A compost of peat and sand with a little loam in a well-drained pan is necessary and it is also essential to reduce watering after the plants have flowered and withhold water completely for a six months' rest period.

Glossodia major

Diuris pedunculata

Pterostylis curta

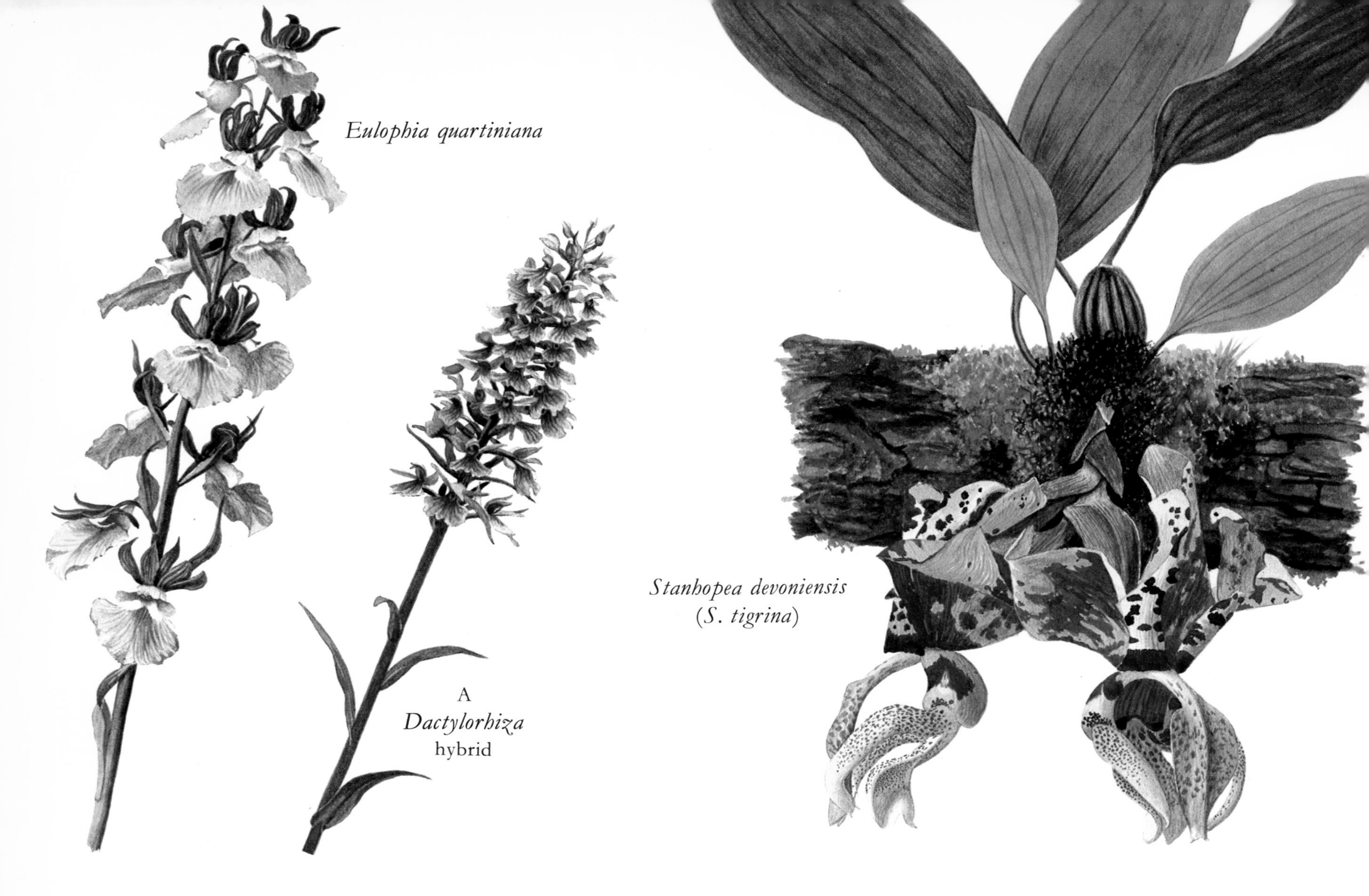

Eulophia quartiniana

A *Dactylorhiza* hybrid

Stanhopea devoniensis (*S. tigrina*)

Lycaste cruenta comes from Guatemala and has tough pseudo-bulbs, strap-shaped leaves 38–45 cms ($1\frac{1}{4}$–$1\frac{1}{2}$ ft) long and deep orange-yellow, long-lasting flowers with greenish-yellow sepals. The lips have blood-red spots at their bases. *L. aromatica* is somewhat similar and richly fragrant.

Stanhopea devoniensis (*S. tigrina*) is a Mexican Orchid of great beauty with large flowers up to 20 cms (8 ins) across of deep orange-yellow or orange-brown, tiger-flecked with purplish-brown markings. They are sweetly scented and hang in pendulous fashion from green pseudo-bulbs. The broad oval-oblong leaves are plicate and rather leathery. This epiphytic species should be rested (that is, given very little water) in winter and can be grown in a hanging basket of fibrous compost. Winter temperatures should not fall below 13°C (55°F). The Venezuelan and Brazilian *S. grandiflora* has fragrant white flowers marked with purple.

Polystachya pubescens comes from South Africa and is a fragrant epiphytic Orchid with terminal racemes carrying several small, inverted flowers. These are bright yellow streaked with red, bearded on the inside with long hairs.

The distinctive flavouring of Vanilla—used in confectionery and perfumes, also some tobaccos—is obtained from the fermented (cured) seed pods of a tall tropical American climbing Orchid called *Vanilla planifolia* (*V. fragrans*).

It first attracted the attention of the western world when in 1520 Bernal Díaz, one of Hernán Cortés' Spanish conquistadores observed the Emperor Montezuma drinking chocolate in Mexico. This beverage was flavoured with ground Vanilla pods.

Appreciating the flavour of this new drink (which had been known for centuries by the Aztecs) the Spanish took *Vanilla* back to Europe. Soon factories were established to manufacture Vanilla-flavoured chocolate and for over three centuries Mexico was the sole supplier of the raw material. Attempts to grow *Vanilla* in Asia and other places always failed because of pollination difficulties—the plants simply would not set fruit.

Habenaria rhodocheila

Lycaste cruenta

Polystachya pubescens

The reason for this was established when it was realized that the flowers in Mexico were fertilized by certain species of bees and hummingbirds unknown in other parts of the world. In 1836 a Belgian botanist called Charles Morren came up with a solution—hand pollination. This is still practised today, especially in Madagascar and the Seychelles which produce most of the world's natural supply (it is also made synthetically).

Although the pale yellow blooms only remain in character for one day the flowering period extends for about two months so that each vine may carry up to 1000 blooms. The plants are reproduced by means of root cuttings.

The species can be grown under glass trained to wires along the roof. Another popular kind is *V. imperialis* from Uganda and West Africa. This has fleshy leaves and spikes of yellow flowers which have yellow-veined, dark purple lips.

Dendrobium nobile from India is one of a large family of epiphytic Orchids from the Old World and Australasia. It has large, fragrant flowers with pink-tipped, white sepals and petals and velvety-crimson blotched lips. This is perhaps the best-known, easiest to grow, and most variable species in commerce and has many varieties. It is illustrated on the title page.

Anoectochilus, as used in gardens, is a name applied to plants from several genera. These are noteworthy for their beautiful leaves which are frequently netted and patterned with contrasting shades. Although the flowers are freely produced they are often insignificant and growers usually nip them out in order to increase the vigour of the foliage. The plants should be grown in a position away from direct sunlight, either on a moss or fibrous base.

Macodes sanderana (*Anoectochilus sanderanus*) is a terrestrial New Guinea plant with olive or brown-green oval leaves up to 10 cms (4 ins) long. These are longitudinally veined in lighter green and have squarish areoles of copper-red and golden lines. The margins are white. *Macodes* are known as Jewel Orchids.

Eulophia quartiniana from tropical Africa is generally reckoned one of the most beautiful of this large genus of approximately 200 species. It has two or three plicate leaves springing from short pseudo-bulbs and ten to twenty flowers on long spikes. These have spreading, greenish-pink sepals and petals and large bright pink labellums with darker veins and white spurs.

Miltonias are tropical Mexican epiphytes with large, flat, Pansy-like flowers of velvety texture. Two Colombian species, *Miltonia vexillaria* and *M. roezlii* have, as the result of careful hybridization, produced some remarkable hybrids. These range in colour from white and pink to deepest crimson as in Red Knight 'Grail'. The strap-shaped leaves are flat and often glaucous.

M. roezlii with white and purple-red flowers is named after Benedict Roezl, a Czech gardener's son, who collected in Mexico, California and Colombia. Colombia is rich in Orchids and Roezl sent back tens of thousands of plants although many were lost. In California Roezl found *Lilium humboldtii* and named it for A. von Humboldt because it was found on September 14th, 1869, the hundredth anniversary of the great naturalist's birth.

Vanilla imperialis

Habenaria rhodocheila is an eastern Asiatic Ground Orchid 15–38 cms (6–15 ins) high with linear leaves and racemes of ten to twelve green sepalled flowers (united into a hood) with bright orange or red lips. The plants have small tubers which should be covered with 1.3 cm ($\frac{1}{2}$ in.) of compost. The growing plants must be protected from strong sunlight and rested in winter in a temperature of 7–10°C (45–50°F).

The genus *Dactylorhiza,* numbering about thirty species, is widespread throughout Europe and just extends into the surrounding countries of Asia and Africa. Several species can grow in close juxtaposition and because of this hybrids are very frequently encountered. The plant shown here is one such plant and as with so many of these natural hybrids its parentage is unresolved. However it is certain that *D. fuchsii* is one of its parents.

Laelia gouldiana is a beautiful Mexican epiphyte with two or three leaves, conical pseudo-bulbs and erect spikes of 30–45 cms (1–1$\frac{1}{2}$ ft) carrying between three and eight flowers. These have light purple petals which are darker at the tips and three-lobed, deep crimson lips. *L. anceps,* another Mexican, is the parent of many garden varieties in a wide range of colours. Laelias cross readily with Cattleyas producing a race of hybrids known as *Laeliocattleya*.

Cattleya also crosses with *Brassavola* to form a group known as *Brassocattleya*. They need similar cultural treatment to *Cattleya* and show a wide range of colours, and they can cross with

Jewel Orchids

Macodes sanderana   *Anoectochilus* species

Miltonia
Red Knight 'Grail'

Calanthe vestita
'Baron Schroeder'

Brassolaeliocattleya
June Moore

Laelia gouldiana

Laelias to form Brassolaeliocattleyas. June Moore has a white flower with a golden throat.

Calanthe is a widely distributed genus of about 120 tropical species, some evergreen others deciduous. All have pseudo-bulbs —small and rounded in the case of the evergreens and more conical in the deciduous kinds. For ornamental purposes hybrids are usually cultivated in preference to species as these are superior in colouring and larger in bloom. The flowers are carried on tall graceful spikes; the one illustrated although technically called by its correct specific name *C. vestita* is actually a hybrid between two forms of the species. All the following 'hybrid' names correctly belong to *C. vestita*: 'Baron Schroeder', 'Excellens', 'William Murray', 'Darbleyana' and 'Bryan', and show a wide range of colours. The plants need good drainage and a winter temperature of 13–16°C (55–60°F).

Few oriental Orchids have captured the imagination more than Vandas, particularly *Vanda coerulea,* the Blue Orchid. This was first seen by a Dr William Griffith in 1837 in the Khasia Hills of Assam and sent to England. Unfortunately it was soon lost and not reintroduced until 1849. *V. coerulea* has up to twenty flowers, 13 cms (5 ins) across, of deep blue with paler blue sepals and petals on spikes of 60–90 cms (2–3 ft).

Hooker's description of the commercial possibilities attached to collecting Blue Orchids attracted the attention of Calcutta businessmen and caused widespread despoiling of natural habitats. Eventually the governments of India and Burma had to step in and prohibit the export of Orchids.

Vandas come in other colours and are usually fragrant, the flowers being poised like butterflies on the slender stems. Those illustrated here are Emma van Deventer, a hybrid between *V. teres* and *V. tricolor* made in 1926, a fine white with purple spots and striations and a brown and yellow labellum; Tan Chay Yan (*V. dearei* x Josephine van Brero) registered in 1952 and having a multicoloured flower of ginger-brown, mauve, yellow and purple; and Pride o' Lanka (*V.* Burgeffii x Miss Joaquim) registered in 1948, orchid-purple with a brown labellum.

The nomenclature of hybrid Orchids is a complicated and confusing affair to the uninitiated. In most particulars it is controlled by the International Code of Botanical Nomenclature

(1966 Edition) and the International Code of Nomenclature for Cultivated Plants (1969 Edition). Many intergeneric hybrid Orchids have been raised but neither of the Codes deals with the collective (grex) names and epithets thoroughly.

A Handbook on Orchid Nomenclature and Registration was published in 1969 by the International Orchid Commission in collaboration with the Royal Horticultural Society, who act as International Registration Authority for Orchids. This sets out the procedure for naming and registering new Orchid hybrids.

The formation of names for genera, species and intergeneric— and interspecific natural hybrids is governed by the Codes cited previously.

Thus the name x *Laeliocattleya leeana* has been given to an intergeneric natural hybrid between *Laelia pumila* and *Cattleya loddigesii*, both names being Latinized, with the sign x in front of the formula to indicate hybridity between two genera.

If natural hybrids occur between species of the same genus the name of the genus is followed by the sign x and a collective (or grex) epithet in Latinized form. *Cymbidium* x *ballianum* is a natural hybrid between *C. eburneum* and *C. mastersii*.

When such hybrids are artificial in origin the collective (or grex) epithet is a word or phrase in a modern language, a 'fancy' name, although the name of the genus is still in Latin. Formerly such collective epithets were also in Latin form but since January 1, 1959, new names for collective epithets must be in a modern language. As examples, x *Vandaenopsis* Hawaii is an artificial cross between *Vanda amesiana* and *Phalaenopsis sanderana*, whilst *Dendrobium* Venus is an artificial hybrid of *D. falconeri* and *D. nobile*; and *Miltonia* Red Knight is a cross between *Miltonia* Piccadilly and *Miltonia* Mrs J. B. Crum. These collective or grex names cover all the individuals of a particular cross and are written with an initial capital letter but without single quotation marks. Individuals within such a grex may vary considerably and often one which is outstanding may be selected, propagated vegetatively and given a cultivar (or clonal) name. Such a name is also given an initial capital letter but is distinguished by being placed in single quotation marks. It may be used in association with the generic name alone—Miltonia 'Grail'; or more often to avoid possible confusion, with both the generic name and the grex epithet, thus *Miltonia* Red Knight 'Grail' where *Miltonia* is the generic name, Red Knight the grex epithet and 'Grail' the cultivar or clonal epithet.

Most of these rules also apply to hybrids generally, but, although collective or grex names are normal in Orchid nomenclature, in the majority of other genera they are seldom used.

Vanda
Emma van Deventer

Vanda
Tan Chay Yan

Vanda
Pride o' Lanka

Oxalidaceae

3 genera and 875 species

This is a family of dicotyledonous plants mostly of tropical and subtropical distribution, with their richest representation in the Southern Hemisphere.

The simple flowers are regular and bear some resemblance to those of Flax (*Linum*) and like them have five sepals, five petals, ten stamens and a superior, five-locular ovary. Although these flowers produce nectar and are regularly visited by bees and other insects, some of the spring-flowering species, such as the European Wood Sorrel (*Oxalis acetosella*), are rarely cross-fertilized: seeds develop from an entirely different set of flowers throughout the summer. These *never open*, but pollen grains germinate inside the bloom and travel down the styles to fertilize the ovules. This method of pollination (which also occurs in Violaceae) is known as cleistogamy.

The seeds have a fleshy aril and when ripe any small disturbance causes the five-edged capsules to invert and shoot out the seeds. This is done instantaneously. This propensity, together with the habit of some to multiply and spread by means of bulbils, which form underground during the growing season, makes certain species noxious weeds. It is important to keep these out of gardens—some of the most troublesome are *Oxalis pes-caprae* (*O. cernua*), *O. latifolia* and *O. corymbosa*—and either concentrate on species and varieties of compact habit or grow kinds likely to be killed off in winter.

Several *Oxalis* (the most important genus) are pretty, long-flowering perennials suitable for rock pockets, pot culture, window boxes and similar containers.

Among the rock-garden species are *O. adenophylla* from Chile and Argentina with grey-green, crinkled leaves and soft lilac flowers on stems 5–8 cms (2–3 ins) high; *O. enneaphylla,* a plant 5 cms (2 ins) high from the Falkland Islands suitable for partial shade, and spring blooming with white, pale or deep pink, wide-open florets; and the autumn-flowering, golden-yellow South American *O. lobata,* which is the same height and also needs partial shade. *O. magellanica* from Bolivia is white flowered and a mere ground carpeter.

The Brazilian species include *O. rubra* (commonly but incorrectly known as *O. floribunda*) which flowers from spring until frost cuts it back if planted in full sun. It is very gay with plenty of deep pink flowers on stems 15–22 cms (6–9 ins) tall. *O. rosea* from Chile is also attractive, sometimes 30 cms (1 ft) high with masses of pink flowers which have white throats.

O. tetraphylla with violet and *O. deppei* with rose-red flowers are Mexican; both have four-parted leaves which are sometimes sold as lucky charms. The tubers of *O. deppei* are edible and were once cultivated for food in parts of Europe, especially France and Belgium. *O. pes-caprae* from South Africa is also edible and the tubers of *O. tuberosa* (*O. crenata*), a Colombian species, have been consumed as a vegetable under the name of Oca since early times in Peru.

O. hedysaroides from Venezuela, Colombia and Ecuador is unusual in the genus in being a shrub or sub-shrub. It reaches 60–90 cms (2–3 ft) in height and is densely clothed with red-purple leaves, each consisting of three ovate stalked leaflets. The bright yellow bunches of typical *Oxalis* flowers contrast well with the deep colour of the foliage. The cultivated plant is probably a variant of the species which in the wild is recorded as having green leaves.

Ornamental greenhouse species grown in Europe and the US include *O. ortgiesii* from Peru with yellow flowers and large reddish-green leaves which have wedge-shaped inverted cuts, and *O. chrysantha,* another yellow species from Brazil.

Biophytum sensitivum is a tropical plant with sensitive pinnate leaves which fold together at night or when touched (like *Mimosa sensitiva*). It has yellow flowers on stems of 15 cms (6 ins).

Paeoniaceae

1 genus and 33 species

This family contains a single genus of dicotyledonous perennials, previously included under Ranunculaceae. There are about thirty-three species, all N temperate, with rhizomatous or tuberous roots; the majority are herbaceous plants but several are of a shrubby nature.

The deeply cut, often biternate leaves are smooth and alternately arranged on the stems; in some cases they assume pleasing tints before dying down at the end of the year. The flowers are large and regular, often very showy with five rounded sepals, five large

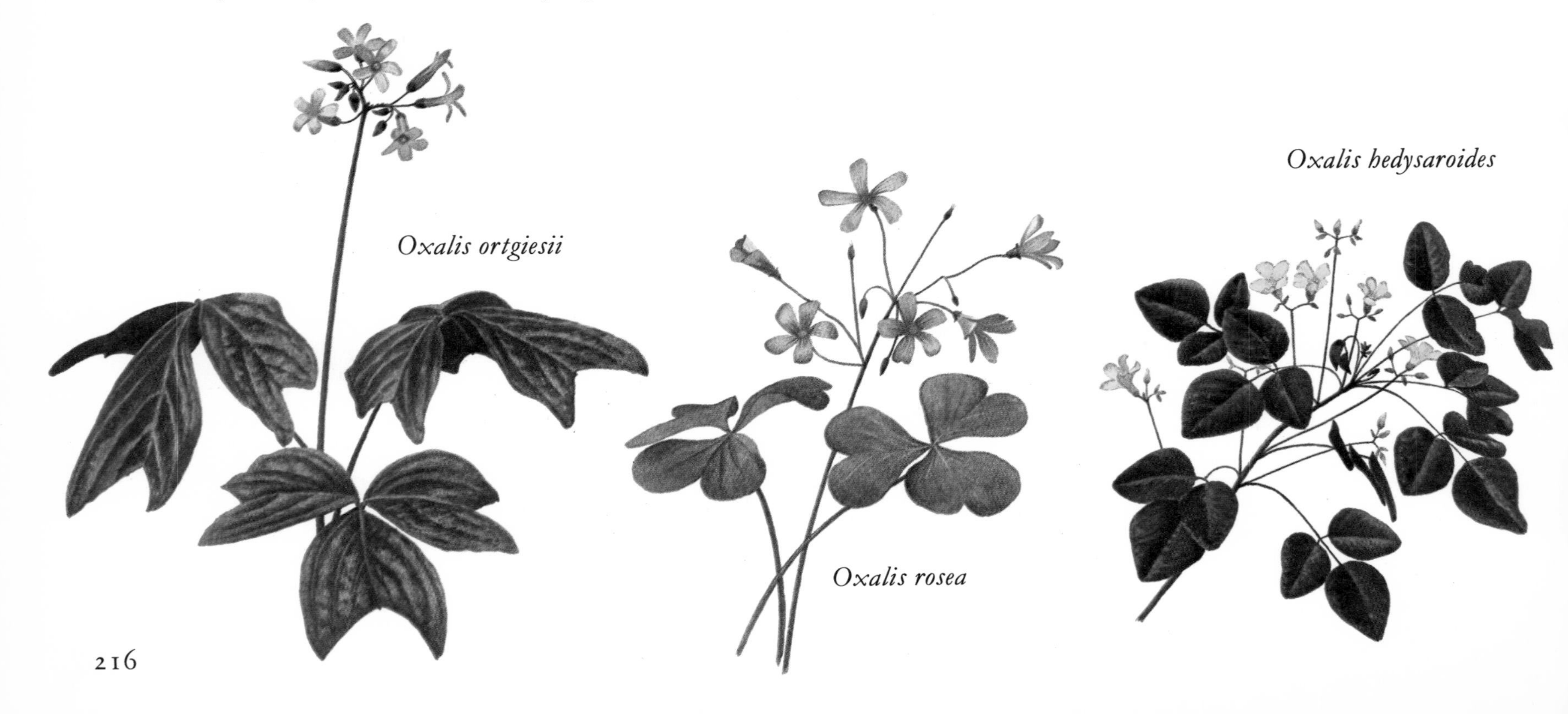

Oxalis ortgiesii

Oxalis rosea

Oxalis hedysaroides

incurving petals and a central boss supporting the stigma surrounded by many stamens. Frequently the blossoms are rich in nectar which attracts bees and various flying insects. Some species are beetle pollinated.

The ovary is superior with a fleshy disc, the fruits consisting of large leathery seed pods containing big seeds with fleshy albumen. The infertile seeds are red and the ripe ones shiny black, becoming extremely ornamental when both immature and ripe seeds are revealed in the same pod.

For many garden purposes Paeonies possess considerable attributes: the blooms appear early and are substantial in size and texture, the colours are rich and in a wide range of shades, and the flowers last well in water. Their perennial nature makes them ideal subjects for permanent planting and since the foliage is attractive and naturally neat, there are no problems of dowdiness after flowering.

Paeonies are also accommodating in the garden and will grow freely in most soils, either in full sun or partial shade. A position which catches the early morning sun is not always advisable, as late frosts can blacken the buds if these thaw out too quickly. Possible frost damage to expanding buds can be counteracted by covering them with wet newspaper for a few hours, so that they thaw out slowly or they can be planted at the fringe of a shrubbery or beneath tall trees (provided there is a cool, rich root run), so that the sun does not reach them until mid-morning or afternoon. The warmer the climate the more important shade becomes. Paeonies like plenty of moisture during the flowering period, otherwise the blooms pass their best very quickly.

The herbaceous Paeonies need to be grown in deep fertile soil, for the thick fleshy roots go well down and their permanent nature demands good nourishment. For this reason an occasional mulch of rotted stable dung, compost or old leaf soil enriched with bonemeal will prove beneficial. Moisture should be supplied during hot, dry spells. Given such treatment the plants respond year after year with masses of tall flower stems topped by large plate-like blooms. Clumps of Paeonies fifty to sixty years old are not uncommon in Britain and once established they should be left alone as much as possible.

Propagation of named cultivars is carried out in early autumn or spring, preferably from young plants. Breaking up an old specimen sometimes delays flower production for several years. Occasionally this may be due to overdeep replanting—the crown buds should be left near the surface, not buried—but often old plants are very woody and do not readily produce new flowering shoots when the roots are disturbed.

The common Paeony of gardens is *Paeonia officinalis,* a native of southern Europe from France to Albania. It is a sturdy plant with red flowers on stems 90–150 cms (3–5 ft) tall and has three double forms, all well known in cultivation. The commonest of these, a rich blood-crimson, is among the oldest of garden plants. It was introduced to Britain before 1548 and seems to have been much cultivated in Tudor times not only for the flowers but on account of its roots. A small piece of the latter, worn round the neck as an amulet, was thought to protect the wearer against evil spirits; bead necklaces, made from dried Paeony root strung on leather, were also worn by infants to prevent convulsions and aid teething.

Of the other varieties, one is dark rose which gradually pales with age and the other, a poor white, which opens flesh-pink and fades to dingy cream. It is sometimes known as 'Alba Plena'. Although many Paeonies have a delightful fragrance, these three have a strong unpleasant odour which, according to the Victorian language of flowers, signifies 'Bashful Shame'—apparently the flower is supposed to blush for its noxious scent!

P. officinalis is mentioned by Theophrastus, a friend and pupil of Plato and Aristotle, in his *Enquiry into Plants* (370 B.C.). He calls it 'the paeony which some call glykyside' and advised that the roots (reputed to cure wounds) should be dug at dead of night for if the operation were viewed by a woodpecker the digger risked attack and possibly 'the loss of his eyesight'. Like the Mandrake (see page 286) it was recommended that the ceremony be carried out with the aid of a hungry dog—tied to a string and enticed by the smell of roast meat, for the groans of the plant as its roots were torn up would, according to the Ancients, prove fatal to all who heard it!

Paeonia officinalis
double form

In the first century A.D. Dioscorides, the Greek herbalist, and Pliny the Elder repeat the woodpecker story and also describe male and female Paeonies: 'the male has leaves like the Royal nut tree. The female has its leaves divided like the gum tree . . . the root of the male is about a finger's thickness and a span in length. It is astringent to the taste and white. The root of the female has seven or eight acorn-shaped off-shoots like the Asphodel'. *P. mascula* is thought to be the male plant referred to (a rosy-red species, also European) and *P. officinalis,* the female.

In Tudor times (1485–1603) contemporary writers mention Paeonies frequently, often under the popular names of the day. These include such delightful terms as Chesses, Hundred-bladed Rose, Sheep-shearing Rose, Rose Royale, Nan Pie, Pie Nanny and Marmaritin.

The most popular Paeonies for garden purposes are cultivars derived from *P. lactiflora* (*P. albiflora, P. edulis*), a fragrant white-flowered species native to Siberia, Mongolia and the northern parts of Manchuria, Tibet and China. These are usually known as Chinese Paeonies, possibly because the first hybrids were raised in that country. Named varieties include singles and doubles, and also Anemone-centred and frill-petalled forms in various colours—white, cream, light to dark pink, red, crimson and purple—or combinations of colours. Those illustrated include the double 'Sarah Bernhardt', an old cultivar raised by Victor Lemoine of Nancy in 1906 but still one of the best for cutting in soft apple-blossom pink; 'Eugene Verdier', even older, dating from 1864 and a fine, silver-edged, soft pink; the single 'Mikado', brilliant red with golden staminodes; and 'Bowl of Beauty', fuchsia-rose, a good early-flowering semi-double sort.

The Daurans and Mongols were reputed to boil the roots of *P. lactiflora* in soup and introduce the ground seeds into their tea. The plant is also used medicinally in some parts of China.

P. mlokosewitschii is the Lemon Paeony, a beautiful Caucasian species with handsome, broad, glaucous foliage and large, single, citron-yellow flowers on stems 45 cms (1½ ft) tall. This delicate shade is accentuated by a central boss of rich golden stamens. Later in the season the seed pods split to reveal a double row of jewel-like seeds of shiny blue-black, interspersed with numbers of brilliant scarlet, infertile ovules.

Other yellow-flowered kinds are the Caucasian *P. wittmanniana* and its variety *nudicarpa,* both growing about 90 cms (3 ft) tall, but neither is as richly coloured as *P. mlokosewitschii.*

P. tenuifolia, from E Europe to the Caucasus, is the Fringed Paeony, with deeply cut leaves having a Fennel-like appearance. The cup-shaped flowers, 8 cms (3 ins) in size, are deep crimson and there is a double form 'Plena'. Growing 30–60 cms (1–2 ft) high this is probably the most suitable species for the rock garden.

P. emodi from the Himalayas is a beautiful plant with single flowers, 10–15 cms (4–6 ins) across, of pure white with prominent yellow stamens. It is further distinguished by having several flowers to a stem.

The Tree Paeonies come from China and the Chinese-Tibetan border. They are shrubs rather than trees and unlike the herbaceous kinds should never be cut down. Beyond an occasional shaping and the removal of weak or damaged wood, pruning is unnecessary.

The main species are *P. delavayi, P. lutea, P. potaninii* and *P. suffruticosa* (*P. moutan*). Hybrids from these and particularly from *P. suffruticosa* often carry very large flowers, sometimes the size of dinner plates, but as they are borne on slender stems these are easily snapped in gales or broken down by weight of snow. For this reason Tree Paeonies should always be planted in a sheltered position and since they are martyrs to late spring frosts, set them in a place which does not receive sun before the afternoon. The plants are lime-tolerant and gross feeders, appreciating a mulch of rotted manure with a handful of bonemeal after flowering.

Cultivars from *P. suffruticosa* may be single or double and many have a rich Rose-like fragrance. Good doubles include 'Fragrans Maxima Plena', salmon-pink; 'Bijou de Chusan', white, shaded pink; and 'Reine Elizabeth', rose-scarlet. Among the singles are

Paeonia
herbaceous cultivars

'Sybil Stern', deep cherry-red with lemon stamens; 'Black Pirate', deep red; and 'Mrs George Warre', pink.

These named sorts are occasionally grafted on *P. suffruticosa* but more generally root grafted on herbaceous sorts such as *P. lactiflora* or *P. officinalis*. This last technique has been practised in China since about A.D. 1000 and is favoured because in time the variety develops its own roots and does not cause annoyance by suckering (as can happen with grafts on *P. suffruticosa* seedlings). Occasionally the odd plant is layered, but only species can reliably be reproduced true to type from seed.

P. delavayi is a species which will grow in sun or shade. Its blooms are deep maroon-red or practically black in some forms but much smaller than *P. suffruticosa*. It grows about 1.8 m (6 ft) high.

P. lutea is a smaller plant, up to 13.5 m (4½ ft) or 1.8 m (6 ft) or more in the variety *ludlowii*, which increases by means of sucker growths. It has yellow flowers, often with red filaments, which provide striking colour contrast. 'L' Espérance', a large semi-double is one of the best hybrids in these shades. 'Souvenir de Maxime Cornu', a fragrant double, is deep golden-yellow with carmine edging; 'Chromatella', also double, is sulphur-yellow; and 'Argosy' is a large single yellow.

P. potaninii makes a low-growing (45 cms; 1½ ft) shrub with maroon flowers. Seedlings, however, show considerable variation and there is an albino form *alba* whose golden anthers have green filaments and a taller buttercup-yellow form called *trollioides*. It increases underground by means of stolons.

In the Orient Tree Paeonies have a long history and are more commonly grown than the herbaceous kinds; in Europe and America the reverse is true. The Japanese are expert at grafting named cultivars and are said to eat the flowers as salad in some parts of the country.

In China they must have been cultivated for centuries for written details concerning Paeonies date back to the 6th century. Prior to the year A.D. 600, they were chiefly grown for medicinal purposes but later their aesthetic qualities were appreciated and by A.D. 700 (according to an old Chinese work on 'The Origin of Things and Matters') their culture was well established, with some thirty different sorts, described according to colour or origin.

During the Tang dynasty (A.D. 618–906) they were particularly popular. Emperors placed them under their protection and the plants were not only cultivated but written about, shown, exchanged and even offered in marriage dowries. They were portrayed in ceramics, textiles and pictorial arts—often with Peacocks and Phoenixes. It was considered appropriate that the Paeony (king of flowers) should appear alongside a Phoenix (king of birds). Ming pottery (1368–1644) shows many representations of Paeonies.

At that time the colour range of Tree Paeonies was confined to white, pink, red and lilac; a yellow variety was not known but much desired. Many ingenious devices were practised in order to obtain the desired colour, such as pouring dye over the roots or wrapping white flowers in yellow paper in the hope that they would take up some of the pigment.

Gardeners were so skilled at retarding and forcing that flowers were to be had from spring to autumn. They also grew the shrubs as espaliers and in various ornamental designs. Tree Paeonies reached England in 1787 after Sir Joseph Banks, head of the Royal Botanic Garden at Kew, commissioned a Doctor John Duncan of the East India Company to obtain a specimen for the gardens. They arrived in the United States about 1820.

Tree Paeonies, their stems 1.8 m (6 ft) high hung with dozens of flowers, live a long time in gardens when grown under suitable conditions, sometimes for more than a century. They also take kindly to pot cultivation and can be forced gently in heat for early bloom.

Paeonia suffruticosa
cultivar

Papaveraceae

26 genera and 200 species

This is a family of dicotyledonous, chiefly herbaceous plants from N temperate and subtropical regions. The name *papaver* is said to refer to the noise made when chewing the seeds, but *pap* also means 'thick milk' and is appropriate since most species secrete white or orange latex in their stems. *Dendromecon* and *Bocconia* species may be shrubs or small trees.

The leaves are alternately arranged; the flowers are solitary or in racemes and often very showy. Individually they are perfect and regular with two sepals (sometimes more or less joined), four petals, which are rolled or crumpled when in bud, many stamens, a superior ovary (except in *Eschscholzia*) and a nut fruit or many-seeded capsule. The seeds are oily.

Among the brightest and showiest is *Papaver,* a genus of herbaceous plants of annual, biennial or perennial duration, with bristly or smooth leaves and stems—often glaucous—containing milky or coloured sap. Most species are native to temperate and subtropical regions in Asia, North Africa and Europe, with several in western North America, one in South Africa and one in Australia.

The brilliant but short-lived flowers may be red, pink, violet, yellow or white and have two (rarely three) sepals, four (rarely six) petals and many pollen-rich stamens. Although much visited by bees and other insects foraging for pollen, they do not in fact contain any honey.

Papaver rhoeas, the Corn Poppy, is an erect branching annual 30–60 cms (1–2 ft) tall, abundant in Europe and Asia and also naturalized in North America. Before the days of selective weed-killers it was a common weed of European cornfields, the blood-red flowers gleaming between the golden stalks of the ripening corn or wheat.

The species itself is rarely cultivated but there are several garden strains of which the most important is the Shirley. Shirley Poppies were raised by the Reverend W. Wilks (of Shirley Vicarage, Croydon, London), who was for thirty-two years Secretary of the Royal Horticultural Society. The story of his find appears in *The Garden* (Vol. 57) of 1905: 'In 1880 I noticed in a waste corner of my garden abutting on the fields a patch of the common wild field Poppy (*P. rhoeas*), one solitary flower of which had a very narrow edge of white. This one flower I marked and the seed saved, and so for several years, the flowers all the while getting a larger infusion of white to tone down the red until they arrived at quite a pale pink and one plant absolutely white.

I then set myself to change the black central portions of the flowers from black to yellow or white, and have at last fixed a strain with petals of varying colour from the brightest scarlet to pure white with all shades of pink between and all varieties of flakes and edged flowers also, but all having yellow or white stamens, anthers and pollen and a white base . . . my ideal is to get a yellow *P. rhoeas*. . . .

The Shirley Poppies are (1) single (2) always have a white base with (3) yellow or white stamens, anthers and pollen and (4) never have the smallest particle of black about them. Double Poppies and Poppies with black centres may be greatly admired by some, but they are not Shirley Poppies. It is rather interesting to reflect that the gardens of the whole world—rich man's and poor man's alike—are today furnished with Poppies which are the direct descendants of one single capsule of seed raised in the garden of the Shirley Vicarage as lately as August 1880.'

These Poppies are easily grown from seed, sown in spring or autumn where they are to flower and then thinned to 15 or 25 cms (6 or 10 ins) apart. The petals of *P. rhoeas* are the source of a red pigment used for colouring wine and medicines. An old legend relates that the red Poppies which followed the ploughing of the field at Waterloo, after Wellington's victory over Napoleon, sprang from the blood of soldiers killed in that battle. After the First World War the battlefields of Flanders were similarly scarlet with *P. rhoeas* flowers.

Today seedsmen offer the Flanders Poppy, which is scarlet with black spots at the base of each petal, for naturalizing, and also Begonia-flowered Poppies with large, double flowers in various colours.

P. somniferum is the Opium Poppy, an annual of 30–60 cms (1–2 ft) with smooth glaucous foliage and large, four-petalled, white to purple flowers with urn-shaped capsules. Modern cultivars, especially the doubles, are very striking in white, pink, rose, crimson and purple.

The wild ancestral form is apparently unknown, but this Poppy has been cultivated for so long that it is now widespread all over the civilized world.

That Opium Poppies were grown in Ancient Egypt is evidenced by illustrations on murals and pictorial records of the times. The writings of Hippocrates and Celsus also mention opium as does De Quincey in his famous *Confessions of an English Opium Eater* (1822). The crude product is derived from the maturing seed pods tediously induced to yield their sap by incising the capsules. This is done around sunset, the resultant coagulated drops of exudation being scraped off next day, dried and then kneaded into brownish balls. This is crude opium and under good conditions an acre of land (with approximately 20,000 plants) will produce 25 to 40 lbs in a season.

Linnaeus with exemplary patience counted 32,000 seeds in a single head of *P. somniferum*. At the beginning of the 19th century it was apparently extensively grown in Britain where some 50,000 lbs were annually consumed and, a sad reflection on the times,

Papaver orientale
cultivars

in 1823 a Miss Kent wrote 'the solution of opium in spirits of wine is called laudanum, or loddy and much used instead of tea by the poorer class females in Manchester and other neighbouring towns'.

The raw material, which is often smoked as an intoxicant, is the source of the habit-forming heroin, use of which has brought misery and degradation to millions. Purified opium on the other hand yields morphia which has given relief to countless sufferers, and also codeine, narcotine and other alkaloids. Morphia is included in many cough medicines since it depresses that part of the brain centre involved in coughing.

The Poppy has long been the flower of sleep and oblivion. As Iago says (*Othello*, IV, iii)

> *'Not poppy, nor mandragora*
> *Nor all the drowsy syrups of the world*
> *Shall ever medicine thee to that sweet sleep*
> *Which thou ownd'st yesterday. . . .'*

However, the Poppy seeds used by bakers for decorating fancy breads contain no opium although they are very rich in oil, which is used in the manufacture of soaps, paints, varnishes, salad oils and cattle cake. The decorative dried seed heads find a ready sale for winter flower arrangements and the fresh capsules have been used as fomentations for such troubles as sprains, bruises, neuralgia and toothache.

A variable species popular for flower decoration and garden adornment is the Iceland Poppy, *P. nudicaule*. It is native to northern sub-arctic regions, producing many solitary flowers on leafless, hairy stems 30 cms (1 ft) tall. These are sweetly fragrant and in soft shades of white, cream, yellow or pale orange. The deeply cut, glaucous leaves form ground-hugging tufts. For indoor use Iceland Poppies should be cut in bud and have the ends sealed by dipping the tips for a few seconds in boiling water, or burning them with a taper.

Although perennial the plants can be grown as annuals or biennials, sowing the seed one summer to flower the next. Established specimens resent disturbance.

There are various strains available such as 'Coonara', with stems of 38–45 cms (15–18 ins) carrying salmon, gold, biscuit or different shades of pink flowers. 'Champagne Bubbles' is a name given to a race of F1 hybrids with immense, crimped-petalled blooms in many pastel shades as well as scarlets and bicolours, and 'Yellow Wonder' is an American sort with buttercup-yellow blooms 10 cms (4 ins) in diameter on stems 60 cms (2 ft) tall.

For garden purposes perhaps the most important species in this colourful family is *P. orientale*. Not only are the blooms large and showy but they come when most required—in early summer. The species is native to S Europe, where it follows the coast-line down as far as Armenia. The flowers are brilliant scarlet, 10–15 cms (4–6 ins) across, with a prominent purple-black blotch at the base of each petal. *P. bracteatum* is similar to *P. orientale* (of which it is sometimes considered to be a variety) but differs in having large leafy bracts below the flower.

Oriental Poppies grow readily in any deep, well-drained soil in full sun and last for years in the same position. They reach a height of about 90 cms (3 ft) and for best effect should be grouped; in mixed borders they are best set in a midway position with later flowering perennials such as Asters in front. The latter then hide the untidy, dying down processes of the Poppies. A few of the cultivars show a tendency to sprawl so may need staking.

Although the type species is easily raised from seed, named cultivars should be increased by root cuttings.

Notwithstanding the fact that *P. orientale* had been grown for many years in Britain it was 1905 before the first colour break appeared. The author's late father-in-law, the nurseryman Amos Perry, inspecting a bed of seedlings one day, found that one plant had rose-pink—instead of scarlet—flowers. He showed this at the Royal Horticultural Society Flower Show the following year, naming it 'Mrs Perry' after his wife and then, feeling that other breaks might occur, hand-pollinated the flowers and sowed the seed, hoping for a still paler form or even a white Poppy. But all the seedlings flowered red.

Then in 1912 a client was making a pink border (a fashion of the time) and a miscellany of plants was carefully selected so that the pale flesh tones appeared at the ends of the beds but worked up to deeper shades towards the middle. Poppies were felt to give the brilliance of colour needed for the central focal point. However, in 1913 Amos Perry received an irate letter. His client wrote that the plants had come to bloom but, in the most conspicuous position reserved for the Scarlet Poppies 'a nasty fat white one had appeared'. Disbelieving, he went to see the garden and there sure enough was a white Poppy which, exchanged for some Montbretias, in time became 'Perry's White'.

Later there came other shades and even double forms and several with fringed petals such as 'King George' and 'Lord Lambourne'. When the latter was first shown at the Chelsea Show in 1920 the national press called it the Mystery Bloom, describing it as a fringed Poppy 'slit all round the edge into dainty tongues . . . an enormous expanse of petals in which one could bury one's face, in dye like the old military tunics'. 'It's a monstrosity' said an expert at the Chelsea Show yesterday, reported another. 'I have never seen one with such a fringe before, admitted the Rev. W. Wilks, the late Secretary of the Royal Horticultural Society, but Lord Lambourne, the President, was proud of the fact that it was called 'Lord Lambourne'. All these varieties are still grown, together with a host of modern cultivars.

The genus *Meconopsis* contains many shade-lovers and some exquisitely beautiful plants. Perhaps the most prized of these is the Himalayan Blue Poppy *Meconopsis betonicifolia* (*M. baileyi*), which was rediscovered by the late Frank Kingdon Ward in 1924.

Romneya coulteri

This had been known since 1886 through a solitary herbarium specimen collected in China by the French missionary Jean Delavay. Although many sought it again through the years it remained undiscovered until 1913, when a Lt.-Col. Bailey came across a plant by chance. He, too, only brought back a pressed flower and it was not until 1926 that plants became available in Europe.

Frank Kingdon Ward, after long days of arduous travel high in the Himalayas described his find thus: 'Suddenly I looked up and there like a blue panel dropped from Heaven a clump of blue Poppies—as dazzling as sapphires—gleamed in the pale light.' Later he returned to collect the seed and sent it to London, where the late Tom Hay, Superintendent of Hyde Park, planted it in the Park. Kingdon Ward on his return to London declared 'I found it growing on the roof of the world and now saw it again in the hub of the world.'

Macleaya microcarpa

This beautiful species grows 60–150 cms (2–5 ft) according to locality and the clarity of the atmosphere. In Scotland it does particularly well with branching stems carrying many nearly round, crimped-petalled flowers 5 cms (2 ins) across. At their best these are a delightful sky-blue with an iridescence which varies in intensity in different lights. The boss of golden stamens in the centre of the flower heightens this beautiful blue. Sometimes the blooms have a purplish tinge and are less attractive; such plants should not be reproduced.

Usually, but not invariably, the plants die after flowering, but fresh seed provides a reliable means of increase and the plants bloom the second season.

Other good Asiatic species include *M. grandis* with four- (although sometimes up to nine-) petalled, purple or deep blue flowers on stems 90 cms (3 ft) tall. The oblong-elliptic, toothed leaves are roughly hairy. The Harebell Poppy, *M. quintuplinervia* has nodding flowers of lavender-blue on stems of 30 cms (1 ft) and *M. napaulensis* (*M. wallichii*), the Satin Poppy, produces beautiful, winter-hardy rosettes of deeply cut leaves. These are densely covered with rusty hairs and in early summer the 1–2 m (3–6 ft) spikes carry numerous red, purple, blue or occasionally white flowers. This species is monocarpic and seed is best selected from the blue kinds.

M. regia from Nepal, another monocarpic species, has rosettes of long narrow leaves, 45–50 cms (18–20 ins) in length, thickly covered with bronze-gold hairs. These gleam in the sunlight and persist in winter. Branched stems 120–150 cms (4–5 ft) tall carry many golden flowers in early summer. *M. integrifolia,* the Yellow Chinese Poppy, and *M. paniculata* are also yellow-flowered; the latter makes an excellent counterpart for *M. napaulensis* as they both grow about the same height.

A rosy-red species, *M. sherriffii,* with nodding flowers on stems of 30–60 cms (1–2 ft) may appeal to some, although to my mind it is inferior in every way to the blue species. It is monocarpic and native to Bhutan and SE Tibet.

All the species described associate pleasingly with ferns and Candelabra Primulas. The only British representative is the Welsh Poppy, *M. cambrica*, a perennial with bright green, deeply lobed leaves. The solitary, Poppy-like flowers, varying from rich orange to an acidulous yellow, are borne on leafy stems about 30 cms (1 ft) high. Double forms exist but are generally less pleasing. The Welsh Poppy is an inveterate seeder and soon naturalizes itself in damp situations.

An entirely different kind of Poppy is *Romneya coulteri* the Matilija Poppy from California. This perennial grows to 120 cms (4 ft) with blue-grey, somewhat leathery leaves and large white blossoms with crimpled petals. These remain open for days and have a central pincushion-like boss of golden stamens which accentuate their whiteness. *R. trichocalyx,* another perennial, is quite similar but the flowers have a disagreeable scent.

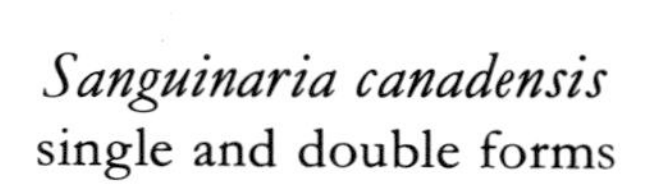

Sanguinaria canadensis
single and double forms

Argemone mexicana

Eschscholzia californica
colour variants

Both need good soil, full sun and a sheltered position. Because of a scarcity of fibrous roots they are difficult to transplant and accordingly are usually propagated from seed or root cuttings.

The two species of *Macleaya,* both Chinese, make good background plants in a mixed border or can be used to mask an untidy garden feature. They thrive in sun or shade and produce branching stems (6–7 ft) tall covered with myriads of small feathery flowers. The large leaves are Fig-shaped, attractively cut and bright silver underneath. When the wind blows this underside is revealed and looks most attractive. *M.* (*Bocconia*) *cordata* is the Plume Poppy with pinkish-buff flowers, which are yellowish in *M. microcarpa.* The stem contains a bright yellow juice which the Chinese use as a dye and also apply as a disinfectant against insect bites and stings.

The related *Bocconia* consists of about ten shrubs native to Asia and Central America. The large, greyish leaves are lobed and the flowers are in terminal panicles. *B. frutescens* is the Mexican Tree Celandine with pinnately divided leaves and greenish flowers. The bark sap contains an alkaloid which has been used as a local anaesthetic by Mexican surgeons.

Sanguinaria canadensis is a charming E North American plant, low growing, with a thick horizontal rootstock which turns blood-red when cut. This circumstance has earned it the name Bloodroot and Red Puccoon; American Indians are said to have painted their faces and bodies with its juice. It was also believed to provide a remedy for rattlesnake stings and early settlers used it as a cough medicine.

The small, heart- to kidney-shaped leaves are lobed or scalloped and silvery beneath; the solitary white flowers about 7 cms (3 ins) across, appear very early in the spring. There is a double form called 'Plena'. Bloodroots will naturalize in grass if the latter is not too closely mown.

Eomecon chionanthum, the Poppy of the Dawn, comes from E China and is a part-shade plant which yet seems to need a generous amount of sunshine together with some root restriction (which means starvation) before it blooms freely. It grows 20–38 cms (8–15 ins) high with panicles of small (4 cms; $1\frac{1}{2}$ ins) very beautiful white flowers and has broad, heart-shaped leaves, cool emerald with scalloped margins.

Argemone is a small genus of about ten species, all American. They are large, robust plants which like light soil with full sun and have prickly stems and leaves. The showy flowers can be yellow, white or purplish and resemble Poppies with four to six petals and many stamens.

A. mexicana, the Prickly Poppy or Devil's Fig, has glaucous sprawling stems and leaves and orange or pale yellow flowers. *A. grandiflora,* with glistening white flowers 5–10 cms (2–4 ins) across, is scarcely spiny.

Both species come from Mexico, where oil from the seeds is used in the manufacture of soap and for lamps.

Similar use is made of oil taken from the seeds of the Yellow Horned Poppy, *Glaucium flavum,* a plant 60–90 cms (2–3 ft) high from Europe, N Africa and W Asia. It is particularly plentiful in the Mediterranean region and usually found on or near sandy sea shores. It accordingly makes a good coastal plant but will grow in any spot where the soil is light and in full sunshine. The species is best treated as biennial and produces large rich yellow flowers and lobed or pinnatifid leaves. The pale orange sap has a disagreeable odour. *G. corniculatum* is the Red Horned Poppy with orange or red blooms.

Eschscholzia (*Eschscholtzia*) *californica* is the orange Californian Poppy, a brilliant hardy annual which is easily grown on most garden soils and flowers over a long period. It needs full sun, the seed being sown where it is to flower in spring or autumn—the latter producing bigger plants.

The usual height is around 30–45 cms (1–$1\frac{1}{2}$ ft); the glaucous grey leaves are deeply cut into narrow segments and the satiny, saucer-shaped, four-petalled flowers open in sunshine. Individually they are 5–7 cms (2–3 ins) across. The buds are pointed, the sepals forming a steeple-like cap, which is pushed off as the flowers open. This characteristic has earned for it the name 'French Bonnets qui tombent' because of this 'irresistible night cap which one is tempted to pull off the awakening flower'.

It is a variable species and in the hands of hybridists has produced single and double cultivars in white, yellow, carmine, vermilion, scarlet and rose-pink.

Californian Poppies are excellent for covering sunny slopes or making large drifts in beds and borders. In San Francisco they gild the grassy hill slopes and were once so abundant that the Spaniards called California 'The Golden West' and 'The Land of Fire'. They dedicated it to San Pascual, whose colour was yellow and described it as 'his altar-cloth spread on the hills'. It is now the State Flower of California.

Years ago Californian Indians ate the leaves as greens, either boiled or roasted on hot stones.

Passiflora vitifolia

Passiflora caerulea

Passifloraceae

12 genera and 600 species

This is an interesting family of shrubs and herbaceous plants, the majority climbers with axillary tendrils, attractive foliage and beautiful flowers.

For gardens the most important genus is *Passiflora* with some 500 species, mainly from tropical and warm temperate parts of America, together with a few species in Asia and Australia.

The leaves are variable in the different species, being occasionally simple, frequently palmate and sometimes bilobed with long pointed, swallow tails. At the base of the leaf stalks there are often extra-floral nectaries. The flowers spring from the same leaf axils as the tendrils, either as solitary blooms or in small cymes. Individually they are bisexual and regular with a cup-shaped receptacle bearing five sepals, five petals, many thread-like appendages forming a central corona, five stamens, three styles with their stigmas and a superior ovary. The attractive blooms with their curious and often spectacular floral attachments have inspired a charming legend.

This relates how Jesuits in the 16th century, following the Conquistadors after their victorious campaigning in South America, sought a favourable omen for their cause as they stepped ashore for the first time. They found it in a trailing, scrambling plant by the sea shore—possibly *P. caerulea*—which they called Passion Flower. The ten sepals and petals of its blooms represented for these men the ten faithful apostles (Judas and the doubting Thomas—some versions say Peter—being omitted); the outer corona signified the countless disciples and the inner corona the Crown of Thorns. The five stamens symbolized Christ's five wounds, the three-parted stigma the nails and the ovary the hammer with which these wounds were made. The curling tendrils were a reminder of the whips with which the Saviour was scourged and the five-parted leaves recalled the clutching hands of the soldiers. When the Jesuits found the natives feasting on the yellow pear-shaped fruits this seemed a sign that the Indians were thirsting for Christianity and that Catholicism would be the religion of this new world.

Passiflora quadrangularis

Most *Passiflora* blooms are flat and poised on short flower stalks, although there are certain species with greatly elongated cylindrical calyx tubes. These are sometimes so extended that the blossoms droop with their own weight and in addition the corona segments are much reduced. The differing appearance of these led early botanists to give them distinct generic rank—as *Tacsonia,* but nowadays they are all collectively grouped under *Passiflora.*

The hardiest Passion Flower for cool temperate countries is *P. incarnata* of south-eastern US as far north as southern New Jersey. Less hardy is *P. caerulea,* a native of S Brazil which can however be grown outside in some parts of the British Isles. This species grows strongly and rapidly with small, five- or seven-lobed leaves and flat, fragrant flowers 7–10 cms (3–4 ins) across. These have white petals and sepals and purple appendages and are succeeded by orange, egg-shaped fruits containing many seeds embedded in pulp. They are edible but less palatable than some other species such as *P. edulis* and *P. quadrangularis.*

Although the individual blooms are of short duration (they only last one day), the flowering season is extended from early summer until autumn. A cultivar called 'Constance Elliott' with ivory-white flowers was very popular in London at the beginning of the 20th century. It makes a good cool house climber.

P. quadrangularis is one of several edible species known as Granadilla. It is an interesting species from tropical America, much cultivated in Australia and the Far East for its fruits. These may be 15–20 cms (6–8 ins) in length and have a sweet acidulous, purple juice. The plant makes a vigorous climber with simple cordate leaves and pink and white flowers 10–13 cms (4–5 ins) across bearing long white tangled filaments, heavily banded with blue and red.

Other species cultivated for their edible fruits are *P. ligularis* from Peru with pale greenish-white flowers and *P. edulis,* a tropical American kind with white, green and purple flowers and yellow or purplish fruits 5 cms (2 ins) long.

Passiflora x *exoniensis,* a splendid climber of hybrid origin, has cordate and downy, three-lobed leaves and bright rosy-red flowers, 10–12 cms (4–5 ins) across, with scarlet-backed sepals and a small white corona. It is sometimes included in *Tacsonia* and needs warm growing conditions.

In Colombia the long-tubed, rosy flowers of *P. mollissima* (*Tacsonia mollissima*) festoon hedgerow shrubs and trees like so many crimson lamps. The yellowish-white, downy fruits are made into a refreshing drink called karuba. This spectacular climber is cultivated outdoors in warm Australia—where it is known as the Banana-Passion Fruit—but has to be grown in heated greenhouses in less favoured climates.

Very similar in appearance is *P. antioquiensis,* another Colombian with larger (10–13 cms; 4–5 ins) and even more brilliant red flowers on long stalks. The leaves are of two shapes, either simple and lanceolate or deeply three-lobed. Stems, flowers and leaf stalks, also the undersides of the leaves, are covered with fine down.

P. racemosa from Brazil is a free-flowering climber with many brilliant, rosy-crimson flowers in pendulous racemes. The narrow petals are rosy and spreading, the outer corona segments purple tipped with white and the inner corona red. A hybrid between this species and *P. quadrangularis,* raised at the John Innes Institute, combines the good qualities of both parents and is known by the name 'John Innes'. It is a strong climber with large, flat, fragrant, blue, white and red flowers which are continuously produced from early summer until autumn, each bloom lasting one day only.

P. vitifolia, native to Central and northern South America, has three-lobed, Vine-shaped leaves with downy stems and large red flowers. The latter are variable, the best forms being almost scarlet.

P. foetida has a wide distribution in tropical South America and can even be found in the Galapagos Islands, where it trails in long strands along the coral sea shores. The neat, downy, three-lobed leaves emit an unpleasant smell when crushed although the small green and white flowers are fragrant. At the back of the blooms the large enwrapping bracts are deeply cut into fine segments and look like green lace.

Passion Flowers are not fastidious as to soil provided they have good drainage and plenty of water during the growing season. Under constricted conditions—as in a greenhouse—the roots need confining, otherwise the plants grow too exuberantly, producing foliage rather than flowers. A soil depth of 30 cms (1 ft) is sufficient, composed of sand, peat and turfy loam.

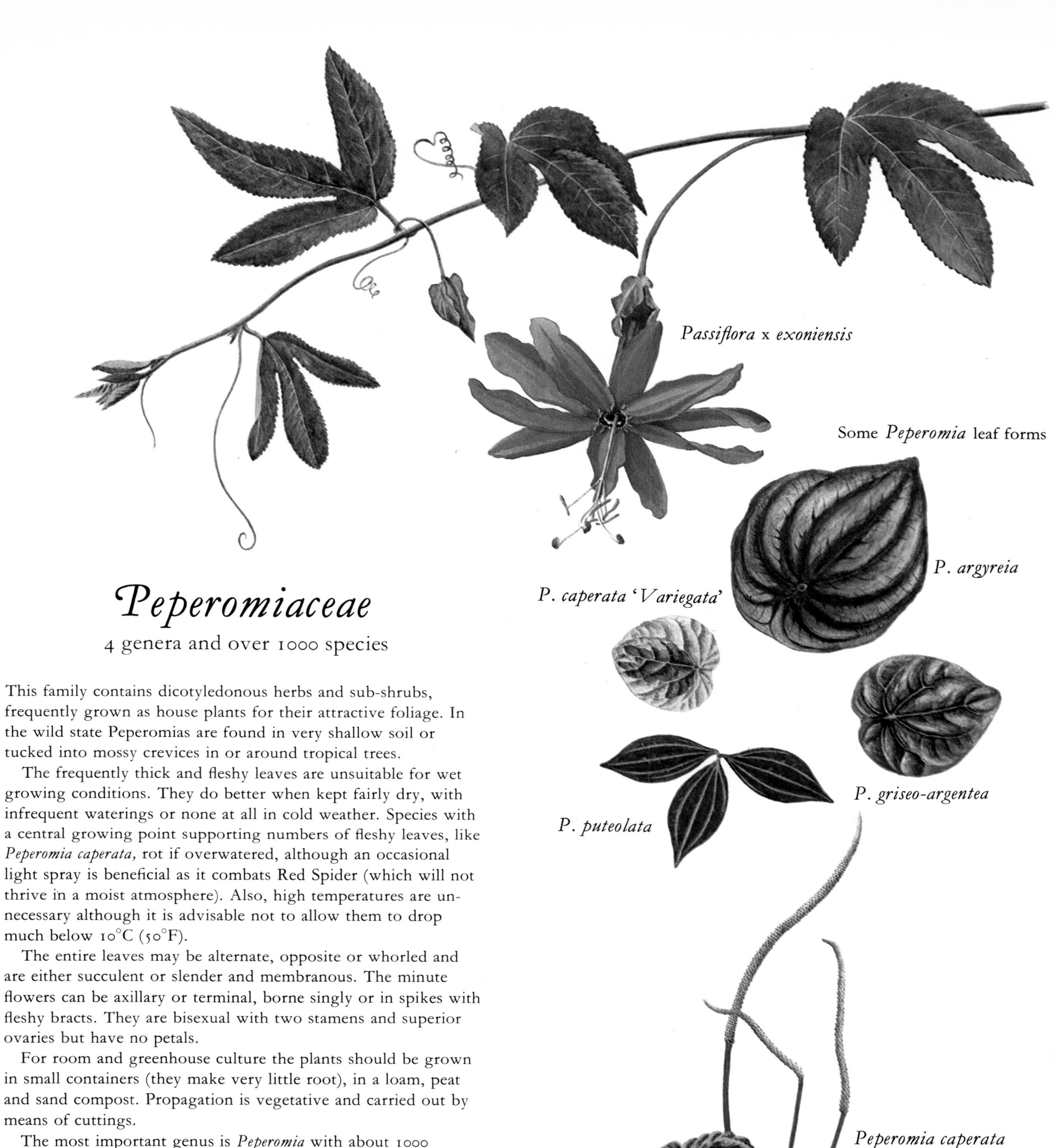

Peperomiaceae

4 genera and over 1000 species

This family contains dicotyledonous herbs and sub-shrubs, frequently grown as house plants for their attractive foliage. In the wild state Peperomias are found in very shallow soil or tucked into mossy crevices in or around tropical trees.

The frequently thick and fleshy leaves are unsuitable for wet growing conditions. They do better when kept fairly dry, with infrequent waterings or none at all in cold weather. Species with a central growing point supporting numbers of fleshy leaves, like *Peperomia caperata,* rot if overwatered, although an occasional light spray is beneficial as it combats Red Spider (which will not thrive in a moist atmosphere). Also, high temperatures are unnecessary although it is advisable not to allow them to drop much below 10°C (50°F).

The entire leaves may be alternate, opposite or whorled and are either succulent or slender and membranous. The minute flowers can be axillary or terminal, borne singly or in spikes with fleshy bracts. They are bisexual with two stamens and superior ovaries but have no petals.

For room and greenhouse culture the plants should be grown in small containers (they make very little root), in a loam, peat and sand compost. Propagation is vegetative and carried out by means of cuttings.

The most important genus is *Peperomia* with about 1000 species—chiefly from South and Central America. Some of these are epiphytic with creeping stems and fleshy leaves.

P. caperata is a popular kind with many small, dark green, heart-shaped leaves which are heavily corrugated. The spikes of pure white flowers contrast pleasingly with these and the whole plant is only a few centimetres high. There is also a form with variegated leaves.

P. argyreia (*P. sandersii*) is one of the finest species; the thick, smooth, rounded leaves are barred with broad stripes of silver and dark green. *P. magnoliifolia* is another popular species, especially in its cream and green leafed form 'Green Gold', and *P. griseo-argentea* and *P. puteolata* are other good foliage plants.

Philadelphaceae

7 genera and 135 species

Several important, mostly summer-flowering shrubs and small trees belong to this family, including plants of easy culture suitable for cool climates. They are dicotyledonous and mostly N temperate but with representatives in the Philippines and Central America.

The simple leaves, with entire or serrated edges grow opposite each other or are verticillate or occasionally alternate. The flowers occur in showy terminal cymes or racemes, or (rarely) are solitary. They are often fragrant and individually are perfect and regular with four or five sepals, five to seven petals, numerous stamens and a superior or inferior ovary.

Philadelphus with some seventy-five species, all N temperate, is one of the most important genera; its conspicuous and richly scented flowers are among the glories of the summer garden. These come to maturity when the great blossoming time of the shrubs is waning and associate splendidly with the Roses, Delphiniums and Lilies of midsummer.

Often, but erroneously, called Syringa (a name which rightly belongs to Lilacs) but also known as Mock Orange, *Philadelphus* are characterized by ovate and opposite, deciduous leaves on slender woody stems and masses of Orange-blossom type flowers, mostly white.

P. coronarius is the oldest-known if not the best species, making a spreading bush 3–4 m (10–12 ft) high. Its habitat is obscure but possibly Caucasian, and plants still found in S Europe are probably naturalized. It has a sweet penetrating fragrance which, however, gives some people hay fever. In Turkey the pithy-centred stems were at one time made into pan-pipes.

For many years *P. coronarius* was the only species known in European gardens, until the 19th century when *P. grandiflorus* and *P. pubescens* arrived from North America, and also several species from the Orient. Crossings were made between the species, M. Lemoine of Nancy being one of the most assiduous hybridists at the turn of the 20th century. To him we owe such splendid plants as *P.* x *lemoinei,* a hybrid between *P. microphyllus* from North America and *P. coronarius*. This in turn became the parent of many fine cultivars such as the doubles 'Virginal' and 'Boule d'Argent' and the singles 'Innocence' and 'Avalanche'.

Philadelphus coronarius

P. x *purpureo-maculatus* is derived from *P.* x *lemoinei* and *P. coulteri* of gardens (not the true Mexican species) and is the parent of several fine cultivars like the sweet-scented 'Belle Étoile', maroon-blotched 'Sybille' and 'Étoile Rose'.

Occasionally *Philadelphus* grow excessively large but may be brought to more manageable proportions by thinning the branches and reducing the rest to vigorous young shoots lower down on the bush. As the plants bloom on the older wood this may cost a season's flowering. Mock Oranges are suitable for most soils, in sun or light shade. They are readily increased from cuttings.

Deutzias are valued for their floriferous habit, neat shape and dainty panicles of starry white or pink flowers. Some, such as *Deutzia gracilis,* can be forced for early bloom. They do well in sun although some will also grow in partial shade. These shrubs like plenty of moisture in the growing season and make small to moderate sized bushes.

Propagation is by means of half-ripe or mature cuttings and

Philadelphus 'Virginal'

Philadelphus 'Belle Etoile'

Carpenteria californica

Deutzia x elegantissima 'Fasciculata'

Lapageria rosea

pruning consists of the removal of overcrowded or worn out shoots soon after flowering.

D. scabra from China and Japan grows 2.5–3 m (8–10 ft) tall, with panicles of white flowers which sometimes have pink on their reverse; 'Pride of Rochester' is one of the showiest cultivars but the double white 'Candidissima' is also good. The white wood of this species is finely grained and used in Japan for mosaic work.

D. gracilis, a Japanese species, grows 90–120 cms (3–4 ft) high, with panicles of pure white, starry flowers. *D.* x *rosea* (a hybrid between the former and *D. purpurascens*) has arching sprays of soft rose, bell-shaped blossoms and *D.* x *elegantissima* 'Fasciculata' is bright rosy-pink outside and paler pink within.

Carpenteria californica is a monotypic genus from California with terminal clusters of fragrant white flowers 5–7 cms (2–3 ins) across, shaped like the yellow Rose of Sharon (*Hypericum calycinum*). Their five-petalled, bowl-shaped blossoms, filled with masses of golden stamens, are really delightful and the evergreen lanceolate leaves are bright green, felted beneath and tapering at each end.

The species is a handsome shrub, 2–5 m (6–15 ft) high, which needs plenty of sun and a warm sheltered situation—for fog and cold ruin the foliage and check flowering. Given these conditions it thrives in most soil types, including chalk, and can be reproduced from seed or cuttings.

Fendlera rupicola, another shrub from the SW United States, also needs sheltered conditions. It is a deciduous species, 1–2 m (3–6 ft) high with opposite, narrowly elliptic to lanceolate leaves which, when happily established, produce masses of large white, rose-tinted, four-petalled flowers. These may be solitary or in groups of three.

Jamesia americana is another deciduous shrub from western North America. It makes a rounded bush of 1.25–2.15 m (4–7 ft) with stiff pithy branches which have peeling bark, opposite, ovate, and toothed leaves and terminal panicles of five-petalled, fragrant, white flowers. It can be propagated from cuttings and likes a sunny situation in well-drained soil.

Philesiaceae

7 genera and 9 species

This is a small group of monocotyledonous shrubs or climbers, all native to the Southern Hemisphere. Occasionally they are epiphytic with fleshy stems and thick simple leaves. The large showy flowers are regular and perfect, always pendent and either solitary or grouped in terminal or axillary inflorescences. Individual flowers have six perianth segments, six stamens and a superior ovary which develops to a berried fruit.

Lapageria rosea is the Chilean Bell Flower, an evergreen climber with smooth, stiff shoots which twine around supports and so hoist themselves to a height of 3–5 m (10–15 ft). The dark, glossy green leaves are alternate and leathery in texture and from their axils or at the ends of the shoots hang one to three tubular flowers. These are about 7 cms (3 ins) long and 5 cms (2 ins) wide with firm, rich crimson petals which have a shiny granulation that gives them an appearance of frozen, red-tinted snow. Later, large fleshy berries appear which are 5 cms (2 ins) long by 2.5 cms (1 in.). These are eaten in their native Chile.

This beautiful climber has long been grown in sheltered parts of Great Britain and Ireland, and more widely in cool greenhouses. In warm climates it should be planted in shade for in the wild it is discovered near river falls in areas of dense vegetation with ferns, mosses and similar plants. A beautiful white form exists and there are several with striped flowers of crimson and white

Philesia magellanica (*P. buxifolia*)

or of shades of light and dark red which look very attractive.

Lapageria rosea is *el copihue,* the National Flower of Chile. It requires plenty of moisture and does best in a peat/sand/loam soil mixture. It flowers freely during summer and autumn and can be propagated by means of seed or layers.

Another Chilean plant is *Philesia magellanica* (*P. buxifolia*), a dwarf evergreen shrub with rich rosy-crimson blossoms which bear some resemblance to those of *Lapageria.* It grows to 90 cms (3 ft) in its native forest habitat, where it is extremely floriferous with Box-like leaves, which are bright green above and greyish-white below. The nodding, tubular flowers have two whorls each of three perianth segments and later develop to round red berries.

Philesia needs much the same soil and cultural conditions as *Lapageria,* although it is not a climber and rarely grows higher than 30 cms (1 ft) in cultivation.

Geitonoplesium cymosum, a twining perennial from the Philippines and Malaysia to Australia, has drooping purplish-green flowers.

Luzuriaga is a genus of three species, two in South America and one in New Zealand.

Luzuriaga parviflora, the Lantern Berry from New Zealand, is a delicate creeping sub-shrub which attaches itself to the stems of Tree Ferns or damp moss-covered banks. It has narrow, alternate shiny leaves which have twisted stalks, and many creamy, bell-shaped flowers up to 2 cms ($\frac{3}{4}$ in.) across which are succeeded by white pear-shaped berries.

L.radicans, a native of Chile and Peru, is often found growing on tree trunks. It is creeping, making an evergreen shrub 5–30 cms (2–12 ins) high with oval, stalkless leaves, set in opposite rows and drooping white flowers on slender stalks. These are succeeded by orange, pea-sized berries. Moist peaty soil makes the best compost for a plant which in cool temperate regions has either to be grown under glass or in sheltered rock-garden pockets.

Eustrephus latifolius, the Wombat Berry or Orange Vine, comes from moist, warm forest regions in Australia, New Guinea and New Caledonia. The weak stems climb without the aid of tendrils or prickles, often to a great height, and are clothed with alternate, glossy, longitudinally veined, ovate-lanceolate leaves. The cream or pinkish flowers hang in clusters of four to six together from persistent stems and are followed by conspicuous, round orange berries.

Some authorities refer *Eustrephus* and *Luzuriaga* to Liliaceae.

Phytolaccaceae

12 genera and 100 species

These are dicotyledonous herbs, shrubs or trees mainly from tropical America and South Africa. They have alternate, entire leaves and inflorescences of regular flowers with four to five sepals and petals, four, five or numerous stamens and usually a superior ovary.

Rivina humilis is the Rouge Plant or Bloodberry from Central America. It derives this name from its showy scarlet berries which are the source of a red dye. It is herbaceous with stems 45–60 cms ($1\frac{1}{2}$–2 ft) high carrying racemes of small pinkish flowers, but the fruits are its main ornament, especially as these remain attractive all winter. The species makes a good pot plant, needing rich heavy soil and warm house conditions in frost-prone climates.

Phytolaccas come from the warmer parts of America, Africa and Asia but several are hardy in the British Isles and warmer zones of the United States.

Phytolacca americana (eastern US), known as Pokeweed, Red Ink Plant and Pigeonberry, makes a striking plant in the wild garden.

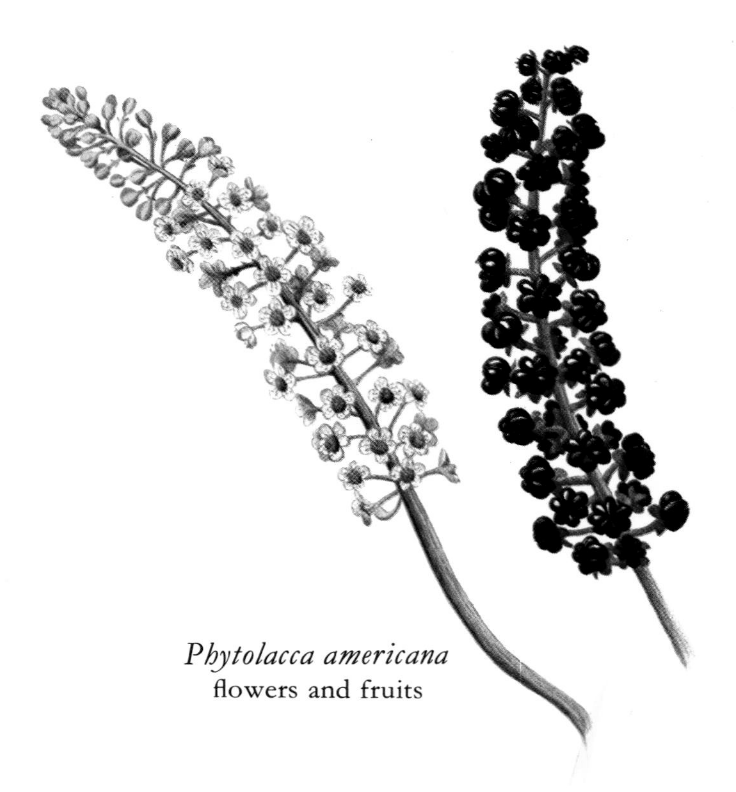

Phytolacca americana
flowers and fruits

It has large, fleshy, very poisonous roots and stems of 1–1.5 m (3–5 ft) with oval leaves. The greenish-white or pinkish flowers are clustered at the tops of these stems and succeeded by dark purple berries which contain a crimson juice. They are said to be poisonous but nevertheless have been used to colour foods and wine, as well as acting as a substitute for red ink. The young shoots can be cooked as greens; country people of Indiana eat the sprouts fried and an excellent pickle can be made from them. Great care should be taken, however, as the old stems and seeds—like the root—are poisonous, even though at one time the sap from the roots was used as an emetic and purgative.

P. clavigera from China has pink and green flowers which are followed by deep black berries. It is somewhat similar to *P. americana* for which it is often mistakenly identified.

Petiveria alliacea from Venezuela, a herbaceous plant with greenish flowers, gives a garlic flavour to milk if eaten by cows. In addition it has numerous medicinal applications and has been used as a fish poison and in the manufacture of one type of curare.

Piperaceae

4 genera and over 2000 species

These are tropical and dicotyledonous climbers, shrubs or small trees, which have simple, alternate leaves and bisexual or one-sexed flowers carried in dense spikes. They have neither sepals nor petals, one to ten stamens if male or one to five stigmas and a superior ovary if female. The fruit is a small berry.

Piper is the most important genus with some 2000 species—mostly climbers. *P. nigrum,* a native of the eastern tropics, is the source of pepper, reputed to be the oldest object of trade between the Orient and Europe. If the berries are gathered and dried before they are ripe black peppercorns result, but if the black outsides are removed by maceration they yield white pepper. Seasoning apart, pepper has been used for countless ailments such as toothache and indigestion and (optimistically!) as a preventative against smallpox, cholera, leprosy, dysentry, typhus, scarlet fever and bubonic plague. It is also added to canned foods, beverages and certain oriental-type perfumes.

P. betle is widely cultivated in many Asiatic countries for use as a masticatory with slices of Betel-nut (*Areca catechu*) and various flavourings. Betle chewing is supposed to stimulate a sense of well-being but turns the saliva red and teeth black. The unripe fruits of the Malayan *P. cubeba* are used in soaps and for flavouring cigarettes and curing throat troubles and in the Pacific Islands a beverage known as Kavakava is prepared from the masticated roots of *P. methysticum.*

One of the most ornamental species is *P. ornatum,* a native of the Celebes off South-east Asia. It is a climber with bronze, heart-shaped leaves pleasantly veined with pink above. As the leaves age these fade to white. The species does well in a warm, moist climate, or in cool countries makes an attractive room plant. If the temperature falls much below 16°C (60°F) the leaves fall. Given a good light—which brings out the pink colouring—it makes a handsome house plant, particularly when trained round and round a moss-filled cylinder of wire netting.

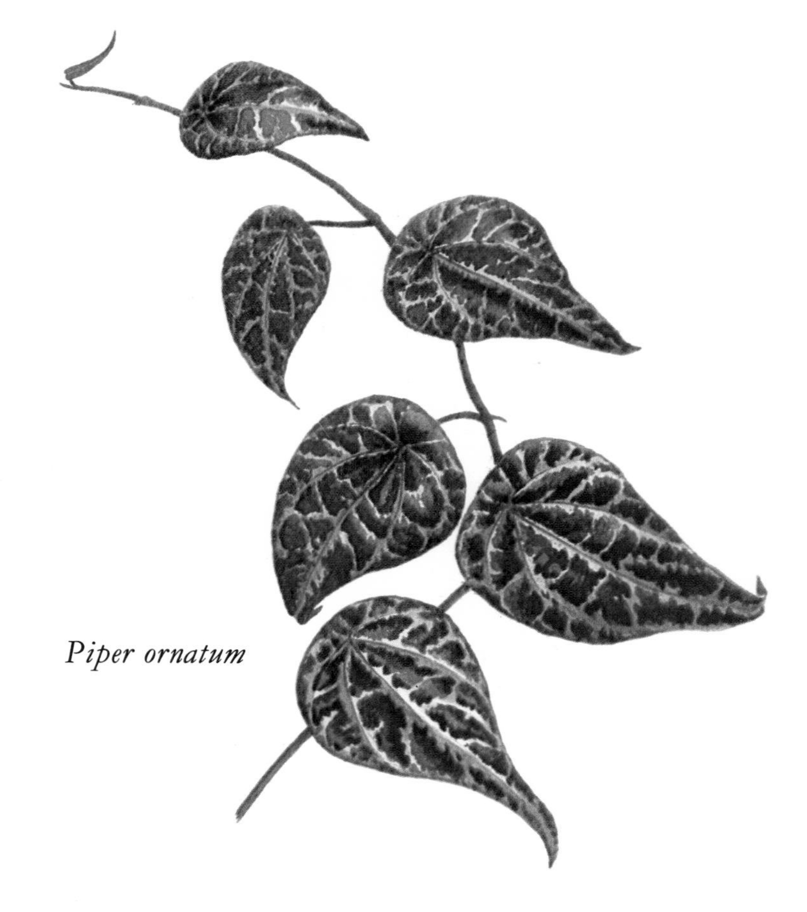

Piper ornatum

Pittosporum tobira

Pittosporaceae

9 genera and 200 species

This is a family of dicotyledonous trees and shrubs of which eight genera are peculiar to Australia. The main characteristics include simple, alternately arranged, leathery evergreen leaves—shiny or woolly—which have a resinous smell when bruised. The perfect flowers are small and bell-like with five sepals, five petals, five stamens and a superior ovary. The fruit is a capsule or berry.

Pittosporum with 150 species has the widest geographical range with representatives in Africa, Australia, New Zealand, various Pacific islands and Asia. Many species have richly perfumed flowers, especially at night. They are handsome evergreens, some yielding useful timber and others such as *P. eugenioides* and *P. tenuifolium* grown for the fresh young foliage. This is esteemed for winter decoration, particularly in northern Europe. *P. eugenioides* is the New Zealand Lemonwood, so called because its pale green, wavy-margined leaves emit a lemon odour when bruised. The species makes a good hedge in mild temperate climates and has yellowish-green, heavily honey-scented flowers.

P. tobira from China and Japan is another good species, especially in its variegated form. It has clusters of white flowers and thick, leathery leaves about 10 cms (4 ins) long in whorls of dark lustrous green (in the type) or speckled with cream in 'Variegata'.

Apart from a few favoured localities Pittosporums are not hardy in Great Britain, nor in the USA, except in very warm areas (zones 8, 9 and 10); elsewhere in these countries they have to be grown in cool greenhouses.

Billardieras are small Australian twining plants with entire, alternate leaves and pendulous flowers on long stalks, succeeded by edible berries. *Billardiera scandens,* the Apple Berry or Dumpling, has cream to purple flowers and *B. longifolia* has yellowish-green blossoms followed by beautiful blue berries.

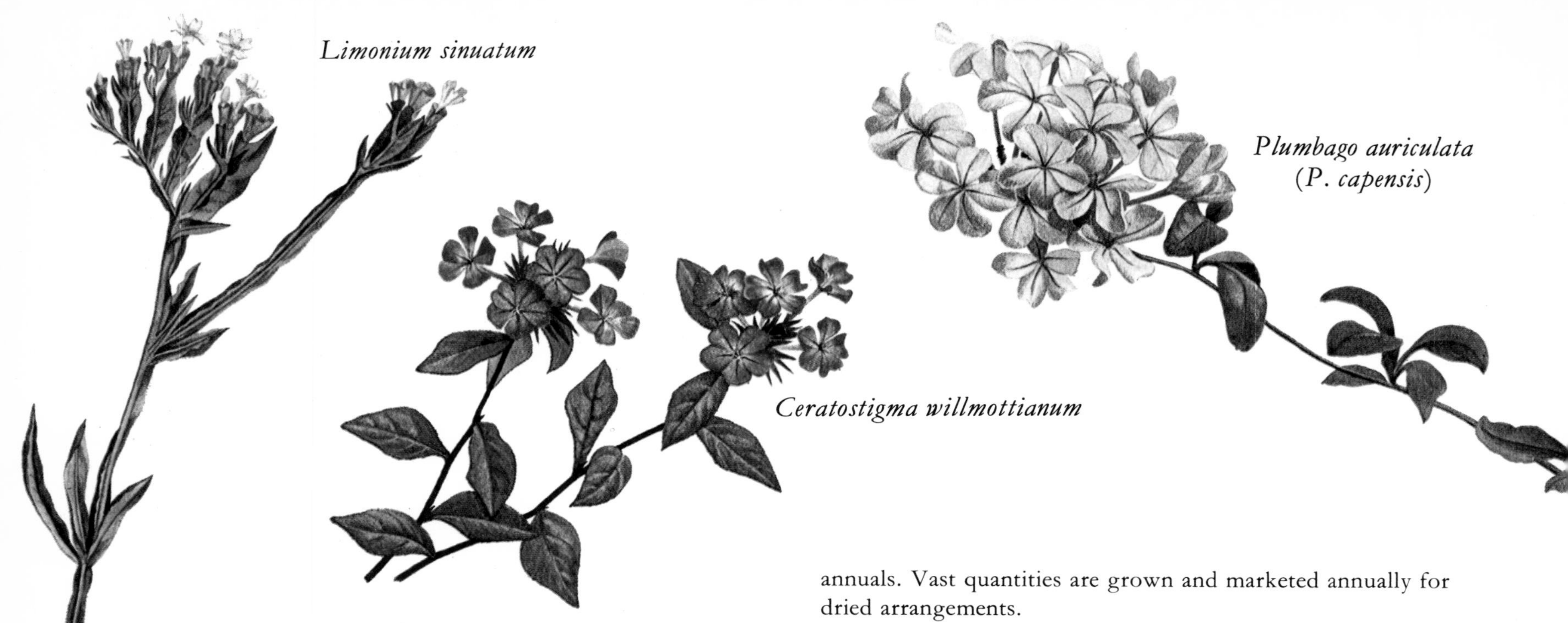

Limonium sinuatum

Plumbago auriculata (*P. capensis*)

Ceratostigma willmottianum

Plumbaginaceae

10 genera and 500 species

These are dicotyledonous plants and shrubs of cosmopolitan distribution, a great number being found in the vicinity of the sea shore. They have simple and alternate leaves, with various types of inflorescence. Individually the blooms are regular and bisexual with five sepals, five petals, five stamens, five styles and a superior ovary.

Plumbago auriculata (*P. capensis*) from South Africa is a slender-stemmed rambling shrub, with neat, evergreen leaves and large trusses of sky-blue, Phlox-like flowers having sticky calyces. It is a great favourite for tropical gardens where it clambers over other shrubs, makes attractives hedges, sprawls down banks or becomes a good wall shrub. In temperate climates it is often planted as a greenhouse climber, but becomes deciduous if the temperature drops much below 7°C (45°F). It flowers long and continuously, thriving in full sun, and is suited by most soils.

P. scandens, the Devil's Herb from the West Indies, is more slender in habit but also white flowered. Juice from the roots and leaves irritates the skin and causes blisters, a circumstance exploited by beggars who create sores to excite pity.

Ceratostigma plumbaginoides from China is hardier and thrives outdoors in England and the US and is valued for its neat habit, vivid blue flowers late in the season and the red autumnal tints assumed by both leaves and stems in the fall. It grows to about 30 cms (1 ft) whereas the somewhat similar *C. willmottianum* from W China reaches 60–120 cms (2–4 ft). This last species is less robust and needs a sheltered position or cool greenhouse treatment.

Limonium (also known as *Statice*) is a genus of some 300 herbs and sub-shrubs of cosmopolitan distribution. Most are found on open sea shores or salt marshes, but in spite of this grow well in ordinary soils—preferring full sun and free drainage. Numerous flowers are carried on the branched inflorescences and the leaves are entire or pinnately lobed. Several species have coloured calyces and are good for cutting or can be dried like everlasting flowers.

L. sinuatum (*Statice sinuata*) from the Mediterranean regions, with blue, mauve, white or rose flowers, and the Algerian *L. bonduelii* with bright yellow blooms are perennials usually cultivated as annuals. Vast quantities are grown and marketed annually for dried arrangements.

L. suworowii, the Candlewick Statice, is a striking annual from Turkestan with stems 45 cms (1½ ft) high terminating in long, dense, cylindrical spikes of rosy-lilac flowers. *L. latifolium* from Bulgaria and Russia is a perennial with rosettes of large leaves topped by stems of 45 cms (1½ ft) packed with cobwebby masses of deep lavender-blue. The flowers of several cultivars, such as 'Violetta' and 'Chilwell Beauty' are a deeper blue.

Armerias are tufted evergreens with round heads of flowers and grassy leaves, suitable for rock-garden pockets and borders. Best known is the European (including British) *Armeria maritima,* a plant 7–15 cms (3–6 ins) high with rich green foliage and small heads of rose, white or red flowers. It is commonly known as Thrift and grows wild on sandy sea shores. *A. plantaginea* (*A. arenaria*) and *A. pseudarmeria,* both from S Europe are similar but taller (up to 45 cms; 1½ ft), their flowers ranging from pale pink to bright ruby red.

Acantholimons are dwarf, tufted evergreens, mostly desert plants, requiring sunny positions and sandy soil. *Acantholimon glumaceum* from Armenia with spikes of bright rose flowers on stems 15 cms (6 ins) high and *A. venustum,* 15–20 cms (6–8 ins), with vivid rose spikelets are typical.

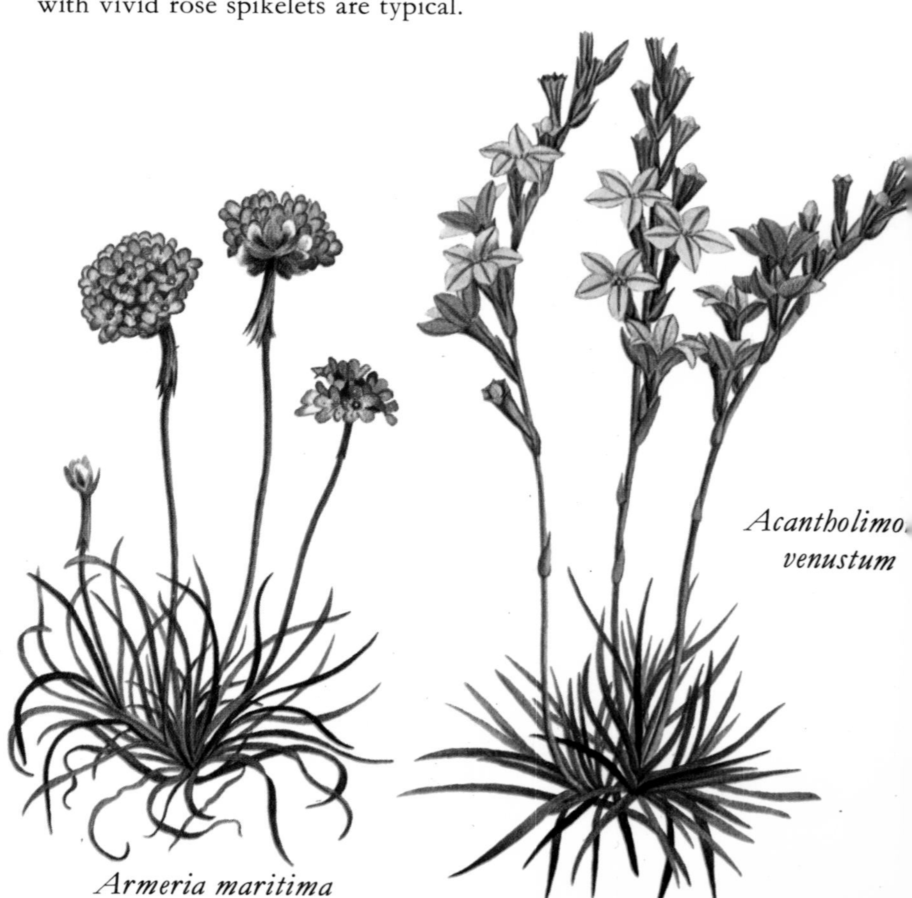

Acantholimon venustum

Armeria maritima

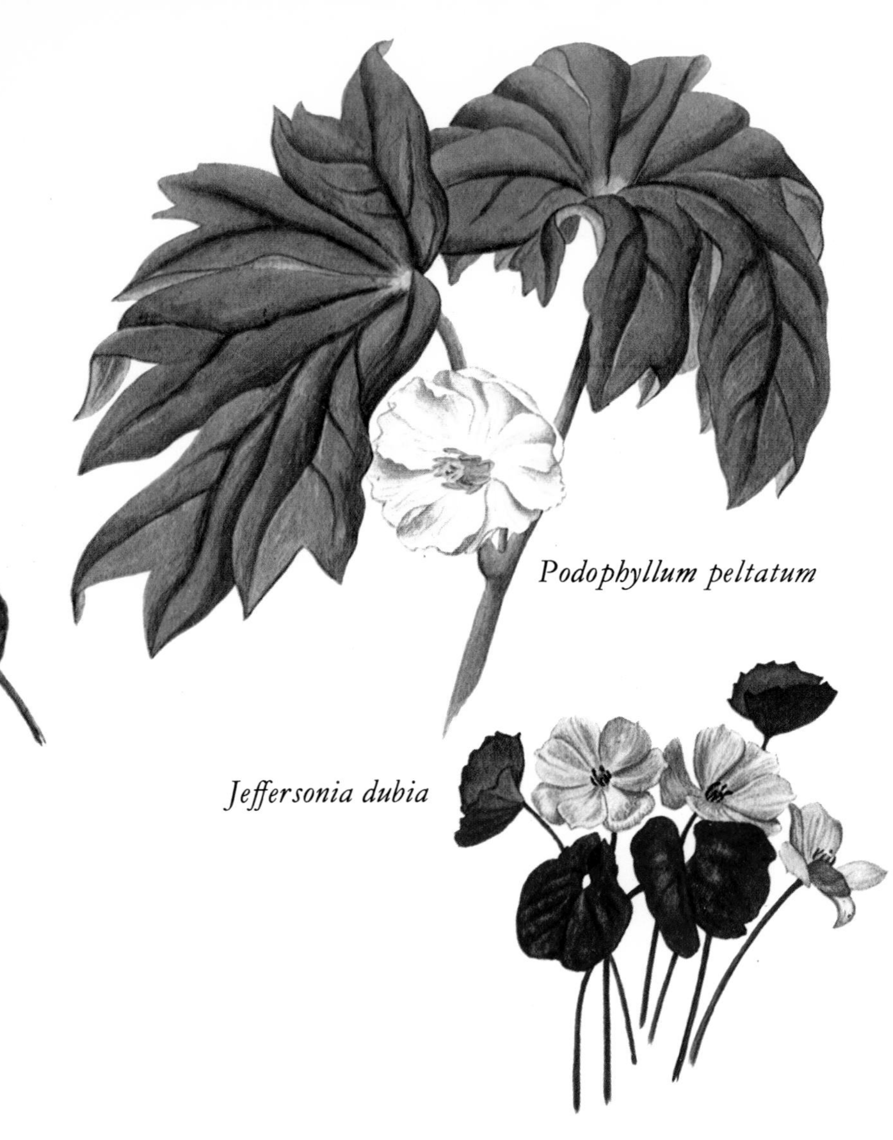

Podophyllum peltatum

Jeffersonia dubia

Podophyllaceae

6 genera and 20 species

These are dicotyledonous N temperate perennials with fleshy roots and palmately lobed or bipartite leaves which spring directly from the root. The flowers may be solitary or borne in cymes or umbels and have four to fifteen sepals, six to nine (rarely ten) petals, four to eight stamens and a superior ovary.

Podophyllum peltatum, the American Mandrake or May Apple, grows to about 30 cms (1 ft) high with stems either terminating in one large centrally poised leaf (balanced like a tea tray) or in a flower stem carrying two large leaves with a single white flower between. Later the blooms give place to oval, yellow edible fruits, which can be made into marmalade and drinks. The roots have medicinal properties but in toxic quantities can cause enteritis. *P. versipelle* from W China has deep crimson flowers and *P. hexandrum* (*P. emodi*) from India, often chocolate-blotched leaves, white or pink flowers and red fruits.

Jeffersonia diphylla, the Twin-leaf or Rheumatism Root, is aptly named because each leaf is neatly cleft down the centre so that there appear to be two leaves on each leaf stalk. The naked flower stems 15 cms (6 ins) tall, carry solitary white flowers about 2.5 cms (1 in.) across. The fruit is a small pod from which the hinged top lifts. *J. dubia*, its Chinese counterpart, has pale to deep lavender-blue flowers and often bronze, young foliage.

Diphylleia cymosa grows taller (30–90 cms; 1–3 ft) and has much the habit of *Podophyllum* with large peltate and lobed leaves 30–60 cms (1–2 ft) across, although the white flowers come in terminal heads instead of singly. They are followed by glaucous blue berries. This species is known in the US as Umbrella Leaf.

Polemoniaceae

15 genera and 300 species

These dicotyledonous plants are herbaceous, although a few do have shrubby bases. Nearly all are native to North America, with slight representation in Chile, Peru, Europe and N Asia. The stems are smooth or slightly hairy, the leaves usually opposite and the flowers in crowded cymes. Individual blooms have five sepals, five petals, five stamens and a superior ovary.

The genus *Phlox* is one of the most important and provides some splendid garden plants, both for the border and rock garden. All but one of the sixty or so species are either North American or Mexican, but British and European hybridists have played an important part in transforming these demure, sometimes insignificant individuals into today's splendid cultivars.

As far as *P. paniculata* is concerned, the first break came when one John Downie of Edinburgh raised an 'eyed' variety, a characteristic which makes so many modern cultivars attractive. In later years M. Prichard and more recently H. B. Symons-Jeune worked on the genus, producing sturdy, free-flowering cultivars in a wide range of good clean colours.

These Phlox are border plants, growing 75–105 cms (2½–3½ ft) tall, with heavy trusses of flower, and should always be grouped for display purposes. They like a fertile soil and must never become dry during the growing season for *Phlox paniculata* is usually the first border perennial to show signs of drought. They will grow in sun or light shade and it is desirable to thin out the weaker shoots in spring in order to obtain better and stronger flowers.

To avoid eelworm infection (the worst pest and one which may be transferred via the leafy shoots), propagation should be by means of root cuttings.

'The Brigadier'

Phlox paniculata hybrids

'Wm Kesselring'

Cantua buxifolia

On December 10th, 1745 'one sod of the fine creeping Lychnis' was sent to Peter Collinson in England by John Bartram of America. This so-called 'Lychnis' was *P. subulata,* the Moss Phlox, one of the glories of the spring rock garden. The mat-like trails sometimes cover a metre or more and when in flower make solid carpets of crimson, pink, mauve or white. Their spectacular appearance when drooping over banks or cascading from rocks, impelled Reginald Farrer (*The English Rock Garden*) to suggest that 'The day that saw the introduction of *Ph. subulata,* ought to be kept as a horticultural festival'. There are cultivars with pink, lavender and red flowers; the two portrayed are 'G. F. Wilson', lavender-blue, and 'Temiskaming' ('Temiscaming'), brilliant magenta. *P. adsurgens,* with large pink flowers, *P. douglasii* 'Boothman's Variety', clear mauve, and *P. amoena*, rich pink, are others for the rock garden.

The summer-flowering annual Phlox are derived from *P. drummondii.* They are beautiful plants for bedding out or sowing in drifts in sunny and rich but well-drained borders. They remain weeks in character with large clustered heads of wide-open white, pink, red, scarlet, purple and violet flowers which generally have conspicuous white or dark eyes. There are three main groups: 'Grandiflora', 25–30 cms (10–12 ins) high; the dwarf 'Nana Compacta' 15–20 cms (6–8 ins); and variants with star-shaped corollas of which 'Twinkling Stars' is an attractive representative.

Polemonium foliosissimum

Phlox drummondii 'Twinkling Stars'

Polemonium caeruleum is a plant of wide distribution throughout the Northern Hemisphere and is conspicuous in Lapland's alpine meadows during the long light days of the arctic summer. It is a perennial, 30–90 cms (1–3 ft) tall, commonly known as Jacob's Ladder or Greek Valerian, with blue (rarely white) open bell-shaped flowers in panicles and smooth ladder-like, pinnate leaves.

P. foliosissimum from the North American Rocky Mountains is deep lavender-blue or white flowered; *P. grandiflorum* from Mexico is lilac, the flowers solitary, or in twos or threes; *P. carneum* is a western North American species with blue, flesh or yellow flowers; and *P. flavum* from Mexico has bunches of yellow blossoms on stems 60–90 cms (2–3 ft) high.

Polemoniums like full sun and will grow in any good garden soil provided this does not dry out in summer. Propagation is by division or seed.

Gilia (in the wide sense) embraces some 120 species, all from temperate and subtropical America. Several of the annuals can be accommodated in the summer border for their clusters of small yellow, white, pink, red or blue funnel-shaped flowers. They favour a light, well-drained, moderately rich soil in sun or light shade but also make charming pot plants in a cold greenhouse.

G. tricolor, the popular Bird's Eye, is a plant of 60–75 cms (2–$2\frac{1}{2}$ ft) with slender stems and variously coloured, bell-shaped flowers. *G. dianthiflorus* (*Linanthus dianthiflorus*) is a small (5–10 cms; 2–4 ins) Californian, forming tufts of thread-like leaves which are almost hidden in summer by countless white, pink or lilac funnel-shaped flowers.

G. achilleifolia (*G. abrotanifolia*), also Californian, is usually pale blue, the flowers in clusters on stems of 60 cms (2 ft).

Among the perennial kinds mention should be made of *G. californica* (*Leptodactylon californicum*), the Prickly Phlox, a low-spreading and branching shrub up to 90 cms (3 ft) with stiff needle-like leaves and large rosy or lilac flowers. It is only hardy in sheltered parts of the British Isles and the warmer zones of North America.

Gilia rubra is the Standing Cypress or Scarlet Gilia, a handsome biennial from the warmer parts of the United States. It is often known as *Ipomopsis rubra* or *Gilia coronopifolia* and has basal rosettes of finely cut leaves and leafy stems 1–2 m (3–6 ft) high, carrying many scarlet, trumpet-shaped flowers, mottled inside with yellow.

Cantuas come from South America, chiefly from the Andean regions of Ecuador, Peru and Bolivia. They are evergreen shrubs and trees with simple, almost stalkless leaves alternately arranged on the branches.

The flowers are occasionally solitary but more frequently arranged in many-flowered corymbs, individual blooms being tubular in shape with five well-defined lobes. In cool temperate countries Cantuas usually have to be grown under glass, although

Phlox subulata

'Temiskaming' 'G. F. Wilson'

stove heat is not necessary. They are no more tender than Fuchsias and grow well in England in cool greenhouses having a winter temperature around 5°C (40°F).

Cantua buxifolia from the Bolivian and Peruvian Andes grows about 120 cms (4 ft) tall and has red and yellow streaked flowers. It is the National Flower of Peru where it may be seen festooning the hedgerows, although of necessity it has to be a greenhouse subject in Europe and nearly all of the US.

C. bicolor from Bolivia 120 cms (4 ft) in height, is a shrub whose flowers are yellow with scarlet lobes and *C. pyrifolia,* an Ecuadorian and Peruvian tree, has yellow and white blooms.

Propagation is effected by seed or cuttings.

Other genera in Polemoniaceae include *Bonplandia,* strong and unpleasantly scented sub-shrubs, about 60 cms (2 ft) high, with pairs of violet blossoms and the somewhat similar *Loeselia* species with two-lipped flowers of lilac or scarlet. These need similar cultural conditions to *Cantua.*

Collomias resemble and should be treated like Gilias. They have dense flower heads; scarlet or rarely yellow in *Collomia biflora,* rose in *C. heterophylla* and yellow fading to red in *C. grandiflora.* The usual height is around 30 cms (1 ft) and the plants are normally treated as annuals.

Polygala myrtifolia

Polygalaceae

12 genera and 800 species

These are dicotyledonous herbs, shrubs or small trees which except for New Zealand, Polynesia and the arctic zone have cosmopolitan distribution.

Distinctive features include simple, entire, opposite, alternate or whorled leaves, sometimes with thorny or scaly stipules; inflorescences which take the form of a spike, raceme or panicle; and bisexual but irregular flowers with five sepals, the two inner often being enlarged and petal-like (three or five petals forming a keel like that of Leguminosae); usually eight stamens and a superior ovary.

Several *Polygala* species are used as rock plants in temperate climates, notably *P. lutea*, an E North American with dense spikes of orange flowers on stems 15–30 cms (6–12 ins) high; *P. chamaebuxus,* the Bastard Box, a European mountain flower with fragrant, cream-tipped purple blossoms and the lime-loving *P. calcarea* or Chalk Milkwort. This is found in Britain and has blunt, compact racemes of six to twelve bright blue, pink or white flowers on stems of 7–15 cms (3–6 ins).

The common Milkwort, *P. vulgaris,* which is also European and similar to *P. calcarea*, is particularly striking in its blue form. Its common name according to Gerard suggests the virtues ascribed to it as an assistant to wet-nursing. The powdered root has been given in cases of pleurisy and infused in boiling water for the relief of coughs.

Several others are grown in a peat/sand/soil compost as early flowering greenhouse subjects. These include the South African *P. myrtifolia,* a densely leafy evergreen shrub of 1.25–2 m (4–6 ft) with rich reddish-purple flowers grouped in clusters at the tips of the branchlets. *P. virgata* is the South African Purple Broom and much resembles *Spartium junceum* because of its green, sparsely leafed, erect stems. The reddish-purple flowers are carried on graceful spikes of 25 cms (10 ins); these last well when cut. It is hardy to light frosts and popular outside in warm temperate climates.

Securidaca longipedunculata is the Violet Tree of South Africa, a deciduous tree of 6 m (20 ft) with long-stalked clusters of bright purple, sweetly scented, pea-shaped flowers in summer. All the *Securidaca* species are tropical; *S. virgata* from the West Indies is a climber with yellow, pink and white fragrant flowers and *S. erecta,* a tropical American shrub, has panicles of red blooms.

Polygonaceae

40 genera and 800 species

This is a large family of dicotyledonous, chiefly N temperate herbs, together with a few shrubs and trees, sometimes twining.

The leaves are usually alternate and simple, rarely toothed; the stipules are united to form a sheath clasping the stem above the leaf base. The flowers are small, regular, perfect and usually grouped with three to six perianth segments, up to nine stamens and a superior ovary.

Certain members of Polygonaceae are coarse perennials out of place in the small garden, but spectacular enough on the banks of large water gardens or in a woodland setting. They include *Rheum palmatum*, the Sorrel Rhubarb from Tibet, with gigantic, deeply cut, five-lobed leaves and stems 1.5 m (5 ft) in height, terminating in showy panicles of creamy flowers; var. *atro-sanguineum* has red flowers. *R. alexandrae* from China and Tibet is a curious species which scarcely resembles a Rhubarb. It has large, yellowish-green, floppy leaves and a stem 90–120 cms (3–4 ft) tall remarkable for the large, reflexed, straw-coloured bracts which sheath the flowers all down the stem like the tiles on a house.

Some of the Knotweeds (Polygonums) are also of giant size, especially *Polygonum cuspidatum* (also known as *Reynoutria japonica*), a handsome but invasive Japanese perennial with smooth, reddish Bamboo-like stems carrying large oval leaves and sprays of feathery white flowers. The variegated form with cream striations on the leaves is the most impressive and makes a handsome picture grown by itself in a bed on a lawn. *P. sachalinense* (*Reynoutria sachalinensis*) from the island of Sakhalin, is another vigorous species, the greenish-white flowers carried in the axils of large, oval-oblong leaves on stems of 2.5–4 m (8–12 ft).

Several small to medium Polygonums are useful in the border and rock garden for their late flowers, especially two mat-forming cultivars of *P. affine,* the Himalayan Fleece Flower from Nepal. They are 'Darjeeling Red' and 'Donald Lowndes', crimson and deep pink respectively; both are free flowering with congested spikes on stems 22 cms (9 ins) high. *P. vacciniifolium,* another Himalayan mat-forming species, is bright rose.

P. campanulatum bears branching stems 90 cms (3 ft) tall with simple oval leaves which are white beneath and has drooping clusters of pale pink, fragrant flowers. This Himalayan plant has a white-flowered Chinese counterpart in the variety *lichiangense*. Both can be propagated by division in spring.

P. bistorta, a N temperate (including British) species likes moist conditions. It has slender stems of 60 cms (2 ft) terminating in dense, cylindrical spikes of white or pink florets 5 cms (2 ins) long, or deep rosy-pink in the clone 'Superbum'. The dried rhizomes have medicinal uses and were once eaten in soups and stews by Cheyenne Indians.

P. "reynoutria" of gardens is a female form of *P. cuspidatum* var. *compactum,* a stiffly erect plant with branching, purple-spotted stems 60 cms (2 ft) high, leathery leaves and axillary spikes of pink flowers—red in the bud—which develop showy enlarged red ovaries. The male plant is less attractive so the female form is usually propagated vegetatively.

Antigonon is a genus of strong perennial climbers with heart-shaped leaves and lacy inflorescences of pretty pink or white flowers. *A. leptopus,* the Corallita, Chain of Love or Coral Vine from Central America, is widely grown in the tropics on banks, hedges, garden fences, trellises and arbours and it even clambers up tall trees. The feathery trails of locket-shaped pink flowers are borne on sprays ending in tendrils. It is the sepals which are coloured—there are no petals. The tubers are edible and have a nutty flavour.

Triplaris are South American trees, strikingly beautiful and often used in parks and gardens in Jamaica and elsewhere in the tropics. One of the most popular is the Long John, *T. surinamensis,* which reaches 20–23 m (60–80 ft) at maturity (occasionally more). It grows conically with dark green, tapering leaves of leathery texture and very large and dense clusters of pure white flowers. As these age they turn first pink and then brown but meantime the calyces elongate like shuttlecocks and change from green to white to bright red. Often flowers and calyces in their various stages can be seen on the same tree, presenting an unforgettable picture.

Antigonon leptopus

Like all *Triplaris,* however, the Long John must be treated with respect. Thousands of fierce stinging ants inhabit the hollowed out stems and rush to attack anyone rash enough to gather the flowers or even shake the tree.

Richard Spruce was the first to describe this phenomenon in *Notes of a Botanist on the Amazon and the Andes* (1908) . . . 'the slender tubular branches, often geniculate at the leaf nodes . . . which are the sallyports of the garrison, whose sentinels are always pacing up and down the main trunk, as the incautious traveller finds to his cost when . . . he ventures to lean his back against a tree.'

Amazonian Indians exploit the activities of the ants and use the ready hollowed out branches for tobacco pipes.

Among other genera in this family mention should be made of the Muehlenbeckias, mostly climbing shrubs with wiry stems and small five-lobed flowers. *Muehlenbeckia complexa,* a deciduous New Zealand climber has greenish-white flowers with a delicious fragrance and twiggy shoots which as they grow become hopelessly entangled. *M. axillaris,* also from New Zealand, makes a dense mat of creeping needle-thick stems and *M. adpressa,* which is Australian and another climber, has white edible berries resembling Gooseberries.

Eriogonum umbellatum from British Colombia to California is the Sulphur Plant, a sub-shrub 30 cms (1 ft) high with silvery plush-textured leaves and umbels of fragrant, soft yellow flowers.

Several members of Polygonaceae are useful to Man, especially Buckwheat (*Fagopyrum esculentum*) with edible seeds used for flour, buckwheat cakes and porridge. The flowers are the source of a commercial honey and the plant is also grown as a green manure. Rhubarb (*Rheum rhaponticum*) has edible leaf stalks used for dessert and wine and other species have roots with medicinal properties.

Polygonum campanulatum

Polygonum affine 'Donald Lowndes'

Pontederiaceae

6 genera and 30 species

This is a small family of tropical, monocotyledonous water plants, often floating and increasing rapidly by means of fleshy stolons. The flowers may be regular or zygomorphic and have six perianth segments, one, three or six stamens and a superior ovary.

Among the most decorative is *Pontederia cordata,* the Pickerel Weed of eastern North America. This is hardy in Britain, where it is a popular aquatic for ornamental ponds and water gardens. Growing about 60 cms (2 ft) high, the species has creeping rootstocks with clusters of heart-shaped leaves on long basal stalks and spikes of funnel-shaped, violet-blue flowers which are hairy on the outside. It thrives in 13–15 cms (5–6 ins) of water. In the SE states of America a form occurs (var. *lancifolia*) with longer flower spikes and more lanceolate foliage. This may grow 1.25–1.5 m (4–5 ft) in height but is less hardy than the type species.

Eichhornia crassipes (*E. speciosa*), the Water Hyacinth of South America, is a striking aquatic with long, trailing roots (which fish use as ova depositories) and rosettes of smooth, light green, cordate leaves. The latter have large, bladder-like petioles, swollen like sausages and filled with spongy tissue which makes them buoyant. The flowers are showy but only last a few hours. Borne on spikes 30 cms (1 ft) tall they are lavender-blue with conspicuous gold and blue peacock markings on the upper petals. Reproduction is by means of stolons and under favourable conditions extremely rapid, so that in many parts of the Old and New World tropics the Water Hyacinth has become a serious pest, choking waterways and impeding navigation. In Mexico the plants are raked out and used as fodder, but have little nutritive value. They are also recommended as a source of cellulose. Frost kills the plant instantly so it is only suitable for tropical pools or glasshouse cultivation in cool climates. *E. azurea*, with spongy but not inflated petioles makes shoots 1.5–2 m (5–6 ft) long which root in the mud. The funnel-shaped flowers are deeper in colour than *E. crassipes* and have yellow markings with the inner segments prettily fringed. It is native to Brazil.

Heteranthera is a genus of about ten tropical and subtropical, moisture-loving plants from America and Africa. They have two kinds of leaves, round or oval floating types and linear submerged ones. Grown completely submerged, a few make valuable subjects for tropical aquaria, although they also thrive outdoors under marsh conditions—for the summer only in frost-prone climates. They can be propagated by division.

The most important are *H. dubia* (*H. graminea*), usually known as the Mud Plantain, with sessile and alternate leaves and yellow axillary flowers; *H. reniformis,* a fleshy plant with rounded leaves above water level, linear beneath and spikes of white or blue flowers; and *H. zosterifolia,* a perennial which grows practically submerged and has slender stems carrying many ribbon-like leaves and light blue flowers. These are all from North or South America.

There are three species of *Monochoria*, all Asiatic, blue-flowered bog or water plants. *M. hastifolia* has fleshy stems terminating in small, broadly heart-shaped leaves, below which are axillary, crowded inflorescences of pale blue to light-violet blossoms. When young the plant is sometimes eaten as a vegetable in some parts of Malaysia.

M. elata (*M. hastifolia elata*) is similar but the inflorescence is elongated instead of bunched and the leaves are much narrower. *M. vaginalis* has spikes of small dark blue flowers. Monochorias can be used in tropical water gardens or under glass elsewhere.

Pontederia cordata

Eichhornia crassipes
(*E. speciosa*)

Portulacaceae

19 genera and 350 species

These are dicotyledonous plants which are frequently prostrate with fleshy, opposite or alternate leaves, circumstances which enable many to thrive in warm, dry regions of the world such as are found in the Cape (South Africa) and South America. The bisexual flowers, usually borne on cymes are regular, with two sepals, four, five or more petals, one or two whorls of five or sometimes many stamens and usually a superior ovary.

The flowers secrete honey and are insect pollinated. Most species have mucilaginous (often acid) stems and foliage. The roots of certain kinds, for example *Claytonia, Talinum* and *Lewisia,* are edible and the leaves of others such as *Portulaca oleracea* and various species of *Calandrinia, Talinum* and *Claytonia* may be eaten as salad or greens.

Lewisias are attractive rock plants for lime-free soils, thriving in rich but well-drained situations. Many are hardy in warm spots in Britain, but in hot dry areas are best planted in partial shade. The succulent, usually narrow leaves are arranged in a rosette at the apex of a thick, fleshy rootstock; the flowers are either solitary or borne on branched panicles. They are variable, in shades of pink, rose and purple and although short-lived maintain themselves by means of self-sown seedlings. *Lewisia tweedyi* and *L. howellii* are particularly fine species.

L. rediviva is the Bitterroot of western North America and the State Flower of Montana. Notwithstanding its diminutive height (2.5–5 cms; 1–2 ins) the rose-purple flowers are large and attractive. It was first discovered in 1805 by Captain Meriwether Lewis while on an expedition to trace the route of the Missouri. He saw the starchy white roots drying in an Indian encampment but found that they 'had a very bitter taste that was naucious'. Later, after finding this handsome plant growing, he gave the name Bitterroot to a river, a valley and a range of mountains.

Some eight years after the expedition had returned home the botanist Pursh—who was writing a flora of North America—examined their herbarium specimens. Discovering signs of life in the Bitterroot he planted some, which grew and flowered the following year. *Lewisia rediviva* thus commemorates not only a great explorer but the unbelievable tenacity of the plant's roots.

Portulaca grandiflora

Other ornamentals belonging to Portulacaceae include *Portulaca grandiflora*, the Rose Moss or Sun Plant of Brazil, whose large flowers range from the original reddish-purple to orange, yellow, salmon and white in the cultivars. *P. lutea* is a large-flowered, yellow, South American sort and *P. oleracea,* the common Purslane from Southern Europe, Africa and Asia is a smaller, yellow-flowered kind of which an improved form *sativa* is sometimes cultivated as a salad vegetable.

The genus *Calandrinia* consists of short annual and perennial succulents, many of which—like the purple to white *C. remota* and magenta-rose *C. polyandra*—are native to Australia. *C. umbellata* from Chile and Peru is called Red Maids on account of its racemes of numerous red flowers and the annual *C. menziesii* is a purple-flowered Californian. Calandrinias and Portulacas only open in sunshine.

Talinums are fleshy plants with flat leaves and terminal clusters (or rarely solitary) flowers. Like others of a similar nature they are sometimes grown in greenhouses in cool countries or in warmer climes as bedding plants. *Talinum caffrum* (*Portulaca caffra*) with yellow flowers, and *T. portulacifolium* (*T. cuneifolium*), magenta, are both native to East Africa.

Lewisia hybrid

Lewisia tweedyi

Calandrinia polyandra

Primula vulgaris

Primulaceae

20 genera and 1000 species

This is a large and important family of dicotyledonous herbaceous plants having cosmopolitan distribution, especially in N temperate zones. The majority favour cool, moist root conditions and will not succeed in intense heat or strong sunshine. They are accordingly plants for temperate gardens, many thriving under woodland conditions or in the vicinity of the water garden.

The roots may be fibrous, rhizomatous or tuberous with the leaves usually (but not always) simple, alternate, opposite or whorled and frequently mealy. The flowers are bisexual and regular, borne on scapes or solitary at the ends of the stems. They have a five-parted calyx, a five-lobed corolla (none in *Glaux*), five stamens, one style and stigma and a superior occasionally semi-inferior ovary which develops to a capsule with few or many seeds.

In some genera such as *Primula, Hottonia* and *Androsace* the styles and stamens are of different lengths (heteromorphic), a device which favours cross-pollination. The family has scant economic importance, although the rhizomes of *Primula* and *Cyclamen* have been used for medicinal purposes and wine is made from the flowers of *Primula vulgaris* and *P. veris*.

Except for Spain and Egypt Cyclamen are known to occur in all countries with a Mediterranean coastline. They are also represented in Switzerland, Austria, Bulgaria, Asia Minor and NW Iran, with their greatest concentration in Greece and its islands, Italy and Yugoslavia.

Some are hardy in N Europe (including Britain) and temperate North America but others are tender or need cool greenhouse treatment. Growing these in an alpine house or frame with frost just excluded is a popular method in Britain, although the small hardy sorts are often colonized in light shade beneath trees or in garden pockets. Some of the latter, notably *Cyclamen hederifolium* (*C. neapolitanum*), live to a great age (more than a century), making large round tubers carrying hundreds of flowers in early autumn.

From a spectacular viewpoint the most arresting are the florists' forms of *C. persicum*. These are usually grown in pots of gritty compost with some good leafmould and a few pieces of chalk or limestone amongst the drainage. The pendulous flowers are large with the petals turned back so that they give the impression of being upright. They come in white and a wide range of pink, salmon-red and purple shades. Sometimes flowers show contrasting eyes of another colour or have frilled or ruffled petals. The type species is native to the Greek islands, Israel and Asia Minor and is sweetly scented with handsome marbled leaves. These characteristics are being bred back into the plants, so that fragrant strains are now available, as too are patterned-leafed sorts such as the Silberblatt (Silver Leaf) group.

Failure with pot Cyclamen can normally be traced to excessive temperatures during cultivation and a dry atmosphere. They need a temperature constant around 16°C (60°F), plenty of light and a buoyant, slightly moist atmosphere. Wet soil can also lead to disaster, causing the buds and leaf bases to rot. Sponging the leaves daily helps maintain humidity in a centrally heated home as does the use of a very fine spray, which however should not touch the blooms.

Outdoor Cyclamen require an annual top dressing of rotted leaf-soil during the dormant period. Although they thrive on chalk they are equally happy in acid, Rhododendron-type soils. Reproduction from seed is easy provided this can be gathered before ants and mice discover the capsules.

Spring-flowering kinds include *C. pseudibericum, C. repandum* and *C. coum* (*C. vernum, C. atkinsii* or *C. orbiculatum*); *C. hederifolium* blooms in autumn and *C. europaeum* in late summer and autumn. The usual height is 5–7 cms (2–3 ins) and the tiny,

Cyclamen persicum variants

Polyanthus hybrids

sometimes fragrant flowers resemble miniature swans poised for flight. The foliage is often beautifully patterned with silver markings.

The genus *Primula* comprises some 500 species, all native to the Northern Hemisphere, where they are frequently found in the vicinity of streams or moist valleys and hillsides. Many come from China, the Himalayas and Tibet. Their main desiderata are constant moisture at the roots (although many need sharp drainage), shade from strong sun and in many cases—but not invariably—a lime-free soil.

Bog Primulas such as *P. rosea, P. yargongensis* and the Candelabra Section (those with whorls of flowers arranged in several tiers on the stems) are both moisture-lovers and lime-haters. These like to feel the influence of water without actually sitting in it, so that bog garden conditions where there is always moisture *beneath* the topsoil are ideal.

Granted these facilities they flourish 'like the green bay tree', the self-set seedlings making splendid patches of colour. *Primula* seed rapidly loses its viability when stored, so should either be sown immediately after harvesting or left to drop and naturalize. If desired the little seedlings can be carefully lifted and transplanted to form new colonies.

The use of Primulas is not restricted to the water garden. Polyanthus, which have been known in gardens since the 17th century and were probably derived from crossing coloured Primroses (*P. vulgaris*) with Cowslips (*P. veris*), are invaluable for spring bedding. Such crosses occur in the wild and the early hybrids, according to Mrs Loudon (*The Ladies Flower Garden*, 1844), were 'always yellow and brown', although the poet Thompson in 1728 wrote of 'the Polyanthus of unnumbered dyes'. Today, thanks to hybridizers in Britain, America and New Zealand especially, they come in a wide range of shades—white, yellow and orange, through pink to purplish-rose and pale blue to deep violet. Double Polyanthus occur occasionally and were known in the 18th century by the comical name Pug-in-a-Pinner.

Many Primulas are used in rock-garden plantings, either in flat pockets as *P. farinosa,* a lilac or purple Eurasian species, or wedged between rocks so that they trail downwards. The lilac-blue *P. marginata* from the Maritime Alps is a good kind for this purpose, and so is *P. allionii,* a small rose, mauve or white species from the same habitat. Temperamental species such as *P. edgeworthii* (*P. winteri*) often rot from excessive cold and wet in winter, but can be kept dry by establishing them under the lee of an overhanging rock. This is a Himalayan plant, pale mauve with a white eye.

Some Primulas are woodlanders, the European (including British) Primrose, *P. vulgaris,* being a noteworthy example. Early in the year lanes, hedgerows, open woodland, streamsides and even railway banks in S England are bright with their pleasantly scented yellow flowers. Each clump of blooms is framed by a rosette of puckered oblong leaves. 'Garryarde Guinevere', the rich crimson variety illustrated, is one of many coloured hybrids of the common Primrose.

Primrose pudding, Primrose vinegar and Primrose wine are still made in country districts but one of the oldest recipes I have is copied from a fifteenth-century cookbook. Called 'Prymerose potage', it entails boiling pounded Primrose flowers with milk of almonds, honey, saffron, rice flour and powdered ginger, garnishing same with Primrose blossoms before serving.

At the death of Disraeli in April, 1881 Queen Victoria sent a wreath of Primroses 'his favourite flower' as a tribute to her old friend and Prime Minister. Later in the century the flower was adopted as the emblem of the Conservative Primrose League.

For damp borders, the wild garden and the waterside single plants are inconspicuous but drifts of Primulas create spectacular effects. Candelabra types such as *P. beesiana, P. bulleyana, P. burmanica, P. pulverulenta* from W China and the Japanese *P. japonica* should always be massed and for the sake of contrast associated with ferns or blue *Iris laevigata, Mimulus* species, *Iris kaempferi* or *Hemerocallis* (Daylilies).

P. beesiana grows about 60 cms (2 ft) with velvety purple flowers but *P. bulleyana,* 75 cms ($2\frac{1}{2}$ ft) high, is more variable, the blooms being orange-scarlet, buff, orange or apricot. The two species cross naturally to produce seedlings showing a wide range of colour. *P. burmanica* has stems of 10 cms (2 ft) with reddish-purple flowers having clearly defined yellow eyes.

P. japonica is the Japanese Primrose, a stout-stemmed species about 45 cms ($1\frac{1}{2}$ ft) high with whorled tiers of white, rose or purplish flowers in the various cultivars.

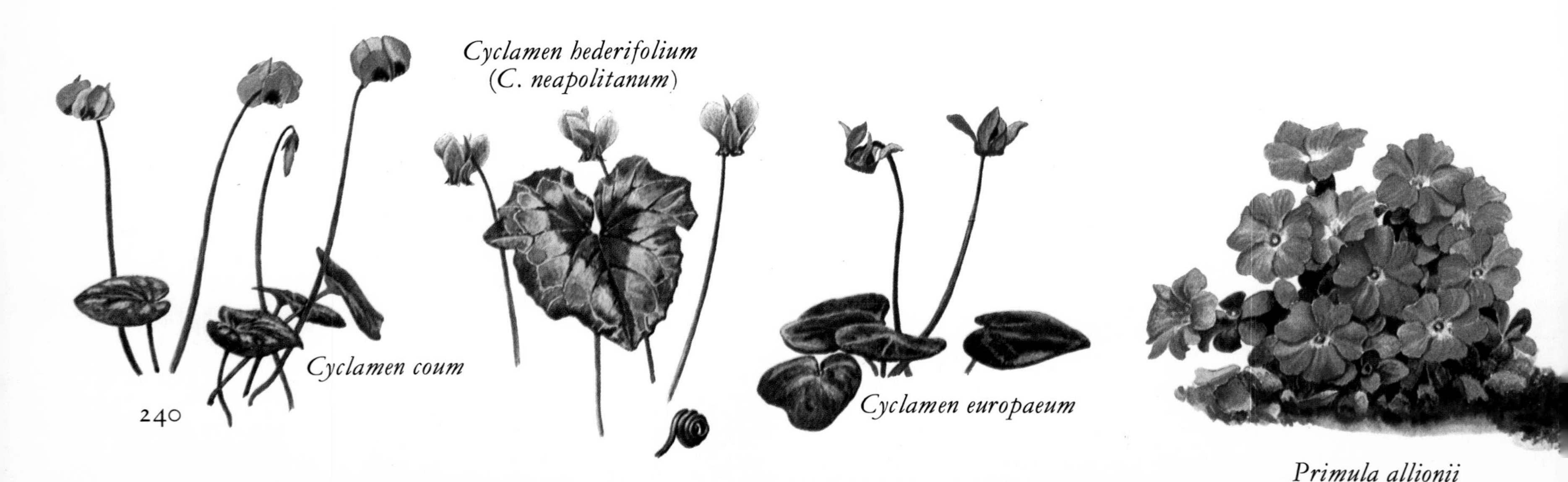

Cyclamen coum

Cyclamen hederifolium (*C. neapolitanum*)

Cyclamen europaeum

Primula allionii

P. pulverulenta, the Chinese Silverdust Primrose, is a robust species with long narrow leaves frequently 30 cms (1 ft) in length and stout, white-powdered, flower scapes carrying several tiers of large deep red, Cowslip-like blossoms. The late G. H. Dalrymple developed a fine strain containing many pink and rose forms which he called Bartley Strain. These, when interplanted with ferns and blue Himalayan Poppies (*Meconopsis*) make an unforgettable fairy-like picture.

Other moisture-loving Primulas include *P. florindae,* the Himalayan Cowslip, a tall species of up to 90 cms (3 ft) from SE Tibet with showy umbels of drooping, fragrant, sulphur-yellow flowers powdered with white inside; *P. helodoxa* from Yunnan with tapering, toothed leaves and whorls of golden flowers on stems of 60–90 cms (2–3 ft); and *P. sikkimensis* from Tibet, another yellow species.

P. rosea from the NW Himalayas is almost the only Primula to tolerate standing water and that for short periods only. It grows about 15 cms (6 ins) high with single brilliant rose flowers, but for garden purposes a form known as 'Micia Visser de Geer' is better as it is taller, brighter in colour and has larger blooms.

The round umbels of *P. denticulata* (another Himalayan species) are about the shape and size of Dandelion seedheads or large golf balls. They come in white, mauve, lavender, deep purple and magenta, the flowers very conspicuous on stems 30 cms (1 ft) tall. Plants come readily from seed but exceptionally fine colour forms are best perpetuated by means of root cuttings.

Primula auricula hybrids

P. juliae is a Caucasian plant only a few centimetres high which makes a good ground carpeter, particularly on heavy soils. The freedom with which it flowers and the brilliant shades of its hybrids provide bold patches of colour in early spring. Birds sometimes attack the flowers otherwise the plants are trouble free and long lasting. Garden forms include 'Old Port', wine-purple; 'Pam', garnet-red; 'Wanda', claret-purple; 'Dorothy', pale yellow; 'Alba', white; and 'Our Pat', dark purple.

P. sieboldii comes from Japan and is one of the best for cutting, the flower stems of 22–30 cms (9–12 ins) terminating in umbels of fragrant white, mauve or rosy-purple flowers. Given a well-drained but moist site, sheltered from hot sun and very bad winter conditions, it proves hardy in the more temperate parts of N Europe and North America; elsewhere it should be grown in a frame and protected with a glass light during the winter months.

Another umbellate species is *P. whitei* from Bhutan, a spring bloomer with blue or bluish-violet flowers which have greenish-yellow eyes. The flower stalk gradually elongates with age.

Countless other Primulas are grown in the world's gardens, including the splendid Auriculas. The type species, *P. auricula* or Dusty Miller, comes from the European Alps and has fragrant yellow flowers on stout mealy stems and powdered oval leaves. The florists' Auriculas may be self-coloured or have a white or yellow eye surrounded by another shade or sometimes have green, grey or white margins to the flowers. They are often very striking but being alpine plants need cool, clear growing conditions—a heavy moist atmosphere is quite unsuitable so that many people overwinter them in frames. Propagation is by offsets (taken in summer) or seed.

Several Asiatic Primulas have become popular pot plants. Large quantities are grown each year for greenhouse and home decoration. Some of these produce their flowers in whorls but *P. obconica* throws up several umbels 15–17 cms (6–7 ins) across, each carrying up to fifteen large Primrose-like flowers. The long-

Primula rosea

Primula denticulata

Primula helodoxa

Primula 'Garryarde Guinevere'

Primula whitei

stemmed rounded leaves are covered with soft glandular hairs which can cause a painful rash on the skin of some persons. The flower colours of modern cultivars range from pale pink to crimson and rich blue to white. *P. obconica* comes from China as does the dainty *P. malacoides*.

The latter is an exquisite little species with many stems 20–25 cms (8–10 ins) high carrying whorls of small mauve, bell-shaped flowers. Under cultivation larger-flowered plants have been developed with single or double mauve, purple, magenta, pink or white blossoms.

P. sinensis seems to have been grown in Chinese gardens for centuries, although no records exist of its being found in the wild. It reached Europe about 1820 and is now a firm favourite, with cultivars ranging from orange-scarlet to deep rose and pink. There are also mauve and white forms and doubles. A less vigorous plant than *P. obconica* with incised hairy leaves, it makes an attractive conservatory subject but should be protected from strong sunlight.

P. x *kewensis* is a hybrid between the Himalayan *P. floribunda* and Arabian *P. verticillata*. It was a chance cross which occurred at Kew in 1898 and was originally sterile, but later mutations have produced fertile seed so that this is now the usual means of increase. The fragrant buttercup-yellow flowers are arranged in six- to ten-flowered whorls on stems 30 cms (1 ft) high, the smooth green leaves being oval-oblong, with waved and toothed edges.

Dodecatheons are small perennials having nodding Cyclamen-like flowers with reflexed petals and rosettes of smooth, oval-oblong, ground-hugging leaves. They require similar conditions to Primroses; that is, light shade and open-textured, leafy soil which will not dry out in summer.

The best known is *Dodecatheon meadia*, the Shooting Star of eastern North America, which has slender, erect stems 30–50 cms (12–20 ins) tall terminating in umbels of up to twenty rosy-purple, backward-pointing flowers. The anthers are bright yellow and form a spearlike tip, giving the bloom a somewhat starry appearance. Others cultivated in gardens are *D. jeffreyi*, reddish-purple with dark purple stamens and *D. hendersonii*, plum-purple with purple anthers and 'rice grain' bulbets on the roots which produce new plants.

Soldanellas are charming little alpines from the mountains of southern and central Europe, with mats of glossy, round or kidney-shaped leaves. The nodding bell-shaped flowers have prettily fringed margins and in their native habitat generate enough heat to melt the surrounding snow and force their way through. They look very charming against a snowy background.

The most commonly cultivated are *Soldanella alpina* which is about 10 cms (4 ins) high and lavender-blue; *S. montana*, one of the finest with round dark green leaves with scalloped margins and stems 15 cms (6 ins) tall carrying between three and ten large lilac-blue bells with out-turned edges; and *S. pindicola*, larger and stouter than *S. montana* but a shyer bloomer. It grows to a height of 20 cms (8 ins) and has rosy-lilac flowers.

Androsace with about 100 annual and perennial species of alpines is constantly acquiring additions by collection and hybridization. Many of these are of easy culture but those species drawn from high altitudes often present cultural problems. These are best left to the specialist. The easier kinds thrive in light, well-drained soil with drainage round the collars—provided by a top dressing of granite chips, pebbles or brick rubble—to prevent rotting at that point. In many situations this is all that is necessary, although protection from excessive winter wet can (if required)

Primula x *kewensis*

Primula malacoides

Primula obconica

be afforded by growing them in the shelter of other plants or fixing a sheet of glass over the crowns during the worst weather.

The most satisfactory for gardens are those in the *Chamaejasme* section. These are tufted with linear or spoon-shaped leaves and umbels of Primrose-like flowers. *A. lanuginosa* in this group has silvery foliage and lavender-rose flowers on stems 5 cms (2 ins) high. It is usually propagated by cuttings which must not be kept too wet and comes from the Himalayas. All the forms of *A. sarmentosa* and *A. primuloides* (also Himalayan) are worth growing, especially *chumbyi*, *watkinsii* and *yunnanense*. The flowers are pale to deep pink on stems of 10 cms (4 ins) and appear in early summer. They make runners and are easily increased by detaching rooted rosettes.

Several species of *Anagallis* are grown as annuals for edging borders in summer or as pot plants in a cold greenhouse. They produce their charming little flowers in great profusion over a period of several months. Light soil and sunshine give the best results—indeed the Scarlet Pimpernel, *A. arvensis,* is sometimes called Poor Man's Weather Glass on account of its extreme sensitivity to atmospheric changes. When rain threatens the flowers close but in fine weather remain open most of the day.

A. arvensis, which is a native of Britain, has brilliant scarlet flowers on branching, almost prostrate stems but it is a variable plant so that bright blue and dark blue flowers are sometimes found. In the Valley of the Nile where it is a common weed it has a local reputation as a medicine in cases of mania and hydrophobia and is also employed as a home remedy for gallstones, lung and other ailments. It is naturalized in the USA.

A. linifolia, a beautiful little plant from the Mediterranean region, has narrow, dark green leaves on spreading stems. These form a compact bushy plant 22–45 cms (9–18 ins) high, smothered with clusters of deep blue flowers which turn crimson with age. The Bog Pimpernel, *A. tenella,* is rosy-pink with deeper red veining. It needs a damp situation and is a W European (including British) native.

Totally aquatic is *Hottonia palustris,* variously known as Water Violet, Water Feather and Featherfoil. It is a European plant frequently established in ornamental ponds as an oxygenator. The submerged foliage is bright green and deeply cut into narrow segments; the flowers in crowded whorls, on stems rising 20–38 cms (8–15 ins) above the water surface. They are lilac-mauve and in flower in early summer. Towards autumn, winter-buds are formed, which drop to the bottom of the pond and remain dormant until the following spring.

Lysimachia comprises a large genus with some 200 species of cosmopolitan distribution. Those in cultivation are plants for damp situations such as the vicinity of a water garden, although several kinds can be naturalized in damp woodland areas.

One of the most widespread is *L. vulgaris,* the Yellow Loosestrife, an erect perennial of 45–90 cms (1½–3 ft), with simple or branched stems carrying the oval-lanceolate leaves in whorls, and axillary and terminal panicles of bright yellow, bell-shaped flowers.

The species is native to temperate Asia and Europe (including Britain), where for centuries its sedative powers were believed to tame fierce animals. The Romans believed that branches of the plant put under the yokes of oxen kept them from quarrelling—probably because flies and other insects were allergic to the herb.

According to Pliny 'it retaineth and keepeth the name *Lysimachia* of King Lysimachus, the sonne of Agathocles, the first finder-out of the nature and vertues of this herbe'. He adds 'it dieth haire yellow, which is not very unlike to be done by reason the flowres are yellow' and that snakes 'craule away at the smell of Loosestrife'. Gerard, who also called the plant Willow-herb, says that 'the smoke of the burned herbe driveth away serpents and killeth flies and gnats in a house'.

L. (*Naumbergia*) *thyrsiflora* from the same geographic regions as *L. vulgaris* tolerates moisture to the extent of sometimes occurring naturally as a bog plant. It is an erect herb of 30–60 cms (1–2 ft) with small yellow, axillary flowers and tapering stalkless leaves.

L. clethroides from China and Japan is distinct with long, nodding, rather dense spikes of small white blossoms on stems 45–90 cms (1½–3 ft) high. Its oval leaves assume brilliant tints in autumn.

L. nummularia is variously known as Creeping Jenny, Creeping Charlie, Moneywort and Herb-twopence. It is a European plant naturalized in the east US with small rounded leaves in opposite pairs on long trailing stems, which root readily in moist soil. The bright golden, cup-shaped flowers appear in the leaf-axils during early summer. When happily situated the plants show a tendency to spread but it is not difficult to control. The golden-leafed form *aurea* is the most decorative and is often used for clothing banks, in hanging baskets, in the bog garden and even submerged as a cold-water aquarium plant.

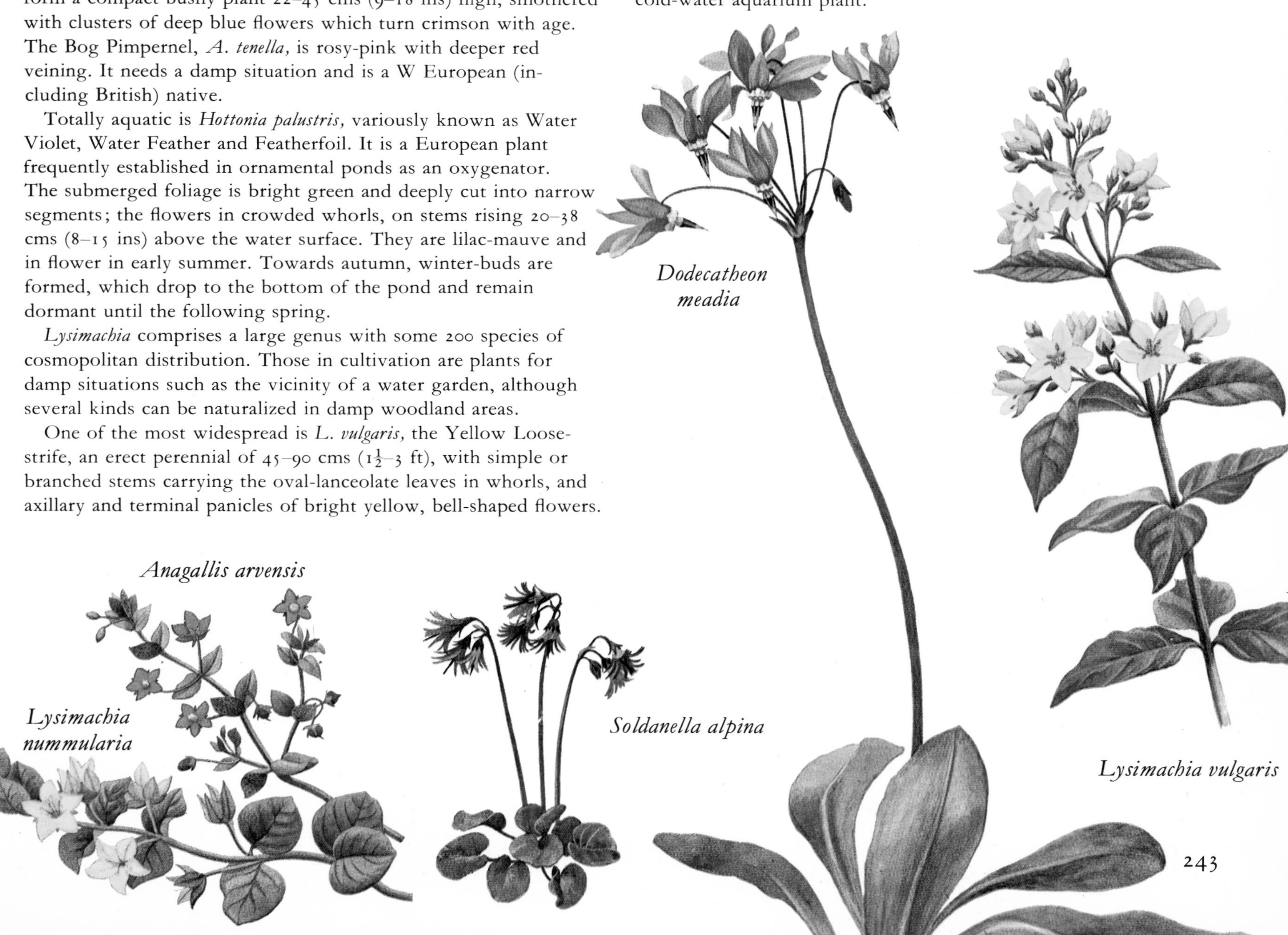

Dodecatheon meadia

Anagallis arvensis

Lysimachia nummularia

Soldanella alpina

Lysimachia vulgaris

Proteaceae

62 genera and 1050 species

This is an interesting family of dicotyledonous trees and shrubs, mostly from tropical areas where a long dry season occurs annually. *Protea*, perhaps the most noteworthy genus is found only in southern and central Africa but other genera are native to South America, Malaya, the Pacific Islands, New Zealand and Australia as well as Africa.

The flowers, insignificant in themselves, are often grouped into large inflorescences (spikes, racemes, heads, and so on) which are among the showiest in the vegetable kingdom. Curiously, although many have their pollen freely exposed in spoon-shaped anthers at the ends of long stamens, they are not wind fertilized but pollinated by birds or sometimes insects. Individually the flowers are bisexual (except in *Leucadendron* and *Aulax*), often zygomorphic, with four perianth segments, four stamens, a superior ovary and one to many seeds (often in hard woody capsules). The leaves are usually tough, and have thick cuticles and frequently a covering of hairs as well to check transpiration.

The only representative grown to any extent in Europe is the Australian Silky Oak, *Grevillea robusta,* which is raised from seed and used as a house plant or on occasions planted outside in summer with other subtropical bedding. It has green, much divided foliage rather like some of the *Asplenium* ferns, but only in the subtropics can one see the handsome clusters of one-sided orange or yellow flowers in large branched racemes. Tea planters at Nandi Hills (Kenya) use the Silky Oak (planted between the tea bushes) as a shade tree and the timber—the species grows to 30 m (100 ft)— is also useful for interior furnishing and cabinet work. Other species have long racemes of white, red, pink or golden flowers or umbel-like racemes of various shades.

G. rosmarinifolia, a bushy shrub of 2–2.15 m (6–7 ft), is often used as a hedging plant in its native Australia. It has narrow, dark green leaves—like Rosemary—and terminal clusters of rosy-red (rarely white or pink) flowers.

The Protea is the National Flower of South Africa where there are over a hundred species. Most of these have proved to be good garden plants in the Republic and would probably succeed in other parts of the world given the requisite conditions of well-drained acid soil with dry windy summers and winter rainfall. Under such circumstances they can tolerate eight to ten degrees of frost and are currently being planted with considerable success in the coastal areas of California, Australia and the Isles of Scilly.

The flowers, massed into striking cup-shaped heads, are surrounded by whorls of coloured bracts which are sometimes hairy and sometimes smooth. The evergreen leaves are simple and either smooth or covered with fine hairs.

Protea cynaroides has the largest flower heads and goes under the appellation of Giant or King Protea. The gorgeous blooms may be 28 cms (11 ins) across, the bracts varying in colour from white to pink or nearly red. These are covered with fine silky hairs which give them a silvery appearance. The true flowers (enclosed by these bracts) come to a snowy peak, which gives the inflorescence some resemblance to the Globe Artichoke, *Cynara*, accounting for its specific name. The green leaves have red leaf stalks and the shrub grows 90–120 cms (3–4 ft) under cultivation or up to 2 m (6 ft) in the wild.

P. compacta is one of the most prolific bloomers with up to two dozen flower heads out at once when established. These are a good clear pink colour and last up to ten days when cut. *P. eximia* (*P. latifolia*) makes an erect shrub of 2–2.5 m (6–8 ft) with rich pink (occasionally paler) outer bracts and spoon-shaped

Protea compacta

Protea eximia (*P. latifolia*)

Protea cynaroides

Leucospermum cordifolium (*L. nutans*)

Grevillea rosmarinifolia

Leucadendron argenteum

inner bracts fringed with white. These in turn enclose countless deep rose flowers. The leaves are broad and silvery-green.

P. barbigera, the Giant Woolly Beard, has blooms with a diameter of 20 cms (8 ins) filled with a mass of soft white hairs which become black at the raised centres of the flower heads. The outer bracts may be soft pink to dark carmine or from pale to a fairly deep yellow in the various local forms.

After blooming Proteas may need a little trimming to keep them shapely and should have the old flowers removed. The plants are normally propagated from seed, germination taking five to eight weeks. Syrupus Protea, a sweet syrup used by early colonists for coughs and pulmonary ailments, is made from the flowers of *P. repens* (*P. mellifera*). This is a shrub or small tree with blush-pink to bright red blooms. In South Africa it is commonly known as Sugarbush; the nectar makes excellent honey.

Another outstanding genus of South African evergreens is *Leucospermum.* This is even more prolific than *Protea,* with an abundance of showy flower heads made up of many tubular, long-styled flowers clustered together so that they resemble pin-cushions. They vary in colour from cream and yellow to pink, scarlet and orange. Leucospermums are often found growing with Proteas but can be distinguished (when out of flower) as the leaves are usually notched at the tips, while those of *Protea* are smooth.

Among the most beautiful is *Leucospermum cordifolium* (*L. nutans*), the Nodding Pincushion, growing up to 120 cms (4 ft) tall and as much in width with more or less globular blooms 10 cms (4 ins) across tightly packed with spiky extruded styles. These vary from pale pink to brick or orange. *L. reflexum* has the brightest flowers and is one of the tallest species (2–2.75 m; 6–9 ft). The velvety blooms consist of a rounded pincushion of thin tubular flowers of salmon-red with yellowish bases. As these age the long styles emerge and curl over backwards and the colour changes to bright crimson. A fully grown shrub of *L. reflexum* can carry almost a thousand flower heads in a season.

The bark of *L. conocarpodendron* (*L. conocarpum*) is used for tanning leather and also for its astringent properties. It has lemon-yellow flowers and is recommended as a drought-resistant, wind-tolerant shrub for seaside planting.

Visitors to Cape Town are always intrigued by the softly hairy, bright silver leaves of the Silver Tree, *Leucadendron argenteum,* which overlap each other so tightly that they mask the thick branches. They are sold as bookmarkers, mats and similar trifles. The species makes an erect tree of 6–8 m (20–25 ft) whose long tapering leaves gleam in the sunshine and ripple in the breeze. The trees are short-lived and rarely last more than twenty years.

All the Leucadendrons have the male and female flowers on separate plants, the former the most attractive as they are fluffy with broad and often colourful bracts. The female flowers resemble small cones.

In *L. discolor* (*L. buekianum*) the male flower heads are orange with yellow bracts and the female flower heads green, also with yellow or yellowish bracts. This shrub 2 m (6 ft) tall has become a popular spring cut flower in South Africa, where it is known as Flame Gold-tips.

Serruria florida is an exquisite evergreen growing about 1.5 m (5 ft) high with delightful nodding flower heads 5 cms (2 ins) across. These are creamy-white flushed with pink, consisting of many papery, petal-like bracts surrounding a mass of pinkish silky hairs—which are the true flowers. The plant is winter-blooming and in great demand in South Africa for wedding and corsage sprays. Its popular name, Blushing Bride, refers to this practice as well as its delicate colouring.

Professor H. Brian Rycroft, Director of Kirstenbosch Botanic Garden, relates a curious story about this plant. Apparently

Serruria florida

although long known to botanical science (it was introduced to Britain in 1824) all the stock disappeared and it was thought extinct. In recent years, however, a single plant was rediscovered in the French Hoek mountains in the Cape. This was propagated at Kirstenbosch and the species has now become a popular South African garden plant.

Embothrium coccineum is another lovely member of Proteaceae. It occurs wild in Chile as an evergreen tree and has proved hardy in sheltered parts of Britain where it blooms when only about a metre high. The brilliant crimson-scarlet flowers are thread-like but massed in showy terminal and axillary racemes, which show up starkly against the dark green, entire and leathery leaves. It requires a sheltered position, moisture and good drainage.

T. Harper Goodspeed in *Plant Hunters in the Andes* described the impact of meeting up with a mass display in the wild as 'hurling back the sunlight as a burning red haze, so powerful and so rich that the onlooker prepares to shield himself from its expected heat'. In view of this the common name Fire Bush seems peculiarly apt. Goodspeed also found Chileans using it as a hedging plant. Where it can be grown outside it should be given lime-free and peaty (Rhododendron-type) soil. Like most members of Proteaceae its life span is short (twenty to twenty-five years) but suckers from the roots may be removed for propagation.

Telopea speciosissima is the Waratah, the National Flower of New South Wales. Making a bushy plant 1–2.15 m (3–7 ft) high it has serrated or entire, leathery, alternate leaves and coral-red flowers like *Grevillea* in large terminal heads surrounded by an involucre of red-crimson bracts. An established plant may bear two hundred or more flower heads in a season.

T. truncata is peculiar to the mountains of Tasmania and has rich crimson flowers with curved perianths. It is hardy in favourable parts of Britain but remains bush-like instead of attaining the 6–8 m (20–25 ft) stature of its native habitat.

Sir Joseph Banks, who for more than fifty years was Honorary Director of King George III's garden at Kew is commemorated by a striking genus of evergreen shrubs and trees calls *Banksia*. Banks is particularly remembered for the important part he played in the botanizing history of the Pacific. He accompanied Captain Cook on his great voyage of exploration to Australia and in the Antipodes is often called the 'Father of Australia'.

There are about fifty species of *Banksia,* mostly Australian but also represented in New Guinea. The leaves are variously shaped, deeply toothed or spiny and frequently have white or reddish-brown undersides. The flowers, borne in pairs in dense cylindrical spikes, are sometimes cone-shaped or like bottlebrushes or the upright ears of Sweet Corn. Like other members of Proteaceae they depend for their beauty on the long styles, which vary in colour above a background which is usually yellow. The flowers are rich in nectar which attracts birds as well as insects.

Unfortunately these striking plants are only suitable for gardens in mild climates, but they tolerate extreme dryness and full sunshine. They are also in character over long periods; the buds take several weeks to open and then remain interesting for months. Their common name in Australia is Honeysuckle Tree or Australian Honeysuckle.

B. baxteri is a tall shrubby species from Western Australia with curious acute triangular leaf segments and globular greenish-yellow flower heads, sometimes up to 7 cms (3 ins) across. *B. burdettii* is also from Western Australia, but has more obovate flower heads, about 10 cms (4 ins) in length, of a rich orange-buff tinged with pink. The leaves are also much narrower, with regularly serrate margins and a white-tomentose undersurface. Other species include *B. serrata*, with purplish wood used for furniture and ship building, green serrated leaves and bluish-grey globose spikes; *B. ericifolia* which has slender cones of yellowish-red flowers ornamented with strands of yellow, silk-like styles; *B. coccinea* with oblong spikes of reddish-grey; and *B. menziesii*, the Honey Flower, one of the best, growing up to 4 m (12 ft) with rounded flower heads of gold and deep pink.

From the Chilean rain forests come several species of *Lomatia,* notably *L. ferruginea* which has proved hardy in Cornwall

Leucadendron discolor
male plant

(England) and has golden-yellow flowers with touches of scarlet on racemes 5–7 cms (2–3 ins) long. It will ultimately grow to 9 m (30 ft), providing a rich close-grained wood which polishes well and is favoured for cabinet work in Chile. Rural settlers also use the dark brown bark sap to dye their ponchos. The fern-like leaves are much divided and about 20 cms (8 ins) long and 10 cms (4 ins) across. *L. obliqua* has ovate leaves and racemes of white flowers.

Another interesting plant is the Firewheel Tree, *Stenocarpus sinuatus*. An Australian evergreen it makes a tall erect tree with shiny green, lobed or entire leaves with wavy edges and bright red blossoms in wheel-shaped inflorescences. These are 10 cms (4 ins) in diameter and come in terminal or axillary clusters of between twelve and twenty flowers.

Edwin Menninger in *Flowering Trees of the World* says that in Florida the flowers 'have a disagreeable foetid odor . . . noticeable 20 feet away when the tree is in full bloom'. He adds that the smell is strongest at night and seems designed to attract night-flying moths for pollination purposes. This curious characteristic, however, seems to be omitted from the Australian floras.

Macadamia or Queensland nuts come from both *Macadamia tetraphylla* with pale lilac, pinkish or white flowers and *M. ternifolia* with cream flowers. They are bushy, evergreen trees from Australia 9–15 m (30–50 ft) high. The flowers are borne in pairs on slender racemes and succeeded by hard-shelled, globular fruits containing one or two white fleshed nuts. These make delicious eating and are grown commercially in Hawaii for shipment to the USA.

There are about 100 different evergreen *Hakea* species in Australia, some of which are cultivated in gardens in warm temperate climates. *Hakea laurina,* the Pincushion Flower, has flat lanceolate leaves which are obliquely twisted and become red before falling; in addition it has globular, sessile balls of pink or scarlet flowers with protruding yellow styles and stigmas.

H. pubescens, also known as *H. gibbosa,* bears quantities of white or pink blossoms and the Needlebush, *H. tenuifolia* (*H. acicularis* or *H. sericea*), which has long needle-like leaves, carries small white blooms and in due course hard woody fruits typical of the genus.

Knightia excelsa has racemes 5 cms (2 ins) long of velvety red flowers with yellowish styles and stigmas which look like bottle-brushes. The species is a New Zealand evergreen tree of columnar habit—like a Lombardy Poplar—and the only one of this shape endemic to New Zealand. Its beautifully marbled wood is much prized by cabinet makers. Native names include Maori Honeysuckle and Bucket-of-Water Tree, the latter referring to its slow rate of combustion. The flowers are much visited by tuis and bell birds.

Gevuina (*Guevina*) *avellana,* a small tree from Chile, has white racemes of flowers and large cherry-red, ball-like fruits. These gradually turn black and have edible nuts known as Chile Hazels. The tree grows up to 12 m (40 ft) in its native country, with pinnate or bi-pinnate, leathery evergreen leaves.

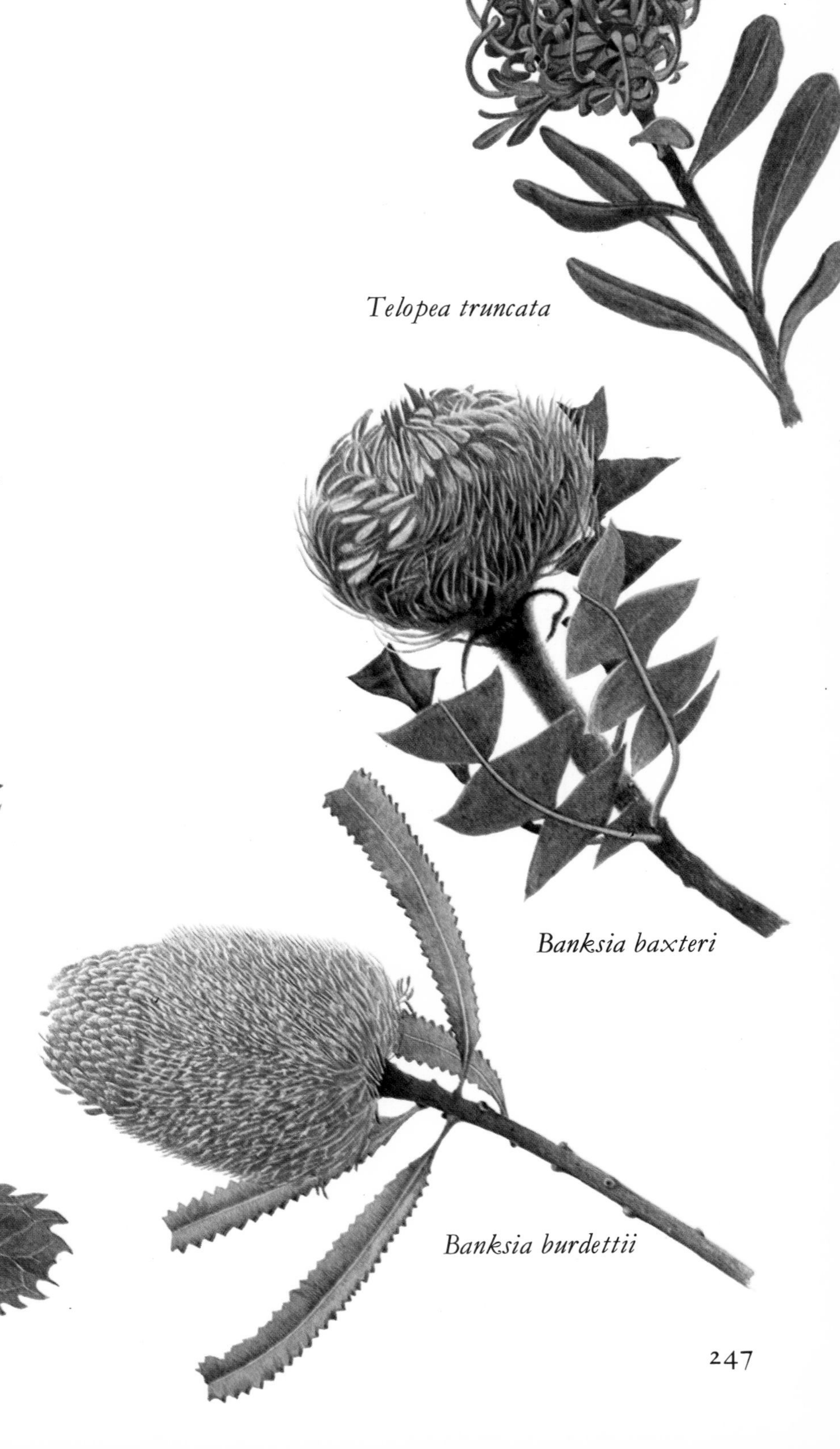

Telopea truncata

Banksia baxteri

Banksia burdettii

Telopea speciosissima

Punica granatum

Punicaceae

1 genus and 2 species

Punica granatum is a small shrubby tree 5–6 m (15–20 ft) high found growing naturally in dense thickets in the Balkans and western Asia but also cultivated and naturalized in the Mediterranean area. It is dicotyledonous and valuable for its bright red and yellow fruits which are the Pomegranates of commerce. These are about the size of an orange, with a tough leathery rind enclosing many seeds, each having a juicy red aril.

The leaves are simple and entire, the flowers brilliant scarlet with crinkled petals and a purplish-red calyx. They have five to eight sepals, five to eight petals and many stamens.

For garden adornment the double red 'Flore Pleno' makes a good wall plant for a sheltered position in cool climates or alternatively can be grown in a cool greenhouse. There is a double white called 'Albo Plena' and a single white, 'Albescens'.

Pomegranates need good but well-drained soil and plenty of sunshine. They can be propagated by cuttings or layers.

In the Mediterranean region the plant is thought to give protection against evil spirits and the fallen blossoms are often threaded on necklaces and hung round the necks of children to alleviate stomach ailments.

There is a belief that every Pomegranate fruit contains one seed which has come from Paradise. It was once thought that a whipping with Pomegranate branches could drive out the devil.

A syrup made from the seeds is sold in Mediterranean regions under the name of Grenadine, and apparently the first sherbert consisted of a preparation of Pomegranate juice mixed with snow. There are several references to the plant in the Bible including mention of its use in King Solomon's Temple (Kings I, 7: 18–20). The silver shekels of Jerusalem (circulated between 143–135 B.C.) bore engravings of Pomegranates.

Ranunculaceae

50 genera and 800 species

This is an important family of dicotyledons—mostly herbaceous perennials and chiefly N temperate in distribution. They include some splendid ornamentals for temperate gardens, also some more suited to warm climates.

The flowers are interesting and very variable in a miscellany of inflorescences. Individually they are bisexual (or rarely unisexual) and regular or irregular in shape, often with the perianth segments in whorls. The genus *Ranunculus* has a distinct corolla and calyx but many other genera bear petaloid sepals of various colours. Usually there are five of these (although there can be as few as three or more than five) with as many petals when present, both sometimes difficult to define as they become honeyglands. There are usually many stamens and the fruits are grouped achenes or follicles containing quantities of small seeds.

Perhaps the most spectacular for the garden are the perennial Delphiniums. There are two main groups: the Garland Larkspurs or Belladonnas, which have branching stems of 90–120 cms (3–4 ft) loosely clothed with spurred, mostly blue flowers; and the large-flowered or Elatum types. The latter are both stately and magnificent, with spikes, 90–215 cms (3–7 ft) in height, of glorious blue semi-double flowers in shades which beggar description. They are characterized by long central racemes which may bear several short branched racemes near their bases and often have contrasting 'eyes' of another colour—usually white or black.

In addition to various shades of blue like gentian-blue, royal blue, heliotrope, cornflower-blue, ultramarine, pastel blue, violet, royal purple and near navy, varieties are available with white, cream, pink, peach, carrot-red and near orange flowers. The Wageningen Horticultural Laboratory in Holland is mainly responsible for the pink and red varieties, which are usually known as University hybrids or 'red Delphiniums' (although a true red colour has still to be raised).

Most Delphiniums are hardy in Britain and over much of the United States. They are greedy plants which require deep, rich and well-manured soil, unlikely to dry out in summer. They are subject to slug damage in spring, like plenty of sun and benefit from an annual mulch and some shoot thinning (which channels all the food into the remaining stems). Propagation is by seed or cuttings (taken when about 7 cms (3 ins) high and rooted in sand). Frequent renewal is advisable with the Elatum type as old plants tend to deteriorate.

Annual Larkspurs are beautiful plants for massing in borders or as cut flowers. The cultivated forms are derived from *Consolida orientalis* (*Delphinium ajacis*), the Rocket Larkspur, and *C. ambigua*. They are of erect compact habit, with dense spikes of flowers in shades of pink, carmine, blue, mauve, lilac and white. Strains are available with double flowers and in a miscellany of heights between 60–135 cms (2–4½ ft).

Several small growing *Delphinium* species are suitable for the garden or greenhouse. They include the Californian *D. cardinale* which is bright red and 60–90 cms (2–3 ft) tall; *D. zalil* (30–60 cms; 1–2 ft), the Yellow Larkspur from Persia; and *D. grandiflorum* (30–90 cms; 1–3 ft), the Bouquet Larkspur from Siberia, and its forms—*album,* white, 'Blue Gem', deep blue, and 'Azure Fairy', pale blue.

Delphiniums have been cultivated since the times of the Pharaohs, although their early importance was entirely due to the supposed ability of the seeds to destroy body vermin! They had other properties such as the power to keep off scorpions, a theory Gerard scoffs at thus '. . . it is set downe . . . that the herbe onely thrown before the Scorpion or any other venomous beast, couseth them to be without force or strength to hurt, insomuch that they cannot move or stirre until the herbe be taken away; with many other such trifling toyes not worth the reading'.

Certainly many have poisonous leaves which (in spite of being harmless to slugs) are left alone by insects. The generic name is derived from the Greek word for a dolphin, *delphinion*, which the flower buds were thought to resemble.

Aconitums are herbaceous perennials with deeply cut palmately lobed leaves and terminal racemes of blue helmet-shaped flowers. They need cool rich soil and will grow in full sun or partial shade. Drought is detrimental so the roots should be kept moist by mulches or watering in dry weather; and the plants should be

Delphinium hybrids

Annual Larkspurs

left alone when well established as they take time to resettle after being disturbed. Aconitums rejoice in a number of common names including Wolf's Bane, Monkshood, Granny's Nightcap, Auld Wife's Huid and Captain over the Garden.

Among the most garden worthy are *Aconitum carmichaelii* (*A. fischeri* of gardens) from Central China, a sturdy species 60–90 cms (2–3 ft) tall with deep purple-blue flowers and the taller var. *wilsonii* which is violet-blue and grows to 2 m (6 ft). The very variable *A. napellus,* the European and Asian Monkshood, has spikes 90–120 cms (3–4 ft) long of broad, helmeted, blue flowers, but the name *A.* x *cammarum* covers a number of presumed hybrids between *A. napellus* and *A. variegatum* such as 'Bicolor', with blue and white flowers. The dark violet-blue 'Sparks Variety' is also referred to *A.* x *cammarum* by some authorities but is more likely to be a form of *A. henryi.*

A. volubile from Altai is a distinct Monkshood inasmuch as its slender stems have a twining habit. When provided with twiggy peasticks—as for Sweet Peas—they elongate and twist to a length of 4–5 m (12–16 ft). The flowers are pale violet.

All *Aconitum* species possess a sinister reputation as poisonous plants so that children should be warned of their dangers. Fatal accidents have occurred in Britain following mistaken identification of the roots of *A. napellus* with Horse-radish or the leaves for Parsley. According to Mrs Lankester (*Sowerby's English Botany*) the narcotic alkaloid Aconitine is even more deadly than pure prussic acid and acts with tremendous rapidity. The name Wolf's Bane refers to its one time use as a poison bait for wolves.

The Himalayan *A. ferox* (still grown in Britain for medicinal purposes) is even more deadly and was once used by the Nepalese to poison wells lying in the path of an approaching British army.

Certain members of Ranunculaceae are very early blooming, particularly the Hellebores, which brave winter's cold and bear shade with a tolerance shown by few other plants. Their foliage is usually evergreen, each leaf deeply cut into segments which are occasionally spiny and the blooms persist for months. This is presumably because the showy part of the flower is the petaloid calyx, the petals being converted to small tubular nectaries which stand just behind the stamens. These sepals may be red, green or white and are sometimes spotted with other colours.

One of the most popular is *Helleborus niger,* the Christmas Rose. A native of Central and S Europe it has deep green leaves divided into oval segments and stems 15–30 cms (6–12 ins) high carrying large white, saucer-shaped flowers filled with bright golden stamens. The species likes lime and needs well-drained soil in partial shade and benefits from an occasional mulch of rotted leaf-soil or peat to conserve moisture. This also protects the early blooms from mud splashes during winter rains. Christmas Roses should be left alone when happily established but may be increased from seed which will produce flowering plants in three years.

H. orientalis, the Lenten Rose, is a variable species about 60 cms (2 ft) tall, from Greece and Asia Minor. Several nodding flowers are carried on each stem, the coloured perianths cream, green, pink, rose, purple or almost black, many of them spotted with contrasting colours. There is also a rare yellow form. Lenten Roses do well at the fringe of trees or between shrubs and associate pleasingly with Primulas, Ferns, Bluebells and similar shade plants.

H. argutifolius (also known as *H. lividus corsicus* and *H. corsicus*) from Corsica and the Balearic Islands is a striking perennial 60–90 cms (2–3 ft) tall carrying many-flowered inflorescences of large apple-green flowers. These come in late winter but persist for months and contrast pleasingly with the dark spiny leaves.

Other green-flowered species are *H. viridis* and *H. foetidus,* both bunch-flowered and both from W and S Europe (including Britain). While the former are pale green and saucer shaped, those of *H. foetidus* have purplish rims and are cup-shaped.

A few hybrids have been raised between certain other species, most of which can be reproduced by careful division of the parent plants soon after flowering.

Hellebores have a romantic history and for centuries were thought to cure madness and counteract witchcraft. Pieces of root were inserted in a hole cut through the ear or dewlap of a sick animal with the idea of warding off the evil spells. On its removal twenty-four hours later the trouble was supposed to be cured. It is certainly a highly poisonous plant with various medicinal applications including that of a heart stimulant. Collectors in ancient times adopted stringent precautions, first circling the plant with a sword and then lifting the roots while chanting prayers to Apollo. At the same time they kept a wary eye open for eagles for, should one of these birds come near, the gatherer would die within the year.

Christmas Roses are still planted near cottage doors, a link with the past when they were often set near the threshold to prevent evil spirits from entering.

Helleborus niger

Eranthis hyemalis

Helleborus orientalis forms

Calthas are early-flowering bog plants, suitable for establishing in wet places near running water or the shallows at the edge of a pond. The flowers, borne on branched inflorescences or sometimes solitary, are usually cup-shaped with yellow sepals (there are no petals). The leaves are smooth and rounded heart-shaped.

Caltha palustris which is widely distributed in the Northern Hemisphere and variously known as Marsh Marigold, Kingcup, Water Cowslip and May-blob, grows 22–38 cms (9–15 ins) high with large Buttercup-like blooms. It has a double variety *plena* (which is the better garden plant) and a white form *alba*.

C. polypetala from the Balkans is taller (up to 90 cms; 3 ft) with branching stems carrying several narrower-segmented, golden-yellow flowers.

White-flowered species include *C. chelidonii* and *C. leptosepala,* both from North America. Early settlers used the leaves of Kingcups (usually *C. palustris*) as a spring vegetable (Cowslip Greens) and the flower buds as a substitute for capers. The blossoms were boiled with a little alum to give a yellow dye to yarn and an infusion of flowers was believed to cure fits in children.

Eranthis are early-flowering plants resembling large Buttercups with ruff-like collars of bright green leaves. They grow about 7 cms (3 ins) high and are suitable for naturalizing in sun or shade. *E. hyemalis,* the Winter Aconite, comes from W Europe, but *E.* x *tubergenii* and 'Guinea Gold' hybrids of this species with *E. cilicicus* (sometimes considered conspecific with *E. hyemalis*) have larger flowers.

Several *Adonis* make pretty, early-flowering plants for the rock garden. They have large Buttercup-like blossoms nearly 5 cms (2 ins) across and soft, finely cut fern-like leaves. *A. amurensis* from Japan has single and var. *plena* double golden flowers on stems 20–30 cms (8–12 ins) high. They bloom in early spring but die down quickly after flowering. *A. aestivalis*, the Pheasant's Eye from S Europe, is an annual with crimson flowers on stems 30 cms (1 ft) tall.

Aconitum x *cammarum* 'Bicolor'

Aconitum henryi 'Sparks Variety'

Helleborus foetidus

Helleborus argutifolius (*H. corsicus*)

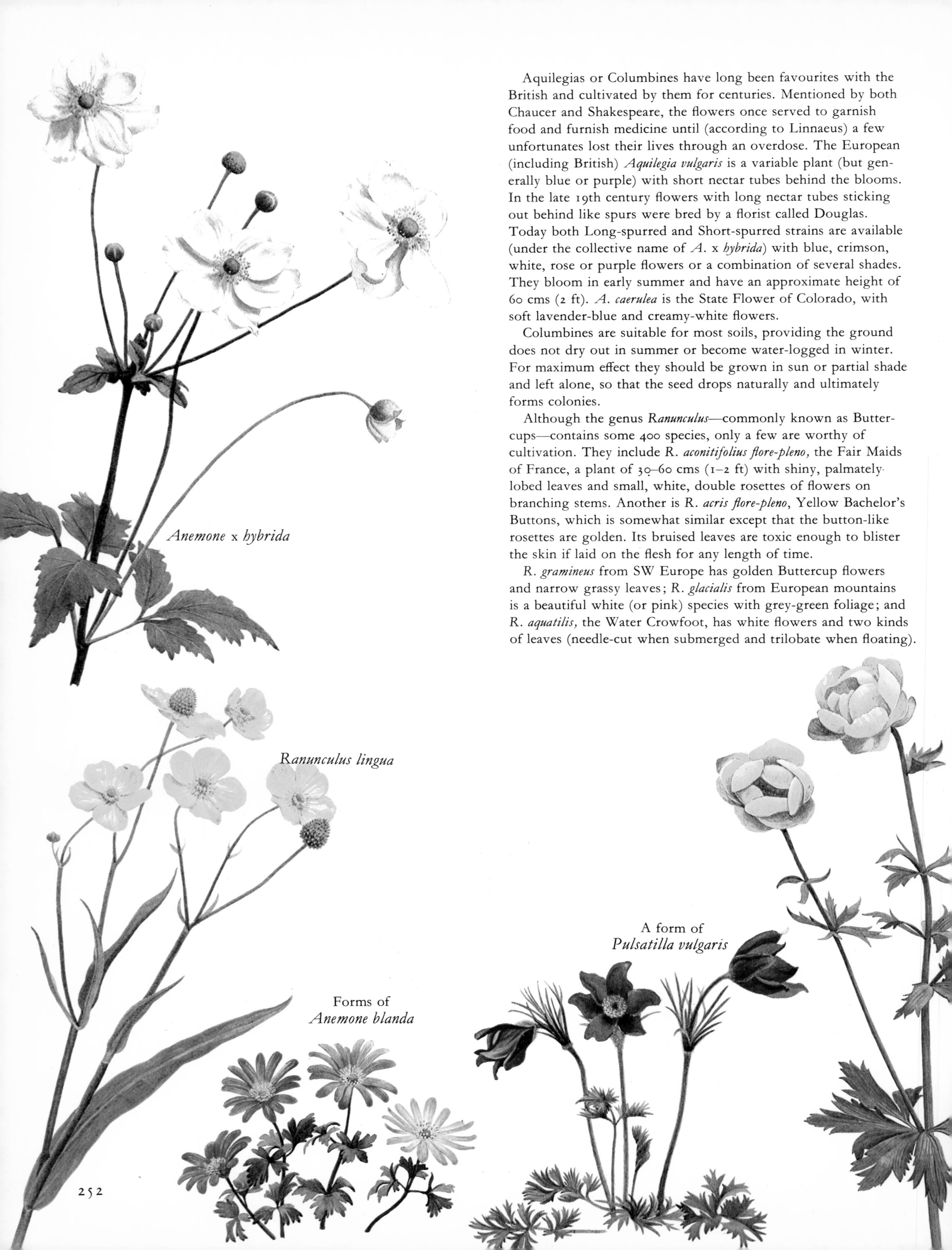

Aquilegias or Columbines have long been favourites with the British and cultivated by them for centuries. Mentioned by both Chaucer and Shakespeare, the flowers once served to garnish food and furnish medicine until (according to Linnaeus) a few unfortunates lost their lives through an overdose. The European (including British) *Aquilegia vulgaris* is a variable plant (but generally blue or purple) with short nectar tubes behind the blooms. In the late 19th century flowers with long nectar tubes sticking out behind like spurs were bred by a florist called Douglas. Today both Long-spurred and Short-spurred strains are available (under the collective name of *A.* x *hybrida*) with blue, crimson, white, rose or purple flowers or a combination of several shades. They bloom in early summer and have an approximate height of 60 cms (2 ft). *A. caerulea* is the State Flower of Colorado, with soft lavender-blue and creamy-white flowers.

Columbines are suitable for most soils, providing the ground does not dry out in summer or become water-logged in winter. For maximum effect they should be grown in sun or partial shade and left alone, so that the seed drops naturally and ultimately forms colonies.

Although the genus *Ranunculus*—commonly known as Buttercups—contains some 400 species, only a few are worthy of cultivation. They include *R. aconitifolius flore-pleno,* the Fair Maids of France, a plant of 30–60 cms (1–2 ft) with shiny, palmately lobed leaves and small, white, double rosettes of flowers on branching stems. Another is *R. acris flore-pleno*, Yellow Bachelor's Buttons, which is somewhat similar except that the button-like rosettes are golden. Its bruised leaves are toxic enough to blister the skin if laid on the flesh for any length of time.

R. gramineus from SW Europe has golden Buttercup flowers and narrow grassy leaves; *R. glacialis* from European mountains is a beautiful white (or pink) species with grey-green foliage; and *R. aquatilis,* the Water Crowfoot, has white flowers and two kinds of leaves (needle-cut when submerged and trilobate when floating).

Anemone x *hybrida*

Ranunculus lingua

A form of *Pulsatilla vulgaris*

Forms of *Anemone blanda*

It is a very variable N temperate species (or group of species) sometimes grown in aquaria and water gardens. The handsome Spearwort, *R. lingua* 'Grandiflora', makes an impressive subject for shallow water in ponds or streams and grows 90 cms (3 ft) high with narrow leaves and golden flowers on branching stems.

Florists' *Ranunculus* are derived from the oriental *R. asiaticus.* They have fanged tubers and should be planted with the claws pointing downwards in well-drained, sunny situations. The large round flower heads—on stems of 22–38 cms (9–15 ins)—may be single or double in shades of scarlet, yellow, rose and orange. The two main sections are the Persian and the Turban, the latter with the larger blooms and hardier.

Most Buttercups have poisonous sap and animals tend to avoid them when grazing. Years ago beggars would ulcerate their feet with the sap to arouse pity and induce charity.

Trollius are known as Globe Flowers because of the shape of the yellow or orange flowers. These are borne on erect and often branching stems 30–60 cms (1–2 ft) tall. They are moisture-loving and will grow anywhere provided the soil is damp, for example in meadows, borders and by ponds. *T. europaeus,* a European (including British) species, is normally pale yellow and has many cultivars in other shades. Some of the best of these are 'Bee's Orange' (orange and gold); 'Canary Bird' (pale lemon); 'Fire Globe' (deep orange); and 'Salamander' (fiery orange).

The species is common in Scandinavia where in the past the dried flowers were used on festive days to strew floors. In Westmorland and Derbyshire (England) Globeflowers were favourites for the traditional well-dressing ceremonies.

Two Actaeas from eastern North America are often grown in light shady borders. *Actaea pachypoda* (*A. alba*), the White Cohosh, is a striking plant when seen in full fruit. Then the white shiny berries with their rosy-red footstalks so weigh down the plants that they become almost prostrate. The species grows about 45 cms ($1\frac{1}{2}$ ft) high and is of branching habit with white flowers. *A. rubra,* the Red Baneberry, is similar except that the berries are scarlet. In the past *Actaea* rhizomes were used to treat skin diseases and asthma, but the berries are poisonous.

Cimicifugas are tall (60–150 cms; 2–5 ft), late summer-flowering perennials with Buttercup-like leaves and erect, feathery plumes of small creamy flowers. Unfortunately these have an unpleasant smell at close quarters so cannot be used for cutting purposes. They even deter insects—hence their common name of Bugbane. They flourish in most soils, being particularly useful for partially shaded borders, and associate pleasingly with *Phlox paniculata* and Red Hot Pokers. Most commonly grown are *Cimicifuga americana* from E and N America; *C. dahurica* which is Central Asian; *C. racemosa,* the Black Snakeroot of E North America; and *C. europaea* (*C. foetida*), an Asian and European species used as an insect deterrent in Russia.

Thalictrums are graceful perennials with attractively cut leaves, in some cases like Maidenhair Fern (*Adiantum*), and branching inflorescences composed of numbers of tiny flowers. This gives them a delicate cobwebby effect. The taller sorts make good background subjects for herbaceous borders and create a pleasant foil for larger flowered plants. They will grow in sun or light shade and are propagated by seed, or division in spring.

Thalictrum dipterocarpum from W China has fine leaves and branching panicles of rosy-mauve flowers with conspicuous yellow stamens. Normally growing about 60 cms (2 ft) it frequently reaches 120–150 cms (4–5 ft) under favourable conditions. For best results mulch the plants annually with well-rotted manure or compost and refrain from cultivating too close to the crowns. The young growths emerge near the parent plant and can be severed by a careless jab with a fork or hoe. There is a double form called 'Hewitt's Double', with rich mauve, round rosettes of flowers. The plants grown in gardens as *T. dipterocarpum* are usually the closely related *T. delavayi.*

Trollius europaeus

Hepatica nobilis

Aquilegia hybrids

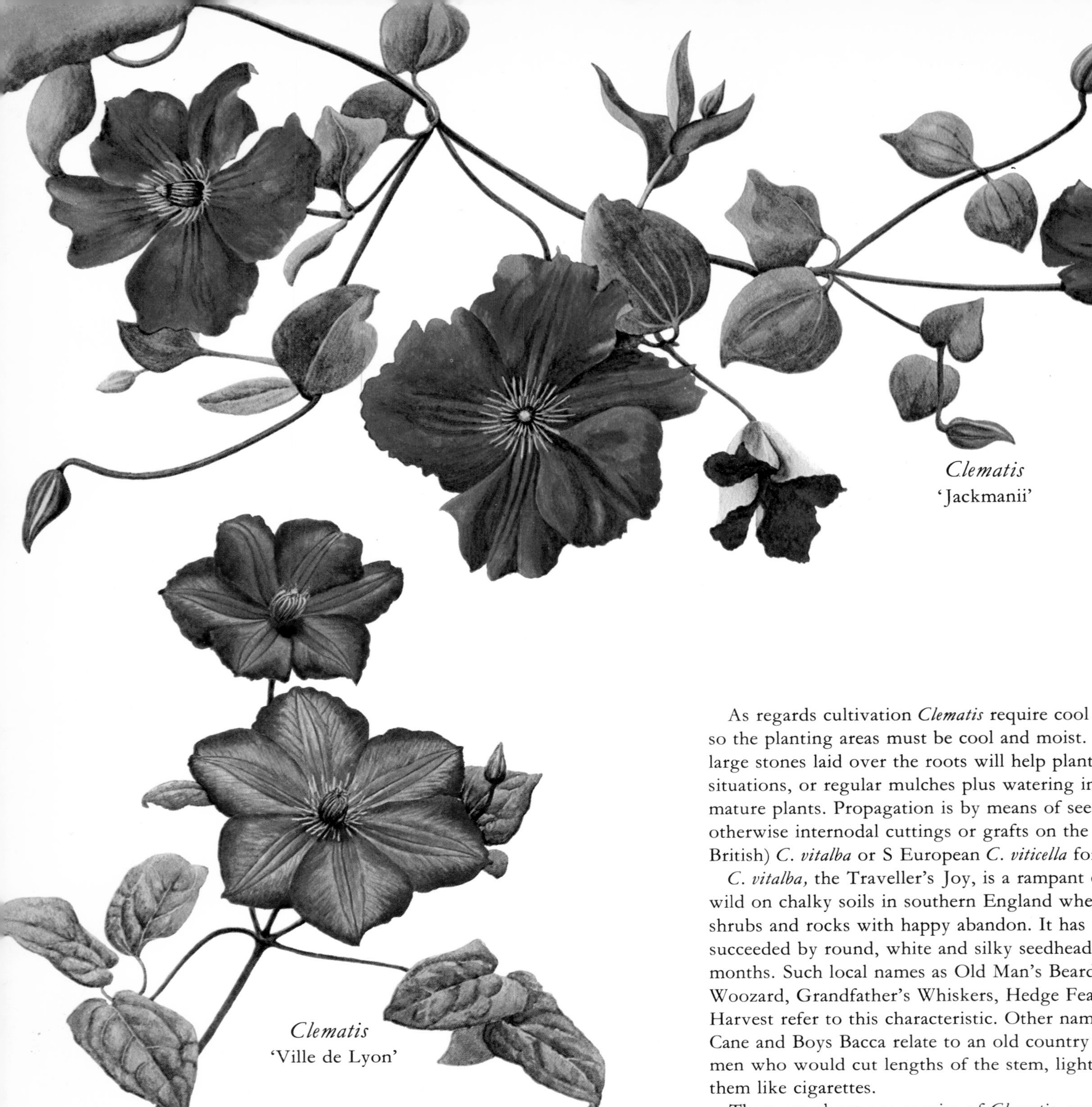

Clematis 'Jackmanii'

Clematis 'Ville de Lyon'

T. flavum, the Yellow Meadow Rue, is a European species with feathery heads of soft yellow flowers; *T. aquilegiifolium* from Europe and North Asia is soft purple with pinnately divided leaves and has a white form *album*; and *T. speciosissimum* (*T. glaucum*), sometimes considered a subspecies of *T. flavum*, the Dusty Meadow Rue from SW Europe, bears glaucous, much cut foliage and large panicles of pale yellow blossoms.

The flowers of *T. flavum* have been used in Britain for dyeing yarn, and in Europe *T. majus* (a form of *T. minus*) was once employed as a cure for jaundice and the plague.

One of the most important genera in Ranunculaceae is *Clematis,* the Queen of Climbers for temperate gardens. These flower freely and for long periods, are easily trained and have a long life. Among a number of uses is that of draping buildings, masking ugly sheds and tree stumps and clambering over trellis and arbours. *Clematis* can also be planted with climbing Roses so that the two intertwine, or set at the feet of old fruit trees to add extra colour. Often the seeds are as attractive as the flowers.

As regards cultivation *Clematis* require cool feet and a hot head, so the planting areas must be cool and moist. A paving slab or large stones laid over the roots will help plants in hot arid situations, or regular mulches plus watering in dry weather assist mature plants. Propagation is by means of seed for the species, otherwise internodal cuttings or grafts on the European (including British) *C. vitalba* or S European *C. viticella* for cultivars.

C. vitalba, the Traveller's Joy, is a rampant climber, growing wild on chalky soils in southern England where it clambers over shrubs and rocks with happy abandon. It has white flowers succeeded by round, white and silky seedheads which persist for months. Such local names as Old Man's Beard, Old Man's Woozard, Grandfather's Whiskers, Hedge Feathers and Snow in Harvest refer to this characteristic. Other names like Smoking Cane and Boys Bacca relate to an old country practice of workmen who would cut lengths of the stem, light one end and smoke them like cigarettes.

There are about 250 species of *Clematis,* among the most noteworthy *C. chrysocoma,* white flowers tinged with pink; *C. lanuginosa,* white to pale lilac; *C. macropetala,* blue or violet-blue with numerous staminodes, so that the nodding flowers appear double; and *C. tangutica* with rich yellow bells. These are all natives of China as is the white-flowered evergreen *C. armandii. C. montana,* a very popular plant for masking ugly features such as a shed, comes from the Himalayas. It has numerous four-sepalled white flowers in spring. There are also pink and rose forms. *C. cirrhosa* (*C. calycina*), a winter-blooming evergreen with deeply cut leaves and red-spotted, yellowish-white flowers, comes from S Europe.

C. orientalis, an Asian plant, is sometimes called Orange or Lemon Peel Clematis on account of its thick, orange or deep yellow sepals. These are succeeded by feathery seed heads and the grey-green leaves are finely cut.

In 1862 the firm of George Jackman of Woking, England, produced a large purple-blue flowered hybrid which it called *C.* 'Jackmanii'. This became the forerunner of many large-flowered cultivars having single and double blooms in shades of purple, mauve, pink, red and white. All the derivatives of 'Jackmanii' flower in late summer on shoots of the present season

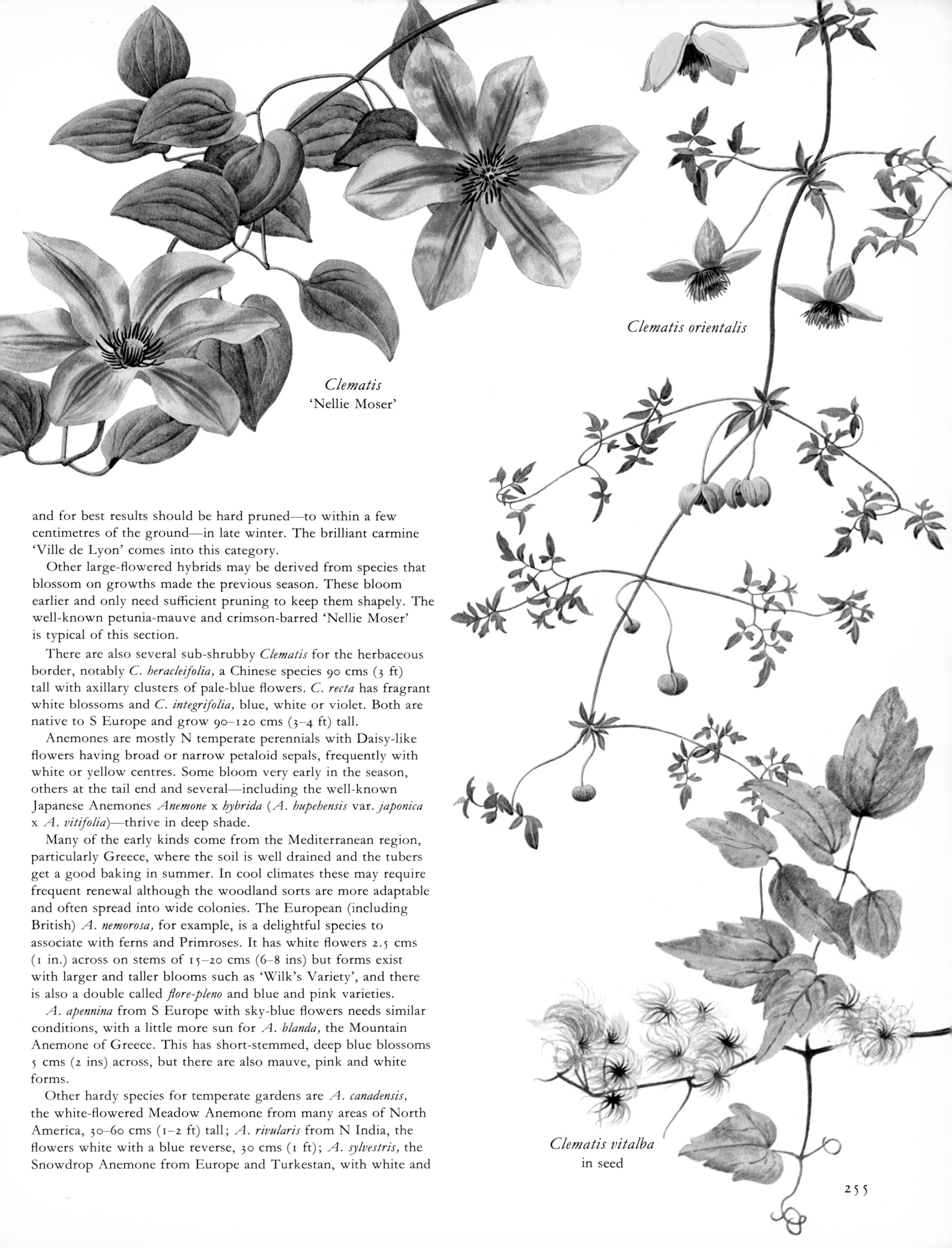

Clematis
'Nellie Moser'

Clematis orientalis

and for best results should be hard pruned—to within a few centimetres of the ground—in late winter. The brilliant carmine 'Ville de Lyon' comes into this category.

Other large-flowered hybrids may be derived from species that blossom on growths made the previous season. These bloom earlier and only need sufficient pruning to keep them shapely. The well-known petunia-mauve and crimson-barred 'Nellie Moser' is typical of this section.

There are also several sub-shrubby *Clematis* for the herbaceous border, notably *C. heracleifolia,* a Chinese species 90 cms (3 ft) tall with axillary clusters of pale-blue flowers. *C. recta* has fragrant white blossoms and *C. integrifolia,* blue, white or violet. Both are native to S Europe and grow 90–120 cms (3–4 ft) tall.

Anemones are mostly N temperate perennials with Daisy-like flowers having broad or narrow petaloid sepals, frequently with white or yellow centres. Some bloom very early in the season, others at the tail end and several—including the well-known Japanese Anemones *Anemone* x *hybrida* (*A. hupehensis* var. *japonica* x *A. vitifolia*)—thrive in deep shade.

Many of the early kinds come from the Mediterranean region, particularly Greece, where the soil is well drained and the tubers get a good baking in summer. In cool climates these may require frequent renewal although the woodland sorts are more adaptable and often spread into wide colonies. The European (including British) *A. nemorosa,* for example, is a delightful species to associate with ferns and Primroses. It has white flowers 2.5 cms (1 in.) across on stems of 15–20 cms (6–8 ins) but forms exist with larger and taller blooms such as 'Wilk's Variety', and there is also a double called *flore-pleno* and blue and pink varieties.

A. apennina from S Europe with sky-blue flowers needs similar conditions, with a little more sun for *A. blanda,* the Mountain Anemone of Greece. This has short-stemmed, deep blue blossoms 5 cms (2 ins) across, but there are also mauve, pink and white forms.

Other hardy species for temperate gardens are *A. canadensis,* the white-flowered Meadow Anemone from many areas of North America, 30–60 cms (1–2 ft) tall; *A. rivularis* from N India, the flowers white with a blue reverse, 30 cms (1 ft); *A. sylvestris,* the Snowdrop Anemone from Europe and Turkestan, with white and

Clematis vitalba
in seed

fragrant drooping flowers filled with golden stamens; and *A. hupehensis* a five-sepalled plant from China introduced between 1902 and 1906. A form of this species *A. hupehensis* var. *japonica* is believed to be one parent of the so-called Japanese Anemones, and was originally sent to England from Japan by Robert Fortune in 1844. The hybrids grow 60–105 cms (2–3½ ft) high with fine vine-shaped leaves and branching stems carrying white, pink or red flowers. Single, semi-double and double sorts are available and usually propagated from root cuttings. They make good late summer perennials for shady spots and are often distributed under the name *A. japonica,* although *A.* x *hybrida* is more correct.

Among the tuberous-rooted kinds are several from the Mediterranean area which may require frequent renewal in cool climates. *A. pavonina*, widespread in central and E Mediterranean countries, has scarlet, pink or purple flowers, often with a peacock 'eye' of yellowish-white. *A. fulgens* from S France is probably a natural hybrid between *A. pavonina* and *A. hortensis*. It has brilliant scarlet blossoms on stems 22–38 cms (9–15 ins) tall.

The Poppy Anemone of gardens has been derived from *A. coronaria* and possibly other species. It has flowers 7 cms (3 ins) across of scarlet, crimson, blue, mauve or white on stems of 15–45 cms (6–18 ins). Frequently these have white centres and there are doubles and semi-doubles as well as singles. The St Brigid Anemones which are mostly semi-double belong here, and also the single De Caen or Giant Flowered French strains.

The latter have a strange legendary history. During the time of the Crusades Bishop Umberto of Pisa, after blessing the soldiers leaving for the war, directed the seamen who were conveying them to bring back soil taken from the Holy Land as ships ballast—instead of the usual sea-shore sand. This was later spread over the Campo Sancta at Pisa 'to bury the honoured dead', but the following spring to everyone's amazement became carpeted with scarlet Anemones. This miraculous event caused the plants to be known as 'Blood Drops of Christ' and under this name they spread across Europe.

Even before the Reformation new shades and varieties started to appear, but in the 17th century a Parisian florist called Maître Bachelieu produced some extra large and colourful varieties. These he refused to sell, exchange or give away and for ten long years kept them to himself. One day, however, the Burgomaster of Antwerp asked to see them in flower. Arriving in his fur-trimmed robes straight from a civic meeting he had the misfortune to drop his cloak, whereupon he called his coachman who took the offending garment back to the carriage. Later the Burgomaster found a few fluffy seeds adhering to the fur and these in due course became the original French Anemones.

Pulsatillas closely resemble Anemones and at one time were included under that genus. *Pulsatilla vulgaris* (*Anemone pulsatilla*) is the Pasque Flower, so called because it blooms in Europe around Easter time. It is also found in Britain, usually near Roman ruins, which leads some authorities to believe that the Romans may have been responsible for its introduction. A variable species, it grows 15–22 cms (6–9 ins) high with large mauve, violet-purple or rarely white flowers having prominent centres full of golden stamens. They are succeeded by fluffy seed heads and the whole plant is covered with silky hairs. Years ago when eggs were dunked (dyed) at Easter *Pulsatilla* flowers were boiled in the water to turn the eggs green. Propagation is by seed as old plants resent disturbance.

Hepatica nobilis (*H. triloba*), a small plant closely related to *Anemone,* has widespread distribution in shady places in Europe and Asia. It grows 10–15 cms (4–6 ins) high with trilobate Ivy-like leaves and deep blue flowers. White and pink forms are also known and so are doubles.

Resedaceae

6 genera and 70 species

This is a small family chiefly Mediterranean in distribution but also represented in Europe, Asia, South Africa and California. The species are dicotyledonous annual or perennial, xerophytic herbs or shrubs with alternate, entire or divided leaves. The flowers come in racemes and are bisexual (rarely monoecious) and irregular with two to eight sepals, petals absent or up to eight, a variable quantity of stamens (between three and forty-five) and a superior ovary.

The only genera of importance in gardens are *Sesamoides* (*Astrocarpus*) and *Reseda*.

Reseda odorata is the sweet-scented Mignonette, a N African annual now naturalized in many countries, which is believed to have achieved much of its early popularity because Napoleon sent home seed gathered during his Egyptian campaign. Josephine grew the little yellow and white flowers as pot plants for the drawing rooms at Malmaison, thus starting a fashion which continued long into the 19th century.

The species has smooth branching stems about 60 cms (2 ft) in height with compact spikes of flowers. Garden forms exist with

A form of
Reseda odorata

orange, yellow, gold, crimson and white blooms. The flowers are the source of an essential oil, formerly used in perfumery but now replaced by various synthetics.

R. alba is a biennial from S Europe (naturalized in Britain) with long slender spikes of small greenish-white flowers on stems up to 90 cms (3 ft). Although not as fragrant as *R. odorata* it has great charm and a long history of cultivation since neolithic times. It was formerly grown in Essex and Yorkshire under the name of Weld for dyeing purposes. Although the seeds are the most productive of a rich yellow colourant, the whole plant was dried and used for this purpose. The dye turned silk, cotton and linen a rich yellow or sometimes blue cloth was dipped into it in order to achieve green.

Linnaeus observed that the nodding flower spikes of *R. alba* followed the course of the sun—pointing towards the east in the morning, the south during the day and the north at night.

Sesamoides pygmaea (*Astrocarpus sesamoides*) is a perennial somewhat similar to *Reseda* with stems 5–20 cms (2–8 ins) high carrying clustered spikes of white flowers. It is native to S Europe.

Ceanothus thyrsiflorus

Rhamnaceae

58 genera and 900 species

This is a dicotyledonous family of cosmopolitan distribution, the species mostly trees and shrubs and often climbing. The leaves are simple and the inflorescences cymose, made up of many individually inconspicuous flowers, which are regular and usually bisexual (rarely unisexual) with four or five sepals, four or five small petals and four or five stamens. Some species are extremely spiny, especially the Chilean *Colletia armata* or Anchor Plant which has rigid, sharply pointed triangular thorns and fragrant white, tubular flowers.

Rhamnus cathartica is the Buckthorn or Rhine Berry, a Eurasian plant which is naturalized in Britain and the United States. It is a deciduous shrub 3–6 m (10–20 ft) tall with opposite and alternate leaves on the same shrub and clusters of small green flowers which are followed by round black fruits. *R. imeritina* from the Caucasus has the largest leaves of the genus. These turn purple in autumn. *R. fallax* from the Alps of SE Europe is also valued for its large, deeply veined leaves and black fruits. The SW European *R. alaternus* is evergreen with black berries and has a striking variegated-leafed form.

Cascara comes from the bark of the Californian *R. purshiana* which years ago was peeled by handcutting and laboriously transported to market on the backs of men and horses.

The most decorative genus is *Ceanothus* with fifty-five species, many native to California where they are known as California Lilac. These can be deciduous or evergreen and range from small trees to the ground-hugging *C. prostratus* or Squaw's Carpet. Unfortunately they are rarely hardy enough to be grown as free-standing bushes in cool temperate climates, but in Britain are often used as wall plants for sunny situations. The masses of small white, grey-blue or deep blue flowers are very showy, especially some of the deeper coloured cultivars which contrast pleasingly with the small, neat leaves.

Among the more commonly cultivated evergreen species are *C. thyrsiflorus* one of the tallest (up to 9 m; 30 ft) and hardiest, the pale to deep blue flowers in panicles of 7 cms (3 ins); *C. dentatus*, deep blue; *C. rigidus*, pale or purplish-blue; and *C.* x *veitchianus*, a good hybrid reaching 3 m (10 ft) with bright blue florets. These all flower in early summer and are all Californian.

C. coeruleus (*C. azureus*) from Mexico is a deciduous shrub up to 2 m (6 ft) with deep blue, late summer flowers. 'Gloire de Versailles' is one of several good cultivars derived from this species. *C. integerrimus* is the Deer Brush, with white, pink or blue flowers. In California the bark is used in the preparation of a tonic. During the American Revolution the leaves of the deciduous *C. americanus* were used as tea and the dried bark and roots as a blood coagulant.

Most Ceanothus transplant badly so cuttings should be struck in small pots and planted out without disturbing the roots.

Rosaceae

100 genera and 2000 species

This is a large and important dicotyledonous family containing many garden favourites and such well-known plants as Apples, Pears, Plums, Strawberries and Raspberries. The species—which have cosmopolitan distribution—are usually perennial and include trees, shrubs and herbs, with simple or compound leaves (generally with stipules) which are nearly always alternately arranged on the stems. Many increase naturally by vegetative means, particularly runners.

The bisexual flowers are in terminal racemes or cymes of various types and have five sepals, five petals, usually many stamens and a superior, or less commonly inferior, ovary.

Rosa is the most important genus, especially in its modern cultivars. These are hardy in cool temperate climates, flower freely and continuously throughout the summer and are easy to grow. They prefer good heavy loam unlikely to dry out in summer, but light soils can be made rich and moisture retentive by digging rotted manure or compost into the ground before planting. In later years mulches of similar material benefit established bushes. The pruning of bush Roses (like hybrid teas and floribundas) is undertaken once a year, removing old, weak or diseased branches and shortening the rest to an outward pointing bud. The more these are reduced by hard cutting the fewer but finer the flowers. Some of the stronger cultivars such as 'Peace' and 'Queen Elizabeth' can be propagated from cuttings, a method also used for species and shrub Roses and many ramblers and climbers. Hybrid tea cultivars are usually budded on strong growing or disease-resistant stocks of *R. canina* the English Briar, *multiflora, rugosa* or *laxa*. *R. odorata* is used in warm climates as it stands heat better than any of the preceding.

Many of the world's museums contain fossil remains of Roses. These have all come from the Northern Hemisphere for Roses have never been discovered growing wild in the Southern Hemisphere. Some of the fossils are very old, especially those found in Asia and in the Miocene deposits of the Baltic provinces. The Oligocene deposits in Colorado (North America) are estimated to be at least thirty-five million years old, so it is fair to assume that Roses have existed on the earth far longer than Man.

Today there are about 250 species and tens of thousands of cultivars. Few temperate gardens are without at least one Rose and they are also widely grown in the subtropics. The Rose is

Rosa canina

probably the most popular garden plant and is the National Flower of England.

The earliest known picture of Roses is that in the House of Frescoes at Knossos, Crete, which dates back to the 16th century B.C. The Island of Rhodes took its name from the flowers, which were portrayed on its coins in 4000 B.C.

In Babylon, too, the Rose was known. Sargon, King of Babylon (2845–2768 B.C.), is reputed to have sent Rose trees to his capital at Akkad, and in Assyrian architecture Roses were used on sill carvings in ceremonial buildings.

The Ancient Greeks and Romans set great store by the flowers. In Homer's *Iliad* Aphrodite anoints the dead Hector with Rose perfume. According to Seneca, the Romans constructed special warm houses, heated by tubes filled with hot water in order to obtain blooms out of season. At that time Roses were used for ceremonial purposes on a scale we can scarcely comprehend today. Not only were they worn on the person in extravagant garlands, but the floors and couches were strewn with petals. Cups of wine were laced with Roses and at one lavish banquet during the reign of Emperor Elagabalus (c. A.D. 218–222) showers of petals were released through apertures in the ceiling, in such quantities that several guests were unable to extricate themselves and suffocated.

In Sybaris (an ancient city of Italy) whose inhabitants were given to luxurious indulgence people slept on mattresses filled with Rose petals. In other words—on a bed of Roses!

The phrase *sub rosa* originated about this period and was linked with Cupid's gift of a Rose (the emblem of love) to Harpocrates, the God of Silence, as a bribe not to reveal the amours of Venus. Accordingly whenever secret matters were discussed a Rose was suspended from the ceiling and what took place beneath it was strictly *sub rosa* (under the Rose). Later carvings took the place of living flowers and for some reason were particularly popular in Victorian times. Incidentally, in the 18th century Jacobites adopted a white Rose as their emblem, for their political leanings and support for Bonnie Prince Charlie had of necessity to be kept secret.

Rosa moyesii 'Geranium'

The extravagant use of Roses in Roman convivialities—and even debaucheries—incensed the Early Fathers of the Christian Church, who would not allow them to be brought into the churches. Later they returned to favour when others saw in them the emblem of martyrs, the five petals representing the five wounds of Christ and the white Rose the virginity of Mary. After this, Roses were commonly used in England in the Middle Ages for decorating shrines, wreathing candles, for rent payments and in heraldry. Beautiful wood carvings of Roses can be seen in many old English churches and chapels.

In Medieval times a Sunday in mid-Lent was known as Rose Sunday and even today, a golden Rose blessed by the Pope may be sent as a mark of outstanding pontifical favour to special personages.

The rosary commemorates a chaplet of Roses supposed to have been bestowed on St Dominick by the Virgin Mary. Originally rosaries were strings of beads made from tightly pressed Rose petals which gave out a pleasing fragrance.

Old Roses

1 'Frau Karl Druschki'
2 'Madame Pierre Oger'
3 *Rosa muscosa*
4 'Rosa Mundi'
5 'Souvenir de la Malmaison'

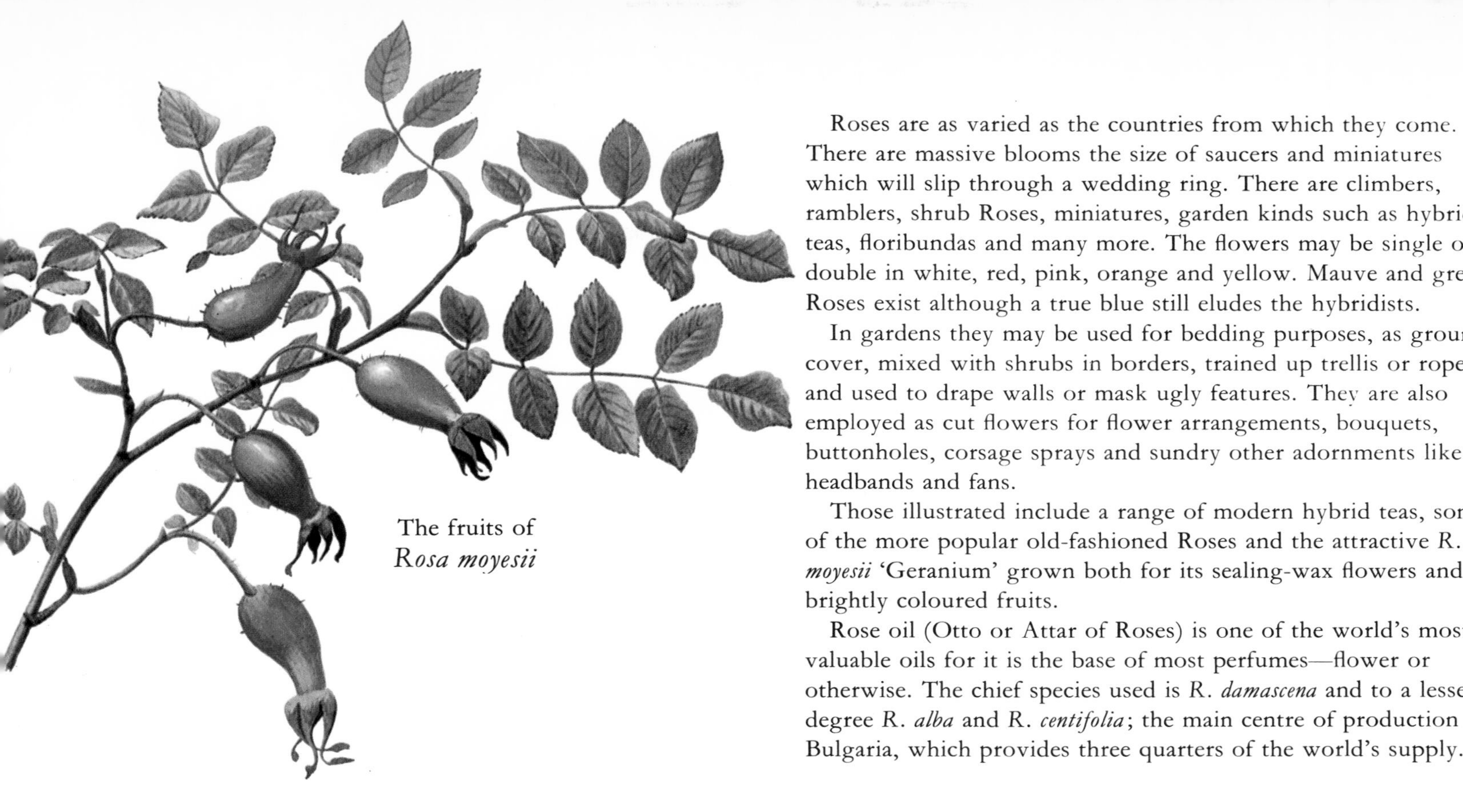
The fruits of *Rosa moyesii*

Roses are as varied as the countries from which they come. There are massive blooms the size of saucers and miniatures which will slip through a wedding ring. There are climbers, ramblers, shrub Roses, miniatures, garden kinds such as hybrid teas, floribundas and many more. The flowers may be single or double in white, red, pink, orange and yellow. Mauve and green Roses exist although a true blue still eludes the hybridists.

In gardens they may be used for bedding purposes, as ground cover, mixed with shrubs in borders, trained up trellis or ropes and used to drape walls or mask ugly features. They are also employed as cut flowers for flower arrangements, bouquets, buttonholes, corsage sprays and sundry other adornments like headbands and fans.

Those illustrated include a range of modern hybrid teas, some of the more popular old-fashioned Roses and the attractive *R. moyesii* 'Geranium' grown both for its sealing-wax flowers and brightly coloured fruits.

Rose oil (Otto or Attar of Roses) is one of the world's most valuable oils for it is the base of most perfumes—flower or otherwise. The chief species used is *R. damascena* and to a lesser degree *R. alba* and *R. centifolia*; the main centre of production is Bulgaria, which provides three quarters of the world's supply.

The Rose has featured prominently in English heraldry, being represented in one form or another in the badges of Edward I, Henry IV, Edward IV, Henry VII, Henry VIII, Edward VI and Queen Elizabeth I. Since 1461 it has been the emblem of Britain following the Wars of the Roses. The Tudor Rose of heraldry is a double bloom with a red outer and white inner row of petals.

The oldest known living Rose tree is at Hildesheim Cathedral in Germany. It is said to be over a thousand years old and is a form of *Rosa canina* with an interesting legendary history. This claims that Kaiser Ludwig the Righteous while out hunting one day became separated from his escort and spent the night in a forest. Returning to his castle the next day he found to his grief that he had lost a sacred relic. This was later found beneath a wild Rose near his sleeping place, but the tree refused to relinquish the relic and Ludwig vowed to build a chapel near the spot. This was completed in A.D. 815, Gunther becoming the first Bishop.

The tree itself has now become a legend. During the Second World War Hildesheim was bombed but miraculously the old tree survived, although all its top growth was destroyed.

Modern hybrid Roses

East of Sofia may be found the Valley of Roses, a place 32 km (20 miles) wide and 160 km (100 miles) long, full of Roses. Here some 200,000 people are employed in tending the bushes, gathering the petals and distilling the oil. It takes a ton of petals to yield a pound of Attar, and production from this area alone averages some three tons per season. Most of this is exported to France, where it is frequently 'stretched' with cheaper oils such as that derived from *Pelargonium capitatum,* the Rose Geranium.

Rose petals at one time were used as scented smoke filters for cigarettes, the petals being rolled and employed in the same fashion as today's cork tips. But the costs were so prohibitive the industry had a short life.

Roses have other uses. *The Grete Herball* of 1526—one of the earliest English herbals—recommends Rose water for the complexion and for the heart 'powdre of perles with sugre of Roses'.

Rose water is still manufactured and Rose-hip syrup has some importance as a tonic, especially for children for it contains four times as much Vitamin C as Blackcurrant juice and twenty times that of Orange juice. Rose-hip marmalade and Rose-petal jelly are popular preserves and Queen Victoria is reputed to have enjoyed a sauce made from Sweet Briar hips and lemon juice with roast mutton.

Rosebuds pickled with white vinegar and sugar are pleasant with cold meat and Rose wine has a delicate flavour and attractive colour. Other items made from Roses include Ointment of Roses, Conserve of Roses (for colds), Rhubarb and Rose-petal Jam, Rose Cold Cream, Rose Pomatum (to combat baldness), candied Rose leaves (for decoration), Pot Pourri, Rose-scented soap and snuff.

Cotoneaster is an important genus which provides first blossom, then berries and often fine autumnal tints in the deciduous species or else evergreen foliage. The species are shrubs or small trees, some almost prostrate like *C. adpressus* from China which makes a weed-suppressing carpeter. Others, such as *C. simonsii* from the Himalayas, make good hedging subjects, or quickly form shelter belts or have their uses in the mixed shrubbery. *C. horizontalis* from China—the Herringbone Cotoneaster—has its branches spread like fishbones. As with the Himalayan *C. microphyllus* it makes a good wall plant but *C. frigidus* and *C. franchetii* var. *sternianus,* both from the Himalayas or the weeping *C. salicifolius* 'Pendulus' may be treated as specimen subjects on lawns.

C. x *watereri,* an evergreen hybrid has spreading branches hung all winter with large and showy clusters of scarlet berries. It is a good plant to grow against trellis, walls or fences.

C. 'Cornubia' is a splendid wide-spreading shrub or small tree with heavy trusses of brilliant red fruits and evergreen leaves.

The small flowers of *Cotoneaster* are white or rose tinted and the berries usually scarlet, less often black or yellow.

Crataegus is the Hawthorn, a genus of deciduous small trees with lobed or pinnatifid leaves, usually spiny branches and clusters of small five-petalled flowers followed by berried fruits (haws). The commonest species is the European (including British) *C. laevigata* (*C. oxyacantha*) with white flowers and dull red haws. This is often used for hedging but left unpruned will make a tree 5–6 m (15–20 ft) high. For garden purposes the double flowered forms are preferable, particularly the white 'Plena', rose-pink 'Rosea Plena' and crimson 'Coccinea Plena'. The leaves have been used medicinally and the fruit pulp was added to flour by some European countries during the Second World War.

C. macracantha from NE North America has diabolical sharply pointed thorns 7–13 cms (3–5 ins) long, also white flowers and large lustrous red haws.

There are countless species and varieties of Spiraeas, all much alike and flowering at the same time, so that the gardener is hard pressed to make a choice. Over the years many have been siphoned off into other genera, notably *Holodiscus, Sorbaria, Filipendula* and *Astilbe* (now in Saxifragaceae).

The remainder are spring- or summer-flowering shrubs which grow freely in sunny situations in most soils. The flowers although small become effective in the sprays and the leaves are neat and usually entire. Popular sorts include *Spiraea* x *arguta* or Bridal Wreath, 2–2.5 m (6–8 ft) tall with slender stems wreathed with dainty white flowers; *S. thunbergii* from China, the earliest to bloom, with white flowers; *S.* x *vanhouttei,* a fine white often used for forcing in Europe and long popular in the United States; and *S.* x *bumalda* 'Anthony Waterer' with large flat, terminal heads of brilliant carmine.

S. prunifolia 'Plena' is a beautiful shrub up to 2 m (6 ft) with double white flowers and red autumnal leaf tints. Although Japanese it was introduced by Robert Fortune from China, where it is frequently planted on graves and has a name which may be literally translated as 'Smile-laugh Flower'.

Holodiscus discolor, the North American Spiraea, is an elegant late summer shrub of loose habit 1.25–2.5 m (4–8 ft) tall, with simple leaves and striking feathery inflorescences which droop under the weight of myriads of small foamy white flowers.

Aruncus dioicus (*A. sylvestris*) is the Goat's Beard from Siberia,

Spiraea x *bumalda* 'Anthony Waterer'

Cotoneaster 'Cornubia' in fruit

Holodiscus discolor

Malus x *purpurea* 'Lemoinei'

a splendid moisture-loving herbaceous perennial growing 1.25–1.5 m (4–5 ft) tall with fine pinnate leaves and impressive plumes of hay-scented, creamy-white flowers.

Filipendula ulmaria (*Spiraea ulmaria*) is the Meadow Sweet, a moisture-loving Asian and European (including British) perennial with three- to five-lobed leaves which are white-downy beneath. The branching stems 60–90 cms (2–3 ft) tall terminate in packed feathery heads of aromatic creamy flowers. There is a double form, another having golden foliage and one with cream and variegated leaves. *F. rubra* from eastern North America grows up to 2.5 m (8 ft), with deep peach flowers, which are carmine-pink in the variant known in gardens as *F. venusta magnifica*.

Many fruits belong to Rosaceae, some genera having ornamental species like the flowering Plums, Pears and Cherries.

Malus includes the Apple and like the orchard fruits the decorative forms appreciate plenty of sunshine and rich well-drained soil. They are often grafted on Crab or Wild Apple stocks and can be grown in a variety of shapes and heights such as Bush, Standard or Half-standard. It is necessary to keep the centres of the standards open by careful pruning.

M. x *purpurea* 'Eleyi' is one of the most outstanding, with wine-red flowers, reddish-copper foliage and small crimson edible fruits. *M. floribunda,* the popular Japanese Crab, has masses of pale pink to white flowers and deep rose buds. There are many other ornamental Crabs—all trees of 6–8 m (20–25 ft) when grown as standards—including 'John Downie' (a hybrid of *M. pumila*) with white flowers and large, conical yellow and red fruits; 'Hillieri' (a hybrid of *M. floribunda*) with semi-double bright pink flowers; 'Golden Hornet' white flowers and bright yellow fruits; and 'Lemoinei' which has dark crimson flowers, bronzy-red foliage and deep purple, Cherry-like fruits. *M.* 'Red Jade', a pendulous freely fruiting form was developed at the Brooklyn Botanic Garden.

The European Medlar, *Mespilus germanica,* is usually cultivated for its flat-topped fruits, which are eaten when bletted—that is, in the early stages of decay—a few weeks after being gathered. The taste is acquired. The flowers resemble those of the Apple and the tree grows up to 8 m (25 ft) tall.

Among a number of Plums grown for garden decoration are forms of *Prunus cerasifera,* a W Asian and E European species. Var. *atropurpurea* (*P. pissardii*) is well known for its rich purple-red leaves and pale pink flowers, and forms of this such as 'Woodii' and 'Nigra' are even finer.

P. spinosa purpurea, the Purple-leafed Sloe, makes a small bushy tree of 2.5–3 m (8–10 ft) or can be used for hedging. Selected forms with pink or white flowers and purple or green leaves are available under such names as 'Purple Flash', 'Blaze' and 'Green-glow' for dwarf edgings 60 cms (2 ft) high.

P. dulcis (*Amygdalus* or *Prunus communis*), the Almond of S Europe, is a well-known early blossoming tree which carries its pink flowers well before the leaves. The nuts produce a pleasant oil which has been valued for centuries.

P. persica from China is the Peach, and ornamental forms include 'Albo-plena', double white, 'Windle Weeping', double pink with a weeping habit, and 'Clara Meyer', double bright pink. Its hybrid with the almond, known as *P.* x *amygdalo-persica* is a valuable March-flowering small tree.

The Apricot of S Europe, *P. armeniaca,* also belongs here. The tree was always plentiful in the Holy Land and its fruits are the 'Apples of gold in pictures of Silver' mentioned in Proverbs 25: 11. Prussic acid is obtained from the kernels.

Japanese Flowering Cherries are frequently considered to be derived from *P. serrulata* and *P. speciosa,* although the Japanese classify them under the name *P. lannesiana* var. *lannesiana*. There are many kinds, some with white flowers like the double 'Shiro-fugen' (*P. serrulata albo-rosea*); pale pink, as is 'Amanogawa', a small erect tree which grows fastigiate like a Lombardy Poplar; deeper pink in 'Yedo Zakura', or yellowish-green in 'Ukon'.

P. subhirtella autumnalis is exceptional in its long flowering period, from autumn to spring in mild climates. The semi-double flowers are pinkish-white. Some species have beautiful shiny bark, especially the Chinese *P. serrula*. *P. laurocerasus,* the common Cherry Laurel, is frequently used in Europe for hedging and shelter belts. It has creamy flower spikes 15 cms (6 ins) long on unpruned trees and glossy, oval-oblong, entire leaves and will grow in deep shade. The odd leaf is sometimes used to give an aromatic flavour to boiled milk but too much can be dangerous as the foliage is poisonous.

P. lusitanica, the Portugal Laurel from Spain and Portugal, makes a tree of 3–6 m (10–20 ft) with long-elliptic evergreen leaves and large racemes of dull white, heavily scented flowers. These are followed by purple fruits.

The best of the ornamental Pears is *Pyrus salicifolia pendula,* an elegant weeping tree of 5 m (15 ft) with clusters of white flowers and long Willow-like, silvery leaves covered with white down.

Chaenomeles are deciduous, spring-flowering, prickly shrubs or small trees often referred to as *Cydonia* or 'Japonicas'. They make good wall shrubs or can be grown in mixed shrubberies, trained over rock banks or used for hedging. They flower best in sun but tolerate some shade and thrive in most soils.

C. japonica (*Pyrus* or *Cydonia maulei*) is a small shrub 90 cms (3 ft) high of spreading habit with spiny, warted twigs and flame-orange flowers. These are succeeded by hard, greenish-yellow, Apple-like fruits with a spicy scent, which are sometimes put amongst linen to impart a pleasant fragrance and also make excellent marmalade or jelly. *C. speciosa* (*C. lagenaria*; *Cydonia japonica*) is the Japanese Quince, a shrub of 2 m (6 ft) or thereabouts having oval, toothed, glossy leaves, spiny stems and orange-scarlet flowers. There are variants in cultivation with white, dark crimson and pink flowers, and also some doubles.

C. japonica and *C. speciosa* are native to Japan, but the small tree *C. cathayensis* is Chinese. It has white or pink flowers and yellow-green, egg-shaped fruits 10–15 cms (4–6 ins) in length. These are used in China for scenting rooms and are often candied.

Pyracantha coccinea from S Europe and Turkey has been grown in Britain since the 17th century, chiefly on account of its glowing scarlet berries which under favourable conditions persist all winter. Since these attract birds which will soon strip the bushes it may be necessary to spray them over with a bitter-tasting substance such as quassia. It is naturalized in the eastern US.

Parkinson called *P. coccinea* the 'ever greene Hawthorne or prickly Corall tree' but nowadays it is commonly known as Firethorn, a literal translation of the Greek generic name. The plant makes an excellent wall shrub—even in shady spots—and can be used for hedging or as a specimen tree (up to 6 m; 20 ft). The leaves are small and entire or toothed and the bunches of small white flowers both fragrant and showy. These are much visited by bees so that normally the branches are heavy with berries in autumn. The cultivar 'Lalandei' is the best for gardens and has bright scarlet fruits.

P. rogersiana from China has red or yellow berries; *P.* x *watereri* makes a good hedging subject with white flowers in summer and masses of red berries later and *P. atalantioides* (*P. gibbsii*), another Chinese species, has the distinction of being the tallest (a good 6 m; 20 ft) and largest-leafed species.

Kerria japonica, a monotypic genus from China, was named after its introducer William Kerr, who sent the double-flowered form known as 'Plena' to Britain in 1805. Kerr was the first resident

Prunus x *amygdalo-persica*

Prunus serrulata 'Shirofugen'

Kerria japonica 'Plena'

Chaenomeles speciosa

Potentilla fruticosa

professional collector to go to China and besides this *Kerria* sent back many new plants including such splendid items as *Pittosporum tobira, Rosa banksiae, Lilium lancifolium* (*L. tigrinum*), *Lonicera japonica* and *Pieris japonica.*

Growing 2 m (6 ft) high it has green stems, small Nettle-like leaves and rounded rosettes 4–5 cms (1½–2 ins) across of golden flowers. It will grow in practically any soil or situation but having weak stems is best set against a wall or fence. The single flowered form of the species (which arrived several years after the double) is less frequently grown.

Most Geums have a clove-like fragrance in their roots and for this reason were formerly used for flavouring wines and ales.

They are easily grown herbaceous perennials much prized for their brightly coloured, chalice-shaped flowers which remain in character throughout the summer. The leaves are irregularly lobed and variously cut. Geums are suitable for sun or light shade in any good garden soil but to ensure constant flowering the roots should be occasionally lifted and split, preferably after blooming.

Geum quellyon (*G. chiloense*) the Scarlet Avens from Chile and Argentina is most important, not for its own sake as it is rarely seen in gardens, but for the fine cultivars raised from it. These include the well-known 'Mrs Bradshaw', a branching plant 60 cms (2 ft) high with large, rich scarlet, semi-double flowers, which was raised in the early 1900's by John Bradshaw of Southgate (London) and comes practically true from seed.

Other desirable sorts are 'Lady Stratheden', with Buttercup-yellow, semi-double flowers (an excellent companion for 'Mrs Bradshaw'); 'Fire Opal', a single having orange flowers overlaid with a warm reddish glow; 'Red Wings', semi-double and rich scarlet; and 'Prince of Orange', bright orange.

G. rivale, the Water Avens from Eurasia and North America, has drooping reddish flowers on hairy stems 30 cms (1 ft) tall, and has given rise to several cultivars, particularly 'Leonard's Variety', which has pink flowers with orange tints.

Geum
'Mrs Bradshaw'

Prunus subhirtella autumnalis

The herbaceous Potentillas are somewhat similar, although the foliage is possibly more attractive. The leaves are finely cut into finger-like segments and often have silver undersides. Some are suitable for edging purposes but others make good long-flowering border plants for well-drained soil, in full sun or partial shade.

Potentilla atrosanguinea, the Himalayan Cinquefoil, has dark reddish-purple flowers but is rarely grown as a garden plant. Instead, a miscellany of cultivars has sprung from the species, in a wide range of shades varying from scarlet and rich orange to yellow. In borders these are most effective when planted in association with *Limonium, Gypsophila* or similar subjects of paler shades, which thus set off the *Potentilla* blooms to better advantage. Good garden sorts are 'Gibson's Scarlet', a single blood-red; 'California', golden-yellow, semi-double; 'Wm Rollinson', semi-double, vermilion and yellow; and 'Yellow Queen', bright yellow. These grow 45–60 cms (1½–2 ft) tall and may be propagated by division in spring or autumn.

Shrubby Potentillas are useful in small gardens as they are of compact habit, require little room and remain in flower for months. The single blossoms are reminiscent of those of the Strawberry (*Fragaria*) and the pinnate leaves are divided into five or seven narrow leaflets. *P. fruticosa,* a hardy little shrub of 60–120 cms (2–4 ft) from the N temperate zone forms a congested mass of twiggy shoots almost hidden in summer under masses of small yellow flowers. Its dried leaves are used for tea in Siberia and by Eskimos in Alaska.

Forms of this species or hybrids with others such as *P. arbuscula* and *P. parvifolia* include 'Katherine Dykes', large primrose-yellow flowers; 'Abbotswood', white; 'Farreri', large buttercup-yellow; 'Tangerine', coppery-yellow; and 'Donard Gold', golden-yellow with orange flecks.

Closely related are *Sarcopoterium* and *Sanguisorba*. *Sarcopoterium* (*Poterium*) *spinosum* comes from Asia Minor and the E Mediterranean. It is a small shrub of 60–90 cms (2–3 ft) with spiny branches, hairy and much divided foliage and spikes of greenish unisexual flowers.

Sanguisorba canadensis (*Poterium canadense*) is the American Burnet, a herbaceous plant with deeply-cut, grey-green foliage and spikes 90 cms (3 ft) long of whitish inflorescences like bottlebrushes. *S. obtusa* from Japan is similar but with rosy-red flowers in packed oblong inflorescences.

S. minor (*Poterium sanguisorba*) with green-purple flowers is the European (including British) Salad Burnet, the leaves of which are sometimes used in soups and salads or in cool drinks. As the generic name implies it was at one time esteemed for staunching wounds and Gerard tells us that 'the leaves steeped in wine and drunken comfort the heart and make it merry and are good against the trembling and shaking thereof'. In former days Salad Burnet was employed in such quantities 'to make the heart merry' that it was frequently called 'Toper's Plant'. It seems certain that even without this prescribed medication wine is still able (as of old) to 'gladden the heart of man'.

Amelanchier canadensis is commonly called Juneberry or Shadbush in its native North America but in Britain (where it or closely related species have become naturalized in some areas) is known as Snowy Mespilus (a name also applied to *A. vulgaris*) or Serviceberry. There is much confusion between the various species in this group and very similar plants may be called *A. lamarckii, A. laevis, A. confusa* and *A. arborea* as well. It makes a tree 6 m (20 ft) or more in height with slender branches, oval, saw-edged leaves and clusters of white flowers in spring, but even quite small saplings carry bloom. The berries which follow are rounded and black-purple in colour, very sweet and pleasant in some forms but almost tasteless in others. Birds are fond of them and help to spread the seeds. Some variation in the autumnal leaf tints occurs and certain trees will have a clear yellow coloration and others a warm red.

Sorbus are deciduous trees and shrubs with simple or pinnate leaves which often assume brilliant autumnal tints and terminal clusters of pink or white Hawthorn-like flowers and small fruits.

Sorbus aucuparia, the Mountain Ash or Rowan, is a tree of 6–15 m (20–50 ft) widely spread over cool temperate parts of Europe and Asia and common in many parts of the British Isles. It likes a cool moist situation, otherwise the leaves scorch, but when well grown the large nodding clusters of bright red fruits weight the branches in autumn and associate pleasantly with the gold and red leaf tints. There are fastigiate, weeping and yellow-fruited varieties.

Mountain Ash is frequently used as a stock plant for grafting other *Sorbus* species and hybrids. The fruits are edible and often made into preserves and have been used as a substitute for coffee. After frosting they are made into a liqueur in Germany and are an ingredient of certain Russian vodkas. The leaves and flowers have been used as an adulterant of tea and the tough wood is suitable for turnery and as a source of cellulose.

S. aria, the Whitebeam, is a fine tree to grow in association with *Prunus cerasifera* var. *atropurpurea.* Its ovate to elliptical leaves are silvery-white beneath and when young look completely argentous, thus providing striking contrast with the rich purple *Prunus* foliage. Later corymbs of heavily scented, white flowers appear and after these bunches of reddish-scarlet oval fruits. Unfortunately unless sprayed with a bitter tasting substance like quassia, the depredations of birds cause them to be of short duration. The dried fruits are used in some countries for coughs and catarrh and the fresh ones made into brandy and vinegar or occasionally baked in bread. The tree, which is a native of the British Isles and widespread in Europe and parts of North America, grows 9–14 m (30–45 ft) high.

Rubus fruticosus

There are many other species of Sorbus including *S. intermedia,* the Swedish Whitebeam; *S. matsumurana,* the Japanese Mountain Ash, a small tree with white flowers and long red fruits; and *S. chamaemespilus,* a shrub of 1.5–2 m (5–6 ft) from the European Alps with rosy-red flowers and scarlet fruits.

Several ornamental Blackberries and Raspberries are cultivated for their coloured stems and flowers rather than their fruits. All make useful semi-shade plants for moist soils and increase naturally by underground stolons. They require a certain amount of pruning, particularly the coloured-stemmed kinds, the older stems of which should be taken back to ground level in spring, when the young shoots (which have the brightest tints) are about 22 cms (9 ins) high.

Rubus biflorus from the Himalayas has white waxy-coated twigs, those of *R. thibetanus* from China are purplish, and the Chinese *R. cockburnianus* carries white waxy stems of 2–2.5 m (6–8 ft). In cool temperate gardens these provide useful sources of colour in winter when so little else is in character.

Another ornamental-stemmed kind is *R. phoenicolasius,* the Japanese Wineberry, which has biennial stems 2.5–3 m (8–10 ft) long, densely covered with bristly red hairs. The terminal racemes of pink flowers are succeeded by edible orange-red fruits.

R. deliciosus from the Rocky Mountains of North America is grown for its single white flowers 5 cms (2 ins) across which resemble Roses and festoon the arched stems 2–3 m (6–10 ft) high. Even finer is *R.* 'Tridel', a hybrid between *R. deliciosus* and *R. trilobus* raised in 1954 by Captain Collingwood Ingram.

R. odoratus is similar but the flowers are purplish-red and sweetly fragrant.

Two good ground carpeters suitable for mild shady spots are *R. irenaeus* and *R. tricolor,* both with white flowers and both Chinese. In cooler areas *R. tricolor* is deciduous.

Mespilus germanica with fruits

Rubus deliciosus

Several of the species in the genus *Rubus* have edible fruits, including the European Blackberries, which are in fact derived from many closely related species under the name R. *fruticosus,* and also R. *idaeus,* the parent of cultivated Raspberries. Many species have cruel prickles which cling to clothing and lacerate the flesh. In New Zealand these prickly *Rubus* are known to colonists as Bush Lawyers—because it is much easier to get into their clutches than out of them!

Alchemilla vulgaris (really a group of closely related species) from Europe (including Britain) Asia and Greenland is the Lady's Mantle, so called from the shape and vandyked edge of the leaves. These are soft grey-green, rounded yet slightly cup-shaped and almost 10 cms (4 ins) across. In wet weather surplus moisture exuded by the leaves beads the edges like drops of quicksilver, or following rain collects on their surfaces. The small greenish-yellow flowers appear in racemes or panicles and the whole plant is softly hairy and much prized by flower arrangers. It thrives in poor dry soils, often seeding itself in such unlikely places as rock crevices or between paving stones, but also makes an attractive under-carpeter or edging plant for Roses. The species commonly grown is *A. mollis* from the Carpathian mountains and Asia Minor.

Gillenia trifoliata is the Bowman's Root, an E North American herbaceous perennial hardy in Britain. It has panicles of white or pinkish flowers 60–90 cms (2–3 ft) tall with five narrow, ribbon-like petals. The almost stalkless leaves are deeply jagged into three parts and have toothed edges. The roots were once used medicinally by Indians and colonists as a mild emetic.

Dryas octopetala is a scrambling, rambling, mat-forming evergreen sub-shrub of the limestone rocks of N Europe (including Britain) and North America. It is popular in rock gardens for its freedom of flower and neat habit. The deeply-toothed, oblong-elliptic leaves are deep green above and downy white beneath, the white (occasionally pinkish) flowers about 4 cms (1½ ins) across on stems 5–13 cms (2–5 ins) high. Later these give place to feathery seed heads. Suitable for most soils, the plant is easily propagated by seed, layers, division or summer-struck cuttings. In the European Alps the leaves are made into a tea called Schweizertee or Kaisertee.

Gillenia trifoliata

Exochorda racemosa

Aronia arbutifolia, the Red Chokeberry, a deciduous shrub from E North America of 1.5–3 m (5–10 ft), is a variable plant with hairy shoots, narrowly-oval, tapering leaves, white or pinkish flowers in corymbs and red globular to pear-shaped fruits. The species thrives in most soils, the foliage colouring well in autumn.

The similar Black Chokeberry, *A. melanocarpa,* is also E North American and has black or black-purple fruits.

Neillia is a genus of E Asian deciduous shrubs allied to *Spiraea.* The inflorescences of small white or pale pink flowers are attractive en masse, the stems often willowy and the leaves alternate, with toothed margins. *N. longiracemosa,* a native of W China, grows 1–2 m (3–6 ft) high with rosy-pink flowers, and *N. sinensis* from central China is characterized by brown peeling bark and striking white calyces to the flowers. It grows 1.5–2 m (5–6 ft) tall.

Exochorda, a small genus of about five species from N Asia contains some of our most beautiful deciduous white-flowered shrubs. It is related to *Spiraea* but has much larger flowers and fruits.

E. racemosa from N China makes a rounded bush up to 3 m (10 ft) with narrow leaves 5–7 cms (2–3 ins) long and erect racemes of pure white, five-petalled flowers, individually up to 4 cms (1½ ins) across. In *E. giraldii* (which may be a form of *E. racemosa*) the blooms are all over 5 cms (2 ins) in diameter, in erect terminal racemes of six to eight flowers. The calyces have red margins.

E. korolkowii from Turkestan has smaller flowers but larger fruits. It has been found at altitudes of 1200–1800 m (4000–6000 ft) and is accordingly very hardy.

Eriobotrya japonica, an evergreen tree, is grown in the subtropics for its woolly-skinned, plum-sized, yellow or orange fruits. These have a flavour reminiscent of apples and pears and contain one or many rather large seeds. They are commonly known as Loquats or in some areas as Japanese Plums or Japanese Medlars.

The species is native to China and Japan and grows 6–9 m (20–30 ft) tall with large and handsome wrinkled leaves 15–22 cms (6–9 ins) long and 7–10 cms (3–4 ins) wide. The yellowish-white flowers are very fragrant and closely packed on stiff terminal panicles of 7–15 cms (3–6 ins).

In cool temperate gardens the tree is only hardy when grown in sheltered places as against a warm wall. The fruits are eaten raw or cooked in preserves, compotes or stewed.

Stranvaesias are evergreen shrubs or small trees with numerous, five-petalled white flowers in panicles and corymbs, narrow leaves and red or yellow haw-like fruits. *Stranvaesia davidiana* and its variants *undulata* and *salicifolia,* all Chinese, are occasionally cultivated.

Acaenas are dwarf herbs or sub-shrubs which make valuable carpeters in the rock garden. They are also grown on graves or as a substitute for lawn grasses. Many are native to New Zealand with other species in South America, Polynesia and Mexico.

Acaena microphylla and *A. inermis,* both from New Zealand, and the Chilean *A. myriophylla* even tolerate the dense shade of Conifers, forming pleasant mats of prostrate stems, small pinnate leaves and bunches of petal-less flowers. The calyces are often bristly with hooked or barbed spines.

A. novae-zelandiae has purplish flower heads, and *A. buchananii* inflorescences of petal-less flowers with yellowish calyx spines.

Rubiaceae

500 genera and 6000 species

This is one of the largest families of dicotyledonous plants, most of them tropical but with some temperate and a few arctic species. They comprise trees, shrubs and herbs, with entire often stipular leaves which, when present, exhibit a great variety of forms. Sometimes the stipules are as large as and look like foliage —so that the plants seem to have whorls of leaves.

The flowers are usually bisexual and regular, with four or five sepals (occasionally one sepal being larger than the others and brightly coloured), four to five (or six) petals, four or five (or six) stamens and an inferior ovary.

Several members of Rubiaceae are important economic plants, especially *Cinchona,* which yields quinine, *Coffea,* coffee and *Cephaelis,* ipecacuanha.

Bouvardias are low-growing, evergreen herbs or small shrubs which are mostly native to Mexico and Central America and have terminal bunches of either bright red, pink, white or yellow tubular flowers. Usually these are single but there are also doubles and a few scented kinds. Bouvardias are valued in the tropics for their long flowering and were once very popular; in Sydney, for example, in the early 1900s more than a quarter of a million plants were raised annually for bedding and show purposes. In cool climates they have to be grown under glass and are sometimes used by florists for sprays, buttonholes and table decorations. They need rich soil and plenty of water during the growing period. After flowering the plants should be cut back to points lower down on the stems showing new bud breaks.

Bouvardia longiflora is one of the most popular on account of its white fragrant flowers, either solitary or in terminal racemes on bushes 60–90 cms (2–3 ft) high. *B. humboldtii corymbiflora* with larger flowers probably belongs here, and there is a rosy-pink sort called *flammea. B. longiflora* has been crossed with other species such as the scarlet *B. leiantha* (both of them Mexican) to produce cultivars with variously coloured flowers. They include a number of doubles such as the deep pink 'President Garfield', the single white 'Alfred Neuner' and the orange-scarlet 'President Cleveland'.

The richly scented *Gardenia jasminoides* (*G. augusta; G. florida*) from S China is frequently cultivated as a pot plant in cool climates or the flowers are sold as buttonholes and sprays. In the tropics the plant makes a bush 2 m (6 ft) high and 1.5 m (5 ft) across. The medium-sized, Camellia-like double white flowers are freely produced over a period of several months. *G. thunbergia* from South Africa grows about 3 m (10 ft) high with large white, strap-petalled, single flowers and round leathery fruits filled with sticky orange pulp. The highly scented blossoms are pollinated by butterflies.

Gardenia flowers of various species are used in perfumery, for scenting tea and coconut oil and also as leis and necklaces. The white wood of the African species is employed for such things as knife handles and spears and the wood ash of *G. thunbergia* in the manufacture of soap and as lye in dyeing. The fruits have been employed for soups and sauces, dyeing purposes (yellow or black according to the species), also in cosmetics and to stupefy fish.

Other outstanding ornamentals are Ixoras with large terminal clusters of brilliantly coloured, four-petalled flowers. *Ixora chinensis,* with a distribution range from India to S China, is one of the most outstanding with rich scarlet flower heads which in West Indian gardens particularly are as arresting as Rhododendrons. These grow on bushes of 2–2.5 m (6–8 ft) and there are hybrids with orange-red, white or yellowish blooms. *I. odorata* from Madagascar has flower trusses 20 cms (8 ins) across, the blossoms richly scented and white or pinkish in colour but maturing to yellow. *I. grandifolia,* from the Malay Archipelago has white flowers. Ixora fruits are black and said to be favoured by peacocks. In India the bush is held sacred to Shiva and Vishnu and offered to the god Ixora—a local Malabar deity. It is usually propagated by layering.

Mussaenda erythrophylla comes from tropical West Africa and is a branched, green-stemmed shrub with long-stalked, evergreen, ovate leaves with red veining and terminal clusters of small creamy-white flowers. The uppermost leaf-like sepals are much enlarged and a splendid velvety red, as are the flower stalks. The inflorescence is thus startlingly conspicuous. *M. luteola* from Arabia and the Sudan has green-veined, yellowish-white calyx lobes. Both species make useful shrubs in the tropics but have to be grown in warm greenhouses elsewhere, in a compost of loam, peat and sand.

An outstanding plant occasionally seen in oriental gardens is *M. phillipica* 'Aurorae'. A bushy shrub growing 2–3 m (6–10 ft) tall, it carries clusters of golden-yellow flowers surrounded by very large, pendent, glistening white bracts. These are very conspicuous for all five sepals of *every* bloom develop these enlarged appendages (not just one) and they persist for practically twelve months. In Singapore this cultivar is known as 'Dona Aurora'. The type species (with a single, white, ovate bract) comes from the Philippines and New Guinea.

Pentas lanceolata is a compact tropical African and Arabian herb (30–60 cms; 1–2 ft) which makes a good bedding subject for sunny subtropical gardens or greenhouse cultivation in cooler areas. The attractive rounded clusters (7–10 cms; 3–4 ins across) of star-shaped lilac, white or red Bouvardia-like flowers unfold from long tubes and bloom all summer, and the oval-oblong leaves are opposite. Although seed propagation is common, good colour forms are best propagated by means of cuttings. *P. coccinea* with bright red flowers prefers a shady situation.

Gardenia jasminoides
double form

The genus *Rubia* from which the family takes its name, is chiefly noteworthy because it contains the Madder, *R. tinctorum.* This species is a weak-stemmed perennial with downward-pointing prickles which enable it to climb and has opposite, entire leaves, yellow flowers and near-black, berry-like fruits. The colouring principle lies in the roots in the form of purpurin and the more important alizarin, which yield madder-red and the brighter more lasting Turkey-red. These have been used since prehistoric times for dyeing purposes, evidenced by the rich colouring of the madder-dyed cloth enshrouding Egyptian mummies and the scarlet fez worn by Muslims. Their importance diminished about 1868 when the first synthetic dyes were manufactured from coal tar. A substitute for Turkey-red was among the first of these substitutes and caused such havoc in the industry that the French army issued its soldiers with madder-dyed scarlet trousers which remained part of their uniform until the First World War. A modest demand for Madder still exists among physicians for therapeutic purposes and for artist's pigments.

The roots of other species also yield red dye. The Indian Madder roots (*R. cordifolia*) are eaten with rice by the Javanese.

Asperula odorata is the Woodruff, a woodland plant 7–10 cms (3–4 ins) high found on calcareous soils in Europe (including Britain). Its small white flowers and whorls of needle-like leaves are very pretty but scentless until dried, when they release a substance called coumarin which smells of new-mown hay. They are then prized for spreading amongst linen. They are also used in Germany to flavour a special wine-based fruit cup.

Burchellia bubalina (*B. capensis*) is a small South African evergreen tree, having gay coral-red starry flowers with long tubes in clusters at the tips of the branches and shiny, dark green leaves. The species is valuable in warm climates because it succeeds in deep shade.

Nertera granadensis (*N. depressa*), the Bead Plant from New Zealand and South America, makes an attractive pot plant when in fruit. It is a dwarf herb with creeping stems, small round shining leaves and inconspicuous flowers which later give place to countless round, crimson berries the size of small Peas.

Cephaelis (*Uragoga*) *ipecacuanha,* a Brazilian herb, is used medicinally to promote sweating and for gastric complaints. The extract comes from the roots which are thickened after the fashion of a row of beads.

Coffee comes from the berries of several species of *Coffea,* the trees being kept down to shrub dimensions by pruning. The branches are laden with small white flowers which have the fragrance of Jasmine and the green berries turn first red and then violet-black. They are then prepared by various methods—some involving drying and others moistening with water. This removes the pulp but the characteristic aroma of coffee is not apparent until the beans are roasted.

Quinine, used in the treatment and prevention of malaria, is obtained from the bark of several species of *Cinchona,* large trees found in the South American Andes. Vast plantations were at one time cultivated for medicinal purposes in Java, from whence they were introduced first by the Dutch in 1854 and then in 1859 by the British. Known as Peruvian or Jesuit's Bark *Cinchona* has lost its importance now there are synthetic anti-malariants.

Mussaenda erythrophylla

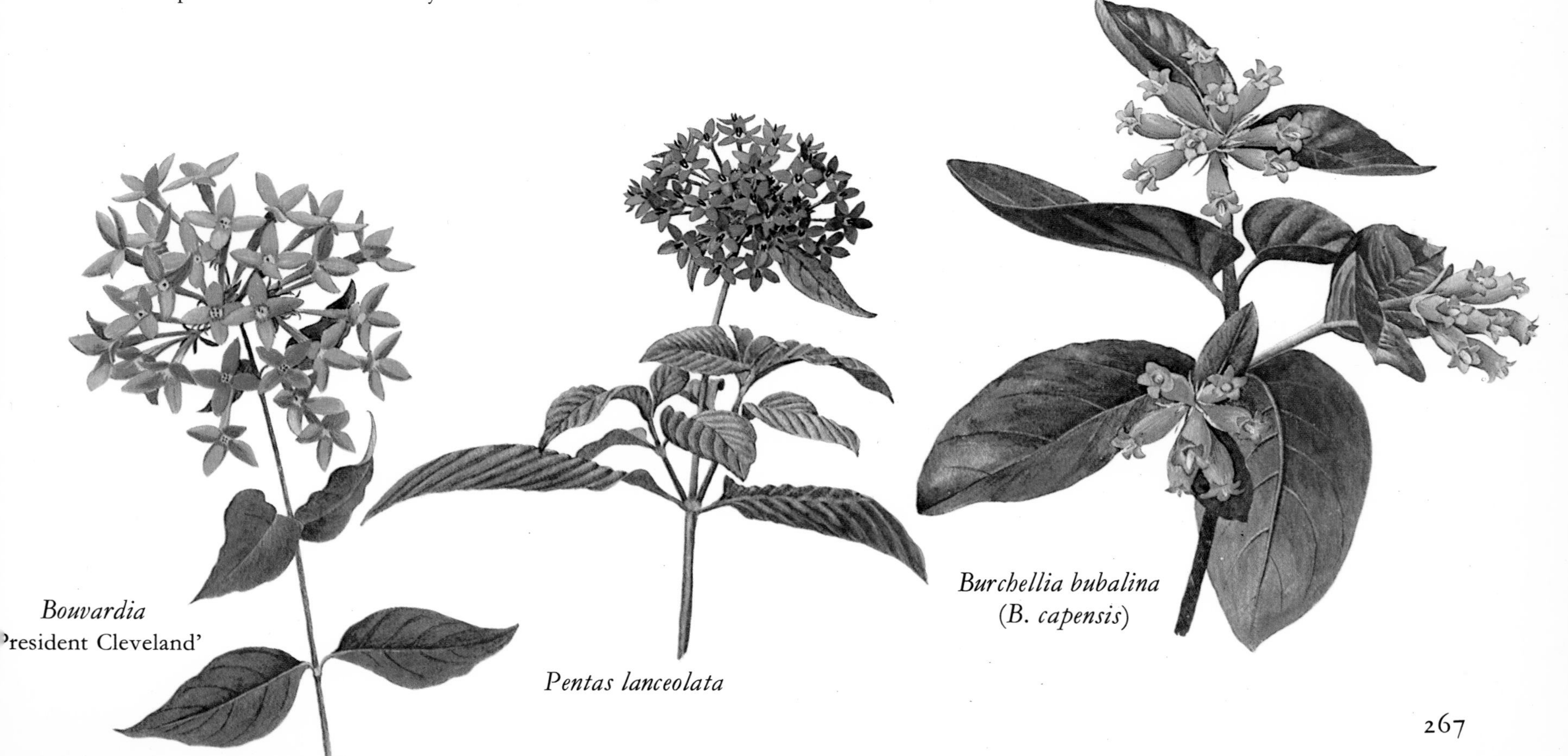

Bouvardia 'President Cleveland'

Pentas lanceolata

Burchellia bubalina (*B. capensis*)

Rutaceae

150 genera and 900 species

This is a family of dicotyledonous plants, mostly trees and shrubs. They come from tropical and temperate parts of the world, particularly South Africa and Australia. The leaves are frequently aromatic and usually compound, arranged alternately or oppositely on the stems.

The inflorescences take various forms with individual flowers normally bisexual (rarely unisexual) and either regular or zygomorphic in shape with four or five sepals and petals, eight or ten stamens and a superior (rarely inferior) ovary.

Several have economic importance, notably Citrus fruits, while others have medicinal application or make useful garden ornamentals.

Citrus, the most important genus, has about twelve species and countless hybrids, including Oranges, Lemons, Grapefruit and Tangerine and in the Orient some at least have been cultivated since very early times. The oldest known reference to 'oranges and pomelos' (*C. sinensis* and *C. decumana*) occurs in an ancient Chinese book called *Yu Kung* or *Tribute of Yu* which relates to an Emperor who reigned around 2000 B.C.

The first *Citrus* to be cultivated in Europe was *C. medica,* the Citron, which probably came overland from the East by the old caravan trade routes through Persia. Theophrastus described its fruits around 300 B.C. and it remained the only *Citrus* known in Europe until the 14th or 15th century A.D.

Christopher Columbus is credited with taking seeds of Lemons (*C. limon*), Citron and Bitter Oranges (*C. aurantiifolia* var. *aurantium*) to the New World during his second voyage of discovery. By the late 16th and mid 17th centuries they had become quite widespread. The Dutch Governor Van Riebeck introduced Lemons and Oranges to the Cape in the 17th century and they reached Australia about 1788.

Sweet Oranges were frequently grown as pot plants in Britain in the 19th and early 20th centuries; the old Orangery at Kew Gardens (now a timber museum) remains as a witness of this cult. The sweetly fragrant, waxy-white flowers were in great demand as bridal wreaths and skilful growers produced good quality fruits.

Citrus sinensis

The famous seedless 'Washington Navel' Orange originated as a bud sport in Bahia, Brazil. It was first grown in America by Mrs Eliza Tibbets in 1873 and a monument in the Eliza Tibbets' Memorial Park at Riverside, California, commemorates the fact.

Poncirus trifoliata (*Aegle sepiaria*), a monotypic genus from N China, is closely related to *Citrus* and hardy enough to grow outdoors in the more temperate regions of Europe (including parts of Britain) and America. It makes a shrub of 4–5 m (12–15 ft) with flattened green branches which have formidable spines 5 cms (2 ins) long. The white flowers 4–5 cms ($1\frac{1}{2}$–2 ins) across are succeeded by small, round, golden, fragrant fruits. The Chinese Orange is sometimes used for hedging in Virginia and Europe.

Dictamnus albus (*D. fraxinella*) is a herb of 45–90 cms ($1\frac{1}{2}$–3 ft) with pinnate leaves like those of the Ash Tree (*Fraxinus*) and showy racemes of white or pink flowers, the latter having deeper red veining. The plant has a variety of names such as Gas Plant, Burning Bush and Candle Plant, all referring to the inflammable nature of a volatile oil which it secretes. This is particularly marked near the old flower heads and on still days can sometimes be ignited. The Fire Worshippers of India considered *Dictamnus* a sacred plant because of this strange phenomenon. When gently rubbed the leaves and stems emit a fragrance like lemon peel.

D. albus is native to central Europe and Mediterranean regions eastward to the Himalayas and N China. The plants should be grown in light, well-drained soil in sheltered situations and when happily established left undisturbed.

Boronias are small Australian shrubs with opposite, simple or compound leaves and four-petalled white, pink, red, brown, yellow or pale blue flowers. These are either solitary or in many-flowered terminal or axillary cymes or umbels. Often the blooms are sweetly scented and the leaves when crushed have a pleasant aromatic scent. A commonly cultivated sort is Brown Boronia, *Boronia megastigma,* a bush of 2–3 m (6–10 ft) with small, soft, narrow leaves and sweetly scented reddish-brown and yellow flowers. The violet-scented Oil of Boronia (an essential oil used in perfumes) comes from this species, which is found naturally in swamps. For garden purposes it is suited by sandy acid soil with plenty of humus so that the ground does not dry out in summer. As the plants are not long lived new stock should be reproduced from time to time by means of seed or cuttings. In cool climates Boronias must be grown under glass.

Eriostemon species are commonly known as Wax-flowers because of their stiff petals. All but one (a species from New Caledonia) are Australian evergreen shrubs. A number are cultivated as greenhouse or garden plants, one of the most floriferous and popular being *E. myoporoides* with short axillary clusters of white flowers (rosy-pink in bud). Under good conditions this makes a rounded bush 90 cms (3 ft) in height, but sometimes grows taller and rather straggly. *E. verrucosus,* the Fairy Wax-flower, grows well on poor soils and has waxy white or pale pink petals. The blue-green leaves are pleasantly aromatic. *E. spicatus* is known as Pepper and Salt and has five-petalled, pink and white flowers in graceful spikes and stiff narrow leaves.

Another Australian genus is *Correa,* the species being attractive small shrubs with opposite leaves and white, pink, yellow or red often tubular flowers with cup-shaped calyces. They have a long flowering period and make good pot plants or can be grown outdoors in sheltered temperate gardens.

Correa reflexa (*C. speciosa*), the Native Fuchsia of Australia, is a very variable plant with heart-shaped or narrow leaves which may be smooth or hairy. It grows 30 cms–2 m (1–6 ft) high and has rich red tubular flowers tipped with green or yellow. Its variety *virens* is a yellowish-green form with rough leaves. *C. alba,* the White Correa, has Box-like leaves (which have been used for tea) and white, four-petalled, bell-shaped flowers.

Skimmias are evergreen shrubs of dwarf compact habit which are useful in cool temperate countries to grow in shade. They have entire, oval-oblong leaves dotted with oil glands which are heavily scented (not always pleasantly) when crushed. The small white flowers occur in close terminal panicles.

Skimmia japonica from Japan is one of the most popular but usually has the male and female flowers on different plants. These are fragrant, the male flowers particularly so, and the female blooms when fertilized give place to round, scarlet berries which often remain in character for months. This species grows 90–120 cms (3–4 ft) tall.

S. reevesiana, a calcifuge from China, has bisexual flowers and crimson berries on stems 60 cms (2 ft) high. *S. laureola* from the Himalayas is less hardy and grows 1.25–1.5 m (4–5 ft) with red fruits and heavily scented foliage when crushed.

Choisya ternata from Mexico is the Mexican Orange, a leathery leaved evergreen with leaves cut like those of *Laburnum.* They have a strong pungent odour when crushed. The large white fragrant flowers are produced in axillary corymbs early in spring, although stray blossoms may be found throughout the year. The bush grows 90–150 cms (3–5 ft) tall and is hardy in parts of Europe (including southern England) and the warmer zones of America. Nevertheless it is advisable to grow it in a sheltered position, or under glass in cool districts. It is easily propagated from cuttings.

Rue, *Ruta graveolens,* belongs to this family and is a branching shrub of 60 cms (2 ft) with green or blue-green, compoundly-pinnate leaves and greenish-yellow flowers. The compact habit and unusual colour of the foliage makes it a good foil for brightly coloured border flowers and there is a green and gold variegated cultivar. 'Jackmans Variety' is a particularly good blue-foliaged variant, which comes more or less true from seed.

The bruised leaves have a pungent odour which Spenser called 'rank-smelling', but as a bitter herb Rue has been valued for centuries. In the Middle Ages it was credited with anti-magical powers and when plagues were abroad people carried sprigs of Rue to ward off infection. It was considered a cure for countless ills from indigestion to bee stings and was often called 'Herb of Grace'.

This herb is one of the few plants to figure in heraldry and is represented in the Collar of the British Order of the Thistle. The finely chopped leaves are eaten in salads and the plants thrive in warm, well-drained soils—particularly where lime is present.

Poncirus trifoliata

Boronia megastigma

Choisya ternata

Correa reflexa var. *virens*

Dictamnus albus

Koelreuteria paniculata

Sarraceniaceae

3 genera and 17 species

This is a small but interesting family of dicotyledonous insectivorous plants, characterized by rosettes of pitcher-shaped, radical leaves which are often strikingly coloured. Frequently the tips of these leaves are expanded into 'lids' or bent over like cowls, a device which partially protects the hollow interiors with their digestive juices against too much dilution by rain. In the case of *Darlingtonia* the hoods terminate in serpent-tongued appendages, a circumstance which explains their common name of Cobra Plants.

These hollow vessels attract insects, primarily because of their colour but later due to the nectar-diffusing glands which are situated near the pitcher openings. In their eagerness to sup this sweetness the insects venture inside and pushing forward slip down into the cavity. Egress is prevented by the presence of downward pointing hairs, so that eventually they drop to the bottom. Here lies a pool of water or acid liquid in which they drown and eventually the products of decomposition are absorbed by the epidermal cells and serve to nourish the plant. In times of drought birds have been observed 'tapping' the pitchers for water and insects. Like *Drosera* and *Dionaea* (see pages 102–3) the members of Sarraceniaceae grow in wet, boggy areas where there is a shortage of soluble nitrates, so the absorption of flies and other insects performs an important dietary function.

The large inverted flowers are carried singly on long stems and consist of rounded, dome-like structures with trailing appendages of sepals and petals. The blooms are bisexual with four, five or six persistent sepals and as many petals (sometimes none at all), also twelve to fifteen or more anthers and a superior ovary.

Darlingtonia californica is a Californian Swamp Plant with green tubular pitchers, matted with white at the tops and having reddish veins. These terminate in fishtail 'lids'. The flowers are pale green with reddish-yellow petals.

The ten species of *Sarracenia* are also North American. At one time they were popular cool greenhouse plants and many hybrids were raised. They all possess some form of fixed protective hood covering. Sarracenias are among the most decorative of the insectivorous plants with brilliant pitchers of reddish-purple, greenish-yellow, green with purple spots, red, white and green striations and similar colour combinations. The flowers are mostly yellow or purple. The hardiest is *S. purpurea* with fat pitchers 10–15 cms (4–6 ins) long, green with red streaks and purple and green flowers. In sheltered bog gardens it makes an unusual feature at the pondside.

S. flava has erect, narrow pitchers of greenish-yellow, individually about 60 cms (2 ft) in height and 5–7 cms (2–3 ins) wide. The yellow flowers, which are 10–13 cms (4–5 ins) across are quite attractive and remain for several weeks in character. *S. leucophylla* (*S. drummondii*) has good trumpet-shaped pitchers with narrow wings which are hairy within and white and green in colour with red-purple veining. A brighter-coloured form known as *rubra* crossed with *S. purpurea* has produced *S.* x *mitchelliana*, a hybrid with wide-mouthed, funnel-shaped pitchers of olive-green profusely striated with fine red veins and heart-shaped, wavy-edged lids.

Sarracenias are also known as Side Saddle Plants, Boots,

Sapindaceae

150 genera and 2000 species

All members of this dicotyledonous family are woody plants, many of them lianas climbing by means of tendrils; the rest are trees or shrubs. The leaves are usually pinnate and the flowers unisexual and either regular or zygomorphic in shape with four or five sepals and petals, four, five, ten or many stamens and a superior ovary.

They include several tropical fruits such as *Litchi chinensis,* the Litchi, which has round Plum-sized fruits with one seed enclosed by sweet white flesh and a brown warty skin. Others include *Nephelium lappaceum,* the Rambutan, with spiny fruits, *N. mutabile,* the Puloesan, *Euphoria longana,* the Longan—all trees from East Asia—and *Blighia sapida,* the W African Akee. Its name commemorates Captain W. Bligh of the ill-fated Bounty and apart from the arils the fruits are poisonous and used in West Africa to stupefy fish.

Koelreuteria paniculata, the Golden Rain Tree from E Asia, has large pinnate leaves and erect, showy panicles of yellow flowers which develop to Walnut-sized, bladder-like fruits. It can grow to a height of 18 m (60 ft) and is hardy in temperate gardens as is *Xanthoceras sorbifolia*, a deciduous Chinese tree of 6–8 m (20–25 ft) with large pinnate leaves and axillary racemes of five-petalled flowers 15–25 cms (6–10 ins) long. These are white with yellow basal blotches which turn to carmine with age. The fruits resemble those of the Horse Chestnut (*Aesculus hippocastanum*).

The tropical American Soapberry, *Sapindus saponaria,* grows to a maximum of 24 m (80 ft) and has white flowers and shiny brown fruits. Their pulp lathers with water and can be used as a soap substitute.

Paullinia pinnata, a Mexican tree which contains the toxic saponin, is used as an arrow poison and to stupefy fish, and also in the Antilles to poison criminals. *P. cupana,* a trailing vine from South America, is rich in caffeine. The seeds are made into guaraná paste, foundation of 'cola'—the stimulating Brazilian drink.

Forefather's Cup, Huntsman's Cup, Soldier's Drinking Cup, Trumpets, Indian Cup, Trumpet Leaf, and Watches. The roots of several species have been used medicinally as a diuretic.

Heliamphora nutans is an interesting perennial from South America. The leaves grow in radical rosettes, each forming a tubular pitcher with a wide mouth and a very small lid. They have wings down the fronts, are hairy inside and green conspicuously veined with red in colour. The white or pale rose flowers are petal-less.

All members of Sarraceniaceae can be grown as cool greenhouse plants if the atmosphere does not become too hot and dry.

Saururaceae

5 genera and 7 species

These moisture-loving, dicotyledonous perennials, which are sometimes planted in the environs of the water garden for their interesting albeit short-lived flowers, are all natives of North America and E Asia. They have leafy stems with simple stipuled leaves, those at the tops of the stalks often coloured or forming an involucre (collar) of bracts.

Individually the blooms are small but being densely packed into spikes or racemes become quite striking. Each flower has three, six or eight stamens, but no sepals or petals, the showy parts being the coloured leaves already mentioned. The ovary may be superior or inferior.

Saururus chinensis is the Lizard's Tail, a bog plant 30–40 cms (12–16 ins) high from China and Japan with oval leaves which have rounded bases and spikes 10–13 cms (4–5 ins) long of yellowish-white flowers. The North American *S. cernuus*, the American Swamp Lily, has dense nodding spikes of fragrant white flowers 10–15 cms (4–6 ins) long and green heart-shaped leaves. In gardens it usually grows to about 30–60 cms (1–2 ft) although taller in its native habitat. In some areas it is also known as the American Lizard's Tail.

The boiled roots were believed to possess healing properties by the Choctow Indians and were applied to wounds in the form of poultices. They still have some medicinal use today, for they are anti-spasmodic, astringent and also act as a sedative.

Houttuynia cordata, a monotypic genus from E Asia, has bluish-green, heart-shaped leaves which later become reddish-crimson, red stems and terminal inflorescences of insignificant flowers. Each is set off by a prominent collar of four white bracts. The plant is of botanical interest inasmuch as it develops seeds parthenogenetically, that is without previous fertilization. It attains a height of around 30–45 cms (1–1½ ft). There is a double form as well which is rather more spectacular in appearance.

Anemopsis californica (*Houttuynia californica*), another monotypic genus, is a Californian bog plant of erect habit with long-stalked, rounded to oblong leaves and conical-shaped inflorescences subtended by whorls of white petaloid bracts. This gives them some resemblance to *Anemone* flowers.

The rootstocks, which are spicily aromatic, have medicinal value and are sold in the drug markets of Mexico for the relief of colds and indigestion and as a blood purifier. They are also strung into necklaces—in the form of beads—as a precaution against malaria and other diseases.

Houttuynia cordata

Darlingtonia californica

Sarracenia x *mitchelliana*

Saxifragaceae

30 genera and 580 species

Saxifragaceae is an interesting family of dicotyledonous, chiefly N temperate herbaceous plants which include many alpines and arctic species of xerophytic habit. This makes them adaptable and able to survive under difficult conditions which is the reason Saxifrages (in particular) are so often used for dry rockery planting. Others, like *Bergenia* and *Heuchera* are tough and persistent so make good border plants in cool climates.

The foliage is extremely variable, sometimes entire, sometimes compoundly divided. It may be smooth or hairy, leathery or tufted, fairly large or drastically reduced in some arctic species.

The flowers are normally bisexual and regular with five sepals, five petals and ten stamens.

One of the most popular genera is *Astilbe,* the species being moisture-loving perennials with attractive compound leaves and large, widely-branching inflorescences of feathery flowers. A single plume may contain hundreds of florets so that the overall effect is often spectacular.

Astilbes are ideal for growing in a woodland setting or close to water in association with Candelabra-type Primulas such as *Primula japonica, P. pulverulenta, P. beesiana* and *P. burmanica.* They also blend pleasantly with *Trollius, Hemerocallis, Iris laevigata* and *I. sibirica* but must have plenty of moisture during the growing season, otherwise the roots perish. A deep, rich soil retentive of moisture and containing plenty of humus constitutes ideal growing conditions.

The most widely used for garden and pot work are a race of cultivars known as *A.* x *arendsii,* which owe their origin to two European growers—Lemoine of France and George Arends of Ronsdorf, Germany. They have evolved from crosses made between the violet-crimson flowered *A. davidii,* introduced from China early in the 20th century, and such Japanese species as *A. thunbergii*, the old *A. japonica* (still known incorrectly in some gardens as *Spiraea japonica*) and *A. astilboides.*

Modern cultivars, with their erect and elegant plumes 60 cms (2 ft) high, ranging in shade from the purest white through tones of salmon, rose and pink, to carmine and deep rose, provide a gorgeous show when generously massed. Representative cultivars include 'Deutschland' and 'Irrlicht', white; 'Bressingham Charm' and 'Rhineland', pink; 'Granat', deep pink; 'William Reeves' and 'Feuer', crimson; and 'Fanal', deep red.

Many Astilbes are easily forced, suitable plants being potted in the autumn then plunged in ashes outdoors until they start to root when they are brought into heat and liberally watered. So treated they make good florists' plants.

Among the species *A. davidii* is one of the tallest, with stems of up to 2 m (6 ft) terminating in spikes 30–60 cms (1–2 ft) long of crimson-magenta flowers. *A. japonica,* the Silver Sheaf, reached Britain in 1908 by chance, in a consignment of Schizocodons from Japan. It has white flowers and there is a variegated-leafed

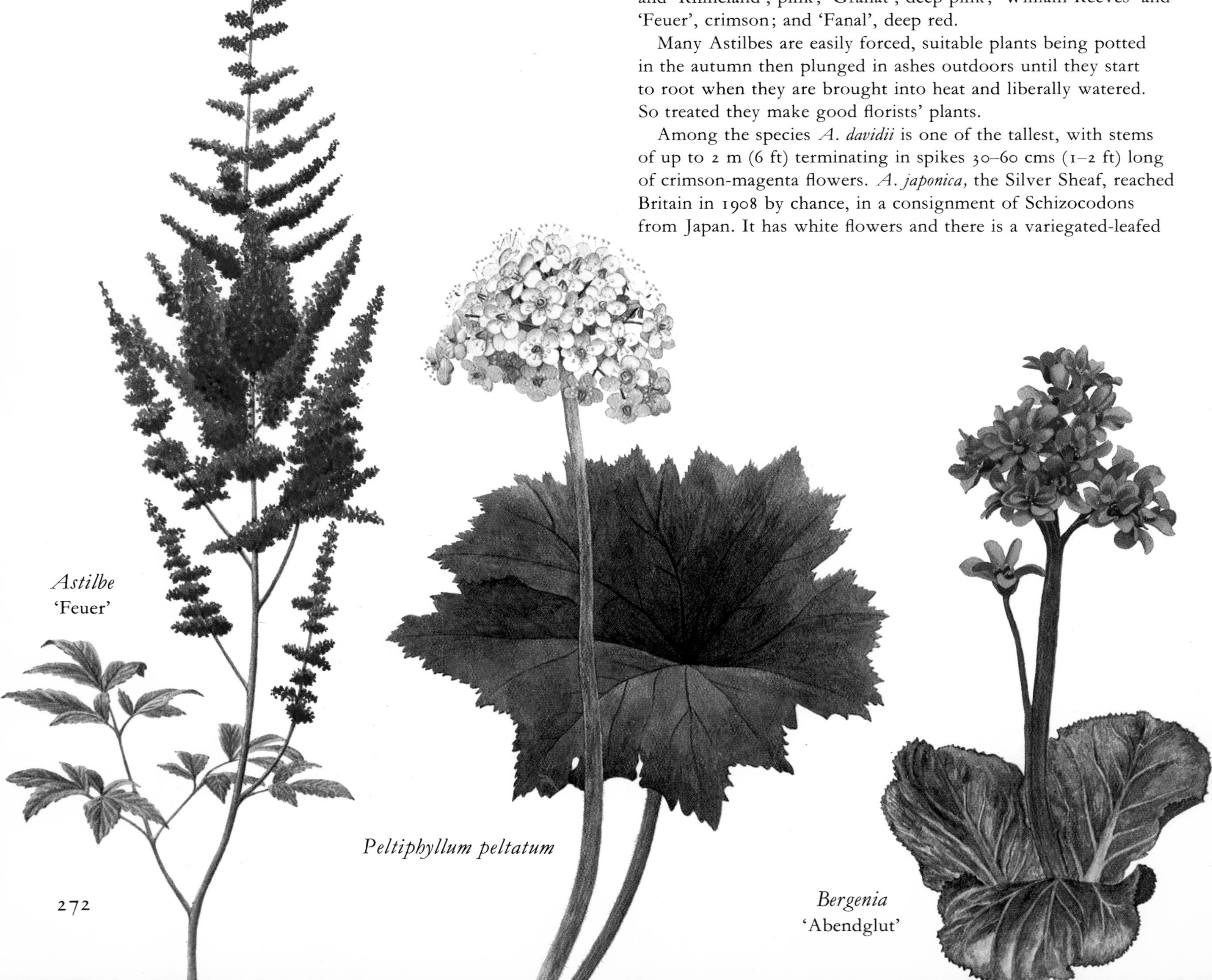

Astilbe 'Feuer'

Peltiphyllum peltatum

Bergenia 'Abendglut'

form. The Japanese *A. thunbergii* with white flowers is particularly useful for forcing and grows about 45 cms ($1\frac{1}{2}$ ft) tall.

The leathery-leafed Bergenias—once known as Megaseas and before that as Saxifrages—are early-flowering perennials with large, rounded, practically evergreen foliage, fleshy rootstocks and dense clusters of white, pink or red flowers. When a leaf of *Bergenia cordifolia* is held between the palms and twisted it makes a noise like a piglet, for which reason the late E. A. Bowles always called it Pig Squeak. This Siberian species has pink flowers and glossy green leaves like those of Water-lilies. These usually persist all winter. There is a red-leafed form and variants with white, deep pink and purplish-pink flowers.

The first species known to European gardeners was *B. crassifolia,* another Siberian, which was sent to Linnaeus in 1760 by the Empress of Russia's gardener, David de Gorter. It has branching panicles of large red flowers on sturdy stems 25–30 cms (10–12 ins) high.

B. stracheyi from the Himalayas has rounded leaves and branched heads of pink flowers, the individual florets of which are fairly large (1.9–2.5 cms; $\frac{3}{4}$–1 in.) and so weigh down the spikes that they appear to droop. There is a white form var. *alba*.

Several cultivars are worth noting, particularly 'Abendglut' with rich crimson, almost double flowers; 'Ballawley' which has bright crimson flowers on arching stems 45 cms ($1\frac{1}{2}$ ft) high and round green leaves which turn crimson-purple in winter; and 'Silberlicht', the best white. The foliage of Bergenias is very popular with flower arrangers.

Peltiphyllum peltatum (previously known as *Saxifraga peltata*) is a Californian moisture-loving perennial. It makes a handsome subject where spectacular foliage effects are desired, with large Lotus-like leaves topping stems of 90–120 cms (3–4 ft). The pale pink flowers appear very early in spring—before the foliage—clustered at the ends of stout stems 60 cms (2 ft) high.

The plant has fat marbled rhizomes which often snake their way above the soil and tenaciously inch their way forward into fresh ground. Deep, rich, moist soils give the best results. The species is popularly known as Umbrella Plant or sometimes as Indian Rhubarb, probably because Californian Indians ate the peeled leaf stalks, either raw or cooked. Propagation is easily effected by division or seed.

Tiarella cordifolia, the Foam Flower, is a plant for shade. It is under 30 cms (1 ft) with green, heart-shaped, hairy, ground-hugging leaves sometimes spotted and veined in red and spikes of foamy white flowers. *T. wherryi* is similar but lacks the stoloniferous habit of *T. cordifolia*. Both species are native to North America.

The fifty species of *Heuchera* are all North American. Being hardy in cool climates they make useful long-flowering perennials for the fronts of borders. They have small bell-shaped flowers hanging from slender, leafless stems 30–60 cms (1–2 ft) tall and heart-shaped, evergreen ground-hugging leaves.

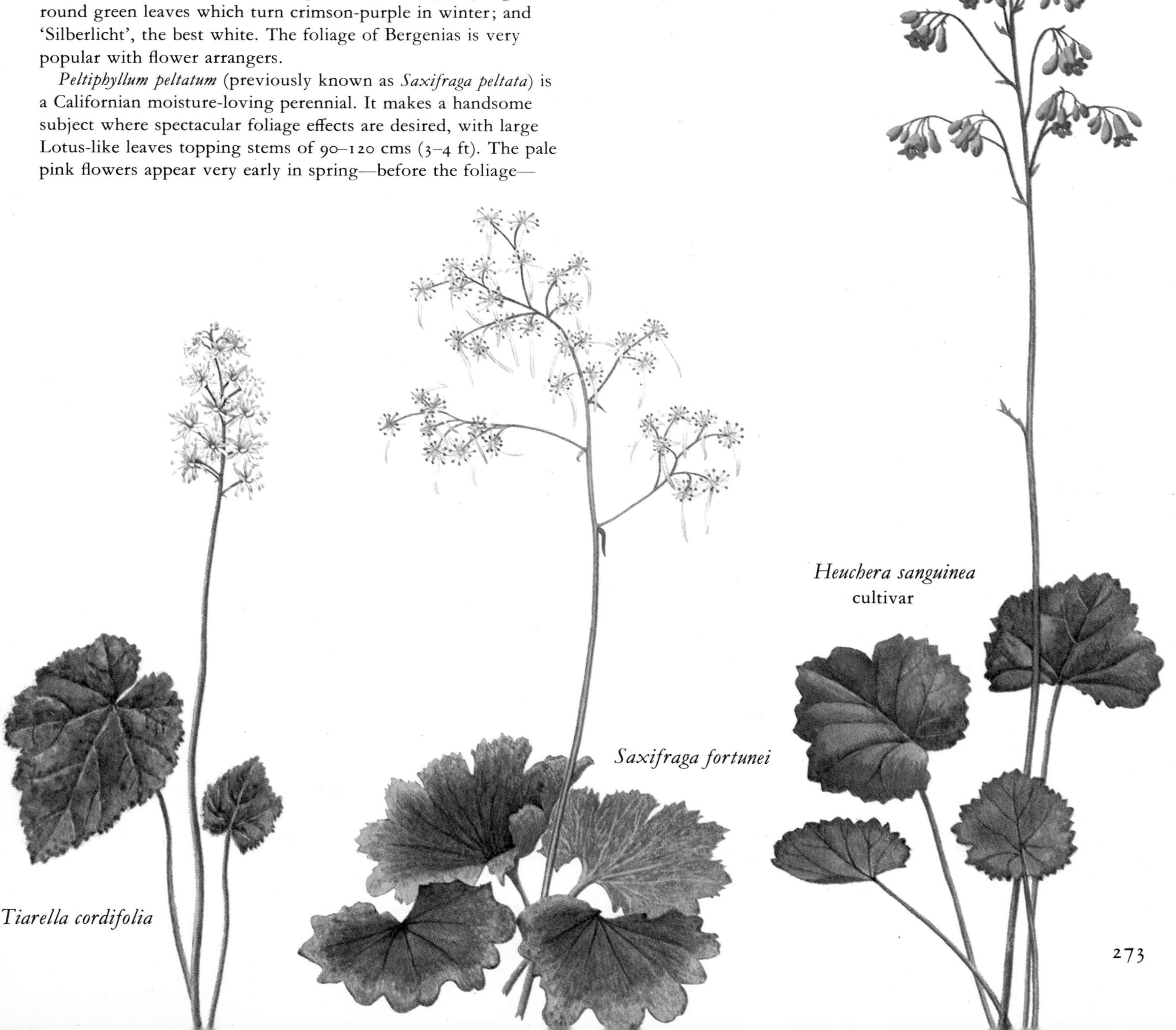

Heuchera sanguinea cultivar

Saxifraga fortunei

Tiarella cordifolia

The commonest is *H. sanguinea* which has slightly hairy, five- or seven-lobed, rounded leaves and spikes of bright red flowers. It reached England from Mexico in 1885, brought by a Dr Murray who took six plants across the Atlantic in an open basket. These, surviving the rigours of what was at that time a long journey, gave such pleasure to English gardeners that hybridists worked on the species and eventually produced many new cultivars. Some of the most interesting are 'Firebrand' (crimson-scarlet), 'Pearl Drops' (pearly-white), 'Apple Blossom' (delicate pink), 'Red Spangles' (scarlet-crimson) and 'Greenfinch' (greenish-yellow).

Another race (sometimes known as *Heuchera brizoides*) has been evolved by crossing *Heuchera* cultivars with *Tiarella cordifolia*. These are grouped under the name *Heucherella,* the cultivar 'Bridget Bloom' being one of the best. It grows 45–50 cms (18–20 ins) high with masses of light pink flowers and fine marbled, lobed leaves. If planted in rich soil and light shade it flowers from spring until autumn.

Rodgersias are named after a United States admiral called John Rodgers who commanded an expedition at the end of the 19th century. During this trip *Rodgersia podophylla* was discovered in Japan. It is one of several species of fine ornamental foliage plants for moist garden conditions, such as the environs of the water garden. Strong winds 'burn' the foliage so they are best afforded a sheltered situation and can then be left undisturbed for years.

Commonly known as Bronze Leaf, *R. podophylla* has large basal long-stalked leaves divided into five toothed leaflets and handsome sprays of yellowish-white flowers, rather like those of *Aruncus dioicus* (*A. sylvestris*) (see page 261). It reaches a height of 90–120 cms (3–4 ft).

R. pinnata, the Feathered Bronze Leaf, is perhaps the best of the genus with emerald-green, sometimes bronze-tinged, large toothed, fan-like leaves. The flowers are borne in large, much-branched panicles and are rosy-pink, which contrasts pleasantly with the foliage. 'Alba' has white flowers, 'Elegans' is warm pink and 'Rubra' deep red. The growth height is 90–120 cms (3–4 ft).

R. tabularis from China has round leaves 30–90 cms (1–3 ft) across poised like circular trays on bristly stems of 90 cms (3 ft). The flowers are creamy-white. *R. aesculifolia*—another Chinese species—has glossy, bronzed and crinkled foliage which is cut up into leaflets rather like those of the Horse Chestnut (*Aesculus hippocastanum*). The fragrant, pinkish-white flowers are clustered on stems 90 cms (3 ft) high.

The largest genus, however, is *Saxifraga* for which this plant family was named. There are about 370 species mostly of an alpine character, often with tufted, closely packed leaves. These exhibit various xerophytic characters such as extreme hairiness or succulence which reduces transpiration, so that they survive in such precarious perches as between vertical rocks or on wind-swept arctic plateaux. Many are vegetatively reproduced by means of offsets or bulbils in the axils of the lower leaves and others possess special glands at the tips or edges of the foliage. These secrete water, containing chalk in solution and as the moisture evaporates the leaves become lime encrusted. Alpine Saxifrages are often called Rockfoils.

Saxifraga umbrosa (like *Arbutus unedo* and several types of insects) is native to western Ireland and Portugal, a circumstance which leads some authorities to suppose the countries were once joined in a common mainland. Its English name, London Pride refers to its ability to survive in a smoggy atmosphere such as was common in nineteenth-century London, although another name St Patrick's Cabbage (also used for the closely related *S. spathularis*) is obviously linked with its Irish habitat. *S.* x *urbium* (*S. spathularis* x *S. umbrosa*) is the plant usually grown as London Pride.

There are some sixteen sections of Saxifrages and hundreds of hybrids. Some are mat-forming and semi-deciduous, others ever-green, the leaves forming basal rosettes. Many of the encrusted Saxifrages (like *S. longifolia* from the Pyrenees) are spectacular when in flower with tall branching stems smothered with loose panicles of flowers. Although the flowering rosettes die afterwards there are usually plenty of young rosettes to grow on and take their place.

The Kabschia types form small, cushion-like hummocks having more or less spiny or spreading foliage with one or several flowers on short stems. They are large in proportion to size and of various colours such as white, pale or deep yellow, pink, red or lilac.

Many Saxifrages are spring-flowering but *S. fortunei* (sometimes considered a variety of *S. cortusifolia*) is one of the more out-standing autumn-blooming sorts. It is a deciduous species, native to China and Japan, with large, rounded, glossy, basal leaves which are reddish beneath and throws up branching stems 30–45 cms (1–1½ ft) high containing loose inflorescences of white flowers. These all have one petal larger than the others, a cir-cumstance which gives the sprays a peculiarly graceful effect. 'Wada's Variety' has purple leaves.

S. grisebachii from Greece and Macedonia forms silver rosettes up to 7 cms (3 ins) across. These eventually make large humped cushions. The flowers are pink on leafy stems 22 cms (9 ins) tall; 'Wisley Variety' is a particularly good darker form.

S. burserana from the European Alps has many spiny silver-grey leaves in basal rosettes and reddish flower buds which open to large white blossoms. These are yellow in 'Major Lutea'. *S.* 'Megasiflora' which is probably derived from *S. burserana* has the same habit with clear rose flowers on stems 5 cms (2 ins) tall.

S. ferdinandi-coburgii, also from Europe, has bright yellow flowers and spiny rosettes and *S. oppositifolia,* a widely distributed species in N Asia, Europe (including Britain) and North America, white, purple, pink or red rounded flowers and tiny glands on its mostly opposite leaves. 'Ruth Draper' is one of its many forms.

Saxifraga oppositifolia 'Ruth Draper' — *Saxifraga ferdinandi-coburgii* — *Saxifraga grisebachii* 'Major Lutea' — *Saxifraga burserana* — *Saxifraga* 'Megasiflora'

Schisandraceae

2 genera and 47 species

This is a small family of dicotyledonous climbing shrubs closely related to Magnoliaceae. They all need warm or sheltered temperate conditions and have simple, alternate, often gland-dotted leaves and more or less regular flowers in which the petals and sepals are similar to one another. These may be unisexual or bisexual and have seven to fifteen perianth segments, four or many stamens and aggregate superior ovaries.

The genus *Schisandra* has some twenty-five species which may be deciduous or evergreen. The dried wood is often delightfully aromatic and the blooms either solitary or several together in axillary clusters. *S. rubriflora* (*S. grandiflora rubriflora*) is a Chinese species with climbing stems 5–6 m (16–20 ft) tall and oblanceolate toothed leaves. The deep crimson flowers swing singly from long red stalks and later give place to red globose berries about the size of Peas. These are very attractive on their stems of 5–13 cms (2–5 ins) and the buds are also showy and look like ripe cherries. The species can be trained up a stout rod or used to cover a pergola. It was introduced by Ernest Wilson in 1908 and has proved hardy in central Europe and Britain and zones 7 to 10 in the United States.

S. chinensis has fragrant, pale rose or white flowers and scarlet berries, and *S. sphenanthera*, orange flowers and red fruits, while the white or yellowish flowers of *S. henryi* are unisexual and on separate plants so that both sexes have to be grown in order to obtain the red, 5–7 cms (2–3 ins), mucilaginous fruits.

Kadsura japonica is the hardiest member of the other genus. It is a Japanese climbing evergreen (up to 4 m; 12 ft) with slender branches carrying oval-lanceolate leaves and solitary creamy-white flowers about 1.9 cms (¾ in.) across. The scarlet berries are clustered in round heads several centimetres across and look very decorative.

Although now transferred to a distinct family (Illiciaceae) *Illicium verum,* the Chinese Star Anise, until very recently was included in Schisandraceae. It is a small evergreen tree with white to yellow flowers when young maturing to rose or purplish-red, star-shaped fruits. It is chiefly important as a specific against various diseases, particularly in Asia; the fruits are also used in perfumery and to give an aromatic flavour to liqueurs, syrups and cordials.

Schisandra rubriflora

Antirrhinum hybrids

Scrophulariaceae

220 genera and 3000 species

This is a large and important family containing many good ornamental plants, chiefly herbs and low shrubs. These have cosmopolitan distribution with a wide range of habitats—from grassy plains and very dry areas to swamps. They include climbers, xerophytes and even semi-parasites like *Bartsia* and *Rhinanthus* which, although possessing green leaves of their own, attach themselves to grasses and derive further nourishment from this source. The leaves may be opposite, alternate or whorled. *Castilleja,* the Indian Paint-brush of western America, has brilliantly coloured upper leaves and bracts.

The inflorescences are extremely varied and may be in the form of a spike, raceme, cyme or complex corymb, although individual axillary flowers are also common. They are bisexual and zygomorphic but vary considerably in structure. Usually however they have four or five sepals, four or five petals (often two-lipped), four stamens and a superior ovary. Normally the blooms are insect pollinated.

A favourite plant in the family is *Digitalis purpurea,* the Foxglove, a species native to W Europe (including Britain) but now naturalized in many parts of the world, even in the Andes and New Zealand (where it is classed as a noxious weed). It is a biennial with greyish-green, oblong to lanceolate leaves and striking spikes of finger-shaped purple flowers 90–150 cms (3–5 ft) tall. These usually have dark purple spots inside although white, pink and rose forms are available, with or without spots. Others have extra wide corollas or are almost double.

Foxgloves are good shade plants and associate pleasingly with ferns and shade-loving Campanulas like *Campanula latifolia* and *C. trachelium*. Other species have yellow flowers, particularly *D. lutea* (60 cms; 2 ft) and *D. grandiflora* (*D. ambigua*), which grows 60–90 cms (2–3 ft) and has brown, net-like striations over the yellow corolla. Both are European.

Foxgloves although poisonous possess important medicinal

properties. The dried leaves contain digitalin which is used in certain heart complaints. Herbalists of the past seem to have employed them for a miscellany of ailments including fevers, liver complaints and the 'King's Evil'.

Antirrhinums are widely planted in many parts of the world. They are annuals or perennials of erect, climbing or prostrate habit with entire leaves and solitary or terminal racemes of sweet-smelling showy flowers. These are tubular with erect two-lobed upper and three-lobed lower lips. When the flowers are open they look like mouths, a circumstance which has earned them such names as Snapdragon, Rabbit's Mouth, Lion's Mouth and Dragon's Mouth.

The most important species is the European *Antirrhinum majus,* garden forms of which are widely used for bedding purposes or grown as pot plants. They also naturalize pleasingly on old walls. The cultivars vary in height from a few centimetres to 60–90 cms (2–3 ft) and the colours range through white, yellow and orange to pink, rose, red, magenta and maroon, frequently displaying several shades on the same flower. Because of the widespread incidence of a disfiguring leaf rust disease these Antirrhinums—although perennial—are usually grown as annuals.

In the past Antirrhinums were valued as a specific against witchcraft and were cultivated in Russia for the sake of the oil in their seeds, which is said to be almost as pure as Olive oil. The blooms form insect traps. They readily yield to pressure from without but once inside the creatures can only find their

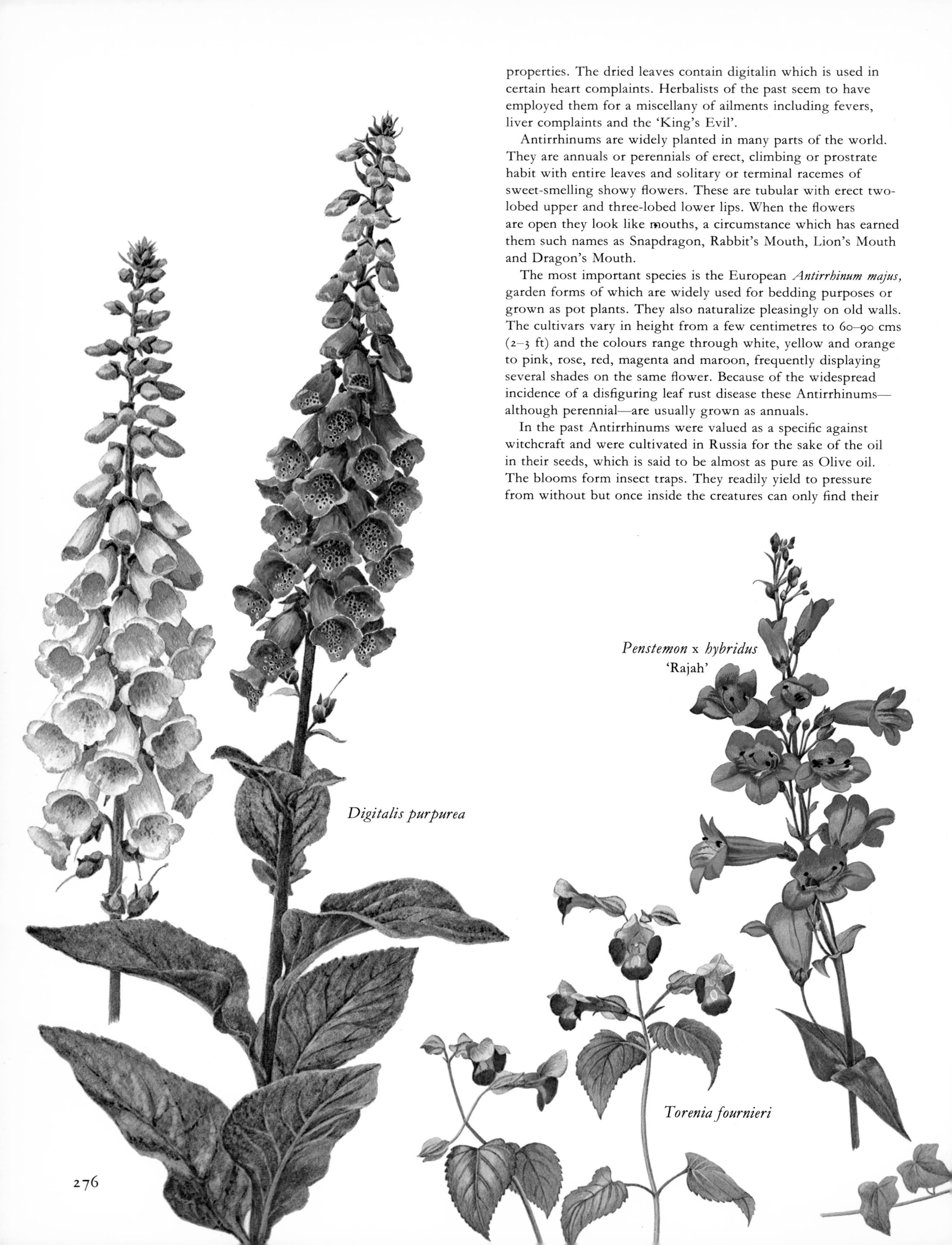

Penstemon x *hybridus* 'Rajah'

Digitalis purpurea

Torenia fournieri

way out again by gnawing through the flowers. *A. asarina* (now known as *Asarina procumbens*) from the Iberian peninsula is a low-growing, scrambling perennial for hot dry places, such as beneath trees. It has five-lobed downy leaves and creamy yellow flowers.

Some authorities consider that *Asarina* and *Maurandya* should be united under the former name. The plant generally grown as *Asarina* (*Maurandya*) *antirrhiniflora* from California to Mexico, a perennial climber of 60 cms–2 m (2–6 ft), is sometimes used as a trailer in hanging baskets or as a greenhouse climber in frost-prone climates. It has trilobate, arrow-shaped leaves and sensitive leaf stalks with which it clings to supports, and rose or purple flowers 2.5 cms (1 in.) across. *Asarina* (*Maurandya*) *barclaiana* from Mexico has larger flowers (up to 7 cms; 3 ins) of various colours from deep purple to white and rose. *Asarina* (*Maurandya*) *scandens* from Mexico is a climbing species having mauve or purple flowers with white throats. Certain species are sometimes included in the genus *Lophospermum* but this is generally reduced to synonymy with either *Maurandya* or *Asarina*.

Torenias are all tropical so in frost-prone climates can only be grown outside during the summer months. The most popular is *Torenia fournieri,* a native of South Vietnam and a profusely flowered annual readily raised from seed. It grows 15–30 cms (6–12 ins) tall and in cool climates particularly makes a popular pot subject on account of its Pansy-hued flowers 4 cms (1½ ins) in diameter. These come in terminal clusters and are funnel-shaped, violet and pale blue with yellow spots. In warm climates the species is often treated as an edging or border plant or set amongst rocks. It flowers long and continuously. *T. baillonii* (*T. flava* of gardens) from the same habitat has bright yellow and purple blooms.

Of the 250 or so species of *Penstemon* all but *P. frutescens* from Asiatic Russia are native to North America, chiefly the western United States.

P. gentianoides, P. hartwegii and *P. cobaea* are rarely seen today except in Botanic Gardens. Their cultivars—grouped as *P.* x *hybridus*—are more showy with stems 60–90 cms (2–3 ft) tall carrying lush, smooth, oval-oblong leaves and spikes or racemes of large scarlet, pink, white or crimson flowers. These frequently have white throats. They make excellent bedding subjects even in cool climates where (although able to withstand a few degrees of frost) they are best overwintered as rooted cuttings under glass.

P. barbatus (*Chelone barbata*) is the Beard Tongue, a species 90 cms (3 ft) high with soft linear-lanceolate leaves and terminal spires of pinkish-red flowers. These have bearded throats and varieties exist with purple, lavender, pink and red blooms.

Among the shrubby species are several splendid plants for the rock garden or small borders. *P. menziesii* grows 30 cms (1 ft) high with dense branches, thick but small leaves and dense racemes of violet-blue or purple flowers. Each blossom is about 2.5 cms (1 in.) long. *P. scouleri* (now considered a variety of *P. fruticosus*) has so many magenta flowers that the bushes (30 cms; 1 ft) become almost hidden beneath the blossom. It has a white-flowered form and another with the same habit is *P. newberryi*, which is rosy-purple with small, round, leathery leaves.

P. heterophyllus is one of the most beautiful. It is slender in growth, about 30 cms (1 ft) high but occasionally more, and has narrow leaves and funnel-shaped flowers 2.5 cms (1 in.) long of soft amethyst, which become clear blue at the mouths. It is easily raised from seed. *P. glaber* has more or less glaucous blue foliage and spikes 30–60 cms (1–2 ft) high of lovely clear blue flowers.

The North American Chelones are near relatives of Penstemons. Their flowers are crowded into short terminal spikes or panicles on sturdy leafy stems and the leaves are opposite and broadly-ovate in shape. They succeed in most soils provided the drainage is good and tolerate light shade.

Chelone obliqua grows 60 cms (2 ft) tall and has rose to red-purple flowers; *C. nemorosa* is violet-purple but rather unpleasantly scented and grows 60 cms (2 ft) high; *C. glabra* which is white or pinkish attains a height of 30–60 cms (1–2 ft); and *C. lyonii* which is 90–120 cms (3–4 ft) tall is rosy-purple.

Calceolarias are mostly native from South America (especially the Andean regions of Chile, Peru and Ecuador) to Mexico.

They are readily recognized by their curious pouched flowers which look rather like inflated slippers. The leaves may be simple or finely cut, with a puckered or hairy texture, and are sometimes sticky.

From a garden viewpoint there are two main groups: the large-flowered, herbaceous species and hybrids which are chiefly used as spring-flowering pot plants and grown as annuals; and the shrubby kinds for bedding out and greenhouse decoration. These are perpetuated vegetatively. The former are reputedly derived from a *Calceolaria corymbosa* and *C. crenatiflora* cross, although other species have doubtless been involved in their background. Their hybrids are known as *C.* x *hybrida* or sometimes as *C.* x *herbeohybrida* and although they rarely come true the seedlings show a tremendously exciting colour range—through cream, yellow and orange to pink and red. In addition they are frequently spotted and blotched with darker shades. They make compact plants about 60 cms (2 ft) in height with large showy blooms which are easily damaged in bad weather or strong sunlight. For this reason they are usually grown under glass, shaded when necessary. To encourage bushiness it is usual to pinch out the growing point when the plants are a few centimetres high.

Calceolarias are 'long-day' plants but earlier flowering can be induced by increasing normal daylight with the aid of artificial illumination. This is done by extending each day's natural light to fifteen hours, using lamps suspended 60 cms (2 ft) above the plants and spaced 120 cms (4 ft) apart.

Shrubby Calceolarias (often known as *C.* x *fruticohybrida*) have much smaller flowers than the large-flowered florists' forms, in rounded, honey-scented trusses which are usually yellow or orange. They are readily propagated from young shoots, rooted in late summer and overwintered in a frost-proof house until all risk of frost is over, when they can be bedded out.

Other interesting species include *C. darwinii,* a dwarf tufted alpine from the Straits of Magellan only 2.5–5 cms (1–2 ins) high. It has small oblong leaves and solitary yellow flowers 2.5 cms

Asarina (*Maurandya*) *scandens*

Phygelius capensis

(1 in.) across which have curious chestnut blotches on their lower pouches. *C. gracilis* (30 cms; 1 ft) is a pretty annual from Ecuador with finely cut, sticky leaves and many creamy-yellow flowers. *C. chelidonioides,* another Ecuadorian annual, has deeply divided succulent leaves and loose clusters of lemon-yellow flowers on stems 30–45 cms (1–1½ ft) tall. *C. arachnoidea* from Chile bears densely white, radical leaves which look as if they were covered with cobwebs, and has panicles of purplish-violet flowers on stems 25 cms (10 ins) in height.

Phygelius capensis is called River Bells in South Africa probably because it grows wild near river banks in various parts of the Republic. It is a shrubby perennial which annually sends up slender, leafy stems 90 cms–2 m (3–6 ft) tall carrying many drooping, coral-red, tubular flowers. There is a deeper red variety. The soft, dark green leaves are ovate and pointed, and larger near the base than at the tops of the stems. *P. aequalis* is similar but carries many more flowers on its stems of 90 cms (3 ft). The flowers are rosy-red with protruding crimson stamens.

Verbascum is another large genus, its 300 odd species all natives of Europe, N Africa and Asia. They are biennials or perennials with thick tap roots and basal rosettes of large leaves which are usually densely covered with white or silver plush-like hairs. The upper leaves are smaller and arranged alternately. These are often pinnate or have toothed edges. The flowers come on stout spikes or racemes. Verbascums make good back-of-the-border subjects and are not fussy about soil provided it is well drained.

A decade ago a decoction of the leaves of *Verbascum thapsus,* the Yellow Mullein, was recommended by physicians to alleviate diarrhoea.

Another common name, Hag's Taper or Higtaper, refers to a country practice of the past when hairs taken from the leaves were twisted into wicks and inserted in mutton fat to act as nightlights. This of course was long before cotton wicks and lamps were invented but even prior to this time the Romans dipped *Verbascum* stems in tallow and used them as torches.

One of the most beautiful garden species is the biennial, *V. bombyciferum* (*V.* 'Broussa') from Asia Minor, 1.25–2 m (4–6 ft) tall. The whole of this plant—its stem, foliage and inflorescence—is densely silvered with short white hairs. The golden flowers gleam through this woolly covering like agates in cotton wool.

Other garden favourites are *V. nigrum,* the Dark Mullein, a European species 60–90 cms (2–3 ft) tall with small yellow flowers and purple stamens; *V. olympicum* from SE Europe, a handsome plant with branching, candelabra-like sprays 1.5–2 m (5–6 ft) tall carrying bright golden flowers; and *V. phoeniceum,* the Purple Mullein, a species from S Europe and Asia 60–120 cms (2–4 ft) tall. The flowers of this last species are very variable in shades of violet, pink and rose and it is probably the parent of such garden varieties as the mauve 'Lilac Domino', white 'Miss Willmott' and rose-pink 'Pink Domino'. These named sorts can be propagated from root cuttings.

For many years *Veronica* embraced a wide diversity of plants including small trees, shrubs, annuals and perennial herbs. The genus has now been split into two main groups—*Hebe* and *Veronica* (although botanists have made some others such as *Veronicastrum*). *Hebe* contains most of the shrubby species and the herbaceous kinds (of Europe, Asia and North America) remain under *Veronica.*

The last are long-flowering plants which require good soil and full sunshine. The best forms have been derived from *V. spicata,* the Spiked Speedwell, a European (including British) perennial of 60 cms (2 ft) with narrow oblong leaves and bright blue flowers in dense spikes. Cultivars with deep blue, violet, rose-pink, clear blue, orchid purple and white flowers exist, most of them growing 30–45 cms (1–1½ ft) tall.

Calceolaria x *hybrida*

V. incana from Russia and China has oblong, silvered, slightly toothed leaves and terminal spikes 30–60 cms (1–2 ft) long of soft blue flowers. More garden-worthy are varieties such as 'Rosea' with pink and 'Saraband' with violet-blue flowers, and also highly silvered foliage forms such as 'Glauca' and 'Argentea'.

For the front of the border *V. teucrium* 'Crater Lake Blue' with short spikes of intense blue has its uses and there are several dwarfs for the rock garden like the blue-flowered *V. armena* from Armenia (10 cms; 4 ins), *V. catarractae* (now *Parahebe catarractae*), a New Zealander with milk-white or blue blossoms (10 cms; 4 ins), and *V. fruticans* (*V. saxatilis*), a European having blue flowers with red eyes (15 cms; 6 ins).

V. gentianoides from the Caucasus is a perennial of 15–30 cms (6–12 ins) with stout oblong leaves and racemes of pale milky-blue flowers. There is also a variegated form.

V. beccabunga, the Brooklime, is a European (including British) bog plant. It is of a creeping nature with blue Forget-me-not-like flowers and fleshy, smooth, oblong leaves. The leaves and young stems were once esteemed as a vegetable and eaten with or instead of Watercress. They are perfectly wholesome and anti-scorbutic. In the past the leaves were applied to burns and bruises.

The 100 to 150 species of *Hebe* are mostly plants of the Southern Hemisphere, particularly Australasia. They are small evergreen trees or shrubs, sometimes with thick, fleshy, simple leaves but particularly in New Zealand often of an alpine nature. New Zealand has no fewer than ninety endemic species, some with their leaves lightly clasping the whippy stems so that they resemble *Cupressus* or other conifers. These are known as Whipcord Hebes.

Noteworthy in this group is *Hebe lycopodioides* with white flowers in dense heads at the tips of the leaf-entwined branches. *H. hectori, H. cupressoides* and *H. ochracea* (usually grown as *H. armstrongii*) are others in the group, none of them with readily recognizable foliage.

The larger-leafed Hebes have showy racemes of blue, pink or white flowers which frequently continue over a period of months. One of the best of these is 'Autumn Glory' with deep violet flowers on bushes of 1.5–2 m (5–6 ft). It is believed to be derived from the large-flowered, fragrant, white and purple *H. elliptica.* This is native to both the Falkland Isles and New Zealand where it often grows up to 6 m (20 ft) high.

Most Hebes do well in coastal areas but are not suitable for very cold climates. *H. speciosa* is one of the most striking, making a bush up to 1.5 m (5 ft) with oblong leaves 2.5–10 cms (1–4 ins) long and dense racemes (7–10 cms; 3–4 ins) of deep purple flowers near the tips of the shoots. White, pink, carmine, crimson, purple and a variegated-leafed form are grown. *H. brachysiphon* has narrow oval leaves and white flowers and *H. salicifolia* has willowy leaves, 5–15 cms (2–6 ins) in length, and long sprays of white, mauve or bluish-purple flowers. They are all native to New Zealand but *H. hulkeana* is considered by many to be the finest of the small shrubby New Zealand Hebes. It has glossy-toothed leaves and panicles of rich blue flowers on stems of 30–45 cms (1–1½ ft). It will only withstand a few degrees of frost so must be given a sheltered situation in cool climates.

In 1880 seed of *Nemesia strumosa* was collected on a sandy bluff overlooking the sea near the Cape and sent to the English firm of Suttons. It is a naturally variable species but Suttons succeeded in breeding plants to separate colours which are now popular annuals for pot work or summer bedding. Wild plants of various colours are still to be found in their native habitat but these—like any batch of mixed hybrids—do not come true because of the necessity of preventing cross-pollination. This is a tiresome chore for laymen so that most gardeners prefer to buy fresh seed each year. The plants are tender to frost and like a sunny, well-drained situation.

The flowers are grouped at the tops of leafy stems 30–38 cms (12–15 ins) tall; the foliage is narrow and tapering with serrated margins. The colours range from white with a yellow spot, lemon and orange to pale pink, crimson, deep red and blue.

Mimulus with some 100 species has a broad cosmopolitan distribution, particularly in America. All are moisture-loving with pouched, often brilliantly coloured flowers and smooth, hairy or sometimes sticky leaves.

M. luteus, the Monkey Musk, is found near streams and water-courses from Alaska to New Mexico but has now become widely naturalized in Britain and parts of Europe. It is a perennial of 30–35 cms (12–14 ins) with hollow stems, smooth, oval-oblong leaves and bright yellow flowers 5 cms (2 ins) across with red dots on their throats. There is a double 'Hose-in-Hose' variety.

The juicy leaves of *M. guttatus* can be eaten as salad greens. The flowers are yellow with red or brown spots on stems of 30–60 cms (1–2 ft).

In 1826 David Douglas discovered *M. moschatus* in California as a creeping plant with glandular hairy, ovate leaves and masses of clear yellow flowers. In Victorian times it was widely planted in damp shady borders or grown as a pot plant on account of its rich scent which, according to those who remember it, persisted for several hours even in a room from which it had been removed.

About 1914 the scent unaccountably disappeared, a circumstance

Hebe speciosa cultivar

Veronica gentianoides

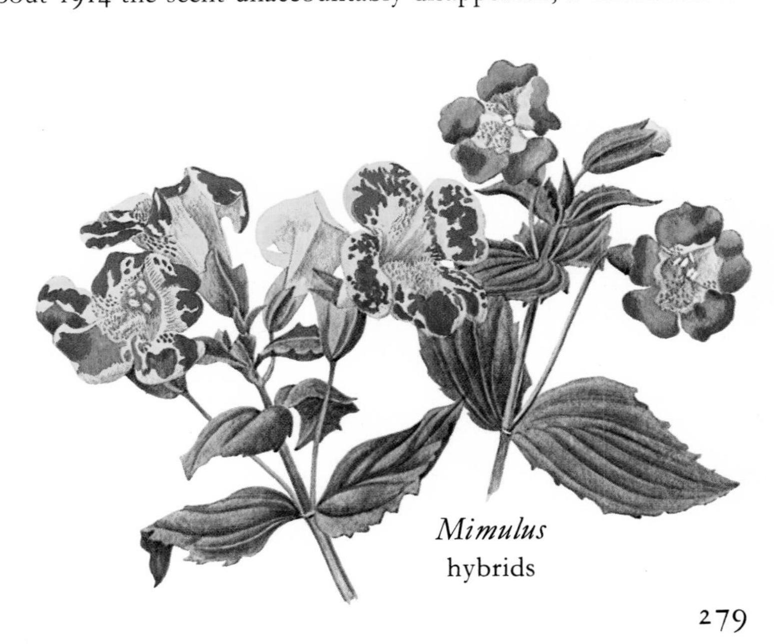

Mimulus hybrids

Rhodochiton atrosanguineus

which has puzzled scientists ever since.* However the plant is still worth growing as a shady ground cover.

M. cardinalis (30–90 cms; 1–3 ft) has rich red flowers and *M. lewisii* red or white blooms; both species come from western North America.

M. cupreus, a low-growing species of 15–25 cms (6–10 ins) from Chile is the parent of such striking bog-garden plants as 'Bees' Dazzler', crimson-scarlet; 'Red Emperor', scarlet; 'Whitecroft Scarlet', vermilion; and 'Leopard', a large yellow with red spots. These only tolerate a degree or so of frost so should be over-wintered in cool climates as rooted cuttings under glass.

Mimulus aurantiacus (*Diplacus glutinosus*) is a shrubby, Californian perennial, 90–150 cms (3–5 ft), which can only be grown outside in frost-free climates. It makes a striking pot plant (which can be plunged in the ground for the summer months), having linear-lanceolate leaves 5–10 cms (2–4 ins) long and orange, buff or salmon flowers. The whole plant is sticky to the touch.

Linarias are the Toadflaxes with small, delicate Antirrhinum-like flowers. Several are grown as summer-flowering annuals outdoors or in pots, particularly varieties of *Linaria maroccana*, the Fairy Flax. These have slender, branching stems 30–45 cms (1–1½ ft) high with narrow pointed leaves and long dense spikes of pale-eyed pink, blue, purple or golden flowers. The species comes from Morocco. *L. reticulata* with deep purple flowers is another annual. It is native to Portugal, Spain and North Africa and grows 60–90 cms (2–3 ft) in height.

L. cymbalaria (sometimes called *Cymbalaria muralis*), the Kenilworth Ivy, is a creeping ground-cover plant which also clambers over old walls or can be used in hanging baskets. It is widely naturalized in Britain and thought to have been introduced from S Europe in the 15th century as a salad plant. The species has smooth, kidney-shaped leaves and pale blue and purple flowers with yellow spots—often so abundant that it is sometimes known as Mother of Thousands. Varieties with white and yellow, pure white, pink and lilac flowers are available and there is one having variegated foliage.

Erinus alpinus is another wall plant and indeed never flowers so well as when growing in old brick crevices or stone cracks. A charming species of 13–15 cms (5–6 ins) from the mountains of western Europe, it has small spoon-shaped leaves and terminal sprays of purple or white flowers. It is best established by mixing the seed with moist loam and peat, pushing the resultant pellets into convenient crannies to germinate.

Rehmannia elata, a striking perennial from China, is referred by some authorities to Gesneriaceae. It has large Foxglove-like flowers of rosy-purple with yellow and red throat markings and clammy, hairy stems 60–90 cms (2–3 ft) tall. It needs the protection of glass in cool climates.

Rhodochiton atrosanguineus (*R. volubile*), Purple Bells, a Mexican climber has sensitive leaf stalks which twine round any convenient support. It has alternate, cordate, sparsely downy leaves and large purplish, tubular flowers with five prominent lobes and pale red, saucer-like calyces.

Russelia juncea from tropical America is a xerophyte with much reduced leaves, arching green stems and many tubular scarlet flowers. It makes a good border plant in dry tropical gardens.

Trees are rare in Scrophulariaceae but *Paulownia tomentosa* (*P. imperialis*) is a splendid representative from China. It makes a large tree of 12–15 m (40–50 ft) with stout branches and very large, plush-like, three to five-lobed leaves. In early spring (before the foliage) the tree is crowned with large terminal panicles of violet-blue, Foxglove-like flowers. Unfortunately frost often catches them in cool climates.

This is one of the imperial plants of old Japan (like the Chrysanthemum) and was used in the crests of the Mikado. In the Orient the wood has a number of applications such as for musical instruments, clogs, boxes and furniture. At one time it was also made into charcoal for gunpowder.

Paulownia tomentosa

* It is probable that the 'scented' form grown was a clone, the constitution of which gradually deteriorated from constant vegetative propagation and slowly died out. The 'loss' of scent may well have been due to stocks having been raised from seed, the resultant seedlings lacking any distinctive scent. Probably the musk scent is the exception rather than the rule in *M. moschatus* and the original clone was grown merely because of this character.

Solanaceae

90 genera and 2000 or more species

This is a large and important family of dicotyledonous herbs, shrubs and small trees of tropical and temperate distribution. Included are several important economic plants like the Potato, Tobacco and Tomato, and also a number with medicinal properties as well as many popular ornamentals.

The non-flowering branches usually have alternate leaves but on the flowering stems they may be alternate or, more frequently, in pairs. Usually (but not invariably) individual leaves are simple.

The flowers can be solitary or grouped in cymes and are generally regular (occasionally zygomorphic) and bisexual. They have five sepals, five petals, five or fewer stamens and a superior ovary which develops to a berry or capsule.

Solanum is the largest genus with about 1700 species, the most important undoubtedly *S. tuberosum,* the Potato. This useful vegetable found its way into Europe from South America towards the end of the 16th century, but although legend persists in linking Sir Walter Raleigh with its discovery and introduction into England there seems little evidence to substantiate this fact. Certainly it was in Spain before Raleigh's time for Mellado in a book of that period (*Diccinario Universal*) describes a gift of Potatoes made to the Pope and King Phillip II. Raleigh went to the New World in 1596 and probably saw Potatoes growing but he could scarcely have brought them from Virginia as is so often stated for they are not native to that part of America. Nor (according to a second version) could he have obtained them from Quito for he was never within a 1000 miles of the Ecuadorian capital.

On the other hand, Sir Francis Drake in 1585—prior to sailing to the rescue of harassed Virginian settlers—certainly visited Cartagena in the north of Colombia and there collected stores and loot. It is quite possible that he took Potatoes on board and thus (perhaps unwittingly) brought them to England. The Germans evidently credit him with the fact for in Offenburg in Baden there is a statue to Drake with the following inscription: 'Sir Francis Drake, who spread the use of the potato in Europe, A.D. 1586.'

However, long before this the Indians of Pre-Colombian times knew the value of Potatoes, not only as food but as the source of an alcoholic beverage. Old tomb patternings portray urns and containers shaped like Potatoes—complete with 'eyes'—or fashioned like men with Potato protuberances round the bases of the pots.

Years before the establishment of the Inca Empire (circa A.D. 1100) certain varieties were made into an edible flour called 'chuño'. To manufacture this the Indians spread Potatoes out on the ground all night to freeze. The next day women, children and men would 'tread' the blackened tubers with their bare feet in order to express the water. This process was repeated for four or five days, at the end of which period the chuño was dried and the flour used for cooking purposes.

Many sorts of Potatoes may still be found in Ecuadorian markets; it is common to see twenty or so sorts thus displayed.

Potatoes have made a marked contribution to the diet of millions, although dependance on them has brought its problems, as Irish peasantry found in 1846 when Potato blight disease caused widespread destruction and misery in Ireland as well as economic collapse.

Strangely, the Potato plant had an unfavourable reputation for many years in France, where people believed it to be responsible for leprosy and other ailments.

Solanum wendlandii

Another important member often included in the genus (although now generally accepted as a separate genus *Lycopersicon*) is *S. lycopersicum* (*S. esculentum, Lycopersicon esculentum* or *L. lycopersicum*), the Tomato. This was also cultivated in Mexico and Peru by the Incas, Mayas, Aztecs and their predecessors long before the arrival of the Spaniards. Soon after the conquest of Mexico the seeds were sent to Spain and Morocco and later the plant was introduced to France under the name of Love Apple (it was thought to be an aphrodisiac) and reached England around 1596.

English colonists carried the Love Apple back to the American continent as an ornamental, for the fruits (actually berries) were not eaten at that time because of their relationship with the Deadly Nightshade (*Atropa belladonna*). Fears of their toxic properties seem to have disappeared by the beginning of the 19th century, for around 1830 Tomatoes were being eaten in many parts of the world. *S.* (*Lycopersicon*) *pimpinellifolium* from Peru, Ecuador and the Galapagos Islands is sometimes grown for the long decorative strings of red or orange currant-sized fruits.

Among the ornamental Solanums *S. wendlandii* is cherished as a beautiful climber with showy heads of lilac-blue flowers. Individually these may be 6 cms (2½ ins) across in cymes of 15 cms (6 ins). The plant is robust, ascending by means of hooked prickles to a height of 6 m (20 ft) or more. On the upper parts of the stems the leaves are three-lobed or heart-shaped but pinnatifid into four or six pairs of leaflets lower down on the plant. A native of Costa Rica the species is only hardy in warm climates but makes an attractive conservatory specimen.

S. crispum from Chile is hardier and will survive a few degrees of frost if planted in a sunny, sheltered position. It has masses of bluish-purple flowers on corymbs of 7–10 cms (3–4 ins) and the entire leaves are crisped at the margins. It is a beautiful subject for covering a wall or low shed and stands hard pruning. 'Glasnevin' is an improved form.

S. jasminoides, a deciduous South American twiner, will clamber to 5 m (15 ft) or more and is called the Potato Vine because of the similarity of the flowers to those of the Potato.

Solanum crispum
'Glasnevin'

Solandra maxima
(*S. hartwegii*)

It has bluish-white flower trusses and there is a more beautiful pure white variety, several free-flowering forms such as 'Floribundum' and one with variegated leaves.

S. aviculare and the closely similar *S. laciniatum* from Australia and New Zealand are known as Kangaroo Apples or in New Zealand as Poro-poro. Both make erect shrubs with entire or lobed leaves and violet-purple flowers followed by large green or yellow fruits. *S. uporo* from Fiji has white flowers which develop to red fruits. Because these were formerly eaten by natives as a side-dish with human flesh this species is commonly known as Cannibal's Tomato.

S. quitoensis comes from Ecuador where it makes a bushy plant of 1.25–1.5 m (4–5 ft) with large, silver-backed spiny leaves and large violet flowers. The golden fruits develop to the size of Tangerines and are made into a refreshing drink called Naranjillo.

S. pseudocapsicum (often confused with and incorrectly grown as *S. capsicastrum*) is highly prized for its round, marble-sized fruits of bright orange-red, which almost mask the shrubs (30–60 cms; 1–2 ft) when in character. The species is commonly cultivated in Europe as a room plant in winter. It has small white flowers and simple oblong-lanceolate leaves. *S. capsicastrum* is closely related but differs in its ovoid, pointed fruits.

S. melongena is the Egg Plant or Aubergine, a well-known vegetable with edible fruits which comes from tropical Africa and S Asia.

One of the world's most spectacular climbing plants is *Solandra maxima* (*S. hartwegii*) which is variously known as Cup of Gold, Chalice Vine and Gold Cup. It is a tropical American species with funnel-shaped blossoms of 22 cms (9 ins) flaring back at the mouths. These are light cream when in bud but after opening become orange-apricot with five brown longitudinal, inner bar markings. The blooms smell like ripe Apricots. *Solandra* leaves are of tough texture and entire.

An interesting genus with medicinal and narcotic properties as well as ornamental features is *Datura*. There are about ten species, one the annual *D. stramonium* or Thorn Apple. A plant of the Old World it is now widespread in many areas, including North America. The Thorn Apple grows about 60 cms (2 ft) high with ovate, toothed smooth leaves and white funnel-shaped flowers which never fully open and later give place to round prickly green fruits superficially like those of Horse Chestnut (*Aesculus hippocastanum*).

The whole plant has a foetid disagreeable smell when bruised and carelessly used can be very poisonous. Notwithstanding this it has various medicinal applications in the treatment of asthma and bronchial complaints and is employed externally as an ointment on bruises and swellings. Several alkaloids including scopolamine (hyoscine) and atropine are found in the seeds, flowers and foliage, scopolamine finding use as a pre-anaesthetic in childbirth and surgery.

Numerous instances are related of fatal or serious consequences following the incautious use of this poisonous plant. In America it is called Jimson Weed a corruption of Jamestown, Virginia, scene of the first English settlement in 1603 where, after its introduction, it grew abundantly.

High doses of Thorn Apple can be dangerous and in India—where it grows abundantly—thieves and assassins administer it to their victims to produce insensibility. Stramonium (from *Datura*) is a common source of 'knockout drops' and produces strange dreams. *D. stramonium* and *D. innoxia* (*D. meteloides*) were at one time used by various Indian tribes during the initiation ceremonies for boys entering manhood.

Historians believe that *Datura* was taken by the priests of Delphos in order to produce the paroxysms attributed to Divine power, a ruse also employed by Peruvians for similar purposes.

Best of the ornamental kinds is probably the double-flowered form of *D. suaveolens*, a tree or large shrub 3–5 m (10–15 ft) high, with strong woody stems, ovate-oblong, velvety leaves often white-downy underneath 15–30 cms (6–12 ins) long, and pendulous, musk-scented flowers of purest white 25 cms (10 ins) across. These hang down like trumpets and flare out at the petal bases like flounced petticoats. They are commonly called Angel's

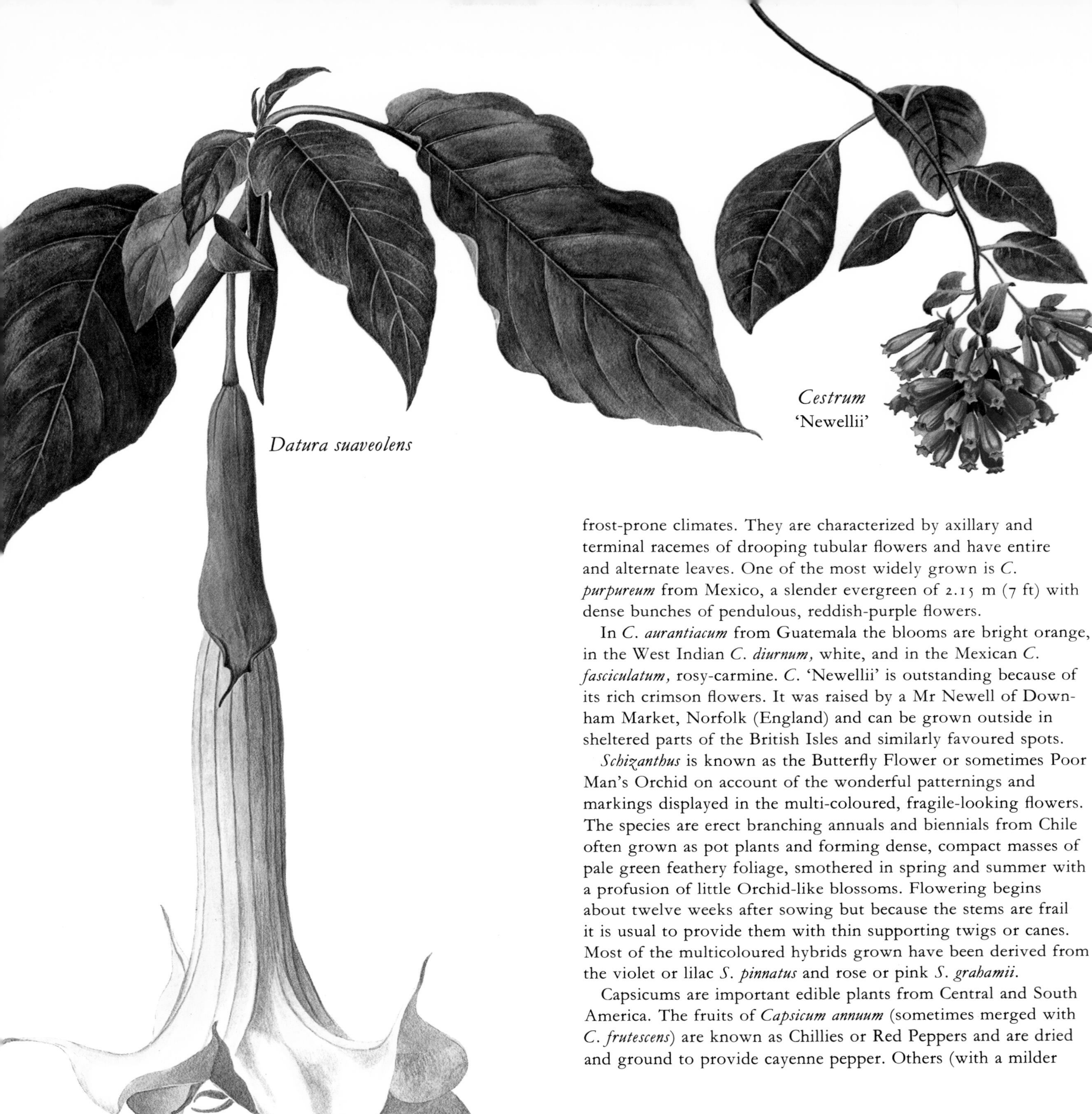

Datura suaveolens

Cestrum 'Newellii'

frost-prone climates. They are characterized by axillary and terminal racemes of drooping tubular flowers and have entire and alternate leaves. One of the most widely grown is *C. purpureum* from Mexico, a slender evergreen of 2.15 m (7 ft) with dense bunches of pendulous, reddish-purple flowers.

In *C. aurantiacum* from Guatemala the blooms are bright orange, in the West Indian *C. diurnum,* white, and in the Mexican *C. fasciculatum,* rosy-carmine. *C.* 'Newellii' is outstanding because of its rich crimson flowers. It was raised by a Mr Newell of Downham Market, Norfolk (England) and can be grown outside in sheltered parts of the British Isles and similarly favoured spots.

Schizanthus is known as the Butterfly Flower or sometimes Poor Man's Orchid on account of the wonderful patternings and markings displayed in the multi-coloured, fragile-looking flowers. The species are erect branching annuals and biennials from Chile often grown as pot plants and forming dense, compact masses of pale green feathery foliage, smothered in spring and summer with a profusion of little Orchid-like blossoms. Flowering begins about twelve weeks after sowing but because the stems are frail it is usual to provide them with thin supporting twigs or canes. Most of the multicoloured hybrids grown have been derived from the violet or lilac *S. pinnatus* and rose or pink *S. grahamii.*

Capsicums are important edible plants from Central and South America. The fruits of *Capsicum annuum* (sometimes merged with *C. frutescens*) are known as Chillies or Red Peppers and are dried and ground to provide cayenne pepper. Others (with a milder

Trumpets. The single form of this species is Mexican where the natives sometimes smoke the foliage and flowers (both poisonous) as an asthma remedy.

By the roadsides in Peru, Ecuador and Colombia one commonly sees small trees weighted with large hanging, white, pink, orange or yellow flowers. These are the species *D. chlorantha* (yellow), *D. sanguinea* (orange-red), and *D. cornigera* (white or cream) or hybrids between them.

All the shrubby Daturas stand hard cutting and make good tub plants in cool climates. South American Indians believe it is courting death to sleep underneath them.

Cestrum comprises a large genus of some 150 ornamental shrubs, most of them requiring winter greenhouse protection in

A *Schizanthus* hybrid

Fruits of *Physalis franchetii*

flavour) are grown for their edible fruits and are used in salads or eaten stuffed with chopped meat. Paprika and Pimentos also come from Capsicums. The fruits assume a variety of shapes and vary in colour from green, yellow, red and violet to dark brown.

Physalis has a widespread distribution with many American representatives. *P. alkekengi,* the Bladder or Winter Cherry, is a perennial herb from central and S Europe and W Asia. It grows to about 30 cms (1 ft), has creeping rhizomes, pointed oval, long-stalked leaves and white flowers which later develop scarlet berries about the size of Cherries. These are edible and surrounded by papery, pointed calyces.

The closely similar *P. franchetii,* the Chinese Lantern (although native to China, Korea and Japan), is often considered only a variant of *P. alkekengi* but is more robust and hardier. It is grown purely for ornamental purposes and is a perennial of 45 cms (1½ ft) with a creeping rootstock (which can become invasive), large long-stalked leaves, white flowers and a showy inflated calyx, which changes through green and yellow to orange-red. It encloses a round orange-red berry. After drying the calyces are frequently used for winter decoration. Cultivars exist with much larger seed pods in 'Gigantea', with canoe-shaped calyx segments in 'Monstrosa', or saucer-shaped calyces in 'Orbiculare' and there is a dwarf called 'Nana'.

P. peruviana, the Cape Gooseberry, is really South American but cultivated in South Africa (and other parts of the world) for its edible fruits. These are normally purplish but yellow in the variety *edulis* which is marketed under the name Golden Berry. *P. ixocarpa* from Mexico is the Jamberberry or Tomatillo, an annual 90–120 cms (3–4 ft) high with yellow flowers and purplish fruits. The latter are made into jam and various preserves.

A monotypic species from Colombia is *Streptosolen jamesonii,* a popular climbing evergreen often used in greenhouses twined round canes in pots or grown up conservatory pillars. Its slender stems reach 2–2.5 m (6–8 ft) and are clothed with sticky, ovate, roughly hairy leaves and panicles of tubular orange flowers.

Several annuals frequently employed for pot work and summer bedding belong to this family. *Salpiglossis sinuata* (*S. variabilis*), the Velvet Trumpet Flower or Painted Tongue from Chile and Peru, is a slender branching species, 30–60 cms (1–2 ft) in height with narrow, sticky leaves and loose sprays of showy flowers about 5 cms (2 ins) long and nearly as much across at the mouths. Modern varieties show a wide colour range through crimson, yellow, scarlet, blue and white. These are usually splashed or marked with gold. They do well in rich deep soil, producing a profusion of blooms which are also good for cutting.

Nicotiana includes *N. tabacum,* the Common Tobacco, from tropical America—the solace of millions.

Long before the Americas were discovered by inhabitants of the Old World, Indians were smoking, chewing and sniffing the leaves of Tobacco. The plant was considered by them to be a magical herb and a cure-all for countless ailments including animal

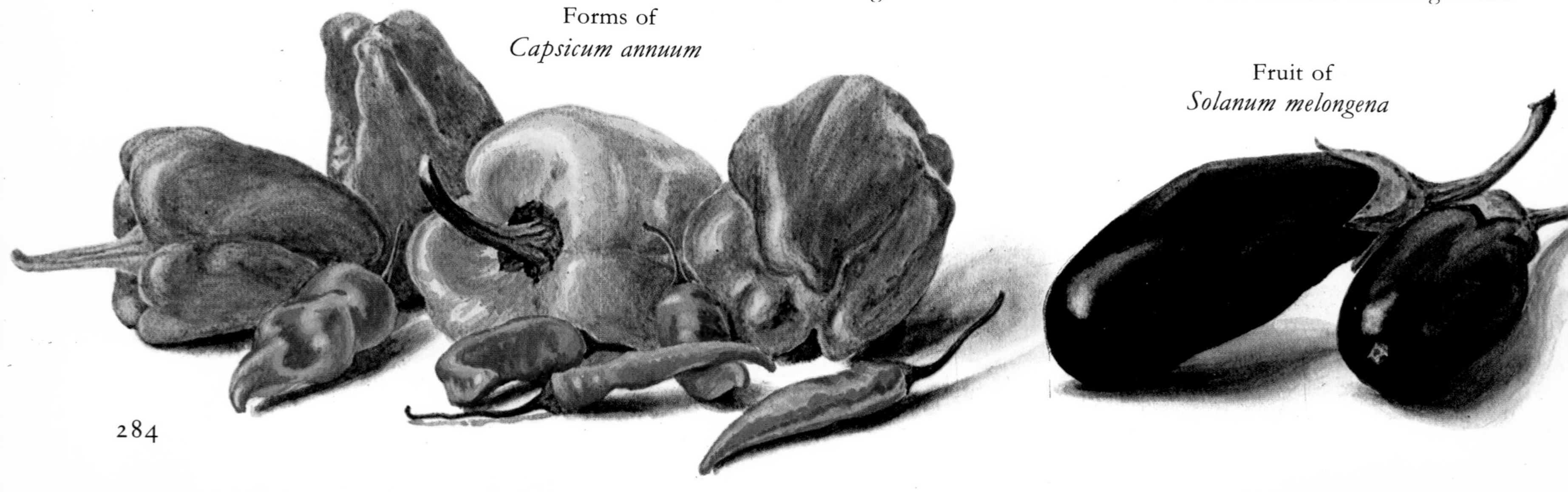
Forms of *Capsicum annuum*

Fruit of *Solanum melongena*

and insect bites, giddiness, headaches, bruises and rheumatism. Indian priests smoked the leaves in connection with religious rites and their medicine men for purposes of magic and healing.

In 1492 two members of Columbus' crew landed in Cuba and later reported having seen natives puffing smoke from their mouths and nostrils. A year later Friar Ramon Pane (who accompanied Columbus on his second voyage of discovery) confirmed this account but added the surprising fact that the Indians also reduced the leaves to a fine powder and sniffed this through a hollow cane to clear their heads. This is the first known record of snuff taking.

Later voyagers (around 1565) discovered natives blending crushed Tobacco leaves and ground sea shells into pellets, which they chewed on long journeys to allay the pangs of hunger and thirst. This is the earliest reference to Tobacco chewing.

In 1560 Jean Nicot de Villemain introduced the Tobacco plant to France and the generic name *Nicotiana* and the alkaloid *nicotine* were later coined from his name. In 1586 Sir Francis Drake and Ralph Lane (first Governor of Virginia) brought plants to England as a present for Sir Walter Raleigh. It was Raleigh who made Tobacco drinking (as pipe-smoking was called at that time) popular and gradually the craze spread all over Europe.

Cigars (always preferred to pipes by the Spanish) take their name from *cigaroo* (cicada) because their shape resembles the bodies of these insects. Over sixty cultivars of Tobacco are now grown.

Nicotiana glauca from southern South America, a yellow-flowered tree 6 m (20 ft) tall, has become naturalized in some Mediterranean areas. It has toxic properties and Indians knew it as marihuana (interpreted by the Conquistadores as Maria Juana or Mary Jane) although the name marihuana is now referred to Indian Hemp (*Cannabis sativa*), the source of hashish or bhang.

N. tabacum, a handsome annual or biennial with erect stems 1.25–2 m (4–6 ft) tall has large, thin and sharply pointed, oblong or lanceolate leaves which are sticky to the touch, and day-blooming, rose or purple, funnel-shaped flowers about 5 cms (2 ins) long in loose clusters or racemes.

The blooms of many ornamental species remain closed until evening but then their rich scent and splendid colours more than compensate for a subdued daytime appearance. Most of them should be grown in a partially shaded situation.

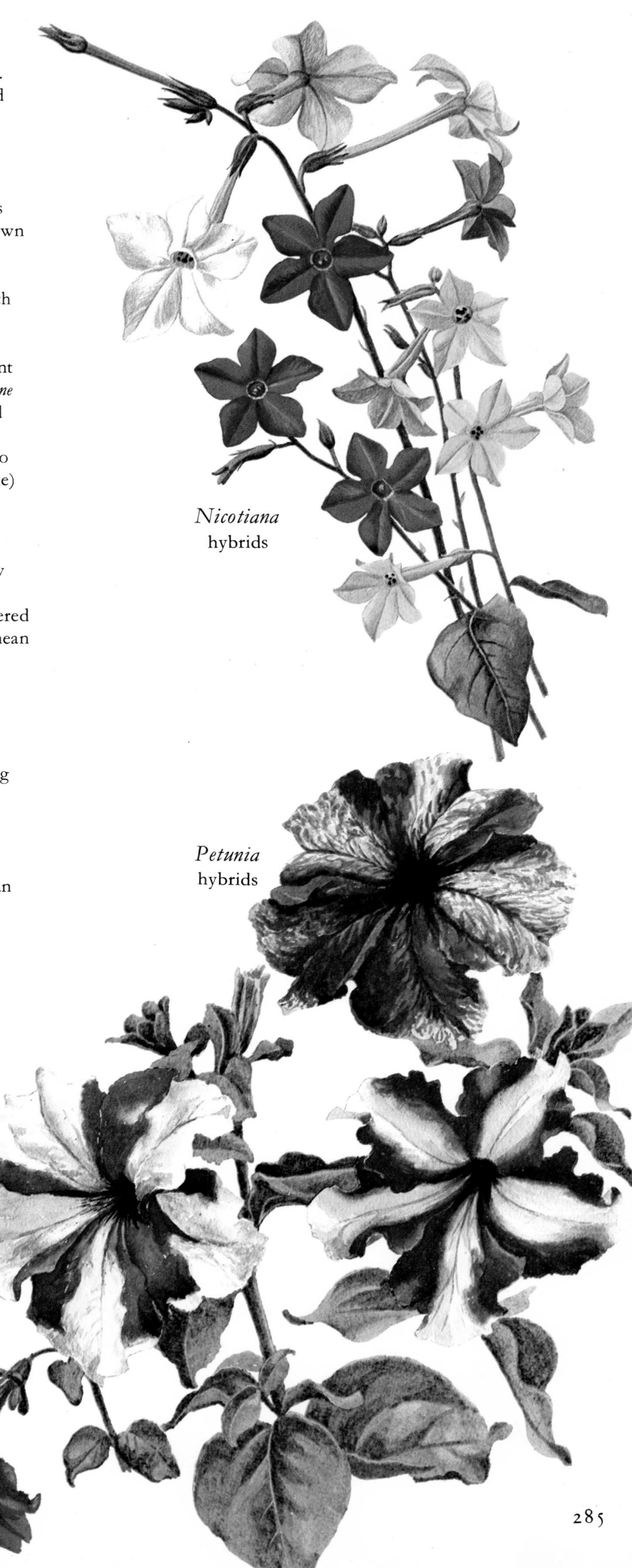

Nicotiana hybrids

Petunia hybrids

Salpiglossis hybrids

The Australian *N. suaveolens* is particularly sweet-scented with nodding, greenish-purple flowers but more garden-worthy is its white form *macrantha* (*N. fragrans*). *N.* x *sanderae*, an annual of garden origin, has flowers 7 cms (3 ins) long in clusters on stems 60–90 cms (2–3 ft) tall. These come in various shades of rose and have greenish-yellow corolla-tubes and wavy-edged leaves.

N. alata from Brazil carries small oblong leaves and clusters of white flowers 7 cms (3 ins) long. It is rarely seen in gardens—most people preferring its large and colourful hybrids, many of which are named. These have mauve, violet, pink, crimson, and scarlet flowers and there is also a cultivar with lime-green blossoms. All are night-blooming and sweetly scented, their heights varying from 30–120 cms (1–4 ft).

Petunias are brilliant, long-flowering plants with large, showy, trumpet flowers. Although most of those grown in gardens are strictly speaking perennials, it is usual to treat them as annuals, raising the seedlings under glass and planting them outside when all fears of frost are past. They are only hardy in a frost-free climate.

The flowers come in mauve, purple, white, yellow and various shades of pink, many being speckled, striped or veined in other colours. Doubles and frilled-petalled sorts are common, also a wide range of F1 hybrids with larger blooms. The usual height is around 45 cms ($1\frac{1}{2}$ ft) but there are dwarf forms known in the trade as *Petunia* 'Hybrida Compacta', about 22 cms (9 ins) high, which are better for bedding, and also trailing sorts for hanging baskets.

Most modern Petunias are derived from the rosy-purple *P. integrifolia* (*P. violacea*) and white *P. axillaris* (*P. nyctaginiflora*), both South American. The latter is fragrant, especially at night, a trait which has been inherited by some of its hybrids.

One very poisonous member of Solanaceae is *Atropa belladonna*, the Deadly Nightshade. It is a vigorous, branching herb 90–120 cms (3–4 ft) high with large ovate leaves and pendulous purple or creamy-yellow flowers. When fertilized these develop to glossy, purple-black fruits (yellow in the cream-flowered form) about the size of cherries. These are sweet-tasting but highly narcotic and many fatal accidents have followed their unwise consumption.

Atropine, formerly used by women to dilate the pupils of their eyes (which gave them a striking effect), is obtained from the roots. Ophthalmologists also used atropine for examination purposes and prior to cataract operations. The dried roots and leaves have a sedative and antispasmodic action which is beneficial when

Brunfelsia calycina macrantha

prescribed by doctors but dangerous used at random. The effect of the poison is to cause diminished sensibility followed by giddiness and delirium and finally death. Strangely animals are rarely affected by the plants. Birds often eat the berries and rabbits graze the foliage seemingly without ill effects.

Atropa belladonna is also known as Dwale, a name previously applied to a concoction of Mandrake, Opium, Henbane, Hemlock and other ingredients, which was used in the Middle Ages as an anaesthetic drug for surgical purposes. Even before this Dwale was a common name for Mandrake (*Mandragora officinarum*), which Pliny says (*Historiae naturalis*) was 'given before incisions or punctures are made in the body in order to ensure insensibility to the patient'. Nowadays Dwale refers to *Atropa,* which is supposed to have been the plant responsible for the poisoning of Marcus Antonius' troops during the Parthian war.

Another highly poisonous plant is *Hyoscyamus niger,* the Henbane, a pretty little European and Asian annual 30–60 cms (1–2 ft) tall with yellow, purple-veined flowers. Hyoscyamine, used in 'twilight sleep' in childbirth, is obtained from the leaves and plant tops. This was the drug used by Dr Crippen to murder his wife in a famous case at the beginning of this century. Crippen was the first murderer ever to be apprehended with the assistance of wireless telegraphy.

Mandragora officinarum or Mandrake is interesting for another reason. The thick roots, often shaped in the human form, were valued by the Ancients for various medicinal and narcotic purposes. Wrapped in a piece of sheet these small 'figures' were supposed to cure a host of maladies, double the amount of money locked up in a box, act as a love charm, keep off evil spirits, foretell the future and render other notable services.

According to Pliny the Mandrake was sometimes conformed like a man, at others like a woman. The male root was white, the female black but gathering the roots was highly dangerous as it had the power to utter sounds when pulled from the ground. Any person hearing these shrieks would immediately be struck dead. To guard against these dangers Josephus (*The Jewish War*) says 'There is one way in which the taking up of the root can be done without danger. This is as follows: They dig all round the root, so that it adheres to the earth only by its extremities. Then they fasten a dog to the root by a string, and the dog, striving to follow his master who calls him away, easily tears up the plant,

Mandragora officinarum

Juanulloa aurantiaca

but dies on the spot; whereat the master can take this wonderful root in his hand without danger.'

Shakespeare describes these shrieks on a number of occasions, as for example in *Romeo and Juliet* IV, iii

'*And shrieks like Mandrakes, torn out of the earth*
That living mortals hearing them run mad'.

They are also mentioned in Genesis, in this case as a cure for sterility. 'And when Rachel saw that she bore Jacob no children, Rachel envied her sister. . . . And Reuben went in the days of wheat harvest, and found mandrakes in the field, and brought them unto his mother Leah. Then Rachel said to Leah, Give me, I pray thee, of thy son's mandrakes. . . . And God remembered Rachel. . . . And she conceived, and bore a son . . . and she called his name Joseph. . . .'

It was long believed that Mandrakes dwelt in the dark places of the earth and thrived under the shadow of the gallows, being nourished by the flesh of criminals executed on the gibbet. However, the plant is not uncommon in European gardens where it is grown for its large and handsome ovate leaves and creamy-yellow or white, cup-shaped flowers heavily veined with purple. In summer the large oval, orange-red, pulpy fruits lie on the ground like a clutch of fowls' eggs. The species grows about 10 cms (4 ins) high and is sometimes known as Devil's Apples.

Another form of this species often known as *M. autumnalis* is smaller with violet flowers in autumn, rounder fruits and oblong, wrinkled and hairy leaves. It is S European.

Brunfelsia calycina, a Brazilian evergreen shrub, is much planted in tropical countries for its neat entire leaves and large and handsome salver-shaped flowers. These are very sweetly scented and produced in profuse intermittent bursts throughout the year. Another characteristic is their ability to change colour—from purple, to mauve to almost white—as the blooms age. Because of this they are known in Kenya as Yesterday, Today and Tomorrow. There are a number of variants including the dwarf, free-flowering, deep violet *floribunda* and the similarly coloured, but larger-flowered *macrantha* (probably only cultivars). Brunfelsias make good pot plants (up to 60 cms; 2 ft) for cool climates.

Juanulloa aurantiaca comes from Peru and is epiphytic on trees and rocks in its native habitat but will grow in well-drained pans of rough peaty soil under glass in cool climates. It has oblong, leathery leaves and many tubular, orange flowers in forked racemes.

Nicandra physalodes from Peru is called the Shoo-Fly Plant because of its reputed power to repel insects. It is a pretty annual growing 60–120 cms (2–4 ft) high with large, salver-shaped, blue flowers and inflated seed pods which are rather like those of *Physalis*.

Nierembergias are frequently grown as summer-bedding annuals or pot plants for their showy white, purple or violet flowers. They are erect or decumbent slender plants with alternate, entire leaves and solitary cup-shaped blooms. *Nierembergia hippomanica* var. *violacea* (*N. caerulea*) from the Argentine and *N. frutescens,* a Chilean species, and their variants are the most important. They grow 30–60 cms (1–2 ft) tall in various colours, the flowers often having yellow throats. Cup Flower is a commonly applied English name.

Other interesting members of Solanaceae include *Cyphomandra betacea,* the Tree Tomato, a branching shrub of 4–4.25 m (12–14 ft) from Brazil with edible fruits, and *Scopolia carniolica*, a neat European plant which will grow in sun or shade in most climates. It has oblong leaves and lurid purple or yellowish-green, nodding flowers on stems 30 cms (1 ft) high in spring. The dried rhizomes are used medicinally and the plant—which is more narcotic than *Atropa belladonna*—is a powerful sleep-inducing hypnotic.

Stachyuraceae

1 genus and 10 species

This is a restricted family of dicotyledonous small trees or shrubs, sometimes climbing, with ovate-lanceolate, simple, serrate leaves arranged alternately on the stems. The flowers appear in the leaf axils as stiff, catkin-like racemes, and are composed of between twelve and twenty waxy, pale yellow blooms which, as the Latin name indicates, look like ears of wheat.

Individually the flowers are regular and bisexual with four sepals, four petals, eight stamens and a superior ovary.

Stachyurus praecox comes from Japan and is hardy in Britain, Europe, New Zealand and America as far north as Massachusetts, particularly when afforded a sheltered situation. Its chief attribute as a garden plant lies in the earliness of the flowers which come before the foliage at the tail end of winter.

The plant needs no regular pruning beyond occasional branch removal and a general trimming to keep the bush shapely. It likes well-drained soil with moisture in summer; this is best afforded by a mulch of leaf-soil, peat or bark fibre. Propagation is by means of half-ripe summer cuttings rooted under glass.

S. chinensis from China is very similar but perhaps a little more vigorous. It usually flowers about two weeks later than *S. praecox*.

Stachyurus praecox

Staphyleaceae

5 genera and 60 species

Staphylea colchica

This is a small family of dicotyledonous trees and shrubs with opposite or alternate, divided leaves having serrated leaflets and regular, usually bisexual flowers in showy panicles. Individually the blooms have five sepals, five petals, five stamens and a superior ovary.

Staphylea is the most important horticultural genus and the species have three to seven-foliate, opposite leaves. *Staphylea colchica* from the S Caucasus is the best and frequently planted in gardens for the beauty of its flowers, foliage and curious inflated fruits.

It is a deciduous shrub 2–3 m (6–10 ft) high and as much across. The three- or five-parted leaves are smooth with bristle-like serrations and the flowers come in spring in showy bunches up to 13 cms (5 ins) long and as much across. Individually they resemble small, white, trumpet Daffodils and have a sweet scent. The fruits consist of inflated capsules 7–10 cms (3–4 ins) long and 5 cms (2 ins) across containing pale brown seeds. The plant needs moist rich soil and sunshine, although specimens occasionally thrive and flower in shady gardens.

S. pinnata from Europe and Asia Minor is known as St Anthony's Nut or Bladder Nut on account of the shape of its inflated fruits. These contain pea-sized seeds and were once thought to be beneficial in bladder ailments. This belief seems to have been short-lived for John Parkinson (*Paradisi in sole Paradisus terrestris*) writes 'Some Quacksalvers have used these nuts as a medicine of rare vertue for the stone but what good they have done, I never yet could learne'. The species grows 3–5 m (10–15 ft) high with pinnate leaves and drooping bunches of white flowers.

S. x *coulombieri,* considered to be a chance hybrid between *S. pinnata* and *S. colchica,* was first noted in 1887 in the grounds of a French nurseryman called M. Coulombier. It is intermediate between the two parents with compact trusses of white blossom.

S. holocarpa from China grows (according to E. H. Wilson who collected it in 1908) to a height of 6–9 m (20–30 ft) in nature. But it is normally a much smaller shrub under cultivation and is deciduous with white or pink flowers.

Sterculiaceae

60 genera and 700 species

This large and handsome family consists in the main of tropical trees and shrubs, although there are also some herbs and vines. General characteristics include simple or digitate leaves and complex cymes of regular flowers, which may be monoecious or bisexual and have three to five sepals, five or no petals, two whorls of stamens, which are often very diverse, and superior ovaries.

The family contains two important economic species in *Theobroma cacao,* source of cocoa, chocolate and cocoa-butter, and *Cola acuminata,* an African shrub which produces Cola Nuts. The latter contain caffeine, a mildly stimulating substance used in drinks and drugs and said to have once been an ingredient of Coca-cola. Years ago natives employed these nuts as money in local trading.

Among the ornamental genera are some beautiful shrubs or small trees, particularly the South African Dombeyas. These have clusters of fragrant, five-petalled, mostly white flowers at the tips of their branches, some very large and resembling *Abutilon. Dombeya pulchra* from the Transvaal has white flowers with pink centres; *D. burgessiae* from Central Africa is white with pink streaks; and *D. natalensis* is called the Wedding Flower on account of its large white, fragrant flower clusters. *D. acutangula* from Mauritius bears long-stalked, red or pink flowers in lateral umbels.

Brachychiton acerifolium, often included in *Sterculia* as *S. acerifolia,* is the Flame Tree of Australia. It makes a beautiful tree with small, brilliant scarlet, bell-like flowers clustered in masses on the leafless branches. Several species are known as Bottletrees as they develop bottle-like trunks. *B. discolor* (*Sterculia discolor*), the Scrub Bottletree or Queensland Lacebark, is widely grown for its large

Brachychiton discolor

Fremontia californica

pink flower bells, which are felted outside. After blooming these carpet the ground with fallen pink blossoms. It has five-lobed leaves which are woolly beneath.

As *Fremontia californica* comes from the western USA it is only suitable for sheltered or completely protected situations when grown in cold climates. Given good drainage and the backing of a warm wall it proves hardy in S England and is quite commonly cultivated in S Europe. Growing 3–5 m (10–15 ft) high, it is an evergreen shrub with bright golden, petal-less flowers 6 cms ($2\frac{1}{2}$ ins) across which look something like Mallow (*Malva*) and heart-shaped foliage on hairy branches. The leaves are thick and leathery and felted beneath. *Fremontia* is known as Flannel Bush in California and was once used by early settlers to relieve sore throats, also to line footwear for extra warmth.

The Mexican *F. mexicana* is similar. Both species resent root disturbance so should be planted out from pots without breaking the soil balls. Propagation is by means of seed or summer cuttings.

The name *Fremontia* was at one time used for the genus *Sarcobatus* in the family Chenopodiaceae. Unless the later (and more common) name of *Fremontia* is retained, the correct name of this genus is apparently *Fremontodendron*.

Another ornamental is *Erythropsis colorata* (*Sterculia* or *Firmiana colorata*) from Asia, a tree of 10.5–19.5 m (35–65 ft) with tubular, coral-red to orange petal-less flowers on leafless branches. The foliage which follows is crowded at the ends of the branches, each leaf having three to five points.

Strelitziaceae

3 genera and 7 species

This is a strange, rather dramatic family of monocotyledonous perennial herbs closely related to the Banana family (Musaceae) Some are of tree-like proportions with unusual flowers and enormous leaves.

Individually the blooms are bisexual with six perianth segments, five or six stamens and an inferior ovary, but are always grouped in striking, terminal or lateral inflorescences.

Ravenala madagascariensis, the Traveller's Tree from Madagascar, is one of the strangest. It is widely planted in the tropics, where it grows 15 m (50 ft) or more, bearing long-stalked leaves which are arranged in two ranks (like an arc), the leaf blades flaring out at the tops. Since the plant is on one plane it thus resembles a feathery fan. The flowers are white and it has edible seeds.

The popular name Traveller's Tree refers to the fact that water collects in hollows formed by the leaf bases. It is possible to drink this liquid and the usual story refers to desperate travellers coming across the plant in the middle of the desert and quenching their thirst from its crystal clear water. However, this *Ravenala* never grows far from marshy ground or springs and the water has an unpleasant vegetable taste.

Strelitzia reginae like all members of its genus comes from South Africa. There it is known as Crane Flower but elsewhere as Bird of Paradise Flower on account of the striking resemblance of the inflorescence to a bird's head. Each inflorescence is about 20 cms (8 ins) in length, poised horizontally on a stout stem 105 cms ($3\frac{1}{2}$ ft) high, the boat-shaped spathe containing several flowers, each with three brilliant orange sepals, with two vivid peacock-blue petals forming a beak and a third small sepal of similar colour. The beak secretes the anthers and when alighted on by birds open and release their pollen. The leaves resemble those of *Canna* and are borne on stems of 90 cms (3 ft). *S. reginae* is the best garden species for outside bedding in the tropics, also for growing under glass in cooler climates. It is the emblem flower of Los Angeles, where it is much planted, and can be propagated by division or separation of the side suckers.

In the dryer parts of South Africa Strelitzias with greatly reduced foliage occur, sometimes with no leaf blades or almost rush-like as in the plant known as *S. parvifolia* var. *juncea*, probably only a variant of *S. reginae*. This lamina reduction is an adaptation to protect the plants against excessive transpiration in the long periods of drought.

Strelitzia reginae

Styrax japonica

Halesia carolina

Taccaceae

2 genera and 31 species

These are monocotyledonous perennial herbs with creeping rhizomes or very large tuberous rootstocks. They all have broad, elliptic or much-lobed leaves springing directly from the ground, usually on very long stalks.

The bisexual and regular flowers are grouped in umbels on scapes and have six perianth segments, six stamens (in two rings) and inferior ovaries. The berried fruits contain many seeds.

The various species of *Tacca* are of tropical distribution, chiefly Asian. It is essential for them to have warm growing conditions, followed by a resting period—during which time they must only be sparingly watered.

T. integrifolia (also known as *T. aspera*) from India, Burma and Malaya grows 45–60 cms (1½–2 ft) tall with oblong-lanceolate, long-stalked leaves and umbels of purplish-brown flowers.

T. artocarpifolia from Madagascar has nodding green blooms with brownish bases and large three-part leaves on stems 60 cms (2 ft) high. *T. pinnatifida* (also referred to *T. leontopetaloides* or in Willis to *T. leontopodioides*), the East Indian Arrowroot, bears attractively dissected leaves, and purplish, funnel-shaped flowers in dense umbels. They have large, fleshy rhizomes which produce a meal equal to the best arrowroot. In some parts of the world this is an important food. *T. pinnatifida* and other species having similar properties are cultivated under the name of Pi in the islands of the Pacific, Malaysia and in India.

Styracaceae

12 genera and 180 species

This family contains dicotyledonous trees and shrubs which have alternate, entire and sometimes leathery leaves and short racemose inflorescences. The flowers are bisexual (rarely dioecious) and regular with four or five sepals and petals and the same number or twice as many stamens. The ovaries may be either superior or inferior.

The most commonly cultivated—especially in cool temperate gardens—are *Halesia* and *Styrax*.

Halesia carolina is the Snowdrop Tree, a big shrub or small tree 3–8 m (10–25 ft) in height from the SE United States. It is very beautiful when in flower in late spring, for the naked branches are then draped with clusters of hanging, white, four-petalled flowers which have a slight resemblance to those of the Snowdrop (*Galanthus*). The smooth ovate to ovate-oblong leaves are minutely toothed.

Styrax officinalis is the Storax native from France to Israel and W Turkey. An aromatic resin extracted from the wood is used (among other things) as incense in churches. It grows 4–6 m (12–20 ft) high, the young leaves underneath and shoots being covered with a whitish down and has ovate leaves and short terminal clusters of fragrant white flowers.

S. hemsleyana from Central and W China is also white-flowered and makes a handsome tree of 6 m (20 ft) or so. *S. japonica* is native to Japan where the fine-grained wood is used in the manufacture of umbrella handles. The seeds are also the source of an oil. The flowers are white and plentiful and very beautiful in early summer.

Most *Styrax* species need sheltered situations, especially in places subjected to wet or cold winters.

Tacca integrifolia
(*T. aspera*)

Tamaricaceae

4 genera and 120 species

The members of this small family are dicotyledonous, temperate and subtropical shrubs, trees or herbs. They have delicate needle-like foliage resembling that of Heaths, a characteristic which reduces transpiration so that they are able to survive in arid or difficult situations, such as the steppes and deserts or in the salt-laden atmosphere of the sea shore.

The flowers may be solitary or more commonly are in large and sometimes showy racemose inflorescences. Individually the blooms are regular and bisexual with four or five sepals, four or five petals, four to five, eight to ten or many stamens and a superior ovary.

Tamarix, the most important genus, has about ninety species. One of these, *T. anglica,* the so-called English Tamarisk (now considered a form of the European *T. gallica*), provides a pleasing midsummer picture around the coasts of SE England with myriads of pale pink, gossamer-like flowers on willowy shoots 90 cms–2.5 m (3–8 ft) long.

The most valuable for garden purposes, however, are the spring-flowering *T. tetrandra* from SE Europe and W Asia and the late summer *T. ramosissima* (*T. pentandra*), another SE European species. Both grow to a height of 3–5 m (10–15 ft) and have showy plumes of rosy-pink florets. They do particularly well in seaside gardens and are easily propagated from cuttings. *T. hispida* is a smaller plant (90–120 cms; 3–4 ft) from the area of the Caspian Sea.

It seems that Tamarisk was first brought to England by Bishop Grindal in the reign of Elizabeth I. It was valued for disorders of the spleen, particularly since the poisonous Yew (*Taxus baccata*) had previously been used for this purpose by English apothecaries. With the introduction of *Tamarix* thousands of plants were cultivated for medical purposes.

Tamarisk was also believed to have magical powers and Pliny recommended the leaves as an ointment for 'nightfoes, or Chilblanes', although adding that the tree was 'unluckie' since it never bore fruit and was accursed, being commonly used to garland criminals.

Two Tamarisks seem to have been brought to England in the 16th century, *T. gallica* from France and *T. germanica* (now more properly *Myricaria germanica*) from Germany. It was probably the latter that Grindal introduced.

John Smith (*A Dictionary of Economic Plants*, 1882) mentions a strange by-product of *T. gallica* and an allied species called *T. mannifera* which is common in the Sinai peninsula. 'Their stems', he says, 'are punctured by a small insect (a scale insect called *Coccus manniparus*) from which a juice exudes, which hardens, and is collected by the Bedouin Arabs and made into cakes and called Manna. It is sweet, and consists of a mucilaginous sugar, and forms a small article of commerce at the present day. It is believed by some to be the Manna of the Israelites.'

While all the species mentioned are of delicate shrub-like proportions there is a tree form in NW India which possesses a remarkable growth rate. This is *T. aphylla* (*T. orientalis*), of which Smith says trees six or seven years of age measure 1.5 m (5 ft) in girth and collapse from old age when about twenty years old.

This same species produces galls which are used by the Arabs for tanning. It gives sheep and goat skins a pretty pinkish colour. The leaves are used against enteritis and the wood as fuel and for construction work in North Africa.

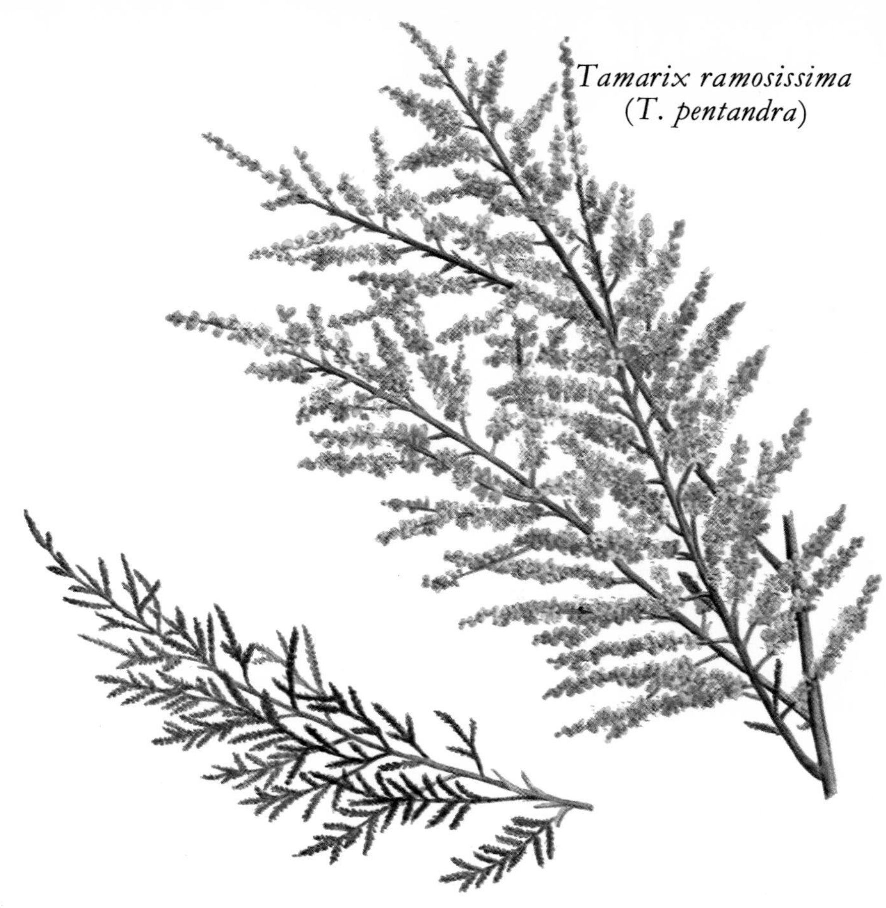

Tamarix ramosissima (*T. pentandra*)

Theaceae

16 genera and 500 species

This family (here taken to include Camelliaceae and Ternstroemiaceae) consists of dicotyledonous trees and shrubs with alternate, leathery and entire or toothed leaves. The usually solitary flowers are bisexual and regular in shape with five to seven sepals, a varying number of petals (but normally four to five), many stamens and a superior (rarely inferior) ovary.

They include several splendid ornamentals of which the most important are the winter- and spring-flowering Camellias.

The genus is named for Georg Kamel (Latinized as Camellus), a Jesuit priest (1661–1706) who was particularly interested in natural science. He collected plants in both the Philippines and China although opinions differ as to whether he brought back Camellia seeds to Europe. Certainly he was responsible for introducing the St Ignatius Bean to Europe (*Strychnos ignatii*), one source of strychnine, and he also wrote a history of the plants of Luzon.

Economically the most important species is *Camellia sinensis,* the Tea Plant, the national beverage of China and one much prized by the English. Many tales and legends surround this plant, an especially charming story referring to an Indian Prince and Buddhist monk called Bodhidharma. This Prince is supposed to have landed on the shores of China in A.D. 510 with the object of converting its natives to Buddhism. To that end he dedicated his life to sleeplessness but one day, after years of wakeful teaching, praying and meditation Ta-Mo (as he was known to the Chinese) fell asleep. Mortified by this weakness of the flesh he cut off his eyelids and threw them to the ground where Buddha caused them to sprout and take root. These became the first Tea plants, the dried leaves of which assume the shape of eyelids and are supposed to represent and induce wakefulness.

Their employment as a beverage is supposed to have been discovered by another Buddhist. He lived the frugal life of a

Camellia
'Cornish Snow'

Camellia
'Mary Christian'

hermit and one day while making up a fire with branches of Tea plant he accidentally dropped some of the leaves in a pot of boiling water. Later he tasted the liquid and finding it exhilarating and pleasant to the palate imparted his discovery to others. And so in course of time the practice of tea-making spread.

Around 1606 the Dutch set up Tea plantations in Java and soon afterwards brought this new beverage to Europe. In 1650 Peter Stuyvesant took it to North America although curiously it was 1652 before Tea reached England.

Among ornamental Camellias (pronounced *mell* not *meel*) is *C. japonica,* a plant from the mountains of Korea and Japan which grows to a height of 9 m (30 ft) among forest trees. It is a splendid but variable species with oval, glossy, evergreen leaves and usually (in the wild) small single, red flowers.

The habitat indicates its needs. It requires shade and shelter and dislikes lime. The first plants grown in England flowered in the Essex garden of Lord Petre in 1739. Petre was a keen plantsman but no one seems to know from whence he obtained his plants. At first they were grown in greenhouses which seems to have been their undoing, for the original plants were soon lost—it is said due to overheating.

However John Slater of the East India Company reintroduced Camellias in 1792. These were two doubles, one white and the other striped.

Camellia japonica
'Lady Vansittart'

Today *C. japonica* is represented in countless cultivars, both single and double, from pure white through various shades of pink to red. Often the flowers exhibit stripes or blotches of other colours. *C. japonica* 'Elegans' is particularly pleasing. It is an anemone-flowered type with bright rose-pink flowers and an attractive centre. It is an old cultivar often known as 'Chandleri Elegans'. Under good conditions most of these plants average about 3 m (10 ft) in height and as much across with hundreds of blooms in early spring.

Their garden value is high for they tolerate temperatures as low as 0°C (32°F); the evergreen foliage is always attractive and the flowers are produced over a long period. They also do well in shade. Bud dropping is a sign of starvation or dryness at the roots while yellow blotches on the foliage or white marks on the flowers may indicate a mineral deficiency or occasionally virus disease.

Other good *Camellia* species are *C. reticulata* from China with semi-double, crimped flowers 15 cms (6 ins) across, and *C. saluenensis,* also Chinese, which was discovered by George Forrest in 1917. This has small but attractive light red flowers borne in profusion. These two species are less hardy than *C. japonica* and unless afforded a sheltered or frost-free situation lose their early blooms and may be badly damaged in severe weather.

The Japanese *C. sasanqua* flowers in winter, with single or double white, pink or red flowers which are slightly fragrant.

There are also many hybrids, a great number of British, American, Australian or New Zealand origin. The late J. C. Williams of Caerhays Castle, Cornwall was particularly successful in raising Camellias, his introductions including 'Cornish Snow', a hybrid between *C. saluenensis* and the white-flowered Chinese *C. cuspidata.* This has dark foliage and many small white single flowers. Another called 'J. C. Williams' (*C. saluenensis* x *C.*

Camellia japonica
'Apple Blossom'

Camellia japonica
'Adolphe Audusson'

japonica) has pale pink, single flowers with the commendable trait of falling as they fade. It is very hardy and free-flowering.

Another of the same parentage raised at Caerhays is 'Mary Christian', a fine pink single with dark green foliage. 'Donation' was bred by the late Colonel Stephenson Clark of Borde Hill, Sussex. It, too, is a *C. saluenensis* and *C. japonica* cross and an extremely fine semi-double, silvery-rose kind of free-flowering habit. It is one of the hardiest and most dependable for general garden purposes. Camellias are often criticized for their refusal to bury their dead, flowering bushes being frequently spoilt by the presence of numbers of dead brown blossoms. In the case of 'Donation', however, they fall as they fade.

Other Camellias illustrated include *C. japonica* 'Apple Blossom' which has pale pink flowers and the semi-double 'Lady Vansittart' which is also a *C. japonica* cultivar originally received by the Caledonian Nurseries in Guernsey among a shipment of plants from Belgium. Its background is obscure although it was originally described as having white flowers lightly striped with pink but variants (sports) occur with light pink and deep rose flowers with darker veinings.

Among *C. japonica* cultivars illustrated are 'Adolphe Audusson', a large, bright crimson semi-double with deeper veining and rounded petals; 'Purity' ('Shiragiku'), a splendid double with firm white petals of various shapes and deep golden stamens; 'Extravaganza', a white, semi-double heavily suffused and blotched with pink and light scarlet; and 'Cinderella', a fimbriated semi-double of pale rose with heavy suffusions of deep rose-pink and a white edging. The last two were raised in the United States of America.

Camellias require little or no pruning. However the young growths of *C. reticulata* and its hybrids—which are naturally wide spreading—may be pinched back occasionally to induce side shoots.

Tea-seed oil (a non-drying oil) is obtained from the seeds of *C. sasanqua* and used for textile purposes in the silk industry. It is also employed in the manufacture of soap. Tsubaki oil, obtained from *C. japonica* seeds, is used as hair oil.

Stuartias (Stewartias) are E North American or E Asian deciduous trees or shrubs with attractive peeling bark, alternate leaves and white flowers (something like single Camellias) filled with numerous stamens.

They have great beauty and the merit of flowering in midsummer when most shrubs have finished, and like Camellias detest lime and prefer a site sheltered from sun during the hottest part of the day. Usually the foliage assumes fine autumnal tints prior to falling. A mild temperate climate is conducive to the best results.

Outstanding among the species is *Stuartia malacodendron,* a shrub or small tree with downy shoots carrying elliptical to

Camellia
'Donation'

Camellia japonica
'Purity'

Camellia japonica
'Extravaganza'

Camellia japonica
'Cinderella'

Stuartia pseudo-camellia

oblong leaves 5–10 cms (2–4 ins) long and solitary, five-petalled flowers up to 8.5 cms (3½ ins) across. These are silky-white with purple filaments with bluish anthers. The species is a native of the SE United States.

S. pseudo-camellia comes from Japan where it makes a tree up to 15 m (50 ft) high. The white flowers are 5–7 cms (2–3 ins) across, the filaments white and anthers golden-yellow. Its leaf tints in the fall are brilliant red and yellow. This species is the most adaptable for English gardens and is also popular in Australia.

S. sinensis from China has fragrant blooms 5 cms (2 ins) across with yellow centres and peeling bark on its trunk. It grows up to 9 m (30 ft).

Gordonia axillaris from China and Formosa is an evergreen shrub or tree with long, smooth, shining leaves and solitary creamy-white flowers 7–13 cms (3–5 ins) across. These are filled with orange-yellow anthers and occur at the ends of the leafy shoots in late winter and spring.

G. chrysandra, also from China blooms in winter, again with white flowers. Both species require a humus type soil and are inclined to be tender in gardens subjected to severe winter wet or cold.

Cleyera japonica 'Tricolor' (*Eurya ochnacea*) from Japan, China and Korea is a shrub grown, especially in Australia, for its patterned Camellia-like leaves. These are green with grey banding, creamy margins and pink tinges. The creamy-white flowers are fragrant and succeeded by red berries which later turn purplish-black and are at their best in the winter. In cool climates the shrub must be grown under glass but small specimens also make good house plants as the leaves remain in fine condition for a long period.

Thunbergiaceae

4 genera and 205 species

This is a family of tropical herbaceous plants or shrubs, including some handsome climbers which are widely planted in warm climates and used for such purposes as festooning trees, draping buildings and masking fences. None of these tolerate frost but nevertheless they are popular in temperate countries for pot culture (twisted round canes) or for growing up trellis or greenhouse supports.

The simple leaves are oppositely arranged and are entire or hastate (arrow-shaped); the flowers are bisexual and regular (or slightly zygomorphic) in axillary or terminal racemes or solitary, with two large, spathe-like bracts, five to ten reduced sepals, five petals, four stamens and a superior ovary. Some authorities refer the family to Acanthaceae.

From a horticultural standpoint the most important genus of this family is *Thunbergia* which contains some 200 species of twining plants, annuals and shrubs. These come mostly from South and tropical Africa, Madagascar and the warmer parts of Asia.

T. alata is called Black-eyed Susan in its native South and tropical East Africa but (possibly because this name is used for *Rudbeckia fulgida* and R. *hirta*) the Black-eyed Clockvine in Europe and the USA. Although perennial it is usually treated as annual in temperate countries and either planted outside for the summer months or grown in large pots with the shoots trained over a balloon-shaped wire or cane framework. It is easily raised from

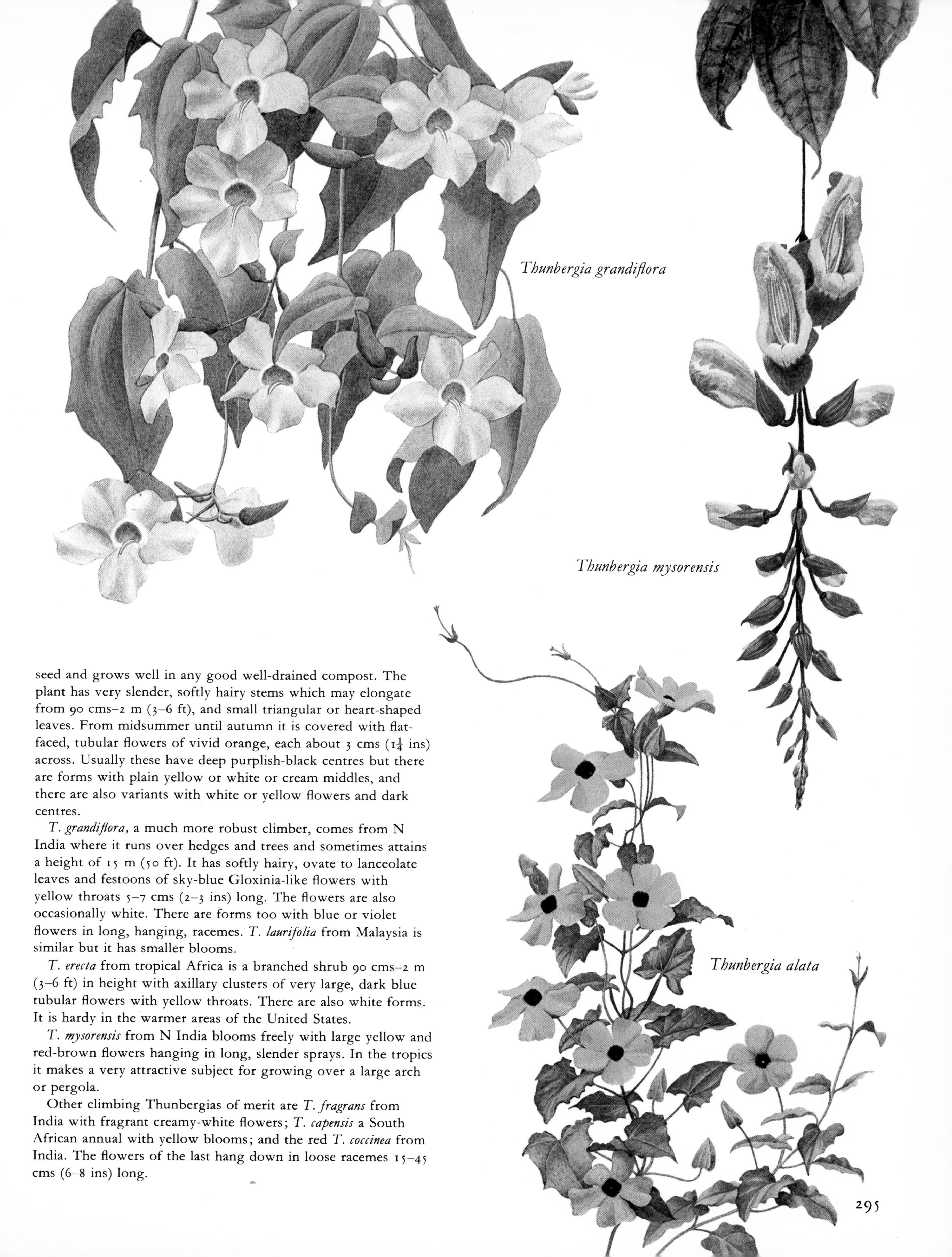

seed and grows well in any good well-drained compost. The plant has very slender, softly hairy stems which may elongate from 90 cms–2 m (3–6 ft), and small triangular or heart-shaped leaves. From midsummer until autumn it is covered with flat-faced, tubular flowers of vivid orange, each about 3 cms (1¼ ins) across. Usually these have deep purplish-black centres but there are forms with plain yellow or white or cream middles, and there are also variants with white or yellow flowers and dark centres.

T. grandiflora, a much more robust climber, comes from N India where it runs over hedges and trees and sometimes attains a height of 15 m (50 ft). It has softly hairy, ovate to lanceolate leaves and festoons of sky-blue Gloxinia-like flowers with yellow throats 5–7 cms (2–3 ins) long. The flowers are also occasionally white. There are forms too with blue or violet flowers in long, hanging, racemes. *T. laurifolia* from Malaysia is similar but it has smaller blooms.

T. erecta from tropical Africa is a branched shrub 90 cms–2 m (3–6 ft) in height with axillary clusters of very large, dark blue tubular flowers with yellow throats. There are also white forms. It is hardy in the warmer areas of the United States.

T. mysorensis from N India blooms freely with large yellow and red-brown flowers hanging in long, slender sprays. In the tropics it makes a very attractive subject for growing over a large arch or pergola.

Other climbing Thunbergias of merit are *T. fragrans* from India with fragrant creamy-white flowers; *T. capensis* a South African annual with yellow blooms; and the red *T. coccinea* from India. The flowers of the last hang down in loose racemes 15–45 cms (6–8 ins) long.

Thymelaeaceae

50 genera and 500 species

The characteristics of this family of dicotyledonous shrubs, trees, lianas and herbs include entire, alternate or opposite leaves and a basically racemose inflorescence. Individually the flowers are bisexual and regular with four to five petal-like sepals, four to five or eight to ten (occasionally many) stamens and a superior ovary.

The most important horticultural genus is *Daphne* with approximately seventy species of evergreen and deciduous shrubs. Many of these come from Europe and various cool parts of Asia which makes them suitable for temperate gardens, particularly when afforded a sheltered situation. They are commonly planted in rock gardens or, in the case of the more fragrant species, close to the house where the rich scent of the flowers can be most enjoyed.

Among the more fragrant is *D. odora* from China, an evergreen shrub up to 2 m (6 ft) tall, with oval-oblong, pointed leaves and crowded heads of small white to reddish-purple flowers. It needs a sheltered position for heavy snow may break the branches and even severe frost damage it irretrievably. There is a form with white margins to the leaves which seems to be slightly hardier. The flowers are used for perfumery purposes in the Orient.

D. mezereum is the European Mezereon, a deciduous shrub of 1.5 m (5 ft) with masses of fragrant, four-sepalled flowers studding its slender, leafless branches in early spring. These vary in colour from pale lilac-pink to deep purplish-red and there are also white forms including a double white. The bright red berries which succeed the flowers are poisonous although birds eat them without ill effect. The whole plant is a powerful irritant and the bark has been used in the treatment of rheumatism and other ailments, The sap affords a yellow dye and according to Linnaeus the berries were at one time used in Sweden to poison wolves and foxes.

The Spurge Laurel, *D. laureola,* a British native but also widespread in central, S and W Europe and W Asia, has shiny, oblanceolate evergreen leaves and pendulous branches of greenish-yellow flowers near the tops of the stems. Towards evening these become markedly fragrant.

D. cneorum, native to S and central Europe, is called the Garland Flower because of the way the bunches of small, pale pink to rose-pink flowers are arranged on the long recumbent branches. These are richly fragrant. The plant is evergreen with small, narrowly oblong leaves and is easily propagated from cuttings. In Spain the bark is thrown into fresh water to stupefy fish, thus ensuring easier netting. It is indigenous to southern and central Europe.

D. blagayana, another evergreen, grows about 30 cms (1 ft) high with spreading branches which can be pegged down to root and make new plants. It comes from Greece and Yugoslavia and has dense heads of fragrant, creamy-white flowers at the ends of the branches. Popular for rock-garden planting, it likes a lime-free soil.

Two good garden hybrids were raised in Britain this century from identical crosses by the brothers Albert and Alfred Burkwood. Working independently they hybridized *D. cneorum* and the white European *D. caucasica* and both produced outstanding cultivars. Alfred's was named 'Somerset' by the firm of Scott & Co. but Albert's was called *D.* x *burkwoodii* by Burkwood & Skipwith who handled the original stock.

'Somerset' is more or less deciduous with pale pink flowers and *D.* x *burkwoodii* is a much-branched, partially evergreen shrub rarely exceeding 90 cms (3 ft) in height with flowers which develop to purplish-pink as they age.

D. pseudo-mezereum from central Japan has oblanceolate leaves and greenish or creamy-yellow unscented flowers in clusters. These are succeeded by red berries. In Japan the fibrous bark is used in the manufacture of paper.

Edgeworthia papyrifera is one of three species of a genus of E Asian shrubs which normally have to be grown under glass in frost-prone districts. It is a deciduous shrub of 1.25–2 m (4–6 ft) from China with alternate, narrowly oval leaves and close-packed heads 2.5–5 cms (1–2 ins) across containing forty to fifty clear yellow, silkily hairy flowers. These are sweetly fragrant. The species has long been cultivated in Japan for paper making as has the very similar *E. gardneri,* the source of Nepal Paper.

Edgeworthia papyrifera

Daphne blagayana

Daphne cneorum 'Eximia'

Daphne odora 'Aureo-marginata'

Daphne laureola

Daphne x *burkwoodii*

Tiliaceae

50 genera and 450 species

This is a large family of dicotyledonous plants, chiefly trees and shrubs, from South-east Asia and South America. The leaves are alternate and often asymmetrical and the inflorescences are cymose and sometimes very complex.

Individual blooms are bisexual and regular with five sepals, five petals (occasionally none), many stamens and a superior ovary.

Included are several well-known trees of the Northern Hemisphere, particularly *Tilia* x *europaea,* the Lime or Linden, and *T. americana,* the North American Bass Wood. Other species in this genus are *T. tomentosa*, the Silver Lime or Linden, and *T. petiolaris*; both come from S Europe. The latter, a tree of 18 m (60 ft), has pendulous branches, fragrant whitish flowers in drooping cymes and heart-shaped, simple leaves which flutter in the wind and reveal striking, silver-felted undersides.

Limes are often planted as shade trees—one of the principal streets in East Berlin is called *Unter den Linden*—but they are also esteemed for the fragrant flowers, which are the source of nectar for honey bees. Lime honey has a delicate flavour and is highly prized; it is also used in medicine and some kinds of liqueurs.

The soft, smooth wood is admirably adapted for carving and many of the magnificent carvings of Stuart times which still adorn so many English churches and palaces are of Lime wood.

Sparmannia africana is the African Hemp, an attractive shrub of 3–6 m (10–20 ft) when grown outdoors, with clusters of pretty white flowers which have large, central puffs of yellow stamens. The latter are touch sensitive and move when disturbed. The soft, evergreen, heart-shaped leaves are downy on both sides and up to 20 cms (8 ins) in length. Where frosts occur the African Hemp must be grown under glass.

Sparmannia africana

Trilliaceae

4 genera and 53 species

This family of monocotyledonous plants is closely related to Liliaceae and was at one time included in that family. The species are all perennial herbs with rhizomatous rootstocks, smooth, erect stems and opposite or whorled leaves of a simple pattern.

The often large flowers are usually solitary but sometimes arranged in umbels; individually they are bisexual and regular with two, three or five (occasionally more) sepals, petals and stamens and a superior ovary. The fruit is a large berry or fleshy capsule.

The largest genus is *Trillium* with about thirty species distributed over North America and in Asia from the western Himalayas to Japan. Most spectacular and best known is *T. grandiflorum,* the Trinity Flower, Wake Robin or Wood Lily of North America. The first name highlights its tripartite character, many parts of the plant occurring in threes or multiples of that number. Thus there are three sepals and three petals, six stamens and a tripartite style to each bloom and a whorl of three leaves on each flowering stem.

The large white blossoms appear at the ends of stems 30–45 cms ($1–1\frac{1}{2}$ ft) tall in early spring, but the plant is variable and pink or reddish forms are not rare. There is also a double white.

The white fleshy roots, once used by Indian herb doctors, are now employed by American Schools of Medicine as remedies for various diseases. The leaves (cooked like greens) provide an emergency food plant. In Nova Scotia *T. grandiflorum* is called Moose Flower because it frequents the habitat of that animal.

T. erectum from eastern North America is known as Purple Trillium or Birth root. The flowers are rich and sombre in colour, a deep red rather than the purple quoted by many writers, although golden-yellow and white flowers are also supposed to occur in the wild. The species, which is widely spread over the whole of Canada and the eastern United States, makes a good companion for *T. grandiflorum*. The flowers have lance-shaped to ovate petals but rather an unpleasant odour.

T. sessile is the Toadshade (sometimes Toad Lily) of the eastern United States. It is one of the first to flower in spring and is characterized by handsome leaves variegated in shades of green and with brown blotches. The stalkless flowers are maroon-purple or sometimes greenish-yellow.

T. undulatum is called the Painted Trillium because of deep red blotches and pale rose streaks on its white, wavy-edged petals. An E North American species, it grows 20–30 cms (8–12 ins)

Trillium grandiflorum
pink form

high. Other Americans are *T. nivale,* the Snow Trillium, with small white flowers on stems 7–15 cms high; *T. chloropetalum* with greenish-yellow flowers (occasionally white to maroon) and mottled leaves; and *T. cernuum,* the Nodding Trillium, which has small, pendent, pink or white blooms on stems 15–60 cms (6–24 ins) high.

T. smallii (*T. apetalon*) from Japan, with small green and purple-red flowers, and the other Asiatic species make less attractive garden plants than the North American forms.

All the Trilliums are woodland subjects, thriving in deep, moist but well-drained leafy or peaty soil. They can be transplanted in summer and propagated by division or seed.

Medeola virginiana, the Indian Cucumber Root, comes from North America and is the sole representative of its genus. It differs from *Trillium* in having two whorls of sessile leaves on the stems of 22–30 cms (9–12 ins). The nodding flowers are greenish-yellow and develop to dark purple berries. The crisp white rhizome tastes like Cucumber.

Paris quadrifolia, the Herb Paris or True Love, is European (including British) and Russo-Asian. It has four stemless leaves in a whorl and umbels of starry, greenish-yellow, terminal flowers followed by dark black berries.

Trillium chloropetalum

Trillium sessile

Tropaeolaceae

2 genera and 92 species

All the plants in this family come from Central and South America and are dicotyledonous, somewhat succulent herbs, often tuberous rooted, many climbing by means of sensitive leaf stalks. The smooth, entire, lobed or palmate leaves are alternate (rarely opposite), with long leaf stalks; the flowers are bisexual but irregular in shape and often very showy. They occur singly in the leaf axils and have five sepals which form a short or long spur below, five irregular petals, eight stamens and a superior ovary.

The first species introduced into Europe was named by the great Swedish botanist Linnaeus. This was *Tropaeolum minus,* a non-climbing Peruvian annual with deep yellow flowers, purple-spotted in the lower petals.

It received its generic name for a singular reason. In ancient times victorious armies after battle would select a convenient tree or set up a tall trophy pole known as the *tropaeum.* On this they draped the armour and equipment of the vanquished foe as an emblem of victory. When Linnaeus saw the little Peruvian annual he called it *Tropaeolum* because the round, peltate leaves resembled soldier's shields and the red and yellow flowers reminded him of the blood-stained helmets of the fallen.

The name Nasturtium which is commonly applied to members of this genus can be literally translated as 'nose-tormentor'. The description refers to its biting peppery taste which reminded our ancestors of Watercress (*Rorippa nasturtium-aquaticum,* but at that time called *Nasturtium officinale*). Tropaeolums were once cultivated as salad plants and they are supposed to contain ten times as much Vitamin C as lettuce. The leaves are tasty between thin slices of bread and butter, the flowers may be used in salads and the green seed pods cured in spiced vinegar make a tolerable substitute for Capers (*Capparis spinosa*). They were extensively used for this purpose in Britain during the Second World War, when genuine Capers were hard to come by.

Nasturtiums are also known as Indian Cress because at the time of their introduction the Spanish possessions in South America were known by the general name of Indies, and also Lark's Heel because 'unto the backe-part (of the flowers) doth hang a taile or spurre, such as hath the Larkes heele, called in Latine *Consolida Regalis*' (Gerard, *Of the Historie of Plants*).

Linnaeus recorded that his daughter Elizabeth Christina in 1762 'observed the flowers of the Nasturtium emit spontaneously, at certain intervals, sparks like electric ones, visible only in the evening' (Richard Folkard Jr. *Plant Lore, Legends and Lyrics*). This phenomenon seems also to have been observed by Goethe and others, although modern science rather discredits these assertions.

Helen Philbrick and Richard Gregg (*Companion Plants*) recommend sowing Nasturtiums in greenhouses to combat White Fly, or under Apple Trees to repel Woolly Aphis and near Broccoli to deter Blackfly.

The species most commonly cultivated today is *Tropaeolum majus,* a strong-growing, smooth-stemmed climber from Peru introduced to Britain around 1686. It has large flowers, orange in the wild plant but very variable in cultivation. Non-climbing sorts known as *nanum* or Tom Thumb Nasturtiums appeared in the 19th century and there are doubles in a wide range of shades from cream, yellow and orange to pink, rose and deep red. The scented Gleam strain originated in California, where 'Golden Gleam' appeared spontaneously in a cottage garden. Nasturtiums are useful summer-flowering annuals as they will grow in poor soil and also in part shade.

T. peregrinum, the Canary Creeper or Canary Bird Flower, is an ornamental climbing annual for growing over fences, hedges and trellis. It comes from Peru where it has been grown since time immemorial in the gardens of Lima and other cities and has slender smooth stems 2.15–2.5 m (7–8 ft) long, bearing long-stalked, rather small lemon-yellow flowers. These have fringed lower petals and five-lobed deeply cut leaves. The common name is thought to refer to the fact that plants were first grown in the Canary Islands.

T. speciosum, the Flame-Flower, is a temperamental jade from Chile which does well in some gardens but fails completely in others. Success is largely linked with the possession of a cool moist soil and heavy dews at flowering time in late summer. A useful establishment trick is to take out a planting pocket 30 cms (1 ft) deep, place the roots at the bottom and barely cover them with soil. As growth progresses gradually add more soil until the ground reaches its normal level. The roots must be in deep shade, although the slender climbing shoots soon wend their way up into the light. The species does particularly well in Scotland where it threads in and out of evergreen hedges, Holly trees and the like. The brilliant scarlet flowers and small five- to six-lobed leaves make this species one of the most beautiful and desirable.

Tropaeolum speciosum

Tropaeolum peregrinum

Tropaeolum polyphyllum

Nasturtiums
(*Tropaeolum majus*)

Tropaeolum tuberosum

T. polyphyllum, also South American (Chile and Argentina), sends its roots down deeply into the soil, so that it is not only a late emerger but difficult to lift. This circumstance, however, makes for hardiness in temperate climates where it is a long-lasting perennial if grown in well-drained soil and a sunny situation. It possesses a long rhizome from which appear prostrate stems which trail over the ground. These are densely clothed with silvery, deeply lobed leaves and large, long-stalked, spurred, golden-yellow flowers in great profusion. After blooming the top growth rapidly disappears.

T. tuberosum is native to Peru where it was originally discovered by European collectors among broken rocks. A climbing species, it has red and yellow, long-spurred flowers on long stalks and five- to seven-lobed leaves. The roots produce a quantity of small, edible, Potato-like tubers which are cultivated and marketed as a vegetable in South America. At the time of the Spanish Conquest these tubers (called *añu*) provided, with Potatoes and Oca (*Oxalis tuberosa* and other *Oxalis* tubers), the staple diet of Indians in the high Andes. They have a disagreeable smell when raw but this disappears with baking. The stalks and leaves may also be eaten and have a slightly acrid taste. Frost kills the plant so it is usual to winter store the tubers in cool climates.

The climbing *T. leptophyllum* (*T. edule*) from Bolivia, Chile and Peru also has edible tubers; the flowers are orange, yellow or pinkish-white.

T. azureum, a tender climber from Chile, is sometimes cultivated under glass in Europe for the sake of its flowers which (although variable) usually come in a brilliant shade of intense blue.

Magallana, the other genus in the family, has only two species which are both native to temperate South America.

Umbelliferae

275 genera and 2850 species

This is a large family of dicotyledonous, mostly N temperate plants of wide distribution. The species are mainly herbs, often with hollow stems and alternate, sheathing, frequently much-divided foliage. The inflorescence is usually unmistakable, being composed of umbels of many small flowers which have petioles of varying lengths so that the blooms are brought to the same flat, plate-like level, although occasionally variations occur as with *Eryngium,* where the head is elongated and dome-shaped.

Although the inflorescences are sometimes spectacular many members of Umbelliferae are so alike in structure that it is difficult to distinguish between them without reference to special (and often inconspicuous) botanical details. The flowers are regular and bisexual with five (rarely none) very small sepals, five petals (rarely none), five stamens and an inferior ovary.

A number of important economic plants occur in the family including the Carrot (*Daucus carota*), Parsnip (*Pastinaca sativa*), Celery (*Apium graveolens*), Fennel (*Foeniculum vulgare*), and sundry herbs like Angelica (*Angelica archangelica* or *Archangelica officinalis*), Chervil (*Anthriscus cerefolium*), Samphire (*Crithmum maritimum*), Dill (*Anethum graveolens*), Caraway (*Carum carvi*), Lovage (*Levisticum officinale*), Parsley (*Petroselinum crispum*), Anise (*Pimpinella anisum*) and Coriander (*Coriandrum sativum*).

The ornamentals although comparatively few in number stand out among a host of rather indifferent species. A few have spectacular foliage, especially certain Eryngiums with spiny, sword-shaped or Yucca-like leaves. These include the Falkland Island and New Zealand *Azorella* (sometimes placed in Hydrocotylaceae), which forms tight cushions 70 cms (3 ft) in diameter and 45 cms (1½ ft) high (a xerophytic adaptation); *Ferula* with thousands of segments; and *Heracleum* which has enormous leaves topped by monster flower heads.

Eryngiums are Thistle-like plants known as Sea Hollies: Sea because *Eryngium maritimum* (a European, including British native) grows naturally on sandy sea shores and Holly on account of its spiny bracts and sharp-toothed leaves. Its flowers are pale blue and the plant grows 15–45 cms (6–18 ins) high.

According to Linnaeus the young flowering shoots of this species can be boiled and eaten like Asparagus and the leaves have a slight aromatic pungency. Gerard says the roots 'condited or preserved with sugar are exceedingly good to be given to old and aged people that are consumed and withered with age' adding that they have the property of 'nourishing and restoring the aged, and amending the defects of nature in the yonger'.

For garden decoration, however, there are more worthy species than *E. maritimum,* especially *E. giganteum,* a biennial from the

Hacquetia epipactis

Caucasus. This is of branching habit, 60–75 cms (2–2½ ft) tall, the green leaves and stems and bracts being heavily overlaid with silver with a touch of blue on the flowers. Although it dies after flowering self-set seedlings normally ensure continuity. Suitable for most soils the species thrives in full sun or partial shade.

E. x *oliveranum* (a hybrid possibly of natural origin with *E. giganteum* as one parent), is more perennial. It has three- to five-lobed leaves and many rich blue, Teasel-like flowers and steel blue upper stems. Deeper blue and light navy cultivars exist and the plants are very spiny.

Two interesting species for key positions in a sunny, well-drained garden are *E. pandanifolium* from Brazil and Uruguay, which grows 3–4 m (10–12 ft) high, and *E. agavifolium* from the Argentine, of lower stature at 1.5–2 m (5–6 ft). The former has spiny-edged, sword-shaped leaves and large branching stems carrying many small, round, purple-tinted, white flower heads. *E. agavifolium* is similar but bears greenish-white flower heads.

Astrantia major, the Masterwort, widespread in Europe, is a fascinating and variable plant 15–60 cms (6–24 ins) in height with branching leafy stems carrying many small inflorescences. The flowers are white or blush-pink, each head set off by an involucre of white bracts which have green mid-ribs and reddish tinges. There are also most attractive variants with deep purplish-red flower heads known erroneously in gardens as *A. carniolica* var. *rubra*.

A. carniolica, a woodland plant from the SE Alps, is usually of dwarfer habit with white or pinkish flowers and five- to seven-lobed radical leaves.

Bupleurums (sometimes called Hare's Ear) are shrubby evergreens or herbs which are occasionally planted in European gardens. They need warm, sheltered conditions or, in the case of *Bupleurum fruticosum,* tolerate an exposed position near the sea shore. This species (which is native to S Europe) grows 1.5–3 m (5–10 ft) high with attractive smooth, narrowly oval and simple leaves of bluish-green. The small yellow flowers, grouped in terminal umbels 5–10 cms (2–4 ins) across, hold great attraction for wasps and bluebottles. *B. fruticescens* is shorter (45–90 cms; 1½–3 ft) with stiff grassy leaves and zigzag umbels of yellow flowers.

Hacquetia epipactis (*Dondia epipactis*), the lone representative of its genus, comes from central Europe where it is found in open wood and scrub in the E Alps. It blooms very early in the year and is a smooth perennial 5–7 cms (2–3 ins) high with bright green, long-stalked leaves which are deeply cut into toothed segments. The tiny yellow flowers set in umbels, are surrounded by a spreading involucre (collar) of green to yellow bracts.

Ferula communis, a noble herbaceous perennial with beautiful foliage, is suitable for such key positions in the garden as the edge of a shrubbery, the bank of a pool or a corner site by a gate. The stem often attains a height of 2.5–3 m (8–10 ft) and a diameter of 5–7 cms (2–3 ins), with huge leaves which are cut into thousands of bright green, needle-like segments. These come through the ground very early in the year. The plant tolerates snow, frost and similar wintry conditions surprisingly well considering its S European origin. When well established it blooms, the huge branching scape (which has a white glaucous bloom) carrying many umbels of yellow flowers. These are set off by extremely large bracts edged with green, leafy fringes.

The species, although commonly known as Giant Fennel, should not be confused with the true culinary Fennel (*Foeniculum vulgare*) although in many respects it resembles this plant.

F. tingitana from Spain, Portugal and North Africa is another fine species for the garden. It grows 2–2.5 m (6–8 ft) tall with deeply cut, shiny leaves and yellow flowers.

Several members of the genus have economic uses, especially *F. assa-foetida* and *F. narthex* which yield a gum-resin called asafoetida when incisions are made in the roots and rhizomes. The drug is used as a stimulant in medicine.

Astrantia major

Eryngium giganteum

F. communis var. *brevifolia* is the source of gum ammoniac and the medicinal gum galbanum comes from *F. rubricaulis* and *F. galbaniflua*.

The stems of *F. communis* are full of white pith, which when dry ignites like tinder. When once alight they burn very slowly without injury to the tube of the stem, a circumstance which accounts for their use as torches in Sicily and other parts of Europe. John Smith (*Dictionary of Popular Names of Economic Plants,* 1882) says this practice of carrying fire from place to place 'is of great antiquity, and serves to explain the passage in Hesiod, where, speaking of the fire Prometheus stole from heaven, he says 'he brought it in a Ferula'. The stems are very light, and Bacchus, the God of Wine, recommended that his votaries should carry them, so that if they quarrelled from the effects of too much wine, they could strike one another without inflicting injury.

Another spectacular umbellifer is *Heracleum mantegazzianum* from the Caucasus, an impressive perennial up to 2–2.5 m (6–8 ft) in height with huge round flower heads 30 cms (1 ft) or more across made up of many small white florets. They look like round tea-trays. The *Royal Horticultural Society Dictionary of Gardening* estimates up to 10,000 flowers in a single inflorescence so that the common name of Cartwheel Flower seems very appropriate.

This *Heracleum* is effective when isolated in a copse or wood or one or two plants make an impressive feature on the banks of a stream or lake, but it is far too large for mixed flower borders. A painful rash may be caused by handling this species.

Among the aromatic or culinary herbs are several grown for ornamental purposes. Fennel (*Foeniculum vulgare*), a European plant (naturalized in Britain) is one such, especially in its dark-leafed form (the so-called 'Black Fennel'). This makes an attractive foil for red and blue flowers in a mixed border. Even the normal type is pretty with its narrowly divided, three or four times pinnate bright green leaves. Fennel has been grown in herb gardens since very early times. The leaves are commonly strewn on fish dishes and the seeds were used in the 16th and 17th centuries to flavour soups, fruit pies and even bread. Unfortunately the plant seeds so freely as to become a nuisance unless the golden flower heads are regularly removed when they fade.

The European (including British) *Myrrhis odorata* or Sweet Cicely is a stately plant for a shady border, which looks especially fine when associated with the scarlet Bergamot (*Monarda didyma*). It has fleshy, carrot-like roots, thrice-pinnate, fern-like leaves and umbels of white flowers which are attractive to bees. The leaves were at one time used in salads although they taste like anise or paregoric. The seeds are full of oil and have a pleasant taste and the roots at one time were boiled as a vegetable or candied. Another past use of the crushed seeds was for polishing wood.

The smell of the plant attracts bees and the insides of empty hives are often rubbed with it before placing them near swarms to induce them to enter.

A few members of Umbelliferae are highly poisonous, notably *Oenanthe crocata,* the Water Dropwort, and *Conium maculatum,* the Common Hemlock—which Socrates took when he decided to depart this life.

Urticaceae

45 genera and 550 species

This is a family of generally rather nondescript dicotyledonous herbs or low shrubs with both tropical and temperate distribution. Some have strong stem fibres which are used for fishing nets, cordage and other purposes and several, notably *Urtica, Urera* and *Laportea,* possess painful stinging hairs.

The leaves may be opposite or alternate and the flowers are unisexual and regular with four or five calyx segments (rarely none at all), no petals, four or five stamens and a superior ovary.

The most commonly cultivated—and even then primarily for their foliage—are *Pellionia, Urera* and several species of *Pilea.*

Pilea cadierei, a native of Indo-China (Annam), is popularly known as the Aluminium Plant because of the deep silvery patches which occur between the veins on its green, oval-oblong, simple leaves. It makes a good house plant growing about 25 cms (10 ins) high, with inconspicuous brownish-green flowers. *P. involucrata* has deeply quilted, fleshy green leaves which on exposure to strong light assume coppery-red overtones. The tiny rosy-red flowers are clustered in the leaf axils.

P. microphylla (*P. muscosa*) from tropical America is the Artillery Plant, Pistol Plant or Gunpowder Plant; all these names refer to its habit of shooting out clouds of pollen when handled. It has a fern-like appearance due to the intricate branching of the green stems (15 cms; 6 ins) and its many minute leaves. The flowers are practically invisible but the general appearance and dwarf habit make it a popular edging plant on greenhouse benches or in mixed house plant arrangements in the home.

Pellionia repens (*P. daveauana*) is a trailer from Indo-China, Burma and Malaya with succulent stems and alternate, roundish-elliptic leaves 5 cms (2 ins) long. These are bronze-green with a tinge of violet, having a central band of bright green in var. *repens* or white blotches on the plain green leaves in var. *viridis.*

Urera baccifera is the Cow-itch of the West Indies and tropical and Central America. Its popularity is lessened by the stinging hairs on the leaves, contact with which has been likened to that of electric shock followed by several hours of intense pain. It makes a shrub of 2–6.25 m (6–21 ft) with large leaves and axillary, much-branched inflorescences of pink or red flowers, followed by the attractive fruiting panicles of white, waxy fruits. In tropical America it is used as a cattle hedge.

Urtica dioica, the Stinging Nettle of Europe, was once used by doctors to flog patients as a counter-irritant in cases of skin trouble (urtication). The young tips are eaten like Spinach as a tonic and the fibre used for many purposes including cloth fabrics. Others in the family (as *Laportea*) have medicinal uses or application as fish poisons.

Pilea cadierei

Pellionia repens

Pilea involucrata

Valerianaceae

13 genera and 400 species

Most members of this dicotyledonous family are annual or perennial herbs, and frequently strongly scented. They bear both basal and stem leaves, the latter being opposite and either simple or pinnate. The flowers occur in compound cymose or capitate inflorescences and may be bisexual or unisexual and regular or irregular. They have five petals (rarely three or four), one to three minute sepals, one to four stamens and an inferior ovary.

Valeriana officinalis, a strongly scented European perennial, has great attraction for cats and also rats, which have been trapped by using pieces of the root as bait. It grows about 90 cms (3 ft) high with inflorescences of pinkish flowers. Medicinally the plant acts as a powerful nerve sedative and in medieval times was used as a condiment. Although Europeans find the scent disagreeable it is valued as a choice perfume in certain oriental countries. A few dwarf species of *Valeriana,* particularly *V. supina* and *V. saliunca*, both European with pinkish fragrant flowers, make low-growing plants of carpeting habit for the rock garden.

Centranthus ruber is a European plant, naturalized in Britain, where it often occurs on railway banks and similar well-drained, sunny situations. It attains a height of 45–90 cms ($1\frac{1}{2}$–3 ft), with smooth, pointed leaves and bold clusters of red, pink or white flowers. In S Europe and Sicily the leaves are eaten in salads and the seeds of this and other species were at one time used for embalming the dead. Its chief disadvantage is an obnoxious catty smell which, however, only becomes apparent when the leaves and stems are trodden on or bruised.

The Bible has several references to a valuable unguent called Spikenard, as in Mark 14:3 where we read of a woman who brought 'an alabaster box of ointment of Spikenard very precious' to Jesus. This costly perfume comes from *Nardostachys jatamansi,* a little Indian plant of 10–15 cms (4–6 ins) with oblanceolate leaves and terminal heads of pale rosy-pink flowers. The lower parts of the flower stems are covered with shaggy hairs and it is from this area that the exquisite perfume is obtained.

Valerianella locusta, a European (including British) native with minute pale lilac flowers on stems of 15–30 cms (6–12 ins), is often eaten as salad when very young. It is called Corn Salad or Lamb's Lettuce from its 'appearing about the time when lambs are dropped' and is particularly valued in France.

Verbenaceae

75 genera and 3000 species

This large family of dicotyledonous herbs, shrubs, trees and climbers includes some which are xerophytic often with sharp thorns. The majority are of tropical or subtropical distribution. Usually the leaves are opposite (sometimes alternate or whorled) and entire or divided. The cymose or variably racemose heads of flowers are frequently set off by an involucre of coloured bracts; individual blooms being bisexual and zygomorphic with five sepals (occasionally four or eight), five petals forming a narrow tube at their base and often two-lipped and four stamens (rarely five or two).

Among a number which possess spiked inflorescences is *Aloysia triphylla* (*Aloysia* or *Lippia citrodora*), the Lemon Verbena of Chile and Argentina. A deciduous shrub or small tree, it has pale lilac, insignificant flowers and narrow, lanceolate leaves arranged in threes, which are delightfully scented and yield an aromatic oil.

Verbenas are others with spiked inflorescences. *Verbena patagonica* (*V. bonariensis*), a small South American perennial, produces branching stems 90 cms (3 ft) high with many violet, purple or blue flowers. It is a great colonizer and may be seen in quantity by the roadsides of Kenya as an alien, but in cooler climates (as in Britain or North America) is usually treated as annual. It is naturalized in California and the SE United States.

V. corymbosa, also South American, is hardy in cool temperate countries provided the soil is moist and it receives full sun. The purplish-blue flowers are densely crowded into terminal, short-spiked panicles.

V. x *hybrida* is a name given to a race of beautiful hybrids 15–45 cms (6–18 ins) high, derived from several South American species (*V. peruviana, V. incisa, V. platensis* (*V. teucroides*) and *V. phlogiflora*). Treated as annuals they make good bedding plants, having a bushy habit with showy, flat clusters of red, pink, yellow, blue, lilac, crimson, scarlet, purple or white flowers. Many have been given cultivar names and some come reasonably true from seed.

Callicarpa bodinieri giraldii is a Chinese shrub which, in temperate gardens subject to severe frost, must be afforded a sheltered situation. It is one of the hardiest of a genus of mainly tropical plants, noteworthy for their fine fruits. The species grows 2–2.75 m (6–9 ft) high with oval-lanceolate, opposite leaves and rounded clusters of lilac flowers followed by globose, glossy berries of a striking, pale bluish-lilac shade which are deep purple in *C. bodinieri giraldii.*

Centranthus ruber

Nardostachys jatamansi

Verbena hybrids

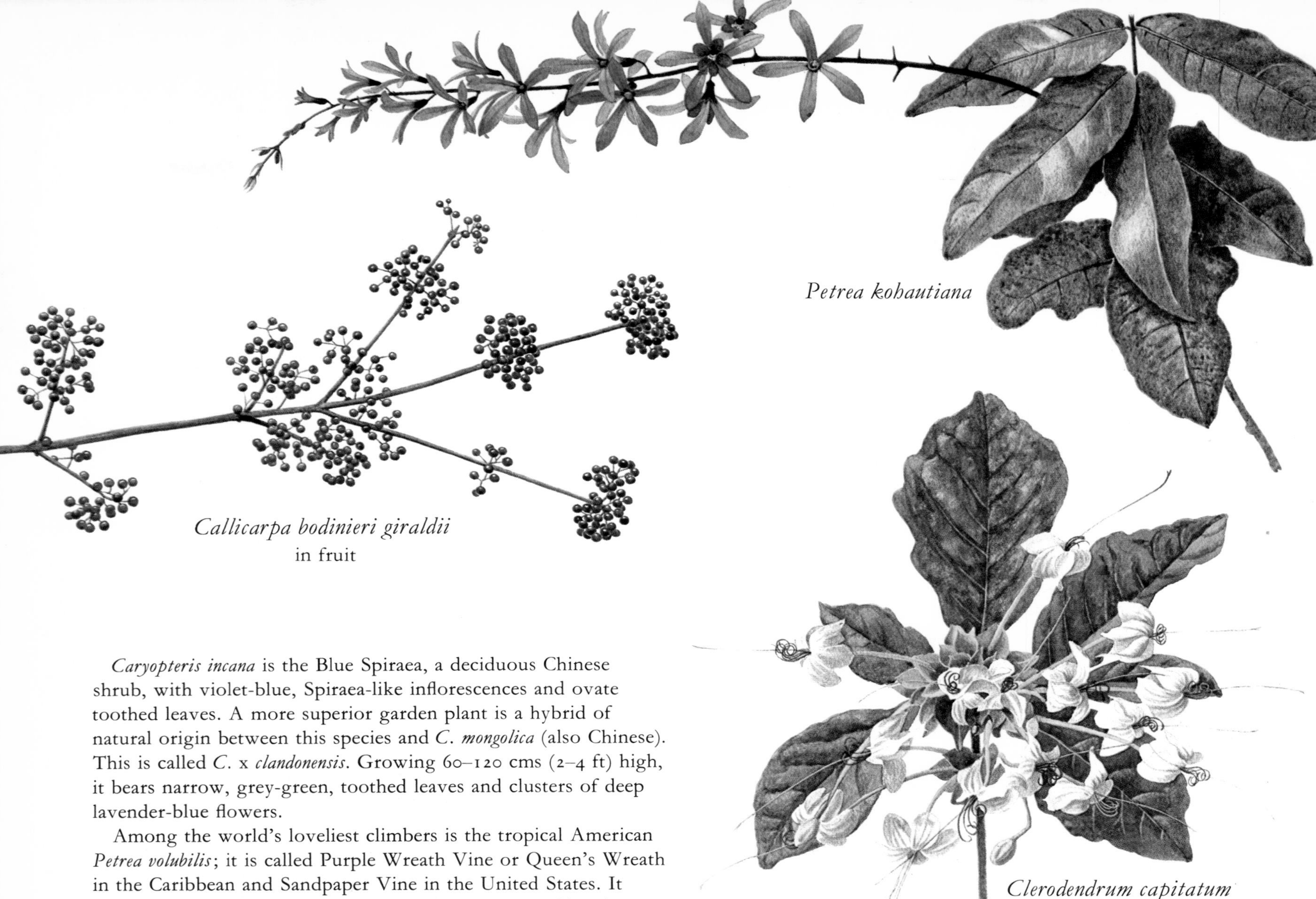

Petrea kohautiana

Callicarpa bodinieri giraldii in fruit

Clerodendrum capitatum

Caryopteris incana is the Blue Spiraea, a deciduous Chinese shrub, with violet-blue, Spiraea-like inflorescences and ovate toothed leaves. A more superior garden plant is a hybrid of natural origin between this species and *C. mongolica* (also Chinese). This is called *C.* x *clandonensis*. Growing 60–120 cms (2–4 ft) high, it bears narrow, grey-green, toothed leaves and clusters of deep lavender-blue flowers.

Among the world's loveliest climbers is the tropical American *Petrea volubilis*; it is called Purple Wreath Vine or Queen's Wreath in the Caribbean and Sandpaper Vine in the United States. It blooms profusely, the sprays (20–30 cms; 8–12 ins) of lovely bluish-violet flowers cascading from a stem which (given a convenient support) will reach a height of over 9 m (30 ft). The sepals are lighter in colour than the corolla and outlast them on the shrub leaving it a lavender-mauve colour which gradually fades to greyish-brown. The plant can be clipped to any convenient height and indeed specimens of 90 cms (3 ft) flower well under glass in pots. The plant often cultivated under this name is *P. kohautiana* from the West Indies which differs in its terminal, not axillary inflorescences.

Duranta repens (*D. plumieri*), the Pigeon Berry or Sky Flower of tropical America is a strongly branched shrub 90 cms–5 m (3–15 ft) high, with square—often drooping—twigs. These carry oblong leaves, and erect or pendent racemes with numerous, small, trumpet-shaped flowers which only appear on one side of the stems. They may be white, blue or purple and have yellow anthers, also darker stripes on some of the petal lobes. The shiny orange berries which follow hang in long strings like Currants, creating a striking effect with the flowers.

Lantana camara, another tropical American shrub, is a prickly stemmed evergreen of 2–3 m (6–10 ft) with ovate, pointed and toothed, aromatic leaves and many small rounded heads of axillary flowers. These may be pink or yellow and change colour with age, to red or orange. There are also forms with white and saffron blooms, which again deepen in colour with maturity.

Lantana sellowiana (*Lippia montevidensis*) with rosy-lilac, yellow-eyed flowers in showy, rounded bunches is a popular mat-forming trailer for rock gardens and borders in the tropics. It comes from Montevideo.

Clerodendrum (*Clerodendron*) contains many popular ornamentals which are grown in sheltered situations or under glass in cool temperate climates, or as border subjects in the tropics or sub-tropics. The species are trees and shrubs, or sometimes climbers, many with fragrant flowers but unpleasantly scented leaves (usually only apparent when these are bruised). A characteristic of the flowers is the long projecting stamens, which can extend 5 cms (2 ins) beyond the corolla and turn inwards at their tips.

Among the commonest in tropical gardens is *C. paniculata*, the Pagoda Flower of Malaya. It has large, lobed, cordate leaves and compact, pyramidal heads of small red, occasionally white or yellowish, flowers. When clipped it makes a neat shrub of 1.25–1.5 m (4–5 ft). The related *C. speciosissimum* (*C. fallax*) has smaller inflorescences (4–5 cms; 1½–2 ins) but larger and more brilliant scarlet flowers. It comes from Java, grows 90–120 cms (3–4 ft) tall and has unpleasant smelling leaves.

Another widely planted species is the Malayan *C. nutans* which grows 1.25–3 m (4–10 ft) with pendulous sprays 25 cms (10 ins) long of pure white flowers having long projecting stamens. *C. capitatum* from tropical Africa forms dense terminal heads (up to 30 cms; 1 ft across), carrying many creamy-white, long-tubed flowers. It suckers freely and can become invasive.

The plant well known in gardens as *C. fragrans,* a native of Japan and China, makes a deciduous shrub (90–150 cms; 3–5 ft) with rounded heads of fragrant rose-pink or white blossoms and cordate-ovate leaves. There is a double form called *pleniflorum* which is, in fact, the type and should correctly be called *C. philippinum*. *C. trichotomum,* a Japanese species, possesses large, ovate to oval leaves and large heads of white flowers with pink calyces followed by bright blue fruits subtended by crimson calyces. *C. t. fargesii* (*C. fargesii*) also has white flowers 10–15 cms (4–6 ins) across and porcelain blue berries set off by persistent

Clerodendrum splendens

Clerodendrum thomsoniae

green calyces which become pink as they age. Both plants have noxious-smelling, ovate leaves.

C. bungei (*C. foetidum*) from China, with fragrant purplish-red flowers is—like the two preceding species—reasonably hardy in the British Isles. *C. thomsoniae* from West Africa, a beautiful evergreen climber, has axillary and terminal, stalked clusters of flowers with locket-shaped white calyces and red corollas—a delightful combination. *C. ugandense,* the Blue Butterfly bush from tropical Africa, grows up to 2.5 m (8 ft) tall and 2 m (6 ft) across, with dainty flowers in two shades of blue (pale blue and deep violet) and long blue stamens.

C. splendens, the Glorybower from Senegal to Angola, grows 1.25–1.5 m (4–5 ft) high—taller in shade—with ovate to oblong, cordate leaves and large clusters of brilliant scarlet, long-stamened flowers.

Among the few plants of economic significance in this family is the E Asian *Tectona grandis* which furnishes Teak, a valuable hard and durable timber. This sinks in water unless thoroughly dried. The tubers of *Priva laevis* from the Argentine are edible.

Clerodendrum trichotomum fargesii

Violaceae

22 genera and 900 species

This family has cosmopolitan distribution and contains many popular annual and perennial herbs like Pansies and Violas, as well as a number of small shrubs. The plants are dicotyledonous with alternate or basal, usually simple leaves, and bisexual and generally zygomorphic flowers arranged singly on the stems or in racemose or fascicled inflorescences. There are five sepals, petals and stamens to each bloom and the ovary is superior.

The most important genus for the garden is *Viola* with some 500 species. Perhaps the most popular is *Viola odorata,* the Sweet Violet, a British native plant but also widespread in parts of Europe, North Africa and Asia. It grows only a few centimetres high with deeply cordate, crenate leaves and deep violet, lilac or white, sweetly scented flowers 2 cms ($\frac{3}{4}$ in.) across. Seed is seldom produced from these blooms but separately, from closed, self-pollinated flowers (cleistogenes).

Under cultivation *V. odorata* has produced larger-flowered forms in a wider range of colours, and also several doubles. Interest in the species is said to have been stimulated by Napoleon who, although not greatly interested in flowers, liked Violets because they reminded him of the Corsican woods in which he played as a child.

Josephine wore Violets on her wedding day and was invariably given a bunch by Napoleon to mark each anniversary. But it was really in defeat that Violets became important to France for as the Emperor was taken away to begin a life of exile in Elba, he told his followers he would return with the Violets in spring.

Viola odorata

Parma Violet

This promise was fulfilled with his escape, when many loyal friends and adherents came forward wearing Violets or violet-coloured dresses and cloaks. But their jubilation was short-lived for soon afterwards Napoleon met his final defeat at Waterloo. Before leaving for St Helena, however, he asked his captors for permission to visit Josephine's grave where he plucked violets, which after his death were found in a locket around his neck.

The large florists' varieties are derived from several species: *V. odorata, V. cyanea, V. alba* and *V. suavis* (or *V. pontica,* considered by some authorities a form of *V. odorata*). All are European but the history of Parma or Neapolitan Violets is more obscure. These are thought to be of oriental origin and were grown at Malmaison by the Empress Josephine and later became a great favourite of Queen Alexandra, who at one time had some 5000 cultivated plants in frames at Windsor.

All these Violets are shade-lovers and appreciate cool moist soil. They are increased by runners, normally removed and potted up or planted in early autumn.

V. hederacea (*V. reniforme; Erpetion hederaceum*) is the Australian Violet, a tufted and creeping species with reniform or orbicular leaves and small blue and white flowers. It is less hardy than *V. odorata* and may need glasshouse protection in winter.

America has many charming small Violas, including *V. papilionacea* (E United States) which has large violet flowers with white hairs in the throat; *V. pedunculata* or Johnny-Jump-Up (California and Mexico) with yellow and brown flowers; and *V. pedata* or Bird's Foot Violet (E United States) with dark violet and lilac flowers (although variable) and three to five cleft leaves. The last grows best under scree conditions.

V. tricolor is the Heartsease or Wild Pansy of John Parkinson (*Paradisus Terrestris,* 1629) and Milton's 'pansy freak'ed with jet' which in folklore had such fanciful names as Tittle-my-Fancy, Kiss-me-at-the-garden-gate and Love-in-Idleness. It is an Asian and European (including British) plant, very variable both as to colour and size of flower which, under the fostering care of the gardener has assumed larger proportions and more brilliant colours.

The Garden Pansy is probably derived from *V. tricolor, V. lutea,* and possibly *V. altaica.* The quaintly marked 'faces' had great attraction for English gardeners from about 1810 onwards and there are still Viola and Pansy Societies which specialize in these flowers, which are usually annual or biennial with very short spurs behind the large flat, circular blooms. These in turn have a well defined, central blotch and should not be less than 5 cms (2 ins) in diameter.

A Lord Gambier (who was ignominiously discharged from the Royal Navy at the beginning of the 19th century and took up gardening to forget his troubles) is often credited with the development of the first Pansies, but it was actually his gardener Mr T. Thompson (later called the Father of the Heartsease) who was the real pioneer. Garden Pansies are now labelled *V.* x *wittrockiana* after Professor Wittrock, a Swedish botanist who wrote a learned treatise on their history in *Viola Studier* in 1896.

Viola 'Irish Molly'

Pansies (*Viola* x *wittrockiana* hybrids)

A particularly charming, small, near black *V. tricolor* seeds itself freely in old gardens. It is known as Bowles' Black although the late E. A. Bowles said his original stock came from Dr Lowe 'who told me it always bred true, and so it does if kept to itself but I know it readily influences other Violas, for its dusky charms appear in Mulattoes, Quadroons and Octaroons all over the place'. He also mentions seeing it at the Chelsea Show labelled 'Viola Black Bowles'. 'I am not so black as I was painted on that label, so I altered it'. (*My Garden in Spring*). It is a charming little weed with a wonderfully friendly and cheerful look in the yellow Cyclopian eye in the middle of its almost black face.

A close relative of *V. tricolor* is *V. saxatilis aetolica,* a dwarf from eastern Europe with pretty little yellow flowers which makes a charming subject for the rock garden.

The Garden Violas are a race of flowers derived from crossing

Viola hederacea

Viola saxatilis aetolica

Viola tricolor

Viola 'Bowles' Black'

Viola cornuta

Show Pansies with *V. cornuta* and *V. lutea*. These were first raised by James Grieve about 1863 and William Robinson tried to call them Tufted Pansies but somehow the name never stuck and *V.* x *williamsii* is now their official title. They are more perennial than Pansies, with longer spurs and more compact growth, often self-coloured. There are countless named varieties of which the one illustrated 'Irish Molly', has most unusual colouring. Violas and Pansies flower long and continuously if the old blooms are regularly removed and make good edging plants or carpeters among taller subjects. They like moist soil and sun or light shade.

V. cornuta from the Pyrenees has fragrant, violet or light mauve, elongated flowers about 2.5 cms (1 in.) across profusely borne on long stems. The petals are rather long and angular, the bottom one slightly broader than the others. There is also a white form.

Violettas are derived from *V. cornuta* and a bedding Pansy. These are small-flowered and lack the usually conspicuous rays or guide lines so prominent in many Pansies.

Vitidaceae

12 genera and 700 species

This family is also known as Vitaceae. Although it includes a few upright shrubs most of its members are dicotyledonous climbers, which cling by means of tendrils or suckers. Frequently they attain great heights and some of the more ornamental kinds are used to drape walls, cover old ugly buildings or mask tree stumps and similar features. Only a proportion is hardy enough for cool areas as most of the species come from warm climates, although some of the latter are popular as house plants.

The leaves may be simple or compound and are arranged alternately on the stems; the deciduous kinds often assuming rich autumnal tints before falling. The small flowers come in cymose inflorescences which are frequently very complex and may be bisexual, monoecious or dioecious. The regular blooms have

Parthenocissis henryana

Ampelopsis brevipedunculata maximowiczii

Cissus discolor

four to five sepals, petals and stamens and a superior ovary.

The genus *Vitis* is economically important for it includes *V. vinifera,* the Grape Vine. Apart from having edible fruits, the dried Grapes become Raisins and a seedless form provides Sultanas. Currants come from a Corinthian variety. In addition the seeds are the source of Grape-seed oil used in the manufacture of soap, paints and (after refining) in certain foods. The waste products are used as cattle food, tannin and Cream of Tartar.

The climate, soil, aspect, variety of Grape Vine and the nature of the yeast all influence the resultant wines. Red wines come from purple Grapes (or are artificially coloured); white wine from white Grapes. In dry wines the fruit sugar is almost entirely fermented whereas in sweet wines the process is stopped before all the sugar is converted. The use of ripe or immature fruit also influences the result.

Fortified wines are of high alcoholic content due to added alcohol or brandy and medicinal wines have iron compounds, quinine and similar tonics added. Among the many drinks linked with the Grape industry are ports and sherries, champagne, claret, hock, madeira, tokay, chianti, burgundy, vermouth and various table wines and liqueurs.

Among the hardier ornamentals is *V. coignetiae*, the Crimson Glory Vine of Japan, a deciduous climber which will reach 18–24 m (60–80 ft) with very large, deeply cordate leaves that have pointed short lobes and are rust-downy beneath, especially when young. The small green flowers are highly fragrant and in autumn the foliage turns crimson.

V. flexuosa from Japan, Korea and China is slender-stemmed with ovate-cordate to triangular, smooth leaves and black fruits, and the woolly-stemmed, Chinese *V. wilsoniae* has wavy-margined, ovate-cordate leaves, very woolly when young, especially on the new shoots and undersides of the leaves. The latter assume rich autumn colours before falling and the black fruits have a purple bloom.

Parthenocissus tricuspidata from China and Japan and *P. quinquefolia* from North America are quick-growing, smooth-leafed, self-clinging climbers hardy in the British Isles and popular for draping house walls. In the fall the foliage turns brilliant scarlet before dropping. Both are commonly known as Virginia Creeper in Britain; in the United States *P. tricuspidata* is the Boston Ivy.

P. henryana (*Vitis henryana*) is a vigorous Chinese climber with beautifully marked, tapering leaves. These are dark green with silvery markings along the veins and divided into three or five leaflets. The fruits are dark blue. This species does best on a shady wall.

Ampelopsis brevipedunculata maximowiczii (*A. heterophylla*) from China, Korea and Japan is a showy climber with variable foliage; the leaves are three- or five-lobed or heart-shaped and without lobes, both kinds frequently on the same plant. The small greenish flowers are succeeded by bright blue berries, which have black dots.

Several plants in Vitidaceae are grown as house plants as they are tolerant of fairly starved conditions (which keeps them dwarfed) and subdued light. *Cissus antarctica,* the Kangaroo Vine from Australia, is one of the commonest, its leaves (5 × 10 cms; 2 × 4 ins) being shaped something like those of the Oak. *C. capensis* from South Africa has round green leaves with crenated margins; *C. discolor*, from Java and Cambodia, has its green leaves marbled with white and purple and deep crimson undersides; and *C. striata* from Chile bears small green leaves 2.5 cms (1 in.) long divided into five leaflets.

Rhoicissus rhomboidea comes from Natal and has compound leaves with three leaflets and the vigorous climbing *Tetrastigma voinieranum* from South-east Asia, five leaflets. The young shoots and the undersides of the leaves are silvery-grey.

Winteraceae

7 genera and 120 species

All members of this family are dicotyledonous trees or shrubs, the majority coming from the Southern Hemisphere. They have alternate or subverticillate, entire leaves and regular and bisexual flowers with two to six sepals, two to many petals, fifteen to many stamens and a superior ovary. The foliage is usually copiously dotted with oil-bearing glands.

Drimys winteri, the plant illustrated, has a wide distribution in South America ranging from Mexico to Tierra del Fuego. It is commonly called Winter's Bark, taking the name from a Captain William Winter, one of Sir Francis Drake's officers at the time of his journey round the world (1577–80). Winter's ship, *Elizabeth,* was the only one of five to survive the rounding of Cape Horn. Scurvy—the scourge of Elizabethan seamen—was an ever present problem at that time and mariners going ashore would try to find a cure by sampling native plants.

Drimys winteri

In the Straits of Magellan Winter discovered a tree whose aromatic bark proved most effective against scurvy. He brought samples back to England in 1578 and later Captain Cook put it to similar use during his voyage of discovery in 1772–75. Not until 1827 did living samples of the plant—now commonly known as Winter's Bark—reach Britain. The bark is used as a stimulant in South America (and, by importation, in Europe as well), for gastric disorders and as a condiment. The wood has some local value and is used for making furniture and for house interiors.

The species is an evergreen tree or large shrub up to 8 m (25 ft) with light, bright green, oblong leaves which are aromatic when crushed and numerous clusters of fragrant, milky-white flowers in loose umbels. The young shoots are often reddish. It is only hardy in Britain in sheltered situations.

Even less hardy is *D. aromatica* from Tasmania. This is smaller, growing up to 5 m (15 ft) with white, long-petalled flowers. The leaves have a pungent peppery taste and the dried fruits have been used as a Pepper substitute.

All species of *Pseudowintera* come from New Zealand. *P. axillaris* is the Pepper Tree, with shiny, strongly aromatic leaves, greenish flowers and red berries. The foliage of *P. colorata* assumes striking autumnal tints before leaf fall.

Alpinia speciosa

Zingiberaceae

45 genera and 700 species

This is a rather confused family of monocotyledonous, tropical perennial herbs closely linked with Marantaceae and Cannaceae, and often very striking with spectacular flowers and noble two-ranked leaves. The latter have sheathing bases and short stalks and the inflorescences appear as heads, spikes (sometimes drooping) or cymes. Individually the flowers often present a hooded appearance and have three sepals and petals, six stamens (some modified) and an inferior ovary. Some authorities include Costaceae in this family.

Members of Zingiberaceae frequently possess fleshy, aromatic rootstocks that are rich in volatile oils. Among the best-known is *Zingiber officinale,* the Ginger Plant, a native of Asia which is widely cultivated over most of the world's tropics. The plant—which likes a warm, moist climate—grows 60–120 cms (2–4 ft) high with leafy stems and two-ranked, lanceolate leaves (30 cms; 1 ft or less) and dense spikes of greenish-yellow flowers with three-lobed purple lips.

The Ginger of commerce comes from the fleshy rhizomes or underground stems, which are curiously lobed or fingered so that they are commonly called 'hands' in the trade. These fresh green rhizomes, after cleaning, shaping and boiling make delicious sweetmeats when candied, crystallized or preserved in syrup. But they are also sold dried (stem Ginger) or powdered, when they have various culinary uses as in cakes, pastry, wine, brandy, Ginger ale and Ginger beer. In addition the essential oil is extracted for its medicinal properties and also used to impart an oriental flavour to certain perfumes.

Ginger appears to have been one of the first spices known to the Old World. It was mentioned by Confucius (551–479 B.C.) in his *Analects* and presumably reached Europe in the camel trains of Arab traders, for the Greek physician Dioscorides frequently refers to its virtues.

Before the Norman Conquest of Britain (A.D. 1066) Ginger was included in the Anglo-Saxon medical works known as leech books and according to Frederic Rosengarten (*The Book of Spices*), a pound of Ginger in the 14th century cost about the same as a sheep and was second only in importance to Pepper as a spice.

In the 16th century Henry VIII recommended Ginger as a specific against the plague and gingerbread men and other fancies became popular at court. Queen Elizabeth I is reputed to have been very fond of them and in *Loves' Labour Lost* V, i, Shakespeare's Costard tells Moth 'An I had but one penny in the world, thou shouldst have it to buy ginger-bread'.

Z. spectabile, the species illustrated, comes from Malacca and the Malay peninsula. It is a more robust plant than *Z. officinale*, growing up to 2 m (6 ft) tall with oblong-lanceolate pointed leaves and spikes 20–30 cms (8–12 ins) long of flowers which have yellow or yellowish-brown bracts and two-lipped creamy-yellow, red and purplish-black petals. Some forms of this plant are more brightly coloured than the specimen illustrated and the bracts may turn scarlet as they age.

Elettaria cardamomum, a native of India, is cultivated there and in Ceylon and Central America for the sake of its ovoid green fruits and near black angular seeds which are the source of Cardamon (sometimes called Cardamom). These are highly aromatic when dried when they become an important ingredient of curry powder and are also used for various flavourings (particularly sausages) and in medicine, perfume and incense. In Scandinavia Cardamon is more widely used than Cinnamon, especially in cakes, pastry and fruit pies. It is the world's third most costly spice, second only to Saffron and Vanilla.

Grown outdoors in the tropics, or in soil beds in a warm moist greenhouse this *Elettaria* makes a noble plant up to 3 m (10 ft) in height—5.4 m (18 ft) in some districts—with linear-lanceolate, pointed leaves 60 cms (2 ft) long and spreading panicles 60 cms (2 ft) across of pale yellow-green flowers with white and blue lips.

In Europe it is grown as a pot plant, but it seldom if ever flowers and only attains a height of about 30 cms (1 ft). It has creeping rhizomes and Bamboo-like, leafy stems carrying dark green, oval-oblong leaves about 15 cms (6 ins) long and 5 cms

(2 ins) across. These often have silvery stripes and release a pungent, Cinnamon odour when handled.

In Arabic countries Cardamon Coffee is a common beverage and a symbol of hospitality. Frederic Rosengarten says 'It is said that a poor man in Saudi Arabia would rather forgo his rice than give up his cardamon. This unique popularity is rather difficult to explain, but there are several possible reasons. One is the belief that cardamon has a cooling effect on the body, especially important during the extreme heat of the summer months, when the temperature may reach 69°C (125°F). Second, it is supposed to be good for the digestion. Third, many Arabs have a traditional, deep-rooted confidence in cardamon's great power as an aphrodisiac.'

Amomum kepulaga from Java is one of several species used as Cardamon substitutes. The leafy stems grow 90 cms (3 ft) high and the flowers are yellow and red. *A. aculeatum* from Penang is taller (4 m; 12 ft) and has orange and red flowers.

One of the most popular ornamentals in Zingiberaceae is *Alpinia speciosa* (also known as *A. nutans*), the Shell Flower or Shell Ginger of South-east Asia. It is much cultivated in tropical gardens for its prolific shell-like flowers, which spill out from the tops of stems sometimes 4 m (12 ft) but more frequently 1.5–2.5 m (5–8 ft) tall. They are white, tipped with a delicate pink, opening to reveal the yellow and scarlet interior, and of an almost porcellanous texture. The broad, strap-like leaves are about 60 cms (2 ft) long and the rhizomes aromatic.

Other attractive species cultivated in warm greenhouses in cool climates or as bedding plants in the tropics include *A.* (or *Catimbium*) *mutica,* a variable species 1.25–2 m (4–6 ft) tall with upright panicles of flowers, usually white-petalled with orange-yellow lips which are spotted or lined with red; *A. malaccensis,* which grows to 3 m (10 ft) and has large white flowers variegated in yellow and red; and *A. purpurata,* the Red Ginger or Ostrich Plume Ginger, which has very showy, upright racemes covered with long, waxen-red bracts. Although these look like flowers the true blossoms, which are white and rather small, nestle between the bracts. All these species come from South-east Asia and have strap-like leaves.

Various *Alpinia* species have slight economic importance. The rhizomes of *A. officinarum* are the source of Galingale, used as a condiment. They also have aromatic, stimulant and carminative properties. The flowers of other species are occasionally eaten raw or pickled, particularly in Java, and a decoction of leaves is sometimes employed (in tropical countries) as a stimulant and aromatic in warm baths.

Hedychium coronarium is the Sweet-scented Garland Flower or Ginger Lily of India. It is a fleshy-rooted perennial which produces many offsets and has smooth, leafy stems 90 cms (3 ft) tall bearing alternate, lanceolate leaves with clasping footstalks. The erect, many-flowered, terminal spikes have very fragrant, white or cream (occasionally) pale yellow, long-tubed flowers with narrow petal segments and long protruding filaments. *H. gardneranum* from the Himalayas has similar leaves but the flowers are yellow and smaller with long stamens.

Hedychiums need similar cultural treatment to Cannas, that is, rich soil with plenty of water during the growing season and a rest in winter. In cool temperate countries they may be grown in tubs or large pots placed outdoors in summer but brought under cover and rested during the winter.

Roscoea is a small genus of Asiatic herbs with clusters of fleshy roots and zygomorphic hooded flowers at the ends of leafy stems 30 cms (1 ft) high. Individual blooms are 4–6 cms ($1\frac{1}{2}$–$2\frac{1}{2}$ ins) long and funnel-shaped, the modified stamen forming a large, petal-like lip about 2.5 cms (1 in.) long. The upper segments arch to form a helmet-like structure. They are hardy in Britain and those parts of Europe, North America and New Zealand where the winter climate is reasonably mild. They should be planted 10–13 cms (4–5 ins) deep in sandy loam soil. *R. humeana* is deep violet-purple; *R. cautleoides,* clear pale-yellow; *R. purpurea,* purple with white lines, and *R. alpina,* white, rose or red-purple.

European greenhouses sometimes display species of *Kaempferia,* especially *K. galanga* from South-east Asia, a tuberous, stemless perennial with two spreading, horizontally placed leaves. These may be 7–13 cms (3–5 ins) long and between them appear up to twelve white flowers with lilac-purple spots on the lips. Similarly coloured, fragrant *K. angustifolia* from the E Himalayas has many ascending leaves, and *K. secunda,* an Indian species, has three to six oblong-lanceolate leaves and red-purple flowers with light purple and white lips. In tropical gardens Kaempferias are often planted beneath trees as ground cover.

Curcuma domestica (often cultivated incorrectly as *C. longa*) yields Turmeric, the main colouring ingredient of curry and a dye used in the East to give a golden tint. It is also applied at times to people, especially at Indonesian weddings, when both bride and groom paint their arms with Turmeric. Turmeric-tinted rice also plays an important part in Indonesian marriage ceremonies and natives of the Pacific Islands paint their bodies and cheeks with it as a decoration and beauty treatment. The effect is very similar to Saffron and in medieval times Turmeric was called 'Indian Saffron' by Europeans. The plant has a large aromatic tuber which is deep yellow throughout and pale yellow flowers in cone-shaped spikes, tipped with pale pink and white and long, thin leaves 45 × 15 cms (18 × 6 ins) broad. It is a native of India but widely cultivated in the tropics.

Roscoea humeana

A form of
Alpina mutica
Hedychium gardneranum
Zingiber spectabile

Index

Page numbers in italics refer to illustrations

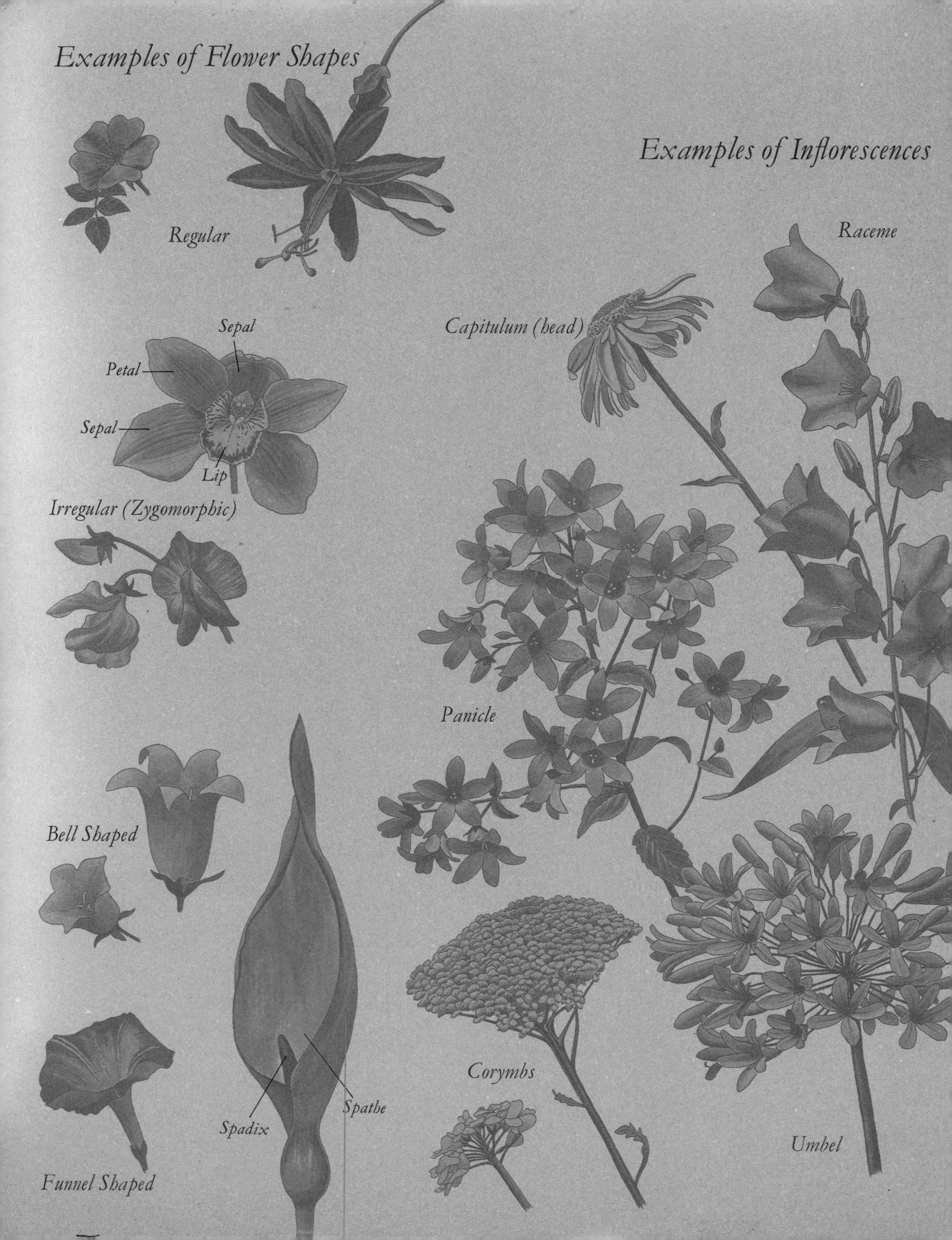
Examples of Flower Shapes
Regular
Sepal
Petal
Sepal
Lip
Irregular (Zygomorphic)
Bell Shaped
Funnel Shaped
Spadix
Spathe
Examples of Inflorescences
Raceme
Capitulum (head)
Panicle
Corymbs
Umbel